Student Solutions Manual

Intermediate Algebra

with Applications

SEVENTH EDITION

Aufmann / Barker / Lockwood

Emily Keaton

Houghton Mifflin Company BOSTON NEW YORK

Publisher: Richard Stratton
Senior Sponsoring Editor: Lynn Cox
Senior Marketing Manager: Katherine Greig
Marketing Associate: Naveen Hariprasad
Associate Editor: Noel Kamm
Editorial Assistant: Laura Ricci
Senior Project Editor: Kerry Falvey
Editorial Assistant: Joanna Carter
Cover Design Manager: Anne S. Katzeff
New Title Project Manager: Susan Brooks-Peltier

Copyright © 2008 by Houghton Mifflin Company. All rights reserved.

No part of this work may be reproduced or transmitted in any form or by any means, electronic or mechanical, including photocopying and recording, or by any information storage or retrieval system without the prior written permission of Houghton Mifflin Company unless such copying is expressly permitted by federal copyright law. Address inquiries to College Permissions, Houghton Mifflin Company, 222 Berkeley Street, Boston, MA 02116-3764.

Printed in the U.S.A.

ISBN 13: 978-0-618-81891-4
ISBN 10: 0-618-81891-X

123456789 – VHO – 11 10 09 08 07

Contents

CHAPTER 1 – REVIEW OF REAL NUMBERS	1
CHAPTER 2 – FIRST-DEGREE EQUATIONS AND INEQUALITIES	13
CHAPTER 3 – LINEAR FUNCTIONS AND INEQUALITIES IN TWO VARIABLES	49
CHAPTER 4 – SYSTEMS OF EQUATIONS AND INEQUALITIES	80
CHAPTER 5 – POLYNOMIALS AND EXPONENTS	131
CHAPTER 6 – RATIONAL EXPRESSIONS	166
CHAPTER 7 – RATIONAL EXPONENTS AND RADICALS	209
CHAPTER 8 – QUADRATIC EQUATIONS AND INEQUALITIES	240
CHAPTER 9 – FUNCTIONS AND RELATIONS	299
CHAPTER 10 – EXPONENTIAL AND LOGARITHMIC FUNCTIONS	313
CHAPTER 11 – SEQUENCES AND SERIES	338
CHAPTER 12 – CONIC SECTIONS	362

CHAPTER 1: REVIEW OF REAL NUMBERS

CHAPTER 1 PREP TEST

1. $\dfrac{5}{12} + \dfrac{7}{30} = \dfrac{25}{60} + \dfrac{14}{60} = \dfrac{39}{60} = \dfrac{13}{20}$

2. $\dfrac{8}{15} - \dfrac{7}{20} = \dfrac{32}{60} - \dfrac{21}{60} = \dfrac{11}{60}$

3. $\dfrac{2}{9}$

4. $\dfrac{4}{15} \div \dfrac{2}{5} = \dfrac{4}{15} \cdot \dfrac{5}{2} = \dfrac{2}{3}$

5. 44.405

6. 73.63

7. 7.446

8. 54.06

9. a, c, d

10. a. $\dfrac{1}{2} = 0.5$ C

    b. $\dfrac{7}{10} = 0.7$ D

    c. $\dfrac{3}{4} = 0.75$ A

    d. $\dfrac{89}{100} = 0.89$ B   Section 1.1

Concept Review 1.1

1. Sometimes true
The only exception to this statement is for zero. $|0| = 0$, which is not a positive number.

3. Sometimes true
If x is a positive integer, then $-x$ is a negative integer.
If x is a negative integer, then $-x$ is a positive integer.

5. Never true
$-4 < -2$ but $(-4)^2$ is not less than $(-2)^2$.

7. Always true

Objective 1.1.1 Exercises

1.   −14: c, e
    9: a, b, c, d
    0: b, c
    53: a, b, c, d
    7.8: none
    −626: c, e

3. $-\dfrac{15}{2}$: b, d
    0: a, b, d
    −3: a, b, d
    π: c, d
    $2.\overline{33}$: b, d
    4.232232223: c, d
    $\dfrac{\sqrt{5}}{4}$: c, d
    $\sqrt{7}$: c, d

5. A terminating decimal is a decimal number that has only a finite number of decimal places – for example, 0.75.

7. The additive inverse of a number is the number that is the same distance from zero on the number line but on the opposite side of zero.

9. A number such as 0.63633633363333…, whose decimal notation neither ends nor repeats, is an example of an <u>irrational</u> number.

11. $y \in \{1, 3, 5, 7, 9\}$ is read "y <u>is an element of</u> the set $\{1, 3, 5, 7, 9\}$."

13. −27

15. $-\dfrac{3}{4}$

17. 0

19. $\sqrt{33}$

21. 91

23. Replace x with each element in the set and determine whether the inequality is true.
$x < 5$
$-3 < 5$ True
$0 < 5$ True
$7 < 5$ False
The inequality is true for −3 and 0.

25. Replace y with each element in the set and determine whether the inequality is true.
$y > -4$
$-6 > -4$ False
$-4 > -4$ False
$7 > -4$ True
The inequality is true for 7.

27. Replace w with each element in the set and determine whether the inequality is true.
$w \leq -1$
$-2 \leq -1$ True
$-1 \leq -1$ True
$0 \leq -1$ False
$1 \leq -1$ False
The inequality is true for −2 and −1.

2 Chapter 1: Review of Real Numbers

29. Replace *b* with each element in the set and evaluate the expression.
    −*b*
    −(−9) = 9
    −(0) = 0
    −(9) = −9

31. Replace *c* with each element in the set and evaluate the expression.
    |*c*|
    |−4| = 4
    |0| = 0
    |4| = 4

33. Replace *m* with each element in the set and evaluate the expression.
    −|*m*|
    −|−6| = −6
    −|−2| = −2
    −|0| = 0
    −|1| = −1
    −|4| = −4

35. Yes, negative real numbers are such that $-x > 0$.

Objective 1.1.2 Exercises

37. The union of two sets will contain all the elements that are in either set. The intersection of the two sets will contain only the elements that are in both sets.

39. Two ways to write the set of natural numbers less than five are {1, 2, 3, 4} and $\{n \mid n < 5, \ n \in \text{natural numbers}\}$. The first way uses the <u>roster</u> method, and the second way uses <u>set-builder</u> notation.

41. The symbol ∞ is called the <u>infinity</u> symbol.

43. {−2, −1, 0, 1, 2, 3, 4}

45. {2, 4, 6, 8, 10, 12}

47. {3, 6, 9, 12, 15, 18, 21, 24, 27, 30}

49. {−35, −30, −25, −20, −15, −10, −5}

51. $\{x \mid x > 4, \ x \text{ is an integer}\}$

53. $\{x \mid x \geq -2\}$

55. $\{x \mid 0 < x < 1\}$

65. $A \cup B = \{1, 2, 3, 4, 5\}$

67. $A \cap B = \{6\}$

69. $A \cap B = \{5, 10, 20\}$

71. $A \cap B = \varnothing$

73. $A \cap B = \{4, 6\}$

75. $\{x \mid -1 < x < 5\}$

77. $\{x \mid 0 \leq x \leq 3\}$

79. $\{x \mid x < 2\}$

81. $\{x \mid x \geq 1\}$

83. $\{x \mid x > 1\} \cup \{x \mid x < -1\}$

85. $\{x \mid x \leq 2\} \cap \{x \mid x \geq 0\}$

87. $\{x \mid x > 1\} \cap \{x \mid x \geq -2\}$

89. $\{x \mid x > 2\} \cup \{x \mid x > 1\}$

91. $\{x \mid 0 < x < 8\}$

93. $\{x \mid -5 \leq x \leq 7\}$

95. $\{x \mid -3 \leq x < 6\}$

97. $\{x \mid x \leq 4\}$

99. $\{x \mid x > 5\}$

101. (−2, 4)

103. [−1, 5]

105. (−∞, 1)

107. [−2, 6)

109. (−∞, ∞)

111. (−2, 5)

113. [−1, 2]

115. (−∞, 3]

117. [3, ∞)

119. (−∞, 2] ∪ [4, ∞)

121. $[-1, 2] \cap [0, 4]$

123. $(2, \infty) \cup (-2, 4]$

125. c

Applying Concepts 1.1

127. $A \cup B$ is
$\{x | -1 \leq x \leq 1\} \cup \{x | 0 \leq x \leq 1\} = \{x | -1 \leq x \leq 1\} = A$

129. $B \cap B$ is set B.

131. $A \cap R$ is $\{x | -1 \leq x \leq 1\}$, which is set A.

133. $B \cup R$ is the set of real numbers, R.

135. $R \cup R$ is the set R.

137. $B \cap C$ is $\{x | 0 \leq x \leq 1\} \cap \{x | -1 \leq x \leq 0\}$, which contains only the number 0.

139.

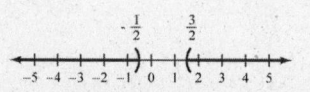

141.

143.

145.

147. $A \cup B = \{x | x > 0, x \text{ is an integer}\}$

149. $A \cap B = \{x | x \geq 15, x \text{ is an odd integer}\}$

151. The answer is b and c. For example:

a. $\dfrac{5-4}{3-2} \leq 0$
$1 \leq 0$
False

b. $\dfrac{2-3}{5-4} \leq 0$
$-1 \leq 0$
True

c. $\dfrac{5-4}{2-3} \leq 0$
$-1 \leq 0$
True

d. $\dfrac{4-5}{2-3} \leq 0$
$1 \leq 0$
False

Section 1.2

Concept Review 1.2

1. Sometimes true
$(-2) + 4 = 2$, a positive number
$(-8) + 4 = -4$, a negative number

3. Never true
$\dfrac{1}{2} + \dfrac{2}{3} = \dfrac{7}{6}$
$\dfrac{1+2}{2+3} = \dfrac{3}{5}$

5. Always true

7. Never true.
The Order of Operations says to work inside parentheses before doing exponents.

9. Always true

Objective 1.2.1 Exercises

1. a. Students should paraphrase the rule: Add the absolute values of the numbers; then attach the sign of the addends.

   b. Students should paraphrase the rule: Find the absolute value of each number; subtract the smaller of the two numbers from the larger; then attach the sign of the number with the larger absolute value.

3. To rewrite $8 - (-12)$ as addition of the opposite, change the subtraction to addition and change -12 to the opposite of -12:
$8 - (-12) = 8 + 12$.

5. $|-8 - (-3)| = |-8 + \underline{3}|$
$= |\underline{-5}|$
$= \underline{5}$

7. $-18 + (-12) = -30$

9. $5 - 22 = 5 + (-22) = -17$

11. $3 \cdot 4(-8) = 12 \cdot (-8) = -96$

13. $18 \div (-3) = -6$

15. $-60 \div (-12) = 5$

17. $-20(35)(-16) = -700(-16) = 11,200$

19. $(-271)(-365) = 98,915$

21. $|12(-8)| = |-96| = 96$

23. $|15 - (-8)| = |15 + 8| = |23| = 23$

25. $|-56 \div 8| = |-7| = 7$

27. $|-153 \div (-9)| = |17| = 17$

29. $-|-8|+|-4| = -8+4 = -4$

31. $-30+(-16)-14-2 = -30+(-16)+(-14)+(-2)$
$= -46+(-14)+(-2)$
$= -60+(-2)$
$= -62$

33. $-2+(-19)-16+12 = -2+(-19)+(-16)+12$
$= -21+(-16)+12$
$= -37+12$
$= -25$

35. $13-|6-12| = 13-|6+(-12)|$
$= 13-|-6|$
$= 13-6$
$= 13+(-6)$
$= 7$

37. $738-46+(-105) = 738+(-46)+(-105)$
$= 692+(-105)$
$= 587$

39. $-442 \div (-17) = 26$

41. $-4897 \div 59 = -83$

43. a. $567+(-812)$ is negative
b. $-259-(-327)$ is positive
c. The product of four positive numbers and three negative numbers is negative.
d. The product of three positive numbers and four negative numbers is positive.

Objective 1.2.2 Exercises

45. a. The least common multiple of two numbers is the smallest number that is a multiple of each of those numbers.
b. The greatest common factor of two numbers is the largest integer that divides evenly into both numbers.

47. 56; 35; 2; 16

49. The sum of 0.773 and -81.5 will have 3 decimal places.

51. $\dfrac{7}{12}+\dfrac{5}{16} = \dfrac{28}{48}+\dfrac{15}{48} = \dfrac{28+15}{48} = \dfrac{43}{48}$

53. $-\dfrac{5}{9}-\dfrac{14}{15} = -\dfrac{25}{45}-\dfrac{42}{45} = \dfrac{-25-42}{45} = -\dfrac{67}{45}$

55. $-\dfrac{1}{3}+\dfrac{5}{9}-\dfrac{7}{12} = -\dfrac{12}{36}+\dfrac{20}{36}-\dfrac{21}{36}$
$= \dfrac{-12+20-21}{36}$
$= -\dfrac{13}{36}$

57. $\dfrac{2}{3}-\dfrac{5}{12}+\dfrac{5}{24} = \dfrac{16}{24}-\dfrac{10}{24}+\dfrac{5}{24} = \dfrac{16-10+5}{24} = \dfrac{11}{24}$

59. $\dfrac{5}{8}-\dfrac{7}{12}+\dfrac{1}{2} = \dfrac{15}{24}-\dfrac{14}{24}+\dfrac{12}{24} = \dfrac{15-14+12}{24} = \dfrac{13}{24}$

61. $\left(\dfrac{6}{35}\right)\left(-\dfrac{5}{16}\right) = -\dfrac{6 \cdot 5}{35 \cdot 16} = -\dfrac{2 \cdot 3 \cdot \cancel{5}}{\cancel{5} \cdot 7 \cdot 2 \cdot 2 \cdot 2 \cdot 2} = -\dfrac{3}{56}$

63. $-\dfrac{8}{15} \div \dfrac{4}{5} = -\dfrac{8}{15} \cdot \dfrac{5}{4} = -\dfrac{8 \cdot 5}{15 \cdot 4} = -\dfrac{2 \cdot 2 \cdot 2 \cdot \cancel{5}}{3 \cdot \cancel{5} \cdot 2 \cdot 2} = -\dfrac{2}{3}$

65. $-\dfrac{11}{24} \div \dfrac{7}{12} = -\dfrac{11}{24} \cdot \dfrac{12}{7}$
$= -\dfrac{11 \cdot 12}{24 \cdot 7}$
$= -\dfrac{11 \cdot 2 \cdot 2 \cdot 3}{2 \cdot 2 \cdot 2 \cdot 3 \cdot 7}$
$= -\dfrac{11}{14}$

67. $\left(-\dfrac{5}{12}\right)\left(\dfrac{4}{35}\right)\left(\dfrac{7}{8}\right) = -\dfrac{5 \cdot 4 \cdot 7}{12 \cdot 35 \cdot 8}$
$= -\dfrac{5 \cdot 2 \cdot 2 \cdot 7}{2 \cdot 2 \cdot 3 \cdot 5 \cdot 7 \cdot 2 \cdot 2 \cdot 2}$
$= -\dfrac{1}{24}$

69. -14.270
$+1.296$
-12.974

$-14.27+1.296 = -12.974$

71. -7.840
$+1.832$
-6.008

$1.832-7.84 = -6.008$

73. $(0.03)(10.5)(6.1) = (0.315)(6.1) = 1.9215$

75. $0.9 \to 9$
$5.418 \to 54.18$

$$\begin{array}{r} 6.02 \\ 9\overline{)54.18} \\ \underline{-54} \\ 01 \\ \underline{-0} \\ 18 \\ \underline{-18} \\ 0 \end{array}$$

$5.418 \div (-0.9) = -6.02$

77. $0.065 \to 65$
$0.4355 \to 435.5$

$$\begin{array}{r} 6.7 \\ 65\overline{)435.5} \\ \underline{-390} \\ 455 \\ \underline{-455} \\ 0 \end{array}$$

$-0.4355 \div 0.065 = -6.7$

79. $38.241 \div [-(-6.027)] - 7.453$
$= 38.241 \div 6.027 + (-7.453)$
$\approx 6.345 + (-7.453)$
≈ -1.11

81. $-287.3069 \div 0.1415 \approx -2030.44$

83.
a. The product is negative.
b. The sum is negative.
c. The quotient is positive.
d. The difference is negative.

Objective 1.2.3 Exercises

85. In the expression $(-6)^3$, -6 is called the <u>base</u> and 3 is called the <u>exponent</u>. To evaluate $(-6)^3$, find the product $\underline{(-6)(-6)(-6)} = \underline{-216}$.

87. $5^3 = 5 \cdot 5 \cdot 5 = 125$

89. $-2^3 = -(2 \cdot 2 \cdot 2) = -8$

91. $(-5)^3 = (-5)(-5)(-5) = -125$

93. $2^2 \cdot 3^4 = (2)(2) \cdot (3)(3)(3)(3) = 4 \cdot 81 = 324$

95. $-2^2 \cdot 3^2 = -(2)(2) \cdot (3)(3) = -4 \cdot 9 = -36$

97. $(-2)^3 \cdot (-3)^2 = (-2)(-2)(-2) \cdot (-3)(-3)$
$= -8 \cdot 9$
$= -72$

99. $4 \cdot 2^3 \cdot 3^3 = 4 \cdot (2)(2)(2) \cdot (3)(3)(3)$
$= 4 \cdot 8 \cdot 27$
$= 32 \cdot 27$
$= 864$

101. $2^2 \cdot (-10)(-2)^2 = (2)(2) \cdot (-10) \cdot (-2)(-2)$
$= 4 \cdot (-10)(4)$
$= -40(4)$
$= -160$

103. $(-3)^3 \cdot 15 \cdot \left(-\dfrac{1}{3}\right)^4 = \dfrac{(-27) \cdot 15}{-81}$
$= -5$

105. $\left(\dfrac{2}{3}\right)^3 \cdot (-3)^4 \cdot 4^5 = \left(\dfrac{8}{27}\right) \cdot (81) \cdot 1024$
$= 24 \cdot 1024$
$= 24{,}576$

107. The expression $(-9)^7$ simplifies to a negative number because it has an odd number of negative factors.

Objective 1.2.4 Exercises

109. We need an Order of Operations Agreement to ensure that there is only one way in which an expression can be correctly simplified.

111. $5^2 - (10 \div 2)^3 \div 5 = 5^2 - (\underline{5})^3 \div 5$
$= \underline{25} - \underline{125} \div 5$
$= 25 - \underline{25}$
$= \underline{0}$

113. $5 - 3(8 \div 4)^2 = 5 - 3(2)^2 = 5 - 3(4) = 5 - 12 = -7$

115. $16 - \dfrac{2^2 - 5}{3^2 + 2} = 16 - \dfrac{4 - 5}{9 + 2}$
$= 16 - \dfrac{-1}{11}$
$= 16 + \dfrac{1}{11}$
$= \dfrac{177}{11}$

117. $\dfrac{3 + \frac{2}{3}}{\frac{11}{16}} = \dfrac{\frac{11}{3}}{\frac{11}{16}} = \dfrac{11}{3} \cdot \dfrac{16}{11} = \dfrac{16}{3}$

119. $5[(2 - 4) \cdot 3 - 2] = 5[(-2) \cdot 3 - 2]$
$= 5[-6 - 2]$
$= 5[-8]$
$= -40$

121. $16 - 4\left(\dfrac{8-2}{3-6}\right) \div \dfrac{1}{2} = 16 - 4\left(\dfrac{6}{-3}\right) \div \dfrac{1}{2}$
$= 16 - 4(-2) \div \dfrac{1}{2}$
$= 16 - (-8) \div \dfrac{1}{2}$
$= 16 - (-8) \cdot 2$
$= 16 - (-16)$
$= 16 + 16$
$= 32$

123. $6[3 - (-4 + 2) \div 2] = 6[3 - (-2) \div 2]$
$= 6[3 - (-1)]$
$= 6[3 + 1]$
$= 6[4]$
$= 24$

125. $\dfrac{1}{2} - \left(\dfrac{2}{3} \div \dfrac{5}{9}\right) + \dfrac{5}{6} = \dfrac{1}{2} - \left(\dfrac{2}{3} \cdot \dfrac{9}{5}\right) + \dfrac{5}{6}$
$= \dfrac{1}{2} - \dfrac{6}{5} + \dfrac{5}{6}$
$= \dfrac{15}{30} - \dfrac{36}{30} + \dfrac{25}{30}$
$= \dfrac{15 - 36 + 25}{30}$
$= \dfrac{4}{30}$
$= \dfrac{2}{15}$

127. $\dfrac{1}{2} - \dfrac{\frac{17}{25}}{4 - \frac{3}{5}} + \dfrac{1}{5} = \dfrac{1}{2} - \dfrac{\frac{17}{25}}{\frac{17}{5}} + \dfrac{1}{5}$
$= \dfrac{1}{2} - \left(\dfrac{17}{25} \cdot \dfrac{5}{17}\right) + \dfrac{1}{5}$
$= \dfrac{1}{2} - \dfrac{1}{5} + \dfrac{1}{5}$
$= \dfrac{1}{2}$

129. $\dfrac{2}{3} - \left[\dfrac{3}{8} + \dfrac{5}{6}\right] \div \dfrac{3}{5} = \dfrac{2}{3} - \left[\dfrac{9}{24} + \dfrac{20}{24}\right] \div \dfrac{3}{5}$
$= \dfrac{2}{3} - \dfrac{29}{24} \div \dfrac{3}{5}$
$= \dfrac{2}{3} - \dfrac{29}{24} \cdot \dfrac{5}{3}$
$= \dfrac{2}{3} - \dfrac{145}{72}$
$= \dfrac{48}{72} - \dfrac{145}{72}$
$= -\dfrac{97}{72}$

131. $0.4(1.2 - 2.3)^2 + 5.8 = 0.4(-1.1)^2 + 5.8$
$= 0.4(1.21) + 5.8$
$= 0.484 + 5.8$
$= 6.284$

133. $1.75 \div 0.25 - (1.25)^2 = 1.75 \div 0.25 - 1.5625$
$= 7 - 1.5625$
$= 5.4375$

135. $25.76 \div (6.54 \div 3.27)^2 = 25.76 \div (2)^2$
$= 25.76 \div 4$
$= 6.44$

137. b
$8^2 - 2^2(5 - 3)^3 = 8^2 - 2^2(2)^3 = 64 - 4(8)$

Applying Concepts 1.2

139. 0

141. No, the multiplicative inverse of zero is undefined.

143. $7^{18} = 1,628,413,597,910,449$
The ones digit is 9.

145. 5^{234} has over 150 digits. The last three are 625.

147. Find b^c. Then find $a^{(b^c)}$.

Section 1.3
Concept Review 1.3

1. Sometimes true
The reciprocal of 1 is 1, a whole number. The reciprocal of 2 is $\dfrac{1}{2}$, not a whole number.

3. Sometimes true
$2xy$ and $3xy$ are like terms with the same variables. $2xy$ and $2x^2y$ are unlike terms with same variables.

5. Always true

Objective 1.3.1 Exercises

1. The fact that two terms can be multiplied in either order is called the <u>Commutative</u> Property of Multiplication.

3. The Multiplication Property of Zero tells us that the product of a number and zero is <u>zero</u>.

5. $3 \cdot 4 = 4 \cdot 3$

7. $(3 + 4) + 5 = 3 + (4 + 5)$

9. $\dfrac{5}{0}$ is undefined.

11. $3(x + 2) = 3x + 6$

13. $\dfrac{0}{-6} = 0$

15. $\dfrac{1}{mn}(mn) = 1$

17. $2(3x) = (2 \cdot 3) \cdot x$
19. A Division Property of Zero
21. The Inverse Property of Multiplication
23. The Addition Property of Zero
25. A Division Property of Zero
27. The Distributive Property
29. The Associative Property of Multiplication
31. When the sum of a number n and its additive inverse is multiplied by the reciprocal of the number n, the result is zero.

Objective 1.3.2 Exercises

33. "Evaluate a variable expression" means replace the variable expression by a numerical value and simplify the resulting expression.

35. $-5; -3; -2; 2; 25; 50$

37. $ab + dc$
$(2)(3) + (-4)(-1) = 6 + 4 = 10$

39. $4cd \div ac$
$4(-1)(-4) \div (2)(-1) = (-4)(-4) \div (2)(-1)$
$= 16 \div (2)(-1)$
$= 16 \div (-2)$
$= -8$

41. $(b - 2a)^2 + c$
$[3 - 2(2)]^2 + (-1) = [3 - 4]^2 + (-1)$
$= [-1]^2 + (-1)$
$= 1 + (-1)$
$= 0$

43. $(bc + a)^2 \div (d - b)$
$[(3)(-1) + 2]^2 \div (-4 - 3) = [-3 + 2]^2 \div (-7)$
$= [-1]^2 \div (-7)$
$= -\dfrac{1}{7}$

45. $\dfrac{1}{4}a^4 - \dfrac{1}{6}bc$
$\dfrac{1}{4}(2)^4 - \dfrac{1}{6}(3)(-1) = \dfrac{1}{4}(16) - \dfrac{1}{6}(3)(-1)$
$= 4 - \dfrac{1}{6}(3)(-1)$
$= 4 - \dfrac{1}{2}(-1)$
$= 4 - \left(-\dfrac{1}{2}\right)$
$= 4 + \dfrac{1}{2}$
$= \dfrac{9}{2}$

47. $\dfrac{3ac}{-4} - c^2$
$\dfrac{3(2)(-1)}{-4} - (-1)^2 = \dfrac{6(-1)}{-4} - (-1)^2$
$= \dfrac{-6}{-4} - (-1)^2$
$= \dfrac{3}{2} - (-1)^2$
$= \dfrac{3}{2} - 1$
$= \dfrac{1}{2}$

49. $\dfrac{3b - 5c}{3a - c}$
$\dfrac{3(3) - 5(-1)}{3(2) - (-1)} = \dfrac{9 - (-5)}{6 - (-1)} = \dfrac{9 + 5}{6 + 1} = \dfrac{14}{7} = 2$

51. $\dfrac{a - d}{b + c}$
$\dfrac{2 - (-4)}{3 + (-1)} = \dfrac{2 + 4}{3 + (-1)} = \dfrac{6}{2} = 3$

53. $-a|a + 2d|$
$-2|2 + 2(-4)| = -2|2 + (-8)|$
$= -2|-6|$
$= -2(6)$
$= -12$

55. $\dfrac{2a - 4d}{3b - c}$
$\dfrac{2(2) - 4(-4)}{3(3) - (-1)} = \dfrac{4 - (-16)}{9 - (-1)}$
$= \dfrac{4 + 16}{9 + 1}$
$= \dfrac{20}{10}$
$= 2$

57. $-3d \div \left|\dfrac{ab - 4c}{2b + c}\right|$
$-3(-4) \div \left|\dfrac{2(3) - 4(-1)}{2(3) + (-1)}\right| = -3(-4) \div \left|\dfrac{6 - (-4)}{6 + (-1)}\right|$
$= -3(-4) \div \left|\dfrac{6 + 4}{6 + (-1)}\right|$
$= -3(-4) \div \left|\dfrac{10}{5}\right|$
$= -3(-4) \div |2|$
$= -3(-4) \div 2$
$= 12 \div 2$
$= 6$

59. $2(d-b) \div (3a-c)$
$2(-4-3) \div [3(2)-(-1)] = 2(-7) \div [6-(-1)]$
$= 2(-7) \div [6+1]$
$= 2(-7) \div 7$
$= -14 \div 7$
$= -2$

61. $-d^2 - c^3 a$
$-(-4)^2 - (-1)^3(2) = -16 - (-1)(2)$
$= -16 + 2$
$= -14$

63. $-d^3 + 4ac$
$-(-4)^3 + 4(2)(-1) = -(-64) + 8(-1)$
$= 64 - 8$
$= 56$

65. 4^{a^2}
$4^{2^2} = 4^4 = 256$

67. $V = LWH$
$V = (14)(10)(6)$
$V = 840$
The volume is 840 in^3.

69. $V = \frac{1}{3} s^2 h$
$V = \frac{1}{3}(3^2)5$
$V = 15$
The volume is 15 ft^2.

71. $V = \frac{4}{3}\pi r^3$ $\qquad r = \frac{1}{2}d = \frac{1}{2}(3) = 1.5$
$V = \frac{4}{3}\pi (1.5)^3$
$V = 4.5\pi$
$V \approx 14.14$
The volume is 4.5π cm^3.
The volume is approximately 14.14 cm^3.

73. $SA = 2LW + 2LH + 2WH$
$SA = 2(5)(4) + 2(5)(3) + 2(4)(3)$
$SA = 40 + 30 + 24$
$SA = 94$
The surface area is 94 m^2.

75. $SA = s^2 + 4\left(\frac{1}{2}\right)bh$
$SA = 4^2 + 2(4)(5)$
$SA = 16 + 40 = 56$
The surface area is 56 m^2.

77. $SA = 2\pi r^2 + 2\pi rh$
$SA = 2\pi (6^2) + 2\pi (6)(2)$
$SA = 72\pi + 24\pi$
$SA = 96\pi$
$SA \approx 301.59$
The surface area is 96π in^2.
The surface area is approximately 301.59 in^2.

79. Because $|b| > |a|$, the denominator will be negative. The numerator will be positive because the product contains an even number of factors. Therefore, the result will be a negative number.

Objective 1.3.3 Exercises

81. If there are two terms with a common variable factor, the Distributive Property allows us to combine the two terms into one term. Add the coefficients of the variable factor, and write the sum as the coefficient of the common variable factor.

83. -5; 25; Commutative; Associative; $4y$

85. $3x + 10x = 13x$

87. $-2x + 5x - 7x = 3x - 7x = -4x$

89. $-2a + 7b + 9a = 7a + 7b$

91. $12\left(\frac{1}{12}\right)x = x$

93. $-3(x - 2) = -3x + 6$

95. $(x + 2)5 = 5x + 10$

97. $-(-x - y) = x + y$

99. $3(x - 2y) - 5 = 3x - 6y - 5$

101. $-2a - 3(3a - 7) = -2a - 9a + 21 = -11a + 21$

103. $2x - 3(x - 2y) = 2x - 3x + 6y = -x + 6y$

105. $5[-2 - 6(a - 5)] = 5[-2 - 6a + 30]$
$= 5[28 - 6a]$
$= 140 - 30a$

107. $5[y - 3(y - 2x)] = 5[y - 3y + 6x]$
$= 5[-2y + 6x]$
$= -10y + 30x$

109. $4(-a - 2b) - 2(3a - 5b) = -4a - 8b - 6a + 10b$
$= -10a + 2b$

111. $-7(2a - b) + 2(-3b + a) = -14a + 7b - 6b + 2a$
$= -12a + b$

113. $2x - 4[x - 4(y - 2[5y + 3])]$
$= 2x - 4[x - 4(y - 10y - 6)]$
$= 2x - 4[x - 4(-9y - 6)]$
$= 2x - 4[x + 36y + 24]$
$= 2x - 4x - 144y - 96$
$= -2x - 144y - 96$

115. $3x + 8(x - 4) - 3(2x - y)$
$= 3x + 8x - 32 - 6x + 3y$
$= 5x - 32 + 3y$

117. $\frac{1}{4}[14x - 3(x - 8) - 7x] = \frac{1}{4}[14x - 3x + 24 - 7x]$
$= \frac{1}{4}[4x + 24]$
$= x + 6$

119. $3[5 - 2(y - 6)] = 3(5) - 3[2(y - 6)] = 15 - 6(y - 6)$

   a. No
   b. Yes

Applying Concepts 1.3

121. $4(3y + 1) = 12y + 4$
The statement is correct; it uses the Distributive Property.

123. $2 + 3x = (2 + 3)x = 5x$
The statement is not correct; it mistakenly uses the Distributive Property. It is in an irreducible statement. That is, the answer is $2 + 3x$.

125. $2(3y) = (2 \cdot 3)(2y) = 12y$
The statement is not correct; it incorrectly uses the Associative Property of Multiplication. The correct answer is $(2 \cdot 3)y = 6y$.

127. $-x^2 + y^2 = y^2 - x^2$
The statement is correct; it uses the Commutative Property of Addition.

129. $3a + 4(b + a)$

   a. $3a + (4b + 4a)$ Distributive Property
   b. $3a + (4a + 4b)$ Commutative Property of Addition
   c. $(3a + 4a) + 4b$ Associative Property of Addition
   d. $(3 + 4)a + 4b$ Distributive Property
      $7a + 4b$

131. $5(3a + 1)$

   a. $5(3a) + 5(1)$ Distributive Property
   b. $(5 \cdot 3)a + 5(1)$
      $15a + 5(1)$ Associative Property of Multiplication
   c. $15a + 5$ Multiplication Property of One

Section 1.4
Concept Review 1.4

1. Never true
   The smaller number is represented by $12 - x$.

3. Never true
   The sum of twice x and 4 is represented by $2x + 4$.

5. Sometimes true
   The square of $-x$ is represented by $(-x)^2$. The only exception is for the number 0.
   $-0^2 = (-0)^2 = 0$

Objective 1.4.1 Exercises

1. ten more than the product of eight and a number

3. the difference between ten times a number and sixteen times the number

5. the unknown number: n
   The sum of the number and two: $n + 2$
   $n - (n + 2) = n - n - 2 = -2$

7. the unknown number: n
   one-third of the number: $\frac{1}{3}n$
   four-fifths of the number: $\frac{4}{5}n$
   $\frac{1}{3}n + \frac{4}{5}n = \frac{5}{15}n + \frac{12}{15}n = \frac{17}{15}n$

9. the unknown number: n
   the product of eight and the number: $8n$
   $5(8n) = 40n$

11. the unknown number: n
    the product of seventeen and the number: $17n$
    twice the number: $2n$
    $17n - 2n = 15n$

13. the unknown number: n
    the square of the number: n^2
    the total of twelve and the square of the number: $12 + n^2$
    $n^2 - (12 + n^2) = n^2 - 12 - n^2 = -12$

15. the unknown number: n
    the sum of five times the number and 12: $5n + 12$
    the product of the number and fifteen: $15n$
    $15n + (5n + 12) = 15n + 5n + 12 = 20n + 12$

17. Let the smaller number be x.
    The larger number is $15 - x$.
    The sum of twice the smaller number and two more than the larger number
    $2x + (15 - x + 2) = 2x + (17 - x) = x + 17$

19. Let the larger number be x.
    Then the smaller number is $34 - x$.
    The difference between two more than the smaller number and twice the larger number
    $[(34 - x) + 2] - 2x = 34 - x + 2 - 2x = 36 - 3x$

10 Chapter 1: Review of Real Numbers

21. c

Objective 1.4.2 Exercises

23. The length of a rectangle is eight more than the width. To express the length and the width in terms of the same variable, let W be the width. Then the length is $W + 8$.

25. The population of Milan, Italy: P
    The population of San Paolo, Brazil: $4P$

27. Distance from Earth to the moon: d
    Distance from Earth to sun: $390d$

29. Amount of the first account: x
    Amount of the second account: $10,000 - x$

31. Flying time between San Diego and New York: t
    Flying time between New York and San Diego: $13 - t$

33. The measure of angle B: x
    The measure of angle A is twice that of angle B: $2x$
    The measure of angle C is twice that of angle A:
    $2(2x) = 4x$

Applying Concepts 1.4

35. The sum of twice a number and three.

37. Twice the sum of a number and three.

39. One-half the acceleration due to gravity: $\frac{1}{2}g$

    Time squared: t^2

    The product: $\frac{1}{2}gt^2$

41. The product of A and v^2: Av^2

CHAPTER 1 REVIEW EXERCISES

1. $\frac{3}{4} \cdot -\frac{3}{4} + \frac{3}{4} = 0$

2. Replace x with the elements in the set and determine whether the inequality is true.
    $x > -1$
    $-4 > -1$ False
    $-2 > -1$ False
    $0 > -1$ True
    $2 > -1$ True
    The inequality is true for 0 and 2.

3. $p \in \{-4, 0, 7\}$
    $-|p|$
    $-|-4| = -4$
    $-|0| = 0$
    $-|7| = -7$

4. $\{-2, -1, 0, 1, 2, 3\}$

5. $\{x | x < -3\}$

6. $\{x | -2 \leq x \leq 3\}$

7. $A \cup B = \{1, 2, 3, 4, 5, 6, 7, 8\}$

8. $A \cap B = \{2, 3\}$

9. $[-3, \infty)$

10. $\{x | x < 1\}$

11. $\{x | x \leq -3\} \cup \{x | x > 0\}$

12. $(-2, 4]$

13. $-10 - (-3) - 8 = -10 + 3 + (-8) = -7 + (-8) = -15$

14. $-204 \div (-17) = 12$

15. $18 - |-12 + 8| = 18 - |-4| = 18 - 4 = 14$

16. $-2 \cdot (4^2) \cdot (-3)^2 = -2 \cdot 16 \cdot 9 = -32 \cdot 9 = -288$

17. $-\frac{3}{8} + \frac{3}{5} - \frac{1}{6} = -\frac{45}{120} + \frac{72}{120} - \frac{20}{120}$
    $= \frac{-45 + 72 - 20}{120}$
    $= \frac{7}{120}$

18. $\frac{3}{5}\left(-\frac{10}{21}\right)\left(-\frac{7}{15}\right) = \frac{3 \cdot 10 \cdot 7}{5 \cdot 21 \cdot 15}$
    $= \frac{3 \cdot 2 \cdot 5 \cdot 7}{5 \cdot 3 \cdot 7 \cdot 3 \cdot 5}$
    $= \frac{2}{15}$

19. $-\frac{3}{8} \div \frac{3}{5} = -\frac{3}{8} \cdot \frac{5}{3}$
    $= -\frac{3 \cdot 5}{8 \cdot 3}$
    $= -\frac{5}{8}$

20. $-4.07 + 2.3 - 1.07 = -1.77 - 1.07 = -2.84$

21. $-3.286 \div (-1.06) = 3.1$

Copyright © Houghton Mifflin Company. All rights reserved.

22. $20 \div \dfrac{3^2 - 2^2}{3^2 + 2^2} = 20 \div \dfrac{9-4}{9+4}$
$= 20 \div \dfrac{5}{13}$
$= 20 \cdot \dfrac{13}{5}$
$= 52$

23. $2a^2 - \dfrac{3b}{a} = 2(-3)^2 - \dfrac{3(2)}{-3}$
$= 2(-3)^2 - \dfrac{6}{-3}$
$= 2(-3)^2 - (-2)$
$= 2(9) - (-2)$
$= 18 + 2$
$= 20$

24. $(a - 2b^2) \div ab$
$(4 - 2(-3)^2) \div (4)(-3) = (4 - 2(9)) \div (4)(-3)$
$= (4 - 18) \div (4)(-3)$
$= -14 \div [(4)(-3)]$
$= -14 \div -12$
$= \dfrac{-14}{-12}$
$= \dfrac{7}{6}$

25. 3

26. y

27. (ab)

28. 4

29. The Inverse Property of Addition

30. The Associative Property of Multiplication

31. $-2(x-3) + 4(2-x) = -2x + 6 + 8 - 4x = -6x + 14$

32. $4y - 3[x - 2(3 - 2x) - 4y] = 4y - 3[x - 6 + 4x - 4y]$
$= 4y - 3[5x - 6 - 4y]$
$= 4y - 15x + 18 + 12y$
$= 16y - 15x + 18$

33. The unknown number: x
The sum of the number and four: $x + 4$
$4(x + 4) = 4x + 16$

34. The unknown number: x
The difference between the number and two: $x - 2$
Twice the difference between the number and two: $2(x - 2)$
$2(x - 2) + 8 = 2x - 4 + 8 = 2x + 4$

35. Let x be the smaller of the numbers. Then the larger number is $40 - x$.
The sum of twice x and five more than $40 - x$.
$2x + (40 - x + 5) = x + 45$

36. Let x be the larger number.
Then the smaller number is $9 - x$.
The difference between three more than twice $(9 - x)$ and one more than x.
$[2(9 - x) + 3] - (x + 1) = 18 - 2x + 3 - x - 1$
$= -3x + 20$

37. The width of the rectangle: W
The length is 3 feet less than $3W$.
The length is $3W - 3$.

38. Let the first integer be x.
The second integer is five more than four times x.
$4x + 5$ is the magnitude of the second integer.

CHAPTER 1 TEST

1. 12

2. Replace x with each element in the set and determine whether the inequality is true.
$-1 > x$
$-1 > -5$ True
$-1 > 3$ False
$-1 > 7$ False
The inequality is true for -5.

3. $2 - (-12) + 3 - 5 = 2 + 12 + 3 + (-5)$
$= 14 + 3 + (-5)$
$= 17 + (-5)$
$= 12$

4. $(-2)(-3)(-5) = (6)(-5) = -30$

5. $-180 \div 12 = -15$

6. $|-3 - (-5)| = |-3 + 5| = |2| = 2$

7. $-5^2 \cdot 4 = -25 \cdot 4 = -100$

8. $(-2)^3 (-3)^2 = (-8)(9) = -72$

9. $\dfrac{2}{3} - \dfrac{5}{12} + \dfrac{4}{9} = \dfrac{24}{36} - \dfrac{15}{36} + \dfrac{16}{36}$
$= \dfrac{24 - 15 + 16}{36}$
$= \dfrac{25}{36}$

10. $\left(-\dfrac{2}{3}\right)\left(\dfrac{9}{15}\right)\left(\dfrac{10}{27}\right) = -\dfrac{2 \cdot 3 \cdot 3 \cdot 2 \cdot 5}{3 \cdot 3 \cdot 5 \cdot 3 \cdot 3 \cdot 3}$
$= -\dfrac{4}{27}$

11. $4.27 - 6.98 + 1.3 = -2.71 + 1.3 = -1.41$

12. $-15.092 \div 3.08 = -4.9$

12   Chapter 1: Review of Real Numbers

13. $12 - 4\left(\dfrac{5^2 - 1}{3}\right) \div 16 = 12 - 4\left(\dfrac{25 - 1}{3}\right) \div 16$
$= 12 - 4\left(\dfrac{24}{3}\right) \div 16$
$= 12 - 4(8) \div 16$
$= 12 - 32 \div 16$
$= 12 - 2$
$= 10$

14. $8 - 4(2 - 3)^2 \div 2 = 8 - 4(-1)^2 \div 2$
$= 8 - 4(1) \div 2$
$= 8 - 4 \div 2$
$= 8 - 2$
$= 6$

15. $(a - b)^2 \div (2b + 1) = (2 - (-3))^2 \div (2(-3) + 1)$
$= (5)^2 \div (-6 + 1)$
$= (5)^2 \div (-5)$
$= 25 \div (-5)$
$= -5$

16. $\dfrac{b^2 - c^2}{a - 2c} = \dfrac{(3)^2 - (-1)^2}{2 - 2(-1)}$
$= \dfrac{9 - 1}{2 - (-2)}$
$= \dfrac{8}{4}$
$= 2$

17. 4

18. The Distributive Property

19. $3x - 2(x - y) - 3(y - 4x)$
$= 3x - 2x + 2y - 3y + 12x$
$= 13x - y$

20. $2x - 4[2 - 3(x + 4y) - 2]$
$= 2x - 4[2 - 3x - 12y - 2]$
$= 2x - 4[-3x - 12y]$
$= 2x + 12x + 48y$
$= 14x + 48y$

21. the unknown number: n
three less than the number: $n - 3$
the product of three less than the number and nine:
$(n - 3)(9)$
$13 - (n - 3)(9) = 13 - 9n + 27 = 40 - 9n$

22. The unknown number: n
The total of twelve times the number and twenty-seven: $12n + 27$
$\dfrac{1}{3}(12n + 27) = 4n + 9$

23. $A \cup B = \{1, 2, 3, 4, 5, 7\}$

24. $A \cup B = \{-2, -1, 0, 1, 2, 3\}$

25. $A \cap B = \{5, 7\}$

26. $A \cap B = \{-1, 0, 1\}$

27. $(\infty, -1]$

28. $(3, \infty)$

29. $\{x \mid x \leq 3\} \cup \{x \mid x < -2\}$

30. $\{x \mid x < 3\} \cap \{x \mid x > -2\}$

Copyright © Houghton Mifflin Company. All rights reserved.

CHAPTER 2: FIRST-DEGREE EQUATIONS AND INEQUALITIES

CHAPTER 2 PREP TEST

1. −4   [1.2.1]
2. −6   [1.2.1]
3. 3   [1.2.1]
4. 1   [1.2.2]
5. $-\dfrac{1}{2}$ [1.2.2]
6. $10x - 5$ [1.3.3]
7. $6(x - 2) + 3 = 6x - 12 + 3$ [1.2.1]
   $= 6x - 9$
8. $3n + 6$ [1.3.3]
9. $0.08x + 0.05(400 - x)$ [1.3.3]
   $= 0.08x + 20 - 0.05x$
   $= 0.03x + 20$
10. $20 - n$ [1.4.2]

Section 2.1

Concept Review 2.1

1. Never true
   An equation must have an equals sign.

3. Sometimes true
   $8 = 4 + 2$ is a false equation. $6 = 6$ is a true equation. Neither equation contains a variable.

5. Never true
   This equation has no solution. Any replacement value for x will result in a false equation.

7. Always true

Objective 2.1.1 Exercises

1. An equation contains an equals sign; an expression does not.

3. The Addition Property of Equations states that the same quantity can be added to each side of an equation without changing the solution of the equation. This property is used to remove a term from one side of an equation by adding the opposite of that term to each side of the equation.

5. $7 - 3m = 4$
   $7 - 3(1) = 4$
   $\quad\quad 4 = 4$
   Yes

7. $6x - 1 = 7x + 1$
   $6(-2) - 1 = 7(-2) + 1$
   $\quad -12 - 1 = -14 + 1$
   $\quad\quad -13 = -13$
   Yes

9. To solve the equation $a - 42 = 13$, use the Addition Property of Equations to add <u>42</u> to each side of the equation. The solution is <u>55</u>.

11. To solve the equation $-\dfrac{2}{5}n = 8$, use the Multiplication Property of Equations to multiply each side of the equation by $-\dfrac{5}{2}$. The solution is <u>−20</u>.

13. $x - 2 = 7$
    $x - 2 + 2 = 7 + 2$
    $\quad\quad x = 9$
    The solution is 9.

15. $a + 3 = -7$
    $a + 3 - 3 = -7 - 3$
    $\quad\quad a = -10$
    The solution is −10.

17. $3x = 12$
    $\dfrac{3x}{3} = \dfrac{12}{3}$
    $x = 4$
    The solution is 4.

19. $\dfrac{2}{7} + x = \dfrac{17}{21}$
    $\dfrac{2}{7} - \dfrac{2}{7} + x = \dfrac{17}{21} - \dfrac{2}{7}$
    $x = \dfrac{17}{21} - \dfrac{6}{21}$
    $x = \dfrac{11}{21}$
    The solution is $\dfrac{11}{21}$.

21. $\dfrac{5}{8} - y = \dfrac{3}{4}$
    $\dfrac{5}{8} - \dfrac{5}{8} - y = \dfrac{3}{4} - \dfrac{5}{8}$
    $-y = \dfrac{6}{8} - \dfrac{5}{8}$
    $-y = \dfrac{1}{8}$
    $(-1)(-y) = \dfrac{1}{8}(-1)$
    $y = -\dfrac{1}{8}$
    The solution is $-\dfrac{1}{8}$.

23. $\dfrac{3}{5}y = 12$

$\dfrac{5}{3}\left(\dfrac{3}{5}y\right) = \dfrac{5}{3}(12)$

$y = 20$

The solution is 20.

25. $\dfrac{3a}{7} = -21$

$\dfrac{7}{3}\left(\dfrac{3}{7}a\right) = \dfrac{7}{3}(-21)$

$a = -49$

The solution is −49.

27. $-\dfrac{5}{12}y = \dfrac{7}{16}$

$-\dfrac{12}{5}\left(-\dfrac{5}{12}y\right) = -\dfrac{12}{5}\left(\dfrac{7}{16}\right)$

$y = -\dfrac{21}{20}$

The solution is $-\dfrac{21}{20}$.

29. $b - 14.72 = -18.45$

$b - 14.72 + 14.72 = -18.45 + 14.72$

$b = -3.73$

The solution is −3.73.

31. $3x + 5x = 12$

$8x = 12$

$\dfrac{8x}{8} = \dfrac{12}{8}$

$x = \dfrac{3}{2}$

The solution is $\dfrac{3}{2}$.

33. $2x - 4 = 12$

$2x - 4 + 4 = 12 + 4$

$2x = 16$

$\dfrac{2x}{2} = \dfrac{16}{2}$

$x = 8$

35. $4x + 2 = 4x$

$4x - 4x + 2 = 4x - 4x$

$2 = 0$

The equation has no solution.

37. $2x + 2 = 3x + 5$

$2x - 3x + 2 = 3x - 3x + 5$

$-x + 2 = 5$

$-x + 2 - 2 = 5 - 2$

$-x = 3$

$(-1)(-x) = (-1)(3)$

$x = -3$

The solution is −3.

39. $2 - 3t = 3t - 4$

$2 - 3t - 3t = 3t - 3t - 4$

$2 - 6t = -4$

$2 - 2 - 6t = -4 - 2$

$-6t = -6$

$\dfrac{-6t}{-6} = \dfrac{-6}{-6}$

$t = 1$

The solution is 1.

41. $2a - 3a = 7 - 5a$

$-a = 7 - 5a$

$-a + 5a = 7 - 5a + 5a$

$4a = 7$

$\dfrac{4a}{4} = \dfrac{7}{4}$

$a = \dfrac{7}{4}$

The solution is $\dfrac{7}{4}$.

43. $\dfrac{5}{8}b - 3 = 12$

$\dfrac{5}{8}b - 3 + 3 = 12 + 3$

$\dfrac{5}{8}b = 15$

$\dfrac{8}{5}\left(\dfrac{5}{8}b\right) = \dfrac{8}{5}(15)$

$b = 24$

The solution is 24.

45. $b + \dfrac{1}{5}b = 2$

$\dfrac{6}{5}b = 2$

$\dfrac{5}{6}\left(\dfrac{6}{5}b\right) = \dfrac{5}{6}(2)$

$b = \dfrac{5}{3}$

The solution is $\dfrac{5}{3}$.

Chapter 2: First-Degree Equations And Inequalities

47. $2x - 9x + 3 = 6 - 5x$
    $-7x + 3 = 6 - 5x$
    $-7x + 5x + 3 = 6 - 5x + 5x$
    $-2x + 3 = 6$
    $-2x + 3 - 3 = 6 - 3$
    $-2x = 3$
    $\dfrac{-2x}{-2} = \dfrac{3}{-2}$
    $x = -\dfrac{3}{2}$
    The solution is $-\dfrac{3}{2}$.

49. $2y - 4 + 8y = 7y - 8 + 3y$
    $10y - 4 = 10y - 8$
    $10y - 10y - 4 = 10y - 10y - 8$
    $-4 = -8$
    The equation has no solution.

51. $9 + 4x - 12 = -3x + 5x + 8$
    $4x - 3 = 2x + 8$
    $4x - 2x - 3 = 2x - 2x + 8$
    $2x - 3 = 8$
    $2x - 3 + 3 = 8 + 3$
    $2x = 11$
    $\dfrac{2x}{2} = \dfrac{11}{2}$
    $x = \dfrac{11}{2}$
    The solution is $\dfrac{11}{2}$.

53. $5.3y + 0.35 = 5.02y$
    $5.3y - 5.3y + 0.35 = 5.02y - 5.3y$
    $0.35 = -0.28y$
    $\dfrac{0.35}{-0.28} = \dfrac{-0.28y}{-0.28}$
    $-1.25 = y$
    The solution is -1.25.

55. Greater than 1. The solution is $b = \dfrac{a}{-5}$. Because $a < -5$, the numerator and denominator are both negative, giving a positive number. The actual value is greater than 1 because $|a| > |-5|$.

Objective 2.1.2 Exercises

57. 42

59. $2x + 3(x - 5) = 15$
    $2x + 3x - 15 = 15$
    $5x - 15 = 15$
    $5x = 30$
    $\dfrac{5x}{5} = \dfrac{30}{5}$
    $x = 6$
    The solution is 6.

61. $5(2 - b) = -3(b - 3)$
    $10 - 5b = -3b + 9$
    $10 = 2b + 9$
    $1 = 2b$
    $\dfrac{1}{2} = \dfrac{2b}{2}$
    $\dfrac{1}{2} = b$
    The solution is $\dfrac{1}{2}$.

63. $3(y - 5) - 5y = 2y + 9$
    $3y - 15 - 5y = 2y + 9$
    $-2y - 15 = 2y + 9$
    $-4y - 15 = 9$
    $-4y = 24$
    $\dfrac{-4y}{-4} = \dfrac{24}{-4}$
    $y = -6$
    The solution is -6.

65. $2x - 3(x - 4) = 2(3 - 2x) + 2$
    $2x - 3x + 12 = 6 - 4x + 2$
    $-x + 12 = 8 - 4x$
    $3x + 12 = 8$
    $3x = -4$
    $\dfrac{3x}{3} = \dfrac{-4}{3}$
    $x = -\dfrac{4}{3}$
    The solution is $-\dfrac{4}{3}$.

67. $-4(7y - 1) + 5y = -2(3y + 4) - 3y$
    $-28y + 4 + 5y = -6y - 8 - 3y$
    $-23y + 4 = -9y - 8$
    $-14y + 4 = -8$
    $-14y = -12$
    $\dfrac{-14y}{-14} = \dfrac{-12}{-14}$
    $y = \dfrac{6}{7}$
    The solution is $\dfrac{6}{7}$.

69.
$$2[4+2(5-x)-2x]=4x-7$$
$$2[4+10-2x-2x]=4x-7$$
$$2[14-4x]=4x-7$$
$$28-8x=4x-7$$
$$28-12x=-7$$
$$-12x=-35$$
$$\frac{-12x}{-12}=\frac{-35}{-12}$$
$$x=\frac{35}{12}$$

The solution is $\frac{35}{12}$.

71.
$$-3[x+4(x+1)]=x+4$$
$$-3(x+4x+4)=x+4$$
$$-3(5x+4)=x+4$$
$$-15x-12=x+4$$
$$-16x-12=4$$
$$-16x=16$$
$$\frac{-16x}{-16}=\frac{16}{-16}$$
$$x=-1$$

The solution is -1.

73.
$$5-6[2t-2(t+3)]=8-t$$
$$5-6(2t-2t-6)=8-t$$
$$5-12t+12t+36=8-t$$
$$41=8-t$$
$$33=-t$$
$$-33=t$$

The solution is -33.

75.
$$3[x-(2-x)-2x]=3(4-x)$$
$$3[x-2+x-2x]=12-3x$$
$$3(-2)=12-3x$$
$$-6=12-3x$$
$$-18=-3x$$
$$\frac{-18}{-3}=\frac{-3x}{-3}$$
$$6=x$$

The solution is 6.

77.
$$\frac{3}{4}t-\frac{7}{12}=\frac{1}{6}$$
$$12\left(\frac{3}{4}t-\frac{7}{12}\right)=12\cdot\frac{1}{6}$$
$$\frac{12\cdot 3t}{4}-\frac{12\cdot 7}{12}=2$$
$$9t-7=2$$
$$9t=9$$
$$\frac{9t}{9}=\frac{9}{9}$$
$$t=1$$

The solution is 1.

79.
$$\frac{1}{2}x-\frac{3}{4}x+\frac{5}{8}=\frac{3}{2}x-\frac{5}{2}$$
$$8\left(\frac{1}{2}x-\frac{3}{4}x+\frac{5}{8}\right)=8\left(\frac{3}{2}x-\frac{5}{2}\right)$$
$$\frac{8\cdot x}{2}-\frac{8\cdot 3x}{4}+\frac{8\cdot 5}{8}=\frac{8\cdot 3x}{2}-\frac{8\cdot 5}{2}$$
$$4x-6x+5=12x-20$$
$$-2x+5=12x-20$$
$$-14x+5=-20$$
$$-14x=-25$$
$$\frac{-14x}{-14}=\frac{-25}{-14}$$
$$x=\frac{25}{14}$$

The solution is $\frac{25}{14}$.

81.
$$\frac{2a-9}{5}+3=2a$$
$$5\left(\frac{2a-9}{5}+3\right)=5\cdot 2a$$
$$\frac{5(2a-9)}{5}+5\cdot 3=10a$$
$$2a-9+15=10a$$
$$2a+6=10a$$
$$6=8a$$
$$\frac{6}{8}=\frac{8a}{8}$$
$$\frac{3}{4}=a$$

The solution is $\frac{3}{4}$.

Chapter 2: First-Degree Equations And Inequalities

83.
$$\frac{2x-1}{4} + \frac{3x+4}{8} = \frac{1-4x}{12}$$
$$24\left(\frac{2x-1}{4} + \frac{3x+4}{8}\right) = 24\left(\frac{1-4x}{12}\right)$$
$$\frac{24(2x-1)}{4} + \frac{24(3x+4)}{8} = \frac{24(1-4x)}{12}$$
$$6(2x-1) + 3(3x+4) = 2(1-4x)$$
$$12x - 6 + 9x + 12 = 2 - 8x$$
$$21x + 6 = 2 - 8x$$
$$29x + 6 = 2$$
$$29x = -4$$
$$\frac{29x}{29} = \frac{-4}{29}$$
$$x = -\frac{4}{29}$$

The solution is $-\frac{4}{29}$.

85.
$$\frac{1}{5}(20x + 30) = \frac{1}{3}(6x + 36)$$
$$15\left[\frac{1}{5}(20x + 30)\right] = 15\left[\frac{1}{3}(6x + 36)\right]$$
$$3(20x + 30) = 5(6x + 36)$$
$$60x + 90 = 30x + 180$$
$$30x + 90 = 180$$
$$30x = 90$$
$$\frac{30x}{30} = \frac{90}{30}$$
$$x = 3$$

The solution is 3.

87.
$$2(y - 4) + 8 = \frac{1}{2}(6y + 20)$$
$$2y - 8 + 8 = \frac{6y}{2} + \frac{20}{2}$$
$$2y = 3y + 10$$
$$-y = 10$$
$$(-1)(-y) = (-1)(10)$$
$$y = -10$$

The solution is −10.

89.
$$\frac{1}{2}x - \frac{3}{5} = \frac{2}{5}x + \frac{1}{2}$$
$$10 \cdot \left(\frac{1}{2}x - \frac{3}{5}\right) = 10 \cdot \left(\frac{2}{5}x + \frac{1}{2}\right)$$
$$\frac{10}{2}x - \frac{30}{5} = \frac{20}{5}x + \frac{10}{2}$$
$$5x - 6 = 4x + 5$$
$$x - 6 = 5$$
$$x = 11$$

The solution is 11.

91.
$$-1.6(b - 2.35) = -11.28$$
$$-1.6b + 3.76 = -11.28$$
$$-1.6b = -15.04$$
$$\frac{-1.6b}{-1.6} = \frac{-15.04}{-1.6}$$
$$b = 9.4$$

The solution is 9.4.

93.
$$x + 0.06(60) = 0.20(x + 20)$$
$$100 \cdot [x + 0.06(60)] = 100 \cdot [0.20(x + 20)]$$
$$100x + 6(60) = 20(x + 20)$$
$$100x + 360 = 20x + 400$$
$$80x + 360 = 400$$
$$80x = 40$$
$$\frac{80x}{80} = \frac{40}{80}$$
$$x = 0.5$$

The solution is 0.5.

95.
$$-3[5 - 4(x - 2)] = 5(x - 5)$$
$$-15 + 12(x - 2) = 5x - 25$$

Objective 2.1.3 Exercises

97. A group of ten friends go to a restaurant. Some people in the group order the all-you-can-eat-buffet, while the rest of the group order the soup-and-sandwich combo. If 2 people order the buffet, then the number of people who order the soup-and-sandwich combo is _8_. If 7 people order the buffet, then the number of people who order the soup-and-sandwich combo is _3_. If n people order the buffet, then an expression that represents the number of people who order the soup-and-sandwich combo is $10 - n$.

99. Strategy
To find the number of bags purchased, write and solve an equation using b to represent the number of bags purchased.

Solution
The first bag purchased cost $10.90. That means that $b - 1$ bags cost $10.50(b - 1)$.
$$84.40 = 10.90 + 10.50(b - 1)$$
$$84.40 = 10.90 + 10.50b - 10.50$$
$$84.40 = 0.40 + 10.50b$$
$$84 = 10.50b$$
$$8 = b$$

The customer purchased 8 bags of feed.

Chapter 2: First-Degree Equations And Inequalities

101. Strategy
To find the charge per hour for labor, write and solve an equation using L to represent the charge per hour for labor.

Solution
The total charge for labor was $4L$ dollars.
$$316.55 = 4L + 148.55$$
$$168 = 4L$$
$$42 = L$$
Labor cost $42 per hour.

103. Strategy
To find the employee's regular hourly rate, write and solve an equation using h to represent the hourly rate and $1.5h$ to represent the overtime rate.

Solution
The wages earned for the first forty hours plus the wages earned for overtime are $642.
$$40h + 9(1.5h) = 642$$
$$40h + 13.5h = 642$$
$$53.5h = 642$$
$$h = 12$$
The regular hourly rate is $12.

105. Strategy
To find the number of each type of ticket, write and solve an equation using g to represent the number of outfield grandstand tickets and $12 - g$ to represent the number of right field box tickets.

Solution
The total price of the tickets was $414.
$$27g + 45(12 - g) = 414$$
$$27g + 540 - 45g = 414$$
$$-18g + 540 = 414$$
$$-18g = -126$$
$$g = 7$$
The fraternity purchased 7 outfield grandstand tickets and 5 right field box tickets.

107. Strategy
To find the number of mezzanine tickets purchased, write and solve an equation using m to represent the number of mezzanine tickets and $7 - m$ to represent the number of balcony tickets.

Solution
The total cost of the tickets was $275.00.
$$50m + 35(7 - m) = 275$$
$$50m + 245 - 35m = 275$$
$$15m + 245 = 275$$
$$15m = 30$$
$$m = 2$$
Two mezzanine tickets were purchased.

Applying Concepts 2.1

109.
$$\frac{1}{\frac{1}{y}} = -9$$
$$y = -9$$
The solution is -9.

111.
$$\frac{10}{\frac{3}{x}} - 5 = 4x$$
$$10\left(\frac{x}{3}\right) - 5 = 4x$$
$$\frac{10}{3}x - 5 = 4x$$
$$\frac{10}{3}x - \frac{10}{3}x - 5 = 4x - \frac{10}{3}x$$
$$-5 = \frac{2}{3}x$$
$$\frac{3}{2}(-5) = \frac{3}{2}\left(\frac{2}{3}x\right)$$
$$-\frac{15}{2} = x$$
The solution is $-\frac{15}{2}$.

113.
$$2[3(x+4) - 2(x+1)] = 5x + 3(1-x)$$
$$2[3x + 12 - 2x - 2] = 5x + 3 - 3x$$
$$2(x + 10) = 2x + 3$$
$$2x + 20 = 2x + 3$$
$$2x - 2x + 20 = 2x - 2x + 3$$
$$20 = 3$$
There is no solution.

115.
$$\frac{4[(x-3) + 2(1-x)]}{5} = x + 1$$
$$\frac{4(x - 3 + 2 - 2x)}{5} = x + 1$$
$$\frac{4(-x - 1)}{5} = x + 1$$
$$\frac{-4x - 4}{5} = x + 1$$
$$5\left(\frac{-4x - 4}{5}\right) = 5(x + 1)$$
$$-4x - 4 = 5x + 5$$
$$-4x + 4x - 4 = 5x + 4x + 5$$
$$-4 = 9x + 5$$
$$-4 - 5 = 9x + 5 - 5$$
$$-9 = 9x$$
$$\frac{-9}{9} = \frac{9x}{9}$$
$$-1 = x$$
The solution is -1.

117.
$$3(2x+2)-4(x-3)=2(x+9)$$
$$6x+6-4x+12=2x+18$$
$$2x+18=2x+18$$
$$2x-2x+18=2x-2x+18$$
$$18=18$$
$$0=0$$
The solution is all real numbers.

Section 2.2
Concept Review 2.2
1. Always true
3. Always true

Objective 2.2.1 Exercises

1.

Coin	Number of coins	•	Value of the coin in cents	=	Total Value of the coins in cents
Nickel	n	•	5	=	$5n$
Quarter	$35-n$	•	25	=	$25(35-n)$

3.
a. There are $20d$ nickels in d dollars.
b. There are $10d$ dimes in d dollars.
c. There are $4d$ quarters in d dollars.

5. Strategy
Number of nickels: x
Number of dimes: $56-x$

Coin	Number	Value	Total Value
Nickel	x	5	$5x$
Dime	$56-x$	10	$10(56-x)$

The sum of the total values of each type of coin equals the total value of all the coins (400 cents).
$5x+10(56-x)=400$

Solution
$$5x+10(56-x)=400$$
$$5x+560-10x=400$$
$$-5x+560=400$$
$$-5x=-160$$
$$x=32$$
$56-x=56-32=24$
There are 24 dimes in the bank.

7. Strategy
Number of twenty-dollar bills: x
Number of five-dollar bills: $68-x$

Bill	Number	Value	Total Value
20-dollar	x	20	$20x$
5-dollar	$68-x$	5	$5(68-x)$

(solution 7 continued) The sum of the total values of each type of bill equals the total value of all the bills (730 dollars).
$20x+5(68-x)=730$

Solution
$$20x+5(68-x)=730$$
$$20x+340-5x=730$$
$$15x+340=730$$
$$15x=390$$
$$x=26$$
The cashier has 26 twenty-dollar bills.

11. Strategy
Number of quarters: x
Number of dimes: $4x$
Number of nickels: $25-5x$

Coin	Number	Value	Total Value
Quarter	x	25	$25x$
Dime	$4x$	10	$10(4x)$
Nickel	$25-5x$	5	$5(25-5x)$

The sum of the total values of each type of coin equals the total value of all the coins (205 cents).
$25x+10(4x)+5(25-5x)=205$

Solution
$$25x+10(4x)+5(25-5x)=205$$
$$25x+40x+125-25x=205$$
$$40x+125=205$$
$$40x=80$$
$$x=2$$
$4x=4\cdot 2=8$
There are 8 dimes in the bank.

13. Strategy
Number of 3¢ stamps: x
Number of 8¢ stamps: $2x - 3$
Number of 13¢ stamps: $2(2x - 3)$

Stamp	Number	Value	Total Value
3¢	x	3	$3x$
8¢	$2x - 3$	8	$8(2x - 3)$
13¢	$2(2x - 3)$	13	$13(2)(2x - 3)$

The sum of the total values of each type of stamp equals the total value of all the stamps (253 cents).
$3x + 8(2x - 3) + 26(2x - 3) = 253$

Solution
$$3x + 8(2x - 3) + 26(2x - 3) = 253$$
$$3x + 16x - 24 + 52x - 78 = 253$$
$$71x - 102 = 253$$
$$71x = 355$$
$$x = 5$$
There are five 3¢ stamps in the collection.

15. Strategy
Number of 18¢ stamps: x
Number of 8¢ stamps: $2x$
Number of 11¢ stamps: $x + 3$

Stamp	Number	Value	Total Value
18¢	x	18	$18x$
8¢	$2x$	8	$8(2x)$
11¢	$x + 3$	11	$11(x + 3)$

The sum of the total values of each type of stamp equals the total value of all the stamps (348 cents).
$18x + 8(2x) + 11(x + 3) = 348$

Solution
$$18x + 8(2x) + 11(x + 3) = 348$$
$$18x + 16x + 11x + 33 = 348$$
$$45x + 33 = 348$$
$$45x = 315$$
$$x = 7$$
There are seven 18¢ stamps in the collection.

Objective 2.2.2 Exercises

17. Integers that follow one another in order are called <u>consecutive</u> integers.

19. Let x represent the smallest integer. Because consecutive integers differ by 1, the next integer can be represented by $x + 1$. The third integer can be represented by $(x + 1) + 1$, or just $x + 2$.

21. Strategy
The smaller integer: n
The larger integer: $10 - n$
Three times the larger integer is three less than eight times the smaller integer.
$3(10 - n) = 8n - 3$

Solution
$$3(10 - n) = 8n - 3$$
$$30 - 3n = 8n - 3$$
$$-11n = -33$$
$$n = 3$$
$10 - n = 10 - 3 = 7$
The integers are 3 and 7.

23. Strategy
The larger integer: n
The smaller integer: $n - 8$
The sum of the two integers is fifty.
$n + (n - 8) = 50$

Solution
$$n + (n - 8) = 50$$
$$2n - 8 = 50$$
$$2n = 58$$
$$n = 29$$
$n - 8 = 29 - 8 = 21$
The two integers are 21 and 29.

25. Strategy
The first number: n
The second number: $2n + 2$
The third number: $3n - 5$
The sum of the three numbers is 123.

Solution
$$n + (2n + 2) + (3n - 5) = 123$$
$$6n - 3 = 123$$
$$6n = 126$$
$$n = 21$$
$2n + 2 = 2(21) + 2 = 42 + 2 = 44$
$3n - 5 = 3(21) - 5 = 63 - 5 = 58$
The numbers are 21, 44, and 58.

27. Strategy
The first integer: n
The second consecutive integer: $n + 1$
The third consecutive integer: $n + 2$
The sum of the integers is –57.
$n + (n + 1) + (n + 2) = -57$

Solution
$$n + (n + 1) + (n + 2) = -57$$
$$3n + 3 = -57$$
$$3n = -60$$
$$n = -20$$
$n + 1 = -20 + 1 = -19$
$n + 2 = -20 + 2 = -18$
The integers are –20, –19, and –18.

29. Strategy
  The first odd integer: n
  The second consecutive odd integer: $n + 2$
  The third consecutive odd integer: $n + 4$
  Five times the smallest of the three integers is ten more than twice the largest.
  $5n = 2(n + 4) + 10$

  Solution
  $5n = 2(n + 4) + 10$
  $5n = 2n + 8 + 10$
  $5n = 2n + 18$
  $3n = 18$
  $n = 6$
  Since 6 is not an odd integer, there is no solution.

31. Strategy
  The first odd integer: n
  The second consecutive odd integer: $n + 2$
  The third consecutive odd integer: $n + 4$
  Three times the middle integer is seven more than the sum of the first and third integers.
  $3(n + 2) = [n + (n + 4)] + 7$

  Solution
  $3(n + 2) = [n + (n + 4)] + 7$
  $3n + 6 = 2n + 11$
  $n = 5$
  $n + 2 = 5 + 2 = 7$
  $n + 4 = 5 + 4 = 9$
  The odd integers are 5, 7, and 9.

33. d

Applying Concepts 2.2

35. Strategy
  Number of nickels: x
  Number of dimes: $2x - 2$
  The number of nickels plus the number of dimes is the total number of coins in the bank (52).

  Solution
  $x + (2x - 2) = 52$
  $3x - 2 = 52$
  $3x = 54$
  $\dfrac{3x}{3} = \dfrac{54}{3}$
  $x = 18$
  $2x - 2 = 2(18) - 2 = 36 - 2 = 34$
  There are 18 nickels in the bank and 34 dimes in the bank. The value of the coins in the bank is equal to the value of the dimes ($.10)(34) plus the value of the nickels ($.05)(18).
  $\$.10(34) + \$.05(18) = \$3.40 + \$.90 = \$4.30$
  The total value of the coins in the bank is $4.30.

37. Strategy
  Number of 5¢ stamps: x
  Number of 3¢ stamps: $x + 6$
  Number of 7¢ stamps: $(x + 6) + 2 = x + 8$

Stamp	Number	Value	Total Value
5¢	x	5	$5x$
3¢	$x + 6$	3	$3(x + 6)$
7¢	$x + 8$	7	$7(x + 8)$

  The sum of the total value of each type of stamp equals the total value of all the stamps (194 cents.)
  $5x + 3(x + 6) + 7(x + 8) = 194$

  Solution
  $5x + 3(x + 6) + 7(x + 8) = 194$
  $5x + 3x + 18 + 7x + 56 = 194$
  $15x + 74 = 194$
  $15x = 120$
  $x = 8$
  $x + 6 = 8 + 6 = 14$
  The number of 3¢ stamps in the collection is 14.

39. Strategy
  First odd integer: n
  Second consecutive odd integer: $n + 2$
  Third consecutive odd integer: $n + 4$
  Fourth consecutive odd integer: $n + 6$
  The sum of the four integers is –64.
  $n + (n + 2) + (n + 4) + (n + 6) = -64$

  Solution
  $n + (n + 2) + (n + 4) + (n + 6) = -64$
  $4n + 12 = -64$
  $4n = -76$
  $n = -19$
  $n + 6 = -19 + 6 = -13$
  The smallest of the four integers is –19.
  The largest of the four integers is –13.
  The sum of the smallest and largest integers is $-19 + (-13)$ or –32.

22 Chapter 2: First-Degree Equations And Inequalities

41. **Strategy**
Units digit: x
Tens digit: $x - 1$
Hundreds digit: $6 - (x + x - 1)$
The value of the number is 12 more than 100 times the hundreds digit.

Solution
$$x + 10(x - 1) + 100[6 - (x + x - 1)] = 100[6 - (x + x - 1)] + 12$$
$$x + 10x - 10 + 100[6 - 2x + 1)] = 100[6 - 2x + 1] + 12$$
$$11x - 10 + 600 - 200x + 100 = 600 - 200x + 100 + 12$$
$$-189x + 690 = -200x + 712$$
$$11x = 22$$
$$x = 2$$

$x - 1 = 2 - 1 = 1$
$6 - (x + x - 1) = 6 - (2 + 2 - 1) = 6 - 3 = 3$
The number is 312.

Section 2.3
Concept Review 2.3
1. Sometimes true
This is true only when the same amounts of gold at each price are mixed.

3. Never true
It takes a time of $t - \frac{1}{2}$ for the cyclist to overtake the runner.

Objective 2.3.1 Exercises
1. **a.** The total value of a 20-pound bag of bird seed that costs $0.42 per pound is $8.40.

b. The cost per ounce of a 24-ounce box of chocolates that has a total value of $16.80 is $0.70.

3. **Strategy**
Cost of mixture: x

	Amount	Cost	Value
Snow peas	20	1.99	39.80
Petite onions	14	39.80	16.66
Mixture	34	x	$34x$

The sum of the values before mixing equals the value after mixing.
$39.80 + 16.66 = 34x$

Solution
$$56.46 = 34x$$
$$1.66 = x$$
The cost per pound of the mixture is $1.66.

5. **Strategy**
Number of adult tickets: x
Number of child tickets: $460 - x$

	Amount	Cost	Value
Adult tickets	x	5.00	$5x$
Child tickets	$460 - x$	2.00	$2(460 - x)$

The sum of the values of each type of ticket sold equals the total value of all the tickets sold (1880 dollars).
$5x + 2(460 - x) = 1880$

Solution
$$5x + 2(460 - x) = 1880$$
$$5x + 920 - 2x = 1880$$
$$3x + 920 = 1880$$
$$3x = 960$$
$$x = 320$$
There were 320 adult tickets sold.

7. **Strategy**
Liters of imitation maple syrup: x

	Amount	Cost	Value
Imitation Syrup	x	4.00	$4x$
Maple Syrup	50	9.50	$9.50(50)$
Mixture	$50 + x$	5.00	$5(50 + x)$

The sum of the values before mixing equals the value after mixing.
$4x + 9.50(50) = 5(50 + x)$

Solution
$$4x + 9.50(50) = 5(50 + x)$$
$$4x + 475 = 250 + 5x$$
$$-x + 475 = 250$$
$$-x = -225$$
$$x = 225$$
The mixture must contain 225 L of imitation maple syrup.

Chapter 2: First-Degree Equations And Inequalities

9. **Strategy**
Number of pounds of nuts used x
Number of pounds of pretzels used $20 - x$

	Amount	Cost	Value
Nuts	x	3.99	$3.99x$
Pretzels	$20 - x$	1.29	$1.29(20 - x)$
Mixture	20	2.37	47.40

The sum of the values before mixing equals the value after mixing.

$3.99x + 1.29(20 - x) = 47.40$
$3.99x + 25.80 - 1.29x = 47.40$
$2.70x + 25.80 = 47.40$
$2.70x = 21.60$
$x = 8$

The mixture must contain 8 pounds of nuts.

11. **Strategy**
Cost per pound of mixture: x

	Amount	Cost	Value
$6.00 tea	30	6.00	6.00(30)
$3.20 tea	70	3.20	3.20(70)
Mixture	100	x	$100x$

The sum of the values before mixing equals the value after mixing.
$6.00(30) + 3.20(70) = 100x$

Solution
$6.00(30) + 3.20(70) = 100x$
$180 + 224 = 100x$
$404 = 100x$
$4.04 = x$

The cost of the mixture is $4.04 per pound.

13. **Strategy**
Gallons of cranberry juice: x

	Amount	Cost	Value
Cranberry	x	4.20	$4.20x$
Apple	50	2.10	$50(2.10)$
Mixture	$50 + x$	3.00	$3(50 + x)$

The sum of the values before mixing equals the value after mixing.
$4.20x + 50(2.10) = 3(50 + x)$

Solution
$4.20x + 50(2.10) = 3(50 + x)$
$4.20x + 105 = 150 + 3x$
$1.20x + 105 = 150$
$1.20x = 45$
$x = 37.5$

The mixture must contain 37.5 gal of cranberry juice.

15. The mixture will cost between $7 and $10 per pounds.
d and f

Objective 2.3.2 Exercises

17. a. Marvin and Nancy start biking at the same time. Marvin bikes at 12 mph and Nancy bikes at 15 mph. After t hours, Marvin has biked <u>12t</u> miles and Nancy has biked <u>15t</u> miles.

b. A plane flies at a rate of 380 mph in calm air. The wind is blowing at 20 mph. Flying with the wind, the plane flies <u>400</u> mph. Flying against the wind, the plane flies <u>360</u> mph.

19. a. Less than

b. Equal to

c. 3 mi

21. **Strategy**
To find the rate of speed solve the equation $d = rt$ for r using $d = 20$ mi, and $t = 30$ min $= 0.5$ h.

Solution
$d = rt$
$20 = r0.5$
$40 = r$
The student drives at 40 mph.

23. **Strategy**
To find the time for the Boeing 737-800, first solve the equation $d = rt$ for r using $d = 1680$ mi, and $t = 3$ h. Then since the Boeing 747-400 is 30 mph faster, subtract 30 mph from the value of r. Finally, solve the equation $d = rt$ for t, using $d = 1680$ mi and r.

Solution
$d = rt$
$1680 = r3$
$560 = r$
$560 - 30 = 530$ mph
$d = rt$
$1680 = 530t$
$3.2 \approx t$
It would take the Boeing 737-800 3.2 h.

25. **Strategy**
Time spent riding: t

	Rate	Time	Distance
1st cyclist	17	t	$17t$
2nd cyclist	19	t	$19t$

The total distance is 54 mi.

Solution
$17t + 19t = 54$
$36t = 54$
$t = 1.5$

The cyclists will meet 1.5 hours after 1 P.M., or at 2:30 P.M.

24 Chapter 2: First-Degree Equations And Inequalities

27. Strategy
t = time for bicyclist
$t + 0.5$ = time for in-line skater

	Rate	Time	Distance
Bicyclist	18	t	$18t$
In-line skater	10	$t + 0.5$	$10(t + 0.5)$

The bicyclist and in-line skater travel the same distance.

Solution
$18t = 10(t + 0.5)$
$18t = 10t + 5$
$8t = 5$
$t = 0.625$
$d = 18t = 11.25$
The bicyclist will overtake the in-line skater after 11.25 miles.

29. Strategy
Rate of the first plane: r
Rate of the second plane: $r + 80$

	Rate	Time	Distance
1st plane	r	1.5	$1.5r$
2nd plane	$r + 80$	1.5	$1.5(r + 80)$

The total distance traveled by the two planes is 1380 mi.
$1.5r + 1.5(r + 80) = 1380$

Solution
$1.5r + 1.5(r + 80) = 1380$
$1.5r + 1.5r + 120 = 1380$
$3r + 120 = 1380$
$3r = 1260$
$r = 420$
$r + 80 = 420 + 80 = 500$
The speed of the first plane is 420 mph.
The speed of the second plane is 500 mph.

31. Strategy
Rate of the second plane: r
Rate of the first plane: $r - 50$

	Rate	Time	Distance
2nd plane	r	2.5	$2.5r$
1st plane	$r - 50$	2.5	$2.5(r - 50)$

The total distance traveled by the two planes is 1400 mi.
$2.5r + 2.5(r - 50) = 1400$

Solution
$2.5r + 2.5(r - 50) = 1400$
$2.5r + 2.5r - 125 = 1400$
$5r - 125 = 1400$
$5r = 1525$
$r = 305$
$r - 50 = 305 - 50 = 255$
The rate of the first plane is 255 mph.
The rate of the second plane is 305 mph.

33. Strategy
Time to the island: t
Time returning from the island: $6 - t$

	Rate	Time	Distance
Going	18	t	$18t$
Returning	12	$6 - t$	$12(6 - t)$

The distance to the island is the same as the distance returning.
$18t = 12(6 - t)$

Solution
$18t = 12(6 - t)$
$18t = 72 - 12t$
$30t = 72$
$t = 2.4$
$d = rt = 18(2.4) = 43.2$
The distance to the island is 43.2 mi.

35. Strategy
Time to the repair shop: t
Time walking home: $1 - t$

	Rate	Time	Distance
To repair shop	14	t	$14t$
Walking home	3.5	$1 - t$	$3.5(1 - t)$

The distance to the repair shop is the same as the distance walking home.
$14t = 3.5(1 - t)$

Solution
$14t = 3.5(1 - t)$
$14t = 3.5 - 3.5t$
$17.5t = 3.5$
$t = 0.2$
$d = rt = 14(0.2) = 2.8$
The distance between the student's home and the bicycle shop is 2.8 mi.

37. Strategy
Time Washington to Pittsburgh train travels: t
Time Pittsburgh to Washington train travels: $t - 1$

	r	t	d
Washington to Pittsburgh	60	t	$60t$
Pittsburgh to Washington	40	$t - 1$	$40(t - 1)$

The trains together cover the distance from Washington to Pittsburgh, 260 miles.
$60t + 40(t - 1) = 260$

Solution
$60t + 40t - 40 = 260$
$100t - 40 = 260$
$100t = 300$
$t = 3$
$t - 1 = 2$
The two trains will pass each other after 2 hours.

Applying Concepts 2.3

39. Strategy
To find the distance solve the equation $d = rt$ for t using $d = 260$ trillion miles $= 260,000,000$ million miles, and $r = 18$ million mph.

Solution
$$d = rt$$
$$260,000,000 = 18t$$
$$t \approx 14,444,444 \text{ h}$$
$$= 14,444,444 \text{ h} \cdot \frac{1 \text{ day}}{24 \text{ h}} \cdot \frac{1 \text{ yr}}{365 \text{ day}}$$
$$\approx 1648 \text{ yr}$$
$$\approx 1600 \text{ yr}$$

It will take about 1600 years.

41. The distance between them 2 min before impact is equal to the sum of the distances each one can travel during 2 min.

$$2 \text{ minutes} \cdot \frac{1 \text{ hour}}{60 \text{ minutes}} = 0.03\overline{3} \text{ hour}$$

Distance between cars
= rate of first car $\cdot 0.03\overline{3}$ + rate of second car $\cdot 0.03\overline{3}$
Distance between cars $= 40 \cdot 0.03\overline{3} + 60 \cdot 0.03\overline{3} = 3.3\overline{3}$

The cars are $3.3\overline{3} \left(\text{or } 3\frac{1}{3}\right)$ miles apart 2 min before impact.

43. Rate during the second mile: x

	Rate	Distance	Time
1st mile	30	1	$\frac{1}{30}$
2nd mile	x	1	$\frac{1}{x}$
Both miles	60	2	$\frac{2}{60} = \frac{1}{30}$

The time traveled during the first mile plus the time traveled during the second mile is equal to the total time traveled during both miles.

$$\frac{1}{30} + \frac{1}{x} = \frac{1}{30}$$

Solution
$$\frac{1}{30} + \frac{1}{x} = \frac{1}{30}$$
$$\frac{1}{x} = 0$$
$$x\left(\frac{1}{x}\right) = 0 \cdot x$$
$$1 = 0$$

There is no solution to the equation. No, it is not possible to increase the speed enough.

45. If a student jogs 1 mi at a rate of 8 mph and jogs back at a rate of 6 mph, then

Total time $= \dfrac{1 \text{ mi}}{8 \text{ mph}} + \dfrac{1 \text{ mi}}{6 \text{ mph}} = \dfrac{7}{24} \text{ h}$

Average rate $= \dfrac{\text{distance}}{\text{time}} = \dfrac{2 \text{ mi}}{\frac{7}{24} \text{ h}} \approx 6.86 \text{ mph}$.

Therefore, the student's average rate is about 6.86 mph, not 7 mph.

Section 2.4

Concept Review 2.4

1. Never true
The amount invested in the other account is $10,000 - x$.

3. Never true
A mixture can never have a greater concentration of an ingredient than the concentrations of both substances going into the mixture.

Objective 2.4.1 Exercises

1. In the equation $I = Pr$, r represents the interest rate given as a decimal, P represents the principal, and I represents the amount of interest earned. If you invest $2000 in a bank that pays 4 percent interest, you can find the interest by evaluating 0.4(2000).

3. The total annual interest earned on the investments in Exercise 2 is $115. Use this information and the information in the table in Exercise 2 to write an equation that can be solved to find the amount of money invested at 6.25%: $\underline{0.0625x + 0.06(x - 5000)}$ $= \underline{115}$.

5. Strategy
To find the total interest solve the equation $I = Pr$ for I using $P = \$2000$, and $r = 5.5\% = 0.055$, solve the equation $I = Pr$ again for I using $P = \$3000$, and $r = 7.25\% = 0.0725$, and add the interest.

Solution
$I = Pr$
$= 2000(0.055)$
$= 110$
$I = Pr$
$= 3000(0.0725)$
$= 217.50$
$110 + 217.50 = 327.50$
Joseph earns $327.50.

Chapter 2: First-Degree Equations And Inequalities

7. Strategy
To find the amount invested in the 8% account, first solve the equation $I = Pr$ for I using $P = \$2000$, and $r = 6.4\% = 0.064$. Then, solve the equation $I = Pr$ for P using I, and $r = 8\% = 0.08$.

Solution
$$I = Pr$$
$$= 2000(0.064)$$
$$= 128$$
$$I = Pr$$
$$128 = P(0.08)$$
$$1600 = P$$
Deon must invest $1600 in the 8% account.

9. Strategy
Amount invested at 6.75%: x
Amount invested at 7.25%: $40,000 - x$

	Principal	Rate	Interest
Amount at 6.75%	x	0.0675	$0.0675x$
Amount at 7.25%	$40,000 - x$	0.0725	$0.0725 \cdot (40,000 - x)$

The sum of the interest earned by the two investments equals the total annual interest earned ($2825).
$0.0675x + 0.0725(40,000 - x) = 2825$

Solution
$$0.0675x + 0.0725(40,000 - x) = 2825$$
$$0.0675x + 2900 - 0.0725x = 2825$$
$$-0.005x = -75$$
$$x = 15,000$$
The amount invested in the certificate of deposit is $15,000.

11. Strategy
Amount invested at 10.5%: x

	Principal	Rate	Interest
Amount at 8.4%	5000	0.084	$0.084 \cdot (5000)$
Amount at 10.5%	x	0.105	$0.105x$
Amount at 9%	$5000 + x$	0.09	$0.09 \cdot (5000 + x)$

The sum of the interest earned by the two investments equals the interest earned by the total investment.
$0.084(5000) + 0.105x = 0.09(5000 + x)$

Solution
$$0.084(5000) + 0.105x = 0.09(5000 + x)$$
$$420 + 0.105x = 450 + 0.09x$$
$$0.015x = 30$$
$$x = 2000$$
$2000 more must be invested at 10.5%.

13. Strategy
Amount invested at 8.5%: x
Amount invested at 6.4%: $8000 - x$

	Principal	Rate	Interest
Amount at 8.5%	x	0.085	$0.085x$
Amount at 6.4%	$8000 - x$	0.064	$0.064 \cdot (8000 - x)$

The sum of the interest earned by the two investments equals the total annual interest earned ($575).
$0.085x + 0.064(8000 - x) = 575$

Solution
$$0.085x + 0.064(8000 - x) = 575$$
$$0.085x + 512 - 0.064x = 575$$
$$0.021x = 63$$
$$x = 3000$$
$8000 - x = 8000 - 3000 = 5000$
The amount invested at 8.5% is $3000.
The amount invested at 6.4% is $5000.

15. Strategy
Amount of additional money at 10%: x

	Principal	Rate	Interest
Amount at 5.5%	6000	0.055	330
Amount at 10%	x	0.10	$0.10x$
Combined Investment	$6000 + x$	0.07	$0.07 \cdot (6000 + x)$

The interest from the combined investment is the sum of the interests from each investment.
$330 + 0.10x = 0.07(6000 + x)$
$330 + 0.10x = 420 + 0.07x$
$0.03x = 90$
$x = 3000$
The additional amount that should be invested at 10% is $3000.

17. Strategy
Amount invested at 4.2%: x
Amount invested at 6%: $13,600 - x$

	Principal	Rate	Interest
Amount 4.2%	x	0.042	$0.042x$
Amount at 6%	$13,600 - x$	0.006	$0.006 \cdot (13,600 - x)$

The interest earned on one investment is equal to the interest earned on the other investment.
$0.042x = 0.06(13,600 - x)$

Solution
$$0.042x = 0.06(13,600 - x)$$
$$0.042x = 816 - 0.06x$$
$$0.102x = 816$$
$$x = 8000$$
$13,600 - x = 13,600 - 8000 = 5600$
The amount that should be invested at 4.2% is $8000.
The amount that should be invested at 6% is $5600.

19. a. 5.5% and 7.2%

b. $6000

Objective 2.4.2 Exercises

21. The label on a 32-ounce bottle of lemonade says that it contains 10% real lemon juice. The amount of real lemon juice in the bottle is 32(0.10) = 3.2 oz.

23. Strategy
To find the amount of juice in the 40-ounce bottle, first solve the equation $Q = Ar$ for r using $Q = 8$ oz, and $A = 32$ oz. Then, solve the equation $Q = Ar$ for Q using r, and $A = 40$ oz.

Solution
$$Q = Ar$$
$$8 = 32r$$
$$0.25 = r$$
$$Q = Ar$$
$$= 40(0.25)$$
$$= 10$$
There is 10 oz of juice in the 40-ounce bottle.

25. Strategy
To find how much more hydrogen peroxide, solve the equation $Q = Ar$ for Q using $A = 750$ ml, and $r = 4\% = 0.04$, solve the equation $Q = Ar$ for Q using $A = 850$, and $r = 5\% = 0.05$, and find the difference.

Solution
$$Q = Ar$$
$$= 750(0.04)$$
$$= 30$$
$$Q = Ar$$
$$= 850(0.05)$$
$$= 42.5$$
$$42.5 - 30 = 12.5$$
There is 12.5 ml of hydrogen peroxide in the 850-milliliter solution.

27. Strategy
Percent concentration of resulting alloy: x

	Amount	Percent	Quantity
60% alloy	15	0.60	9
20% alloy	45	0.20	9
Mixture	60	x	$60x$

The sum of the quantities before mixing is equal to the quantity after mixing.
$9 + 9 = 60x$

Solution
$$9 + 9 = 60x$$
$$18 = 60x$$
$$0.30 = x$$
The resulting alloy is 30% silver.

29. Strategy
Percent concentration of the resulting alloy: x

	Amount	Percent	Quantity
70%	25	0.70	0.70 · (25)
15%	50	0.15	0.15 · (50)
Mixture	75	x	$75x$

The sum of the quantities before mixing is equal to the quantity after mixing.
$0.70(25) + 0.15(50) = 75x$

Solution
$$0.70(25) + 0.15(50) = 75x$$
$$17.5 + 7.5 = 75x$$
$$25 = 75x$$
$$0.3\overline{3} = x$$
the resulting alloy is $33\frac{1}{3}\%$ silver.

31. Strategy
Pounds of 12% aluminum alloy: x

	Amount	Percent	Quantity
12%	x	0.12	$0.12x$
30%	400	0.30	0.30 · (400)
20%	$400 + x$	0.20	0.20 · $(400 + x)$

The sum of the quantities before mixing is equal to the quantity after mixing.
$0.12x + 0.30(400) = 0.20(400 + x)$

Solution
$$0.12x + 0.30(400) = 0.20(400 + x)$$
$$0.12x + 120 = 80 + 0.20x$$
$$-0.08x + 120 = 80$$
$$-0.08x = -40$$
$$x = 500$$
500 lb of the 12% aluminum alloy must be used.

33. Strategy
Liters of 65% solution: x
Liters of 15% solution: $50 - x$

	Amount	Percent	Quantity
65% solution	x	65%	$0.65x$
15% solution	$50 - x$	15%	0.15 · $(50 - x)$
Mixture	50	40%	0.40 · (50)

The sum of the quantities before mixing is equal to the quantity after mixing.
$0.65x + 0.15(50 - x) = 0.40(50)$

Solution
$$0.65x + 0.15(50 - x) = 0.40(50)$$
$$0.65x + 7.5 - 0.15x = 20$$
$$0.5x = 12.5$$
$$x = 25$$
25 L of 65% disinfectant solution and 25 L of 15% disinfectant solution were used.

28 Chapter 2: First-Degree Equations And Inequalities

35. Strategy
Number of quarts of water: x

	Amount	Percent	Quantity
Water	x	0	0
80% antifreeze	5	0.80	4
50% antifreeze	$5 + x$	0.50	$0.50 \cdot (5 + x)$

The sum of the quantities before mixing is equal to the quantity after mixing.
$0 + 4 = 0.50(5 + x)$

Solution
$4 = 2.5 + 0.50x$
$1.5 = 0.5x$
$x = 3$
3 quarts of water should be added.

37. Strategy
Ounces of water to be added: x
Ounces of 5% solution: $60 + x$

	Amount	Percent	Quantity
Water	x	0	0
7.5% solution	60	0.075	4.5
5% solution	$60 + x$	0.05	$0.05 \cdot (60 + x)$

The sum of the quantities before mixing is equal to the quantity after mixing.
$0 + 4.5 = 0.05(60 + x)$

Solution
$0 + 4.5 = 0.05(60 + x)$
$4.5 = 3 + 0.05x$
$1.5 = 0.05x$
$x = 30$
30 oz of water should be added.

39. Strategy
Percent concentration of result: x

	Amount	Percent	Quantity
5% fruit juice	12	0.05	$0.05 \cdot (12)$
Water	2	0	0
Result	10	x	$10x$

The sum of the quantities before mixing is equal to the quantity after mixing.
$0.05(12) + 0 = 10x$

Solution
$0.05(12) + 0 = 10x$
$0.60 = 10x$
$x = 0.06$
The result is 6% fruit juice.

41. Strategy
Percent concentration of the resulting alloy: x

	Amount	Percent	Quantity
54%	80	0.54	$0.54 \cdot (80)$
22%	200	0.22	$0.22 \cdot (200)$
Mixture	280	x	$280x$

The sum of the quantities before mixing is equal to the quantity after mixing.
$0.54(80) + 0.22(200) = 280x$

Solution
$0.54(80) + 0.22(200) = 280x$
$43.2 + 44 = 280x$
$87.2 = 280x$
$0.3114 \approx x$
The resulting alloy is about 31.1% copper.

43. b

Applying Concepts 2.4

45. Strategy
Total amount invested: x
Amount invested at 9%: $0.25x$
Amount invested at 8%: $0.30x$
Amount invested at 9.5%: $0.45x$

	Amount	Percent	Quantity
Amount at 9%	$0.25x$	0.09	$0.09(0.25x)$
Amount at 8%	$0.30x$	0.08	$0.08(0.30x)$
Amount at 9.5%	$0.45x$	0.095	$0.095(0.45x)$

The total annual interest earned is $1785.

Solution
$0.09(0.25x) + 0.08(0.3x) + 0.095(0.45x) = 1785$
$0.0225x + 0.024x + 0.04275x = 1785$
$0.08925x = 1785$
$x = 20{,}000$

$0.25x = 0.25(20{,}000) = 5000$
$0.3x = 0.3(20{,}000) = 6000$
$0.45x = 0.45(20{,}000) = 9000$
The amount invested at 9% was $5000.
The amount invested at 8% was $6000.
The amount invested at 9.5% was $9000.

47. Strategy

Cost per pound of mixture: x

	Amount	Cost	Value
$5.50 tea	50	550	550(50)
$4.40 tea	75	440	440(75)
Mixture	125	x	$125x$

The sum of the quantities before mixing is equal to the quantity after mixing.
$550(50) + 440(75) = 125x$

Solution
$$550(50) + 440(75) = 125x$$
$$27,500 + 33,000 = 125x$$
$$60,500 = 125x$$
$$484 = x$$

The tea mixture would cost $4.84 per pound.

49. Strategy

Grams of water: x

	Amount	Percent	Quantity
Pure water	x	0	0
Pure acid	20	1.00	1.00(20)
25% acid	$20+x$	0.25	$0.25 \cdot (20+x)$

The sum of the quantities before mixing is equal to the quantity after mixing.
$0 + 1.00(20) = 0.25(20 + x)$

Solution
$$0 + 1.00(20) = 0.25(20 + x)$$
$$20 = 5 + 0.25x$$
$$15 = 0.25x$$
$$60 = x$$

60 g of pure water were in the beaker.

51.
  a. The percent increase was greatest in 2000.

  b. The cost of consumer goods and services was highest in 2005.

Section 2.5

Concept Review 2.5

1. Always true

3. Sometimes true
The rule states that when dividing an inequality by a negative integer, we must reverse the inequality.

5. Sometimes true
This is not true for $a = 0$.

Objective 2.5.1 Exercises

1. The Addition Property of Inequalities states that the same number can be added to each side of an inequality without changing the solution set of the inequality. Examples will vary. For instance,
$$8 > 6$$
$$8 + 4 > 6 + 4$$
$$12 > 10$$
and
$$-5 < -1$$
$$-5 + (-7) < -1 + (-7)$$
$$-12 < -8$$

3. $x + 7 \leq -3$
$x \leq -10$
The solution is a, c.

5. remains the same

7. is reversed

9. $x - 3 < 2$
$x < 5$
$\{x | x < 5\}$

11. $4x \leq 8$
$\dfrac{4x}{4} \leq \dfrac{8}{4}$
$x \leq 2$
$\{x | x \leq 2\}$

13. $-2x > 8$
$\dfrac{-2x}{-2} < \dfrac{8}{-2}$
$x < -4$
$\{x | x < -4\}$

15. $3x - 1 > 2x + 2$
$x - 1 > 2$
$x > 3$
$\{x | x > 3\}$

17. $2x - 1 > 7$
$2x > 8$
$\dfrac{2x}{2} > \dfrac{8}{2}$
$x > 4$
$\{x | x > 4\}$

19. $6x + 3 > 4x - 1$
$2x + 3 > -1$
$2x > -4$
$\dfrac{2x}{2} > \dfrac{-4}{2}$
$x > -2$
$\{x \mid x > -2\}$

21. $8x + 1 \geq 2x + 13$
$6x + 1 \geq 13$
$6x \geq 12$
$\dfrac{6x}{6} \geq \dfrac{12}{6}$
$x \geq 2$
$\{x \mid x \geq 2\}$

23. $7 - 2x \geq 1$
$-2x \geq -6$
$\dfrac{-2x}{-2} \leq \dfrac{-6}{-2}$
$x \leq 3$
$\{x \mid x \leq 3\}$

25. $4x - 2 < x - 11$
$3x - 2 < -11$
$3x < -9$
$\dfrac{3x}{3} < \dfrac{-9}{3}$
$x < -3$
$\{x \mid x < -3\}$

27. $x + 7 \geq 4x - 8$
$-3x + 7 \geq -8$
$-3x \geq -15$
$\dfrac{-3x}{-3} \leq \dfrac{-15}{-3}$
$x \leq 5$
$(-\infty, 5]$

29. $6 - 2(x - 4) \leq 2x + 10$
$6 - 2x + 8 \leq 2x + 10$
$14 - 2x \leq 2x + 10$
$14 - 4x \leq 10$
$-4x \leq -4$
$\dfrac{-4x}{-4} \geq \dfrac{-4}{-4}$
$x \geq 1$
$[1, \infty)$

31. $2(1 - 3x) - 4 > 10 + 3(1 - x)$
$2 - 6x - 4 > 10 + 3 - 3x$
$-6x - 2 > 13 - 3x$
$-3x - 2 > 13$
$-3x > 15$
$\dfrac{-3x}{-3} < \dfrac{15}{-3}$
$x < -5$
$(-\infty, -5)$

33. $\dfrac{3}{5}x - 2 < \dfrac{3}{10} - x$
$10\left(\dfrac{3}{5}x - 2\right) < 10\left(\dfrac{3}{10} - x\right)$
$6x - 20 < 3 - 10x$
$16x - 20 < 3$
$16x < 23$
$\dfrac{16x}{16} < \dfrac{23}{16}$
$x < \dfrac{23}{16}$
$\left(-\infty, \dfrac{23}{16}\right)$

35. $\dfrac{1}{3}x - \dfrac{3}{2} \geq \dfrac{7}{6} - \dfrac{2}{3}x$
$6\left(\dfrac{1}{3}x - \dfrac{3}{2}\right) \geq 6\left(\dfrac{7}{6} - \dfrac{2}{3}x\right)$
$2x - 9 \geq 7 - 4x$
$6x - 9 \geq 7$
$6x \geq 16$
$\dfrac{6x}{6} \geq \dfrac{16}{6}$
$x \geq \dfrac{8}{3}$
$\left[\dfrac{8}{3}, \infty\right)$

37. $\dfrac{1}{2}x - \dfrac{3}{4} > \dfrac{7}{4}x - 2$
$4\left(\dfrac{1}{2}x - \dfrac{3}{4}\right) > 4\left(\dfrac{7}{4}x - 2\right)$
$2x - 3 > 7x - 8$
$-5x - 3 > -8$
$-5x > -5$
$\dfrac{-5x}{-5} < \dfrac{-5}{-5}$
$x < 1$
$(-\infty, 1)$

39.
$2 - 2(7 - 2x) < 3(3 - x)$
$2 - 14 + 4x < 9 - 3x$
$-12 + 4x < 9 - 3x$
$-12 + 7x < 9$
$7x < 21$
$\dfrac{7x}{7} < \dfrac{21}{7}$
$x < 3$
$(-\infty, 3)$

41. Only positive numbers

43. Only negative numbers

Objective 2.5.2 Exercises

45.
a. When a compound inequality is combined with *or*, the set operation union is used.
b. When a compound inequality is combined with *and*, the set operation intersection is used.

47. To solve the compound inequality $4x \leq 4$ or $x + 2 > 8$, divide each side of the first inequality by $\underline{4}$, and subtract $\underline{2}$ from each side of the second inequality: $x \leq \underline{1}$ or $x > \underline{6}$.

49. One interval of real numbers

51. Empty set

53.
$3x < 6$ and $x + 2 > 1$
$x < 2 \qquad x > -1$
$\{x | x < 2\} \quad \{x | x > -1\}$
$\{x | x < 2\} \cap \{x | x > -1\} = (-1, 2)$

55.
$x + 2 \geq 5$ or $3x \leq 3$
$x \geq 3 \qquad x \leq 1$
$\{x | x \geq 3\} \quad \{x | x \leq 1\}$
$\{x | x \geq 3\} \cup \{x | x \leq 1\} = (-\infty, 1] \cup [3, \infty)$

57.
$-2x > -8$ and $-3x < 6$
$x < 4 \qquad x > -2$
$\{x | x < 4\} \quad \{x | x > -2\}$
$\{x | x < 4\} \cap \{x | x > -2\} = (-2, 4)$

59.
$\dfrac{1}{3}x < -1$ or $2x > 0$
$x < -3 \qquad x > 0$
$\{x | x < -3\} \quad \{x | x > 0\}$
$\{x | x < -3\} \cup \{x | x > 0\} = (-\infty, -3) \cup (0, \infty)$

61.
$x + 4 \geq 5$ and $2x \geq 6$
$x \geq 1 \qquad x \geq 3$
$\{x | x \geq 1\} \quad \{x | x \geq 3\}$
$\{x | x \geq 1\} \cap \{x | x \geq 3\} = [3, \infty)$

63.
$-5x > 10$ and $x + 1 > 6$
$x < -2 \qquad x > 5$
$\{x | x < -2\} \quad \{x | x > 5\}$
$\{x | x < -2\} \cap \{x | x > 5\} = \varnothing$

65.
$2x - 3 > 1$ and $3x - 1 < 2$
$2x > 4 \qquad 3x < 3$
$x > 2 \qquad x < 1$
$\{x | x > 2\} \quad \{x | x < 1\}$
$\{x | x > 2\} \cap \{x | x < 1\} = \varnothing$

67.
$3x + 7 < 10$ or $2x - 1 > 5$
$3x < 3 \qquad 2x > 6$
$x < 1 \qquad x > 3$
$\{x | x < 1\} \quad \{x | x > 3\}$
$\{x | x < 1\} \cup \{x | x > 3\} = (-\infty, 1) \cup (3, \infty)$

69.
$-5 < 3x + 4 < 16$
$-5 - 4 < 3x + 4 - 4 < 16 - 4$
$-9 < 3x < 12$
$\dfrac{-9}{3} < \dfrac{3x}{3} < \dfrac{12}{3}$
$-3 < x < 4$
$\{x | -3 < x < 4\}$

71.
$0 < 2x - 6 < 4$
$0 + 6 < 2x - 6 + 6 < 4 + 6$
$6 < 2x < 10$
$\dfrac{6}{2} < \dfrac{2x}{2} < \dfrac{10}{2}$
$3 < x < 5$
$\{x | 3 < x < 5\}$

73.
$4x - 1 > 11$ or $4x - 1 \leq -11$
$4x > 12 \qquad 4x \leq -10$
$x > 3 \qquad x \leq -\dfrac{5}{2}$
$\{x | x > 3\} \quad \left\{x | x \leq -\dfrac{5}{2}\right\}$
$\{x | x > 3\} \cup \left\{x | x \leq -\dfrac{5}{2}\right\}$
$= \left\{x | x > 3 \text{ or } x \leq -\dfrac{5}{2}\right\}$

75.
$2x + 3 \geq 5$ and $3x - 1 > 11$
$2x \geq 2 \qquad 3x > 12$
$x \geq 1 \qquad x > 4$
$\{x | x \geq 1\} \quad \{x | x > 4\}$
$\{x | x \geq 1\} \cap \{x | x > 4\} = \{x | x > 4\}$

Chapter 2: First-Degree Equations And Inequalities

77. $9x - 2 < 7$ and $3x - 5 > 10$
$\quad\quad 9x < 9 \quad\quad\quad 3x > 15$
$\quad\quad x < 1 \quad\quad\quad\quad x > 5$
$\{x | x < 1\} \quad \{x | x > 5\}$
$\{x | x < 1\} \cap \{x | x > 5\} = \varnothing$

79. $3x - 11 < 4$ or $4x + 9 \geq 1$
$\quad\quad 3x < 15 \quad\quad 4x \geq -8$
$\quad\quad x < 5 \quad\quad\quad x \geq -2$
$\{x | x < 5\} \quad \{x | x \geq -2\}$
$\{x | x < 5\} \cup \{x | x \geq -2\}$
$= \{x | x \text{ is a real number}\}$

81. $3 - 2x > 7$ and $5x + 2 > -18$
$\quad\quad -2x > 4 \quad\quad 5x > -20$
$\quad\quad x < -2 \quad\quad\quad x > -4$
$\{x | x < -2\} \quad \{x | x > -4\}$
$\{x | x < -2\} \cap \{x | x > -4\} = \{x | -4 < x < -2\}$

83. $5 - 4x > 21$ or $7x - 2 > 19$
$\quad\quad -4x > 16 \quad\quad 7x > 21$
$\quad\quad x < -4 \quad\quad\quad x > 3$
$\{x | x < -4\} \quad \{x | x > 3\}$
$\{x | x < -4\} \cup \{x | x > 3\} = \{x | x < -4 \text{ or } x > 3\}$

85. $3 - 7x \leq 31$ and $5 - 4x > 1$
$\quad\quad -7x \leq 28 \quad\quad -4x > -4$
$\quad\quad x \geq -4 \quad\quad\quad x < 1$
$\{x | x \geq -4\} \quad \{x | x < 1\}$
$\{x | x \geq -4\} \cap \{x | x < 1\} = \{x | -4 \leq x < 1\}$

87. $\frac{2}{3}x - 4 > 5$ or $x + \frac{1}{2} < 3$
$\quad\quad \frac{2}{3}x > 9 \quad\quad\quad x < \frac{5}{2}$
$\quad\quad x > \frac{27}{2}$
$\left\{x | x > \frac{27}{2}\right\} \quad \left\{x | x < \frac{5}{2}\right\}$
$\left\{x | x > \frac{27}{2}\right\} \cup \left\{x | x < \frac{5}{2}\right\}$
$= \left\{x | x > \frac{27}{2} \text{ or } x < \frac{5}{2}\right\}$

89. $-\frac{3}{8} \leq 1 - \frac{1}{4}x \leq \frac{7}{2}$
$-3 \leq 8 - 2x \leq 28$
$-11 \leq -2x \leq 20$
$\frac{11}{2} \geq x \geq -10$
$\left\{x | -10 \leq x \leq \frac{11}{2}\right\}$

Objective 2.5.3 Exercises

91. $n \geq 40$

93. Strategy
the unknown number: x
five times the difference between the number and two: $5(x - 2)$
the quotient of two times the number and three: $2x \div 3$

Solution
five times the difference between a number and two is greater than or equal to the quotient of two times the number and three
$5(x - 2) \geq \frac{2x}{3}$
$5x - 10 \geq \frac{2x}{3}$
$3(5x - 10) \geq 2x$
$15x - 30 \geq 2x$
$13x - 30 \geq 0$
$13x \geq 30$
$x \geq \frac{30}{13}$
$x \geq 2\frac{4}{13}$
The smallest integer is 3.

95. Strategy
the width of the rectangle: x
the length of the rectangle: $4x + 2$
To find the maximum width, substitute the given values in the inequality $2L + 2W < 34$ and solve.

Solution
$2L + 2W < 34$
$2(4x + 2) + 2x < 34$
$8x + 4 + 2x < 34$
$10x + 4 < 34$
$10x < 30$
$x < 3$
The maximum width of the rectangle is 2 ft.

97. Strategy
To find the four consecutive integers, write and solve a compound inequality using x to represent the first integer.

Solution
Lower limit of the sum < sum < Upper limit of the sum
$62 < x + (x + 1) + (x + 2) + (x + 3) < 78$
$62 < 4x + 6 < 78$
$62 - 6 < 4x + 6 - 6 < 78 - 6$
$56 < 4x < 72$
$\frac{56}{4} < \frac{4x}{4} < \frac{72}{4}$
$14 < x < 18$
The four integers are 15, 16, 17, and 18; or 16, 17, 18, and 19; or 17, 18, 19, and 20.

99. Strategy
First side of the triangle: $x + 1$
Second side of the triangle: x
Third side of the triangle: $x + 2$

Solution
The perimeter of the triangle is more than 15 in. and less than 25 in.
$$15 < P < 25$$
$$15 < (x+1) + x + (x+2) < 25$$
$$15 < 3x + 3 < 25$$
$$15 - 3 < 3x + 3 - 3 < 25 - 3$$
$$12 < 3x < 22$$
$$\frac{12}{3} < \frac{3x}{3} < \frac{22}{3}$$
$$4 < x < 7\frac{1}{3}$$
If $x = 5$: $5 + 1 = 6$; $5 + 2 = 7$;
$5 + 6 + 7 = 18 =$ perimeter
If $x = 6$: $6 + 1 = 7$; $6 + 2 = 8$;
$6 + 7 + 8 = 21 =$ perimeter
If $x = 7$: $7 + 1 = 8$; $7 + 2 = 9$;
$7 + 8 + 9 = 24 =$ perimeter
The lengths of the second side could be 5 in., 6 in., or 7 in.

101. Strategy
To find the number of minutes, write and solve an inequality using N to represent the number of minutes of cellular phone time.

Solution
cost of second option $\leq$ cost of first option
$$0.4N + 25 \leq 49$$
$$0.4N \leq 24$$
$$N \leq 60$$
A customer can use a cellular phone 160 min before the charges exceed the first option.

103. Strategy
To find the number of messages for which the AirTouch plan is less expensive, solve an inequality using N to represent the number of messages.
Cost of AirTouch < Cost of TopPage
$$6.95 + 0.10(x - 400) < 3.95 + 0.15(x - 400)$$
$$6.95 + 0.10x - 40 < 3.95 + 0.15x - 60$$
$$33.05 + 0.10x < -56.05 + 0.15x$$
$$23 < 0.05x$$
$$460 < x$$
AirTouch is less expensive for jobs more than 460 messages.

105. Strategy
To find the number of minutes for which a call will be cheaper to pay with coins, solve an inequality using N to represent the number of minutes.
Coins < Calling card

Solution
$$(0.70) + 0.15(N - 3) < 0.35 + 0.196 + 0.126(N - 1)$$
$$0.70 + 0.15N - 0.45 < 0.35 + 0.196 + 0.126N - 0.126$$
$$0.25 + 0.15N < 0.42 + 0.126N$$
$$0.024N < .17$$
$$N < 7.08$$
Using coins will be cheaper for 7 minutes or less.

107. Strategy
To find the number of checks that have to be written for Glendale Federal to cost less than the competitor, solve an inequality using N to represent the number of checks.
Glendale account < Other account
$$8 + 0.12(N - 100) < 5 + 0.15(N - 100)$$
$$8 + 0.12N - 12 < 5 + 0.15N - 15$$
$$0.12N - 4 < 0.15N - 10$$
$$-0.03N < -6$$
$$N > 200$$
The Glendale Federal account will cost less for more than 200 checks.

109. Strategy
To find the range of miles that a car can travel, write and solve an inequality using N to represent the range of miles.

Solution
$$22(19.5) < N < 27.5(19.5)$$
$$429 < N < 536.25$$
The range of miles is between 429 mi and 536.25 mi.

Applying Concepts 2.5

111. $8x - 7 < 2x + 9$
$$6x - 7 < 9$$
$$6x < 16$$
$$x < \frac{16}{6}$$
$$x < \frac{8}{3}$$
$\{1, 2\}$

113. $5 + 3(2 + x) > 8 + 4(x - 1)$
$$5 + 6 + 3x > 8 + 4x - 4$$
$$11 + 3x > 4 + 4x$$
$$3x > -7 + 4x$$
$$-x > -7$$
$$x < 7$$
$\{1, 2, 3, 4, 5, 6\}$

115. $-3x < 15$ and $x + 2 < 7$
$\quad x > -5 \qquad\qquad x < 5$
$\{x | x > -5\} \cap \{x | x < 5\} = \{x | -5 < x < 5\}$
$\{1, 2, 3, 4\}$

117.
$$-4 \leq 3x + 8 < 16$$
$$-4 + (-8) \leq 3x + 8 + (-8) < 16 + (-8)$$
$$-12 \leq 3x < 8$$
$$\frac{-12}{3} \leq \frac{3x}{3} < \frac{8}{3}$$
$$-4 \leq x < \frac{8}{3}$$
$\{1, 2\}$

119. Strategy
To find the temperature range in degrees Celsius, write and solve a compound inequality.

Solution
$$77 < \frac{9}{5}C + 32 < 86$$
$$77 - 32 < \frac{9}{5}C + 32 - 32 < 86 - 32$$
$$45 < \frac{9}{5}C < 54$$
$$\frac{5}{9}(45) < \frac{5}{9}\left(\frac{9}{5}C\right) < \frac{5}{9}(54)$$
$$25 < C < 30$$
The temperature is between 25°C and 30°C.

121. Strategy
To find the largest whole number of minutes the call could last, set up an inequality with N representing the number of minutes and $N - 3$ representing the number of minutes after the first 3 minutes.

Solution
$$156 + 52(N - 3) < 540$$
$$156 + 52N - 156 < 540$$
$$52N < 540$$
$$N < 10.38$$
The largest whole number of minutes the call could last is 10 min.

Section 2.6
Concept Review 2.6

1. Always true

3. Never true

5. Never true
$|x + b| < c$ is equivalent to $-c < x + b < c$.
The absolute value of a number is always positive.

Objective 2.6.1 Exercises

1. $|x - 8| = 6$
$|2 - 8| = 6$
$|-6| = 6$
$6 = 6$
Yes

3. $|3x - 4| = 7$
$|3(-1) - 4| = 7$
$|-3 - 4| = 7$
$|-7| = 7$
$7 = 7$
Yes

5. If $|x| = 8$, then $x = \underline{8}$ or $x = \underline{-8}$.

7. $|x| = 7$
$x = 7 \qquad x = -7$
The solutions are 7 and −7.

9. $|-t| = 3$
$-t = 3 \qquad -t = -3$
$t = -3 \qquad t = 3$
The solutions are −3 and 3.

11. $|-t| = -3$
There is no solution to this equation because the absolute value of a number must be non-negative.

13. $|x + 2| = 3$
$x + 2 = 3 \qquad x + 2 = -3$
$x = 1 \qquad x = -5$
The solutions are 1 and −5.

15. $|y - 5| = 3$
$y - 5 = 3 \qquad y - 5 = -3$
$y = 8 \qquad y = 2$
The solutions are 8 and 2.

17. $|a - 2| = 0$
$a - 2 = 0$
$a = 2$
The solution is 2.

19. $|x - 2| = -4$
There is no solution to this equation because the absolute value of a number must be non-negative.

21. $|2x - 5| = 4$
$2x - 5 = 4 \qquad 2x - 5 = -4$
$2x = 9 \qquad 2x = 1$
$x = \frac{9}{2} \qquad x = \frac{1}{2}$
The solutions are $\frac{9}{2}$ and $\frac{1}{2}$.

23. $|2 - 5x| = 2$
$2 - 5x = 2 \qquad 2 - 5x = -2$
$-5x = 0 \qquad -5x = -4$
$x = 0 \qquad x = \frac{4}{5}$
The solutions are 0 and $\frac{4}{5}$.

Chapter 2: First-Degree Equations And Inequalities   35

25. $|5x+5|=0$
    $5x+5=0$
    $5x=-5$
    $x=-1$
    The solution is -1.

27. $|2x+5|=-2$
    There is no solution to this equation because the absolute value of a number must be non-negative.

29. $|x-9|-3=2$
    $|x-9|=5$
    $x-9=5 \quad x-9=-5$
    $x=14 \quad x=4$
    The solutions are 14 and 4.

31. $|8-y|-3=1$
    $|8-y|=4$
    $8-y=4 \quad 8-y=-4$
    $-y=-4 \quad -y=-12$
    $y=4 \quad y=12$
    The solutions are 4 and 12.

33. $|4x-7|-5=-5$
    $|4x-7|=0$
    $4x-7=0$
    $4x=7$
    $x=\dfrac{7}{4}$
    The solution is $\dfrac{7}{4}$.

35. $|3x-2|+1=-1$
    $|3x-2|=-2$
    There is no solution to this equation because the absolute value of a number must be non-negative.

37. $|4b+3|-2=7$
    $|4b+3|=9$
    $4b+3=9 \quad 4b+3=-9$
    $4b=6 \quad 4b=-12$
    $b=\dfrac{3}{2} \quad b=-3$
    The solutions are $\dfrac{3}{2}$ and -3.

39. $|5x-2|+5=7$
    $|5x-2|=2$
    $5x-2=2 \quad 5x-2=-2$
    $5x=4 \quad 5x=0$
    $x=\dfrac{4}{5} \quad x=0$
    The solutions are $\dfrac{4}{5}$ and 0.

41. $2-|x-5|=4$
    $-|x-5|=2$
    $|x-5|=-2$
    There is no solution to this equation because the absolute value of a number must be non-negative.

43. $|3x-4|+8=3$
    $|3x-4|=-5$
    There is no solution to this equation because the absolute value of a number must be non-negative.

45. $5+|2x+1|=8$
    $|2x+1|=3$
    $2x+1=3 \quad 2x+1=-3$
    $2x=2 \quad 2x=-4$
    $x=1 \quad x=-2$
    The solutions are 1 and -2.

47. $3-|5x+3|=3$
    $-|5x+3|=0$
    $|5x+3|=0$
    $5x+3=0$
    $5x=-3$
    $x=-\dfrac{3}{5}$
    The solution is $-\dfrac{3}{5}$.

49. Two positive solutions

51. Two negative solutions

Objective 2.6.2 Exercises

53. If $|x|>9$, then $x<-9$ or $x>9$.

55. $|x|>3$
    $x>3 \quad$ or $\quad x<-3$
    $\{x|x>3\} \quad \{x|x<-3\}$
    $\{x|x>3\} \cup \{x|x<-3\} = \{x|x>3 \text{ or } x<-3\}$

57. $|x+1|>2$
    $x+1>2 \quad$ or $\quad x+1<-2$
    $x>1 \quad\quad x<-3$
    $\{x|x>1\}$
    $\{x|x<-3\} \quad \{x|x>1\} \cup \{x|x<-3\} = \{x|x>1 \text{ or } x<-3\}$

59. $|x-5|\le 1$
    $-1\le x-5\le 1$
    $-1+5\le x-5+5\le 1+5$
    $4\le x\le 6$
    $\{x|4\le x\le 6\}$

61. $|2-x| \geq 3$
$2-x \leq -3$ or $2-x \geq 3$
$-x \leq -5 \quad\quad -x \geq 1$
$x \geq 5 \quad\quad x \leq -1$
$\{x|x \geq 5\} \quad \{x|x \leq -1\}$
$\{x|x \geq 5\} \cup \{x|x \leq -1\} = \{x|x \geq 5 \text{ or } x \leq -1\}$

63. $|2x+1| < 5$
$-5 < 2x+1 < 5$
$-5-1 < 2x+1-1 < 5-1$
$-6 < 2x < 4$
$\dfrac{-6}{2} < \dfrac{2x}{2} < \dfrac{4}{2}$
$-3 < x < 2$
$\{x|-3 < x < 2\}$

65. $|5x+2| > 12$
$5x+2 > 12$ or $5x+2 < -12$
$5x > 10 \quad\quad 5x < -14$
$x > 2 \quad\quad x < -\dfrac{14}{5}$
$\{x|x > 2\} \quad \left\{x\Big|x < -\dfrac{14}{5}\right\}$
$\{x|x > 2\} \cup \left\{x\Big|x < -\dfrac{14}{5}\right\} = \left\{x\Big|x > 2 \text{ or } x < -\dfrac{14}{5}\right\}$

67. $|4x-3| \leq -2$
The absolute value of a number must be non-negative. The solution set is the empty set, $\varnothing$.

69. $|2x+7| > -5$
$2x+7 > -5$ or $2x+7 < 5$
$2x > -12 \quad\quad 2x < -2$
$x > -6 \quad\quad x < -1$
$\{x|x > -6\} \quad \{x|x < -1\}$
$\{x|x > -6\} \cup \{x|x < -1\} = \{x|x \text{ is a real number}\}$

71. $|4-3x| \geq 5$
$4-3x \geq 5$ or $4-3x \leq -5$
$-3x \geq 1 \quad\quad -3x \leq -9$
$x \leq -\dfrac{1}{3} \quad\quad x \geq 3$
$\left\{x\Big|x \leq -\dfrac{1}{3}\right\} \quad \{x|x \geq 3\}$
$\left\{x\Big|x \leq -\dfrac{1}{3}\right\} \cup \{x|x \geq 3\} = \left\{x\Big|x \leq -\dfrac{1}{3} \text{ or } x \geq 3\right\}$

73. $|5-4x| \leq 13$
$-13 \leq 5-4x \leq 13$
$-13-5 \leq 5-5-4x \leq 13-5$
$-18 \leq -4x \leq 8$
$\dfrac{-18}{-4} \geq \dfrac{-4x}{-4} \geq \dfrac{8}{-4}$
$\dfrac{9}{2} \geq x \geq -2$
$\left\{x\Big|-2 \leq x \leq \dfrac{9}{2}\right\}$

75. $|6-3x| \leq 0$
$6-3x = 0$
$-3x = -6$
$x = 2$
$\{x|x = 2\}$

77. $|2-9x| > 20$
$2-9x > 20$ or $2-9x < -20$
$-9x > 18 \quad\quad -9x < -22$
$x < -2 \quad\quad x > \dfrac{22}{9}$
$\{x|x < -2\} \cup \left\{x\Big|x > \dfrac{22}{9}\right\} = \left\{x\Big|x < -2 \text{ or } x > \dfrac{22}{9}\right\}$

79. All negative solutions

Objective 2.6.3 Exercises

81. The desired dosage of a particular medicine is 50 mg, but it is acceptable for the dosage to vary by 0.5 mg from the desired dosage. The number 0.5 is called the <u>tolerance</u> for the desired dosage of 50 mg. The value $50 + 0.5 = \underline{50.5}$ mg is called the <u>upper</u> limit for the dosage and the value $50 - 0.5 = \underline{49.5}$ mg is called the <u>lower</u> limit for the dosage.

83. Strategy
Let b represent the diameter of the bushing, T the tolerance, and d the lower and upper limits of the diameter. Solve the absolute value inequality $|d-b| \leq T$ for d.

Solution
$|d-b| \leq T$
$|d-1.75| \leq 0.008$
$-0.008 \leq d-1.75 \leq 0.008$
$-0.008 + 1.75 \leq d - 1.75 + 1.75 \leq 0.008 + 1.75$
$1.742 \leq d \leq 1.758$
The lower and upper limits of the diameter of the bushing are 1.742 in. and 1.758 in.

85. **Strategy**
Let p represent the prescribed amount of medication, T the tolerance, and m the lower and upper limits of the amount of medication. Solve the absolute value inequality $|m-p| \leq T$ for m.

Solution
$|m-p| \leq T$
$|m-2.5| \leq 0.2$
$-0.2 \leq m-2.5 \leq 0.2$
$-0.2+2.5 \leq m-2.5+2.5 \leq 0.2+2.5$
$2.3 \leq m \leq 2.7$
The lower and upper limits of the amount of medicine to be given to the patient are 2.3 cc and 2.7 cc.

87. **Strategy**
Let v represent the prescribed number of volts, T the tolerance, and m the lower and upper limits of the amount of voltage. Solve the absolute value inequality $|m-v| \leq T$ for m.

Solution
$|m-v| \leq T$
$|m-110| \leq 16.5$
$-16.5 \leq m-110 \leq 16.5$
$-16.5+110 \leq m-110+110 \leq 16.5+110$
$93.5 \leq m \leq 126.5$
The lower and upper limits of the amount of voltage to the computer are 93.5 volts and 126.5 volts.

89. **Strategy**
Let r represent the length of the piston rod, T the tolerance, and L the lower and upper limits of the length. Solve the absolute value inequality $|L-r| \leq T$ for L.

Solution
$|L-r| \leq T$
$\left|L-10\frac{3}{8}\right| \leq \frac{1}{32}$
$-\frac{1}{32} \leq L-10\frac{3}{8} \leq \frac{1}{32}$
$-\frac{1}{32}+10\frac{3}{8} \leq L-10\frac{3}{8}+10\frac{3}{8} \leq \frac{1}{32}+10\frac{3}{8}$
$10\frac{11}{32} \leq L \leq 10\frac{13}{32}$
The lower and upper limits of the length of the piston rod are $10\frac{11}{32}$ in. and $10\frac{13}{32}$ in.

91. The desired diameter is 5 in. The actual diameter can vary by 0.01 in.

93. **Strategy**
Let M represent the amount of ohms, T the tolerance, and r the given amount of the resistor. Find the tolerance and solve $|M-r| \leq T$ for M.

Solution
$T = (0.10)(15,000) = 1500$ ohms
$|M-r| \leq T$
$|M-15,000| \leq 1,500$
$-1,500 \leq M-15,000 \leq 1,500$
$-1,500+15,000 \leq M-15,000+15,000 \leq 1,500+15,000$
$13,500 \leq M \leq 16,500$
The lower and upper limits of the resistor are 13,500 ohms and 16,500 ohms.

95. **Strategy**
Let M represent the amount of ohms, T the tolerance, and r the given amount of the resistor. Find the tolerance and solve $|M-r| \leq T$ for M.

Solution
$T = (.05)(56) = 2.8$
$|M-r| \leq T$
$|M-56| \leq 2.8$
$-2.8 \leq M-56 \leq 2.8$
$-2.8+56 \leq M-56+56 \leq 2.8+56$
$53.2 \leq M \leq 58.8$
The lower and upper limits of the resistor are 53.2 ohms and 58.8 ohms.

Applying Concepts 2.6

97. $\left|\frac{3x-2}{4}\right| + 5 = 6$

$\left|\frac{3x-2}{4}\right| = 1$

$\frac{3x-2}{4} = 1 \qquad \frac{3x-2}{4} = -1$
$3x-2 = 4 \qquad 3x-2 = -4$
$3x = 6 \qquad 3x = -2$
$x = 2 \qquad x = -\frac{2}{3}$

The solutions are 2 and $-\frac{2}{3}$.

99. $\left|\dfrac{2x-1}{5}\right| \leq 3$

$$-3 \leq \dfrac{2x-1}{5} \leq 3$$
$$5(-3) \leq 5\left(\dfrac{2x-1}{5}\right) \leq 5(3)$$
$$-15 \leq 2x-1 \leq 15$$
$$-15+1 \leq 2x-1+1 \leq 15+1$$
$$-14 \leq 2x \leq 16$$
$$\dfrac{-14}{2} \leq \dfrac{2x}{2} \leq \dfrac{16}{2}$$
$$-7 \leq x \leq 8$$
$$\{x \mid -7 \leq x \leq 8\}$$

101. $|y+6| = y+6$

Any value of y that makes $y+6$ negative will result in a false equation because the left side of the equation will be positive and the right side of the equation will be negative. Therefore, the equation is true if $y+6$ is greater than or equal to zero.
$$y+6 \geq 0$$
$$y \geq -6$$
$$\{y \mid y \geq -6\}$$

103. $|b-7| = 7-b$

Any value of b that makes $7-b$ negative will result in a false equation because the left side of the equation will be positive and the right side of the equation will be negative. Therefore, the equation is true if $7-b$ is greater than or equal to zero.
$$7-b \geq 0$$
$$b \leq 7$$
$$\{b \mid b \leq 7\}$$

105. $|x-2| < 5$

107.
a. $|x+y| \leq |x|+|y|$

b. $|x-y| \geq |x|-|y|$

c. $||x|-|y|| \geq |x|-|y|$

d. $\left|\dfrac{x}{y}\right| = \dfrac{|x|}{|y|}, \ y \neq 0$

e. $|xy| = |x||y|$

CHAPTER 2 REVIEW EXERCISES

1. $x+4 = -5$
$$x+4-4 = -5-4$$
$$x = -9$$
The solution is -9.

2. $\dfrac{2}{3} = x + \dfrac{3}{4}$
$$\dfrac{2}{3} - \dfrac{3}{4} = x + \dfrac{3}{4} - \dfrac{3}{4}$$
$$\dfrac{8}{12} - \dfrac{9}{12} = x$$
$$-\dfrac{1}{12} = x$$
The solution is $-\dfrac{1}{12}$.

3. $-3x = -21$
$$\dfrac{-3x}{-3} = \dfrac{-21}{-3}$$
$$x = 7$$
The solution is 7.

4. $\dfrac{2}{3}x = \dfrac{4}{9}$
$$\dfrac{3}{2}\left(\dfrac{2}{3}x\right) = \dfrac{3}{2}\left(\dfrac{4}{9}\right)$$
$$x = \dfrac{2}{3}$$
The solution is $\dfrac{2}{3}$.

5. $3y-5 = 3-2y$
$$3y+2y-5 = 3-2y+2y$$
$$5y-5 = 3$$
$$5y-5+5 = 3+5$$
$$5y = 8$$
$$\dfrac{5y}{5} = \dfrac{8}{5}$$
$$y = \dfrac{8}{5}$$
The solution is $\dfrac{8}{5}$.

6. $3x-3+2x = 7x-15$
$$5x-3 = 7x-15$$
$$5x-3-7x = 7x-15-7x$$
$$-2x-3 = -15$$
$$-2x-3+3 = -15+3$$
$$-2x = -12$$
$$\dfrac{-2x}{-2} = \dfrac{-12}{-2}$$
$$x = 6$$
The solution is 6.

Chapter 2: First-Degree Equations And Inequalities 39

7.
$$2(x-3) = 5(4-3x)$$
$$2x - 6 = 20 - 15x$$
$$2x - 6 + 15x = 20 - 15x + 15x$$
$$17x - 6 = 20$$
$$17x - 6 + 6 = 20 + 6$$
$$17x = 26$$
$$\frac{17x}{17} = \frac{26}{17}$$
$$x = \frac{26}{17}$$
The solution is $\frac{26}{17}$.

8.
$$2x - (3 - 2x) = 4 - 3(4 - 2x)$$
$$2x - 3 + 2x = 4 - 12 + 6x$$
$$4x - 3 = -8 + 6x$$
$$4x - 3 - 6x = -8 + 6x - 6x$$
$$-2x - 3 = -8$$
$$-2x - 3 + 3 = -8 + 3$$
$$-2x = -5$$
$$\frac{-2x}{-2} = \frac{-5}{-2}$$
$$x = \frac{5}{2}$$
The solution is $\frac{5}{2}$.

9.
$$\frac{1}{2}x - \frac{5}{8} = \frac{3}{4}x + \frac{3}{2}$$
$$8\left(\frac{1}{2}x - \frac{5}{8}\right) = 8\left(\frac{3}{4}x + \frac{3}{2}\right)$$
$$8\left(\frac{1}{2}x\right) - 8\left(\frac{5}{8}\right) = 8\left(\frac{3}{4}x\right) + 8\left(\frac{3}{2}\right)$$
$$4x - 5 = 6x + 12$$
$$4x - 5 - 6x = 6x + 12 - 6x$$
$$-2x - 5 = 12$$
$$-2x - 5 + 5 = 12 + 5$$
$$-2x = 17$$
$$\frac{-2x}{-2} = \frac{17}{-2}$$
$$x = -\frac{17}{2}$$
The solution is $-\frac{17}{2}$.

10.
$$\frac{2x-3}{3} + 2 = \frac{2-3x}{5}$$
$$15\left(\frac{2x-3}{3} + 2\right) = 15\left(\frac{2-3x}{5}\right)$$
$$\frac{15(2x-3)}{3} + 15(2) = \frac{15(2-3x)}{5}$$
$$5(2x-3) + 30 = 3(2-3x)$$
$$10x - 15 + 30 = 6 - 9x$$
$$10x + 15 = 6 - 9x$$
$$10x + 15 + 9x = 6 - 9x + 9x$$
$$19x + 15 = 6$$
$$19x + 15 - 15 = 6 - 15$$
$$19x = -9$$
$$\frac{19x}{19} = \frac{-9}{19}$$
$$x = -\frac{9}{19}$$
The solution is $-\frac{9}{19}$.

11.
$$3x - 7 > -2$$
$$3x > 5$$
$$\frac{3x}{3} > \frac{5}{3}$$
$$x > \frac{5}{3}$$
The solution is $\left(\frac{5}{3}, \infty\right)$.

12.
$$2x - 9 < 8x + 15$$
$$2x - 8x - 9 < 8x - 8x + 15$$
$$-6x - 9 < 15$$
$$-6x - 9 + 9 < 15 + 9$$
$$-6x < 24$$
$$\frac{-6x}{-6} > \frac{24}{-6}$$
$$x > -4$$
The solution is $(-4, \infty)$.

Copyright © Houghton Mifflin Company. All rights reserved.

Chapter 2: First-Degree Equations And Inequalities

13.
$$\frac{2}{3}x - \frac{5}{8} \geq \frac{5}{4}x + 3$$
$$24\left(\frac{2}{3}x - \frac{5}{8}\right) \geq 24\left(\frac{5}{4}x + 3\right)$$
$$16x - 15 \geq 30x + 72$$
$$16x - 30x - 15 \geq 30x - 30x + 72$$
$$-14x - 15 \geq 72$$
$$-14x - 15 + 15 \geq 72 + 15$$
$$-14x \geq 87$$
$$\frac{-14x}{-14} \leq \frac{87}{-14}$$
$$x \leq -\frac{87}{14}$$
The solution is $\left\{x \mid x \leq -\frac{87}{14}\right\}$.

14.
$$2 - 3(x - 4) \leq 4x - 2(1 - 3x)$$
$$2 - 3x + 12 \leq 4x - 2 + 6x$$
$$-3x + 14 \leq 10x - 2$$
$$-3x - 10x + 14 \leq 10x - 10x - 2$$
$$-13x + 14 \leq -2$$
$$-13x + 14 - 14 \leq -2 - 14$$
$$-13x \leq -16$$
$$\frac{-13x}{-13} \geq \frac{-16}{-13}$$
$$x \geq \frac{16}{13}$$
The solution is $\left\{x \mid x \geq \frac{16}{13}\right\}$.

15.
$$-5 < 4x - 1 < 7$$
$$-5 + 1 < 4x - 1 + 1 < 7 + 1$$
$$-4 < 4x < 8$$
$$\frac{-4}{4} < \frac{4x}{4} < \frac{8}{4}$$
$$-1 < x < 2$$
The solution is $(-1, 2)$.

16. $5x - 2 > 8$ or $3x + 2 < -4$
$\quad\; 5x > 10 \qquad\quad 3x < -6$
$\quad\;\; x > 2 \qquad\qquad x < -2$
$\{x \mid x > 2\} \qquad \{x \mid x < -2\}$
$\{x \mid x > 2\} \cup \{x \mid x < -2\} = \{x \mid x > 2 \text{ or } x < -2\}$
The solution is $(-\infty, -2) \cup (2, \infty)$.

17. $3x < 4$ and $x + 2 > -1$
$\quad x < \frac{4}{3} \qquad\qquad x > -3$
$\left\{x \mid x < \frac{4}{3}\right\} \quad \{x \mid x > -3\}$
$\left\{x \mid x < \frac{4}{3}\right\} \cap \{x \mid x > -3\} = \left\{x \mid -3 < x < \frac{4}{3}\right\}$

18. $3x - 2 > -4$ or $7x - 5 < 3x + 3$
$\quad 3x > -2 \qquad\qquad 4x - 5 < 3$
$\quad \frac{3x}{3} > \frac{-2}{3} \qquad\qquad 4x < 8$
$\quad\quad x > -\frac{2}{3} \qquad\qquad \frac{4x}{4} < \frac{8}{4}$
$\qquad\qquad\qquad\qquad\;\; x < 2$
$\left\{x \mid x > -\frac{2}{3}\right\} \quad \{x \mid x < 2\}$
$\left\{x \mid x > -\frac{2}{3}\right\} \cup \{x \mid x < 2\} = \{x \mid x \text{ is any real number}\}$

19. $|2x - 3| = 8$
$2x - 3 = 8$ or $2x - 3 = -8$
$\quad 2x = 11 \qquad\quad 2x = -5$
$\quad\; x = \frac{11}{2} \qquad\quad x = -\frac{5}{2}$
The solutions are $\frac{11}{2}$ and $-\frac{5}{2}$.

20. $|5x + 8| = 0$
$\quad 5x + 8 = 0$
$\quad\quad 5x = -8$
$\quad\quad\;\; x = -\frac{8}{5}$
The solution is $-\frac{8}{5}$.

21. $6 + |3x - 3| = 2$
$\quad\;\; |3x - 3| = -4$
There is no solution to this equation because the absolute value of a number must be non-negative.

22. $|2x - 5| \leq 3$
$\quad -3 \leq 2x - 5 \leq 3$
$-3 + 5 \leq 2x - 5 + 5 \leq 3 + 5$
$\quad\quad 2 \leq 2x \leq 8$
$\quad\quad \frac{2}{2} \leq \frac{2x}{2} \leq \frac{8}{2}$
$\quad\quad 1 \leq x \leq 4$
The solution is $\{x \mid 1 \leq x \leq 4\}$.

23. $|4x - 5| \geq 3$

$4x - 5 \geq 3$ or $4x - 5 \leq -3$

$4x \geq 8$ ⠀⠀ $4x \leq 2$

$x \geq 2$ ⠀⠀ $x \leq \frac{1}{2}$

$\{x | x \geq 2\}$ ⠀ $\{x | x < \frac{1}{2}\}$

$\{x | x \geq 2\} \cup \{x | x \leq \frac{1}{2}\} = \{x | x \leq \frac{1}{2} \text{ or } x \geq 2\}$

24. $|5x - 4| < -2$

There is no solution to this equation because the absolute value of a number must be non-negative.

25. **Strategy**
Let b represent the diameter of the bushing, T the tolerance, and d the lower and upper limits of the diameter. Solve the absolute value inequality $|d - b| \leq T$ for d.

Solution
$|d - b| \leq T$
$|d - 2.75| \leq 0.003$
$-0.003 \leq d - 2.75 \leq 0.003$
$-0.003 + 2.75 \leq d - 2.75 + 2.75 \leq 0.003 + 2.75$
$2.747 \leq d \leq 2.753$

The lower and upper limits of the diameter of the bushing are 2.747 in. and 2.753 in.

26. **Strategy**
Let p represent the prescribed amount of medication, T the tolerance, and m the lower and upper limits of the amount of medication. Solve the absolute value inequality $|m - p| \leq T$ for m.

Solution
$|m - p| \leq T$
$|m - 2| \leq 0.25$
$-0.25 \leq m - 2 \leq 0.25$
$-0.25 + 2 \leq m - 2 + 2 \leq 0.25 + 2$
$1.75 \leq m \leq 2.25$

The lower and upper limits of the amount of medicine to be given to the patient are 1.75 cc and 2.25 cc.

27. **Strategy**
The smaller integer: n
The larger integer: $20 - n$
Five times the smaller integer is two more than twice the larger integer.
$5n = 2 + 2(20 - n)$

Solution
$5n = 2 + 2(20 - n)$
$5n = 2 + 40 - 2n$
$5n = 42 - 2n$
$7n = 42$
$n = 6$
$20 - n = 20 - 6 = 14$
The integers are 6 and 14.

28. **Strategy**
First consecutive integer: x
Second consecutive integer: $x + 1$
Third consecutive integer: $x + 2$
Five times the middle integer is twice the sum of the other two integers.
$5(x + 1) = 2[x + (x + 2)]$

Solution
$5(x + 1) = 2[x + (x + 2)]$
$5x + 5 = 2(2x + 2)$
$5x + 5 = 4x + 4$
$x + 5 = 4$
$x = -1$
$x + 1 = -1 + 1 = 0$
$x + 2 = -1 + 2 = 1$
The integers are -1, 0, and 1.

29. **Strategy**
Number of nickels: x
Number of dimes: $x + 3$
Number of quarters: $30 - (2x + 3) = 27 - 2x$

Coin	Number	Value	Total Value
Nickel	x	5	$5x$
Dime	$x + 3$	10	$10(x + 3)$
Quarter	$27 - 2x$	25	$25(27 - 2x)$

The sum of the total values of each type of coin equals the total value of all the coins (355 cents).
$5x + 10(x + 3) + 25(27 - 2x) = 355$

Solution
$5x + 10(x + 3) + 25(27 - 2x) = 355$
$5x + 10x + 30 + 675 - 50x = 355$
$-35x + 705 = 355$
$-35x = -350$
$x = 10$
$27 - 2x = 27 - 2(10) = 27 - 20 = 7$
There are 7 quarters in the collection.

Chapter 2: First-Degree Equations And Inequalities

30. Strategy
Cost per ounce of the mixture: x

	Amount	Cost	Value
Pure silver	40	$8.00	8(40)
Alloy	200	$3.50	3.50(200)
Mixture	240	x	240x

The sum of the values before mixing equals the value after mixing.
$8(40) + 3.50(200) = 240x$

Solution
$$8(40) + 3.50(200) = 240x$$
$$320 + 700 = 240x$$
$$1020 = 240x$$
$$4.25 = x$$
The mixture costs $4.25 per ounce

31. Strategy
Gallons of apple juice: x

	Amount	Cost	Value
Apple juice	x	3.20	3.20x
Cranberry juice	40	5.50	40(5.50)
Mixture	$40 + x$	4.20	$4.20(40 + x)$

The sum of the values before mixing equals the value after mixing.
$3.20x + 40(5.50) = 4.20(40 + x)$

Solution
$$3.20x + 40(5.50) = 4.20(40 + x)$$
$$3.20x + 220 = 168 + 4.20x$$
$$-x = -52$$
$$x = 52$$
The mixture must contain 52 gal of apple juice.

32. Strategy
Rate of the first plane: r
Rate of the second plane: $r + 80$

	Rate	Time	Distance
1st plane	r	1.75	1.75r
2nd plane	$r + 80$	1.75	$1.75(r + 80)$

The total distance traveled by the two planes is 1680 mi.
$1.75r + 1.75(r + 80) = 1680$

Solution
$$1.75r + 1.75(r + 80) = 1680$$
$$1.75r + 1.75r + 140 = 1680$$
$$3.5r + 140 = 1680$$
$$3.5r = 1540$$
$$r = 440$$
$r + 80 = 440 + 80 = 520$
The speed of the first plane is 440 mph. The speed of the second plane is 520 mph.

33. Strategy
Amount invested at 10.5%: x
Amount invested at 6.4%: $8000 - x$

	Principal	Rate	Interest
Amount at 10.5%	x	0.105	0.105x
Amount at 6.4%	$8000 - x$	0.064	$0.064(8000 - x)$

The sum of the interest earned by the two investments equals the total annual interest earned ($635).
$0.105x + 0.064(8000 - x) = 635$

Solution
$$0.105x + 0.064(8000 - x) = 635$$
$$0.105x + 512 - 0.064x = 635$$
$$0.041x + 512 = 635$$
$$0.041x = 123$$
$$x = 3000$$
$8000 - x = 8000 - 3000 = 5000$
The amount invested at 10.5% was $3000.
The amount invested at 6.4% was $5000.

34. Strategy
Pounds of 30% tin: x
Pounds of 70% tin: $500 - x$

	Amount	Percent	Quantity
30%	x	0.30	0.30x
70%	$500 - x$	0.70	$0.70(500 - x)$
40%	500	0.40	0.40(500)

The sum of the quantities before mixing is equal to the quantity after mixing.
$0.30x + 0.70(500 - x) = 0.40(500)$

Solution
$$0.30x + 0.70(500 - x) = 0.40(500)$$
$$0.30x + 350 - 0.70x = 200$$
$$-0.40x + 350 = 200$$
$$-0.40x = -150$$
$$x = 375$$
$500 - x = 500 - 375 = 125$
375 lb of 30% tin and 125 lb of 70% tin were used

35. Strategy
To find the minimum amount of sales, write and solve an inequality using N to represent the amount of sales.

Solution
$$800 + 0.04N \geq 3000$$
$$0.04N \geq 2200$$
$$N \geq 55,000$$
The executive's amount of sales must be $55,000 or more.

36. Strategy
To find the range of scores, write and solve an inequality using N to represent the score on the last test.

Solution
$$80 \leq \frac{92+66+72+88+N}{5} \leq 90$$
$$80 \leq \frac{318+N}{5} \leq 90$$
$$5 \cdot 80 \leq 5 \cdot \frac{318+N}{5} \leq 5 \cdot 90$$
$$400 \leq 318+N \leq 450$$
$$400-318 \leq 318+N-318 \leq 450-318$$
$$82 \leq N \leq 132$$

Since 100 is the maximum score, the range of scores to receive a B grade is $82 \leq N \leq 100$.

CHAPTER 2 TEST

1. $x-2=-4$
$x-2+2=-4+2$
$x=-2$
The solution is -2.

2. $x+\frac{3}{4}=\frac{5}{8}$
$x+\frac{3}{4}-\frac{3}{4}=\frac{5}{8}-\frac{3}{4}$
$x=\frac{5}{8}-\frac{6}{8}$
$x=-\frac{1}{8}$
The solution is $-\frac{1}{8}$.

3. $-\frac{3}{4}y=-\frac{5}{8}$
$-\frac{4}{3}\left(-\frac{3}{4}y\right)=-\frac{4}{3}\left(-\frac{5}{8}\right)$
$y=\frac{5}{6}$
The solution is $\frac{5}{6}$.

4. $3x-5=7$
$3x-5+5=7+5$
$3x=12$
$\frac{3x}{3}=\frac{12}{3}$
$x=4$
The solution is 4.

5. $\frac{3}{4}y-2=6$
$\frac{3}{4}y-2+2=6+2$
$\frac{3}{4}y=8$
$\frac{4}{3}\left(\frac{3}{4}y\right)=\frac{4}{3}(8)$
$y=\frac{32}{3}$
The solution is $\frac{32}{3}$.

6. $2x-3-5x=8+2x-10$
$-3x-3=-2+2x$
$-3x-3+3x=-2+2x+3x$
$-3=-2+5x$
$-3+2=-2+5x+2$
$-1=5x$
$\frac{-1}{5}=\frac{5x}{5}$
$-\frac{1}{5}=x$
$x=-\frac{1}{5}$
The solution is $-\frac{1}{5}$.

7. $2[x-(2-3x)-4]=x-5$
$2[x-2+3x-4]=x-5$
$2[4x-6]=x-5$
$8x-12=x-5$
$8x-x-12=x-x-5$
$7x-12=-5$
$7x-12+12=-5+12$
$7x=7$
$\frac{7x}{7}=\frac{7}{7}$
$x=1$
The solution is 1.

8. $\frac{2}{3}x-\frac{5}{6}x=4$
$\frac{4}{6}x-\frac{5}{6}x=4$
$-\frac{1}{6}x=4$
$-6\left(-\frac{1}{6}x\right)=-6(4)$
$x=-24$
The solution is -24.

44 Chapter 2: First-Degree Equations And Inequalities

9.
$$\frac{2x+1}{3} - \frac{3x+4}{6} = \frac{5x-9}{9}$$
$$18\left(\frac{2x+1}{3} - \frac{3x+4}{6}\right) = 18\left(\frac{5x-9}{9}\right)$$
$$18\left(\frac{2x+1}{3}\right) - 18\left(\frac{3x+4}{6}\right) = 18\left(\frac{5x-9}{9}\right)$$
$$6(2x+1) - 3(3x+4) = 2(5x-9)$$
$$12x + 6 - 9x - 12 = 10x - 18$$
$$3x - 6 = 10x - 18$$
$$3x - 6 - 10x = 10x - 18 - 10x$$
$$-7x - 6 = -18$$
$$-7x - 6 + 6 = -18 + 6$$
$$-7x = -12$$
$$\frac{-7x}{-7} = \frac{-12}{-7}$$
$$x = \frac{12}{7}$$

The solution is $\frac{12}{7}$.

10.
$$2x - 5 \geq 5x + 4$$
$$-3x - 5 \geq 4$$
$$-3x \geq 9$$
$$\frac{-3x}{-3} \leq \frac{9}{-3}$$
$$x \leq -3$$
$$(-\infty, -3]$$

11.
$$4 - 3(x+2) < 2(2x+3) - 1$$
$$4 - 3x - 6 < 4x + 6 - 1$$
$$-2 - 3x < 4x + 5$$
$$-7x < 7$$
$$\frac{-7x}{-7} > \frac{7}{-7}$$
$$x > -1$$
$$(-1, \infty)$$

12. $3x - 2 > 4$ or $4 - 5x < 14$
$$3x > 6 \qquad -5x < 10$$
$$x > 2 \qquad \frac{-5x}{-5} > \frac{10}{-5}$$
$$\qquad\qquad x > -2$$
$$\{x|x > 2\} \qquad \{x|x > -2\}$$
$$\{x|x > 2\} \cup \{x|x > -2\} = \{x|x > -2\}$$

13. $4 - 3x \geq 7$ and $2x + 3 \geq 7$
$$-3x \geq 3 \qquad\qquad 2x \geq 4$$
$$\frac{-3x}{-3} \leq \frac{3}{-3} \qquad \frac{2x}{2} \geq \frac{4}{2}$$
$$x \leq -1 \qquad\qquad x \geq 2$$
$$\{x|x \leq -1\} \cap \{x|x \geq 2\} = \emptyset$$

14. $|3 - 5x| = 12$
$$3 - 5x = 12 \qquad 3 - 5x = -12$$
$$-5x = 9 \qquad\qquad -5x = -15$$
$$x = -\frac{9}{5} \qquad\qquad x = 3$$

The solutions are $-\frac{9}{5}$ and 3.

15. $2 - |2x - 5| = -7$
$$-|2x - 5| = -9$$
$$|2x - 5| = 9$$
$$2x - 5 = 9 \qquad 2x - 5 = -9$$
$$2x = 14 \qquad\qquad 2x = -4$$
$$x = 7 \qquad\qquad x = -2$$
The solutions are 7 and –2.

16. $|3x - 1| \leq 2$
$$-2 \leq 3x - 1 \leq 2$$
$$-2 + 1 \leq 3x - 1 + 1 \leq 2 + 1$$
$$-1 \leq 3x \leq 3$$
$$\frac{-1}{3} \leq \frac{3x}{3} \leq \frac{3}{3}$$
$$-\frac{1}{3} \leq x \leq 1$$
$$\left\{x \Big| -\frac{1}{3} \leq x \leq 1\right\}$$

17. $|2x - 1| > 3$
$$2x - 1 > 3 \text{ or } 2x - 1 < -3$$
$$2x > 4 \qquad\qquad 2x < -2$$
$$x > 2 \qquad\qquad x < -1$$
$$\{x|x > 2\} \cup \{x|x < -1\} = \{x|x > 2 \text{ or } x < -1\}$$

18. $4 + |2x - 3| = 1$
$$|2x - 3| = -3$$
There is no solution because the absolute value of a number is always non-negative.

19. Strategy
To find the number of miles, write and solve an inequality using N to represent the number of miles.

Solution
cost of car A < cost of car B
$$12 + 0.10N < 24$$
$$0.10N < 12$$
$$N < 120$$
It costs less to rent from Agency A if the car is driven less than 120 mi.

20. Strategy
Let p represent the prescribed amount of medication, T the tolerance, and m the lower and upper limits of the given amount of medication. Solve the absolute value inequality $|m-p| \leq T$ for m.

Solution
$|m-p| \leq T$
$|m-3| \leq 0.1$
$-0.1 \leq m-3 \leq 0.1$
$-0.1+3 \leq m-3+3 \leq 0.1+3$
$2.9 \leq m \leq 3.1$

The lower and upper limits of the amount of medication to be given to the patient are 2.9 cc and 3.1 cc.

21. Strategy
Number of 15¢ stamps: x
Number of 11¢ stamps: $2x$
Number of 21¢ stamps: $30 - 3x$

Stamp	Number	Value	Total Value
15¢	x	15	$15x$
11¢	$2x$	11	$11(2x)$
21¢	$30-3x$	21	$21(30-3x)$

The sum of the total values of each type of stamp equals the total value of all the stamps (440 cents).
$15x + 11(2x) + 21(30 - 3x) = 422$

Solution
$15x + 11(2x) + 21(30-3x) = 422$
$15x + 22x + 630 - 63x = 422$
$-26x + 630 = 422$
$-26x = -208$
$x = 8$
$30 - 3x = 30 - 3(8) = 30 - 24 = 6$
There are six 21¢ stamps.

22. Strategy
Price of hamburger mixture: x

	Amount	Cost	Value
$2.60 hamburger	100	2.60	2.60(100)
$4.20 hamburger	60	4.20	4.20(60)
Mixture	160	x	$160x$

The sum of the values before mixing equals the value after mixing.
$2.60(100) + 4.20(60) = 160x$

Solution
$2.60(100) + 4.20(60) = 160x$
$260 + 252 = 160x$
$512 = 160x$
$3.20 = x$

The price of the hamburger mixture is $2.20/lb

23. Strategy
Time jogger runs a distance: t
Time jogger returns same distance: $1\frac{45}{60} - t$

	Rate	Time	Distance
Jogger runs a distance	8	t	$8t$
Jogger returns same distance	6	$\frac{7}{4} - t$	$6\left(\frac{7}{4} - t\right)$

The jogger runs a distance and returns the same distance.
$8t = 6\left(\frac{7}{4} - t\right)$

Solution
$8t = 6\left(\frac{7}{4} - t\right)$
$8t = \frac{21}{2} - 6t$
$14t = \frac{21}{2}$
$\frac{1}{14}(14t) = \frac{1}{14}\left(\frac{21}{2}\right)$
$t = \frac{3}{4}$

The jogger ran for $\frac{3}{4}$ hour.

$8t = 8 \cdot \frac{3}{4} = 6$

The jogger ran a distance of 6 mi one way. The jogger ran a total distance of 12 mi.

24. Strategy
Amount invested at 7.8%: x
Amount invested at 9%: $12,000 - x$

	Principal	Rate	Interest
Amount invested at 7.8%	x	0.078	$0.078x$
Amount invested at 9%	$12,000 - x$	0.09	$0.09 \cdot (12,000 - x)$

The sum of the interest earned by the two investments equals the total annual interest earned ($1020).
$0.078x + 0.09(12,000 - x) = 1020$

Solution
$0.078x + 0.09(12,000 - x) = 1020$
$0.078x + 1080 - 0.09x = 1020$
$-0.012x = -60$
$x = 5000$
$12,000 - x = 12,000 - 50000 = 7000$
The amount invested at 7.8% was $5000.
The amount invested at 9% was $7000.

25. Strategy
Ounces of pure water: x

	Amount	Percent	Quantity
Pure water	x	0	0
8% salt	60	0.08	0.08(60)
3% salt	$60 + x$	0.03	$0.03(60 + x)$

The sum of the quantities before mixing is equal to the quantity after mixing.
$0 + 0.08(60) = 0.03(60 + x)$

Solution
$$0 + 0.08(60) = 0.03(60 + x)$$
$$4.8 = 1.8 + 0.03x$$
$$3 = 0.03x$$
$$100 = x$$
There are 100 oz of pure water.

CUMULATIVE REVIEW EXERCISES

1. $-2^2 \cdot 3^3 = -(2 \cdot 2)(3 \cdot 3 \cdot 3) = -(4)(27) = -108$

2. $4 - (2-5)^2 \div 3 + 2 = 4 - (-3)^2 \div 3 + 2$
$= 4 - 9 \div 3 + 2$
$= 4 - 3 + 2$
$= 1 + 2$
$= 3$

3. $4 \div \dfrac{\frac{3}{8}-1}{5} \cdot 2 = 4 \div \dfrac{-\frac{5}{8}}{5} \cdot 2$
$= 4 \div \left(-\dfrac{5}{8} \cdot \dfrac{1}{5}\right) \cdot 2$
$= 4 \div \left(-\dfrac{1}{8}\right) \cdot 2$
$= 4 \cdot (-8) \cdot 2$
$= -32 \cdot 2$
$= -64$

4. $2a^2 - (b-c)^2 = 2(2)^2 - (3-(-1))^2$
$= 2 \cdot 4 - (3+1)^2$
$= 2 \cdot 4 - 4^2$
$= 2 \cdot 4 - 16$
$= 8 - 16$
$= -8$

5. The Commutative Property of Addition

6. $A \cap B = \{3, 9\}$

7. $3x - 2[x - 3(2-3x) + 5] = 3x - 2[x - 6 + 9x + 5]$
$= 3x - 2[10x - 1]$
$= 3x - 20x + 2$
$= -17x + 2$

8. $5[y - 2(3-2y) + 6] = 5[y - 6 + 4y + 6]$
$= 5[5y]$
$= 25y$

9. $4 - 3x = -2$
$4 - 3x - 4 = -2 - 4$
$-3x = -6$
$\dfrac{-3x}{-3} = \dfrac{-6}{-3}$
$x = 2$
The solution is 2.

10. $-\dfrac{5}{6}b = -\dfrac{5}{12}$
$\left(-\dfrac{6}{5}\right)\left(-\dfrac{5}{6}\right)b = \left(-\dfrac{6}{5}\right)\left(-\dfrac{5}{12}\right)$
$b = \dfrac{1}{2}$
The solution is $\dfrac{1}{2}$.

11. $2x + 5 = 5x + 2$
$2x + 5 - 5x = 5x + 2 - 5x$
$-3x + 5 = 2$
$-3x + 5 - 5 = 2 - 5$
$-3x = -3$
$\dfrac{-3x}{-3} = \dfrac{-3}{-3}$
$x = 1$
The solution is 1.

12. $\dfrac{5}{12}x - 3 = 7$
$\dfrac{5}{12}x - 3 + 3 = 7 + 3$
$\dfrac{5}{12}x = 10$
$\left(\dfrac{12}{5}\right)\left(\dfrac{5}{12}\right)x = \left(\dfrac{12}{5}\right)10$
$x = 24$
The solution is 24.

13. $2[3 - 2(3 - 2x)] = 2(3 + x)$
$2[3 - 6 + 4x] = 6 + 2x$
$2[-3 + 4x] = 6 + 2x$
$-6 + 8x = 6 + 2x$
$-6 + 8x - 2x = 6 + 2x - 2x$
$-6 + 6x = 6$
$-6 + 6x + 6 = 6 + 6$
$6x = 12$
$\dfrac{6x}{6} = \dfrac{12}{6}$
$x = 2$
The solution is 2.

14.
$$3[2x - 3(4-x)] = 2(1-2x)$$
$$3[2x - 12 + 3x] = 2 - 4x$$
$$3[5x - 12] = 2 - 4x$$
$$15x - 36 = 2 - 4x$$
$$15x - 36 + 4x = 2 - 4x + 4x$$
$$19x - 36 = 2$$
$$19x - 36 + 36 = 2 + 36$$
$$19x = 38$$
$$\frac{19x}{19} = \frac{38}{19}$$
$$x = 2$$
The solution is 2.

15.
$$\frac{3x-1}{4} - \frac{4x-1}{12} = \frac{3+5x}{8}$$
$$24\left(\frac{3x-1}{4} - \frac{4x-1}{12}\right) = 24\left(\frac{3+5x}{8}\right)$$
$$\frac{24(3x-1)}{4} - \frac{24(4x-1)}{12} = \frac{24(3+5x)}{8}$$
$$6(3x-1) - 2(4x-1) = 3(3+5x)$$
$$18x - 6 - 8x + 2 = 9 + 15x$$
$$10x - 4 = 9 + 15x$$
$$10x - 4 - 15x = 9 + 15x - 15x$$
$$-5x - 4 = 9$$
$$-5x - 4 + 4 = 9 + 4$$
$$-5x = 13$$
$$\frac{-5x}{-5} = \frac{13}{-5}$$
$$x = -\frac{13}{5}$$
The solution is $-\frac{13}{5}$.

16.
$$3x - 2 \geq 6x + 7$$
$$3x \geq 6x + 9$$
$$-3x \geq 9$$
$$x \leq -3$$
$$\{x | x \leq -3\}$$

17.
$5 - 2x \geq 6$ and $3x + 2 \geq 5$
$-2x \geq 1$ $3x \geq 3$
$x \leq -\frac{1}{2}$ $x \geq 1$
$\left\{x \mid x \leq -\frac{1}{2}\right\}$ $\{x | x \geq 1\}$
$\left\{x \mid x \leq -\frac{1}{2}\right\} \cap \{x | x \geq 1\} = \emptyset$

18.
$4x - 1 > 5$ or $2 - 3x < 8$
$4x > 6$ $-3x < 6$
$x > \frac{3}{2}$ $x > -2$
$\left\{x \mid x > \frac{3}{2}\right\}$ $\{x | x > -2\}$
$\left\{x \mid x > \frac{3}{2}\right\} \cup \{x | x > -2\} = \{x | x > -2\}$

19.
$|3 - 2x| = 5$
$3 - 2x = 5$ $3 - 2x = -5$
$-2x = 2$ $-2x = -8$
$x = -1$ $x = 4$
The solutions are -1 and 4.

20.
$3 - |2x - 3| = -8$
$-|2x - 3| = -11$
$|2x - 3| = 11$
$2x - 3 = 11$ $2x - 3 = -11$
$2x = 14$ $2x = -8$
$x = 7$ $x = -4$
The solutions are 7 and -4.

21.
$|3x - 5| \leq 4$
$-4 \leq 3x - 5 \leq 4$
$-4 + 5 \leq 3x - 5 + 5 \leq 4 + 5$
$1 \leq 3x \leq 9$
$\frac{1}{3} \leq \frac{3x}{3} \leq \frac{9}{3}$
$\frac{1}{3} \leq x \leq 3$
$\left\{x \mid \frac{1}{3} \leq x \leq 3\right\}$

22.
$|4x - 3| > 5$
$4x - 3 < -5$ or $4x - 3 > 5$
$4x < -2$ $4x > 8$
$x < -\frac{1}{2}$ $x > 2$
$\left\{x \mid x < -\frac{1}{2}\right\}$ $\{x | x > 2\}$
$\left\{x \mid x < -\frac{1}{2}\right\} \cup \{x | x > 2\} = \left\{x \mid x < -\frac{1}{2} \text{ or } x > 2\right\}$

23. $\{x | x \geq -2\}$

24.

25. The unknown number: n
three times the number: $3n$
the sum of three times the number and six: $3n + 6$
$(3n + 6) + 3n = 6n + 6$

26. Strategy
The first integer: n
Second consecutive odd integer: $n + 2$
Third consecutive odd integer: $n + 4$
Three times the sum of the first and third integers is fifteen more than the second integer.

Solution
$3[n + (n + 4)] = (n + 2) + 15$
$3(2n + 4) = n + 17$
$6n + 12 = n + 17$
$5n + 12 = 17$
$5n = 5$
$n = 1$
The first integer is 1.

27. Strategy
Number of 23¢ stamps: n
Number of 19¢ stamps: $2n - 5$

Stamp	Number	Value	Total Value
19¢	$2n - 5$	19	$19(2n - 5)$
23¢	n	23	$23n$

The sum of the total values of each denomination of stamp equals the total value of all the stamps (393 cents).
$19(2n - 5) + 23n = 393$

Solution
$19(2n - 5) + 23n = 393$
$38n - 95 + 23n = 393$
$61n - 95 = 393$
$61n = 488$
$n = 8$
$2n - 5 = 2(8) - 5 = 16 - 5 = 11$
There are eleven 19¢ stamps.

28. Strategy
Number of adult tickets: n
Number of children's tickets: $75 - n$

	Number	Value	Total Value
Adult tickets	n	2.25	$2.25n$
Children's tickets	$75 - n$	0.75	$0.75(75 - n)$

The sum of the total values of each denomination of ticket equals the total value of all the tickets ($128.25).
$2.25n + 0.75(75 - n) = 128.25$

Solution
$2.25n + 0.75(75 - n) = 128.25$
$2.25n + 56.25 - 0.75n = 128.25$
$1.5n + 56.25 = 128.25$
$1.5n = 72$
$n = 48$
48 adult tickets were sold.

29. Strategy
Slower plane: x
Faster plane: $x + 120$

	Rate	Time	Distance
Slower plane	x	2.5	$2.5x$
Faster plane	$x + 120$	2.5	$2.5(x + 120)$

The two planes travel a total distance of 1400 mi.
$2.5x + 2.5(x + 120) = 1400$

Solution
$2.5x + 2.5(x + 120) = 1400$
$2.5x + 2.5x + 300 = 1400$
$5x + 300 = 1400$
$5x = 1100$
$x = 220$
$x + 120 = 220 + 120 = 340$
The speed of the faster plane is 340 mph.

30. Strategy
Liters of 12% acid solution: x

Solution	Amount	Percent	Quantity
12%	x	0.12	$0.12x$
5%	4	0.05	$0.05(4)$
8%	$x + 4$	0.08	$0.08(x + 4)$

The sum of the quantities before mixing equals the quantity after mixing.
$0.12x + 0.05(4) = 0.08(x + 4)$

Solution
$0.12x + 0.05(4) = 0.08(x + 4)$
$0.12x + 0.2 = 0.08x + 0.32$
$0.04x + 0.2 = 0.32$
$0.04x = 0.12$
$x = 3$
3 L of 12% acid solution must be in the mixture

31. Strategy
Amount invested at 9.8%: x
Amount invested at 12.8%: $10,000 - x$

	Principal	Rate	Interest
Amount at 9.8%	x	0.098	$0.098x$
Amount at 12.8%	$10,000 - x$	0.128	$0.128(10,000 - x)$

The sum of the interest earned by the two investments is equal to the total annual interest earned ($1085).
$0.098x + 0.128(10,000 - x) = 1085$

Solution
$0.098x + 0.128(10,000 - x) = 1085$
$0.098x + 1280 - 0.128x = 1085$
$-0.03x + 1280 = 1085$
$-0.03x = -195$
$x = 6500$
$6500 was invested at 9.8%.

CHAPTER 3: LINEAR FUNCTIONS AND INEQUALITIES IN TWO VARIABLES

CHAPTER 3 PREP TEST

1. $-4(x-3) = -4x + 12$ [1.3.3]

2. $\sqrt{(-6)^2 + (-8)^2} = \sqrt{36 + 64}$ [1.2.4]
   $= \sqrt{100}$
   $= 10$

3. $\dfrac{3 - (-5)}{2 - 6} = \dfrac{8}{-4}$ [1.2.4]
   $= -2$

4. $-2(-3) + 5 = 6 + 5$ [1.3.2]
   $= 11$

5. $\dfrac{2(5)}{5 - 1} = \dfrac{10}{4}$ [1.3.2]
   $= 2.5$

6. $2(-1)^3 - 3(-1) + 4 = 2(-1) + 3 + 4$ [1.3.2]
   $= -2 + 7$
   $= 5$

7. $\dfrac{7 + (-5)}{2} = \dfrac{2}{2}$ [1.3.2]
   $= 1$

8. $3x - 4(0) = 12$ [2.1.1]
   $3x = 12$
   $x = 4$

Section 3.1

Concept Review 3.1

1. Always true

3. Never true
   The first number is the *x*-coordinate and the second number is the *y*-coordinate.

5. Never true
   The point (–2, –4) is in the third quadrant.

Objective 3.1.1 Exercises

1. To graph the point (6, –2), start at the origin and move 6 units <u>right</u> and 2 units <u>down</u>.

3. [graph]

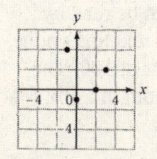

5. $A(0, 3)$
   $B(1, 1)$
   $C(3, -4)$
   $D(-4, 4)$

7. [graph]

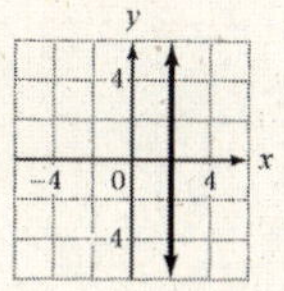

9. [graph]

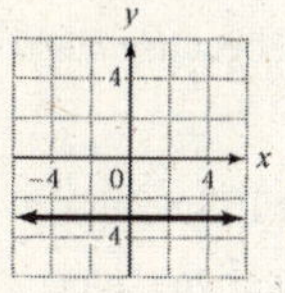

11. $y = x^2$
    Ordered pairs: (–2, 4)
    (–1, 1)
    (0, 0)
    (1, 1)
    (2, 4)
    [graph]

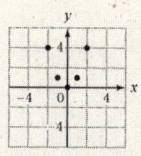

13. $y = |x + 1|$
    Ordered pairs: (–5, 4)
    (–3, 2)
    (0, 1)
    (3, 4)
    (5, 6)
    [graph]

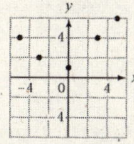

15. $y = -x^2 + 2$
    Ordered pairs: (–2, –2)
    (–1, 1)
    (0, 2)
    (1, 1)
    (2, –2)
    [graph]

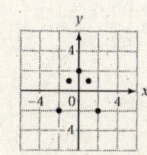

17. $y = x^3 - 2$

Ordered pairs:
(-1, -3)
(0, -2)
(1, -1)
(2, 6)

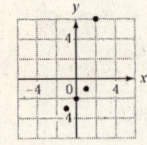

19.
a. (a, b) is in Quadrant I.

b. $(-a, b)$ is in Quadrant II.

c. $(b-a, -b)$ is in Quadrant IV because $b-a$ is positive (since $b > a$).

d. $(a-b, -b-a)$ is in Quadrant III because $a-b$ is negative and $-b-a$ is negative (since $b > a$).

Objective 3.1.2 Exercises

21. For points $P_1(3, -3)$ and $P_2(-1, 4)$, identify each of the following values:
$x_1 = \underline{3}$, $x_2 = \underline{-1}$, $y_1 = \underline{-3}$, and $y_2 = \underline{4}$.

23. $d = \sqrt{(4-3)^2 + (1-5)^2}$
$= \sqrt{1+16}$
$d = \sqrt{17}$
$x_m = \frac{3+4}{2} = \frac{7}{2}$
$y_m = \frac{1+5}{2} = 3$

The length is $\sqrt{17}$ and the midpoint is $\left(\frac{7}{2}, 3\right)$.

25. $d = \sqrt{(-2-0)^2 + (4-3)^2}$
$= \sqrt{4+1}$
$d = \sqrt{5}$
$x_m = \frac{0+(-2)}{2} = -1$
$y_m = \frac{3+4}{2} = \frac{7}{2}$

The length is $\sqrt{5}$ and the midpoint is $\left(-1, \frac{7}{2}\right)$.

27. $d = \sqrt{[2-(-3)]^2 + [-4-(-5)]^2}$
$= \sqrt{5^2 + 1^2}$
$d = \sqrt{26}$
$x_m = \frac{-3+2}{2} = -\frac{1}{2}$
$y_m = \frac{-5+(-4)}{2} = -\frac{9}{2}$

The length is $\sqrt{26}$ and the midpoint is $\left(-\frac{1}{2}, -\frac{9}{2}\right)$.

29. $d = \sqrt{(-1-5)^2 + [5-(-2)]^2}$
$= \sqrt{(-6)^2 + 7^2}$
$d = \sqrt{85}$
$x_m = \frac{5+(-1)}{2} = \frac{4}{2} = 2$
$y_m = \frac{-2+5}{2} = \frac{3}{2}$

The length is $\sqrt{85}$ and the midpoint is $\left(2, \frac{3}{2}\right)$.

31. $d = \sqrt{(2-5)^2 + [-5-(-5)]^2}$
$= \sqrt{(-3)^2 + 0^2}$
$d = 3$
$x_m = \frac{5+2}{2} = \frac{7}{2}$
$y_m = \frac{-5+(-5)}{2} = -5$

The length is 3 and the midpoint is $\left(\frac{7}{2}, -5\right)$.

33. For $P_1(-a, -b)$ and $P_2(c, d)$, the midpoint is $\left(\frac{-a+c}{2}, \frac{-b+d}{2}\right)$. If $a < c$ and $b < d$, then the x-coordinate is positive and the y-coordinate is positive. This corresponds to Quadrant I.

Objective 3.1.3 Exercises

35.
a. The number of calories can be read by looking at the y-axis. Each dotted line represents an increment of 250. The number of calories for a hamburger is 275.

b. The number of milligrams can be read by looking at the x-axis. Each dotted line on the x-axis represents an increment of 100. The number of milligrams of sodium in a Big Mac is 1100.

37. a. Look for 2001 on the *x*-axis. Then look for the first point that lies above the *x*-axis. This point has an *y*-coordinate of $1.8 trillion.

b. Look for 1.9 on the *y*-axis. Then look for the point(s) that lie above the 1.9. This point has an *x*-coordinate of 2003.

c. The points on the graph appear to be increasing in value for both *x*- and *y*-coordinates. We would expect consumer debt 2004 to be greater than in 2003.

39. The lowest *x*-value is <u>1920</u> and the highest *x*-value is <u>2000</u>. The lowest *y*-value is <u>35</u> and the highest *y*-value is <u>97</u>. In a scatter diagram of the data, the *x*-axis must show values from at least <u>1920</u> to <u>2000</u> and the *y*-axis must show values from at least <u>35</u> to <u>97</u>.

41. Label the years on the *x*-axis. The *y*-values represent percent so the values on the *y*-axis should go from 0 to 100. Since some of the values are between multiples of 10, the graph would be easier to read with increments of 5.

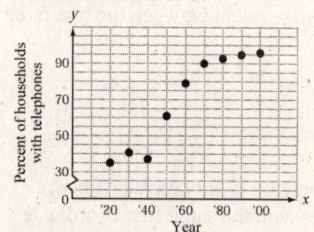

Objective 3.1.4 Exercises

43. a. Strategy
To find the average annual rate of change, divide the change in population in millions (20.9 – 3.0) by the change in years (2000 – 1900).

Solution

$$\text{average rate of change} = \frac{20.9 - 3.0}{2000 - 1900}$$
$$= \frac{17.9}{100}$$
$$= 0.179$$

0.179 million = 179,000
The average annual rate of change in population was 179,000 people.

b. Strategy
To find if the average annual rate of change from 1900 to 1950 was greater than or less than the change from 1950 to 2000, subtract the average annual rate of change in population from 1950 to 2000 from the average annual rate of change in population from 1900 to 1950.

Solution
For 1900 to 1950,

$$\text{average rate of change} = \frac{7.7 - 3.0}{1950 - 1900}$$
$$= \frac{4.7}{50}$$
$$= 0.094$$

For 1950 to 2000,

$$\text{average rate of change} = \frac{20.9 - 7.7}{2000 - 1950}$$
$$= \frac{13.2}{50}$$
$$= 0.264$$

0.094 < 0.264
The average annual rate of change in population from 1900 to 1950 was less than the average annual rate of change in population from 1950 to 2000.

c. Strategy
To find if the average annual rate of change from 1980 to 1990 in Texas was greater than or less than the change from 1980 to 1990 in California, subtract the average annual rate of change in population in Texas from 1980 to 1990 from the average annual rate of change in population in California from 1980 to 1990.

Solution
For Texas from 1980 to 1990,

$$\text{average rate of change} = \frac{17 - 14.2}{1990 - 1980}$$
$$= \frac{2.8}{10}$$
$$= 0.28$$

For California from 1980 to 1990,

$$\text{average rate of change} = \frac{29.8 - 23.7}{1990 - 1980}$$
$$= \frac{6.1}{10}$$
$$= 0.61$$

0.28 < 0.61
The average annual rate of change in population in Texas was less than the average rate of change in population in California.

47. a. Strategy
To find the average annual rate of change, divide the change in fatalities (4727 − 6482) by the change in years (2000 − 1990).

Solution
average rate of change $= \dfrac{4727 - 6482}{2000 - 1990}$
$= -175.5$
The average annual rate of change was −175.5 pedestrians per year.

b. The answer can be considered encouraging because it means the number of pedestrian fatalities is decreasing.

49. a. Strategy
To find the average annual rate of change, divide the change in applications in thousands (375 − 120) by the change in years (2000 − 1990).

Solution
average rate of change $= \dfrac{375 - 120}{10}$
$= 25.5$
The average annual rate of change was 25,500 applications.

b. Strategy
To find the how much greater, compare the average annual rate of change in applications for 1995 to 2000 with that of 1990 to 1995.

Solution
For 1995 to 2000,
average rate of change $= \dfrac{375 - 175}{5}$
$= 40$
For 1990 to 1995,
average rate of change $= \dfrac{175 - 120}{5}$
$= 11$
40 − 11 = 29
The average rate of change was greater from 1995 to 2000 by 29,000 applications per year.

Applying Concepts 3.1

51. Ordered pairs: (−2, 4)
(−1, 1)
(0, 0)
(1, 1)
(2, 4)

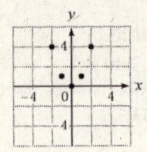

53.

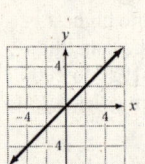

55. The graph of all ordered pairs (x, y) that are equidistant from two fixed points is a line that both is perpendicular to the line that passes through the two fixed points and also passes through the midpoint of the line segment between the two fixed points.

Section 3.2
Concept Review 3.2

1. Sometimes true
By definition, a function cannot have different second coordinates with the same first coordinate.

3. Sometimes true
The function $f(x) = \dfrac{2}{x-3}$ is not defined for $x = 3$.

5. Always true

Objective 3.2.1 Exercises

1. Yes, every function is a relation, but not every relation is a function. A relation and function are similar in that both are sets of ordered pairs, but a function is a specific type of relation. A function is a relation for which there are no two ordered pairs with the same first element.

3. For a function, each element of the domain is paired with <u>exactly one</u> element of the range.

5. {(−3, 1), (−2, 2), (1, 5), (4, −7)}
The relation is a function because each number in the domain is paired with exactly one number in the range.
Domain = {−3, −2, 1, 4}
Range = {−7, 1, 2, 5}

7. {(1, 5), (2, 5), (3, 5), (4, 5), (5, 5)}
The relation is a function because each number in the domain is paired with exactly one number in the range.
Domain = {1, 2, 3, 4, 5}
Range = {5}

9. $\left\{\left(-2, -\frac{1}{2}\right), (-1, -1), (1, 1), \left(2, \frac{1}{2}\right), \left(3, \frac{1}{3}\right)\right\}$
The relation is a function because each number in the domain is paired with exactly one number in the range.
Domain = {−2, −1, 1, 2, 3}
Range = $\left\{-1, -\frac{1}{2}, \frac{1}{3}, \frac{1}{2}, 1\right\}$

11. {(2, 3), (4, 5), (6, 7), (8, 9), (6, 8)}
The relation is not a function because there are two ordered pairs with the same first coordinate and different second coordinates.
Domain = {2, 4, 6, 8}
Range = {3, 5, 7, 8, 9}

13. Yes, the diagram does represent a function because each number in the domain is paired with exactly one number in the range.

15. No, the diagram does not represent a function. The numbers 1 and 4 are paired with two different numbers.

17. Yes, the diagram does represent a function because each number in the domain is paired with exactly one number in the range.

19. a. Yes, this table defines a function because no weight occurs more than once.
    b. $29.25

21. Evaluating a function means finding the element in the range that corresponds to a specific element in the domain. To evaluate $f(x) = 3x$ for $x = 2$, substitute 2 for x:
$f(2) = 3 \cdot 2 = 6$
The value of the function when $x = 2$ is 6.

23. True

25. Given $f(x) = 5x - 7$, find $f(3)$:
$f(x) = 5x - 7$
$f(3) = 5(3) - 7$
$f(3) = 8$

27. $f(x) = 4x + 5$
    a. $f(2) = 4(2) + 5 = 8 + 5 = 13$
    b. $f(-2) = 4(-2) + 5 = -8 + 5 = -3$
    c. $f(0) = 4(0) + 5 = 0 + 5 = 5$

29. $v(s) = 6 - 3s$
    a. $v(3) = 6 - 3(3) = 6 - 9 = -3$
    b. $v(-2) = 6 - 3(-2) = 6 + 6 = 12$
    c. $v\left(-\frac{2}{3}\right) = 6 - 3\left(-\frac{2}{3}\right) = 6 + 2 = 8$

31. $f(x) = -\frac{3}{2}x - 2$
    a. $f(4) = -\frac{3}{2}(4) - 2 = -6 - 2 = -8$
    b. $f(-2) = -\frac{3}{2}(-2) - 2 = 3 - 2 = 1$
    c. $f\left(\frac{4}{3}\right) = -\frac{3}{2}\left(\frac{4}{3}\right) - 2 = -2 - 2 = -4$

33. $p(c) = \frac{1}{2}c - \frac{3}{4}$
    a. $p\left(\frac{7}{2}\right) = \frac{1}{2}\left(\frac{7}{2}\right) - \frac{3}{4} = \frac{7}{4} - \frac{3}{4} = \frac{4}{4} = 1$
    b. $p\left(-\frac{1}{2}\right) = \frac{1}{2}\left(-\frac{1}{2}\right) - \frac{3}{4} = -\frac{1}{4} - \frac{3}{4} = -\frac{4}{4} = -1$
    c. $p\left(\frac{1}{2}\right) = \frac{1}{2}\left(\frac{1}{2}\right) - \frac{3}{4} = \frac{1}{4} - \frac{3}{4} = -\frac{2}{4} = -\frac{1}{2}$

35. $f(x) = 4x - 1$
    a. $f(a + 3) = 4(a + 3) - 1 = 4a + 12 - 1 = 4a + 11$
    b. $f(2a) = 4(2a) - 1 = 8a - 1$

37. $f(x) = 2x^2 - 1$
    a. $f(3) = 2(3)^2 - 1 = 2(9) - 1 = 18 - 1 = 17$
    b. $f(-2) = 2(-2)^2 - 1 = 2(4) - 1 = 8 - 1 = 7$
    c. $f(0) = 2(0)^2 - 1 = 0 - 1 = -1$

39. $h(t) = 3t^2 - 4t - 5$
    a. $h(-2) = 3(-2)^2 - 4(-2) - 5 = 3(4) + 8 - 5 = 12 + 8 - 5 = 15$
    b. $h(-1) = 3(-1)^2 - 4(-1) - 5 = 3(1) + 4 - 5 = 3 + 4 - 5 = 2$
    c. $h(w) = 3w^2 - 4w - 5$

41. $g(x) = x^2 + 2x - 1$

a. $g(1) = (1)^2 + 2(1) - 1 = 1 + 2 - 1 = 2$

b. $g(-3) = (-3)^2 + 2(-3) - 1 = 9 - 6 - 1 = 2$

c. $g(a) = a^2 + 2a - 1$

43. $p(t) = 4t^2 - 8t + 3$

a. $p(-2) = 4(-2)^2 - 8(-2) + 3 = 4(4) + 16 + 3 = 16 + 16 + 3 = 35$

b. $p\left(\dfrac{1}{2}\right) = 4\left(\dfrac{1}{2}\right)^2 - 8\left(\dfrac{1}{2}\right) + 3 = 4\left(\dfrac{1}{4}\right) - 4 + 3 = 1 - 4 + 3 = 0$

c. $p(-a) = 4(-a)^2 - 8(-a) + 3 = 4a^2 + 8a + 3$

45. $f(x) = |x - 3|$

a. $f(-1) = |-1 - 3| = |-4| = 4$

b. $f(5) = |5 - 3| = |2| = 2$

c. $f(3) = |3 - 3| = |0| = 0$

47. $C(r) = 3|r| - 2$

a. $C(-3) = 3|-3| - 2 = 3(3) - 2 = 9 - 2 = 7$

b. $C(4) = 3|4| - 2 = 3(4) - 2 = 12 - 2 = 10$

c. $C(0) = 3|0| - 2 = 0 - 2 = -2$

49. $K(p) = 5 - 3|p + 2|$

a. $K(-2) = 5 - 3|(-2) + 2| = 5 - 3|0| = 5 - 0 = 5$

b. $K(-7) = 5 - 3|(-7) + 2| = 5 - 3|-5| = 5 - 3(5) = 5 - 15 = -10$

c. $K(3) = 5 - 3|(3) + 2| = 5 - 3|5| = 5 - 3(5) = 5 - 15 = -10$

51. For $t = 3$:
$s(t) = -16t^2 + 48t$
$s(3) = -16(3)^2 + 48(3)$
$ = -16(9) + 48(3)$
$ = -144 + 144$
$ = 0$

53. For $x = 2$:
$P(x) = 3x^3 - 4x^2 + 6x - 7$
$P(2) = 3(2)^3 - 4(2)^2 + 6(2) - 7$
$ = 3(8) - 4(4) + 6(2) - 7$
$ = 24 - 16 + 12 - 7$
$ = 8 + 12 - 7$
$ = 20 - 7$
$ = 13$

55. For $p = -3$:
$R(p) = \dfrac{3p}{2p - 3}$

$R(-3) = \dfrac{3(-3)}{2(-3) - 3}$

$ = \dfrac{-9}{-6 - 3}$

$ = \dfrac{-9}{-9}$

$ = 1$

57. $f(x) = 3x - 5$
$f(-3) = 3(-3) - 5 = -14$
$f(-2) = 3(-2) - 5 = -11$
$f(-1) = 3(-1) - 5 = -8$
$f(0) = 3(0) - 5 = -5$
$f(1) = 3(1) - 5 = -2$
$f(2) = 3(2) - 5 = 1$
Range = $\{-14, -11, -8, -5, -2, 1\}$

59. $r(t) = \dfrac{t}{2}$

$r(-3) = \dfrac{(-3)}{2} = -\dfrac{3}{2}$

$r(-2) = \dfrac{(-2)}{2} = -1$

$r(-1) = \dfrac{(-1)}{2} = -\dfrac{1}{2}$

$r(0) = \dfrac{(0)}{2} = 0$

$r(1) = \dfrac{(1)}{2} = \dfrac{1}{2}$

$r(2) = \dfrac{(2)}{2} = 1$

Range $= \left\{-\dfrac{3}{2}, -1, -\dfrac{1}{2}, 0, \dfrac{1}{2}, 1\right\}$

61. $f(x) = x^2 + 3$

$f(-3) = (-3)^2 + 3 = 12$

$f(-2) = (-2)^2 + 3 = 7$

$f(-1) = (-1)^2 + 3 = 4$

$f(0) = (0)^2 + 3 = 3$

$f(1) = (1)^2 + 3 = 4$

$f(2) = (2)^2 + 3 = 7$

$f(3) = (3)^2 + 3 = 12$

Range $= \{3, 4, 7, 12\}$

63. $c(n) = n^3 - n - 2$

$c(-3) = (-3)^3 - (-3) - 2 = -26$

$c(-2) = (-2)^3 - (-2) - 2 = -8$

$c(-1) = (-1)^3 - (-1) - 2 = -2$

$c(0) = (0)^3 - (0) - 2 = -2$

$c(1) = (1)^3 - (1) - 2 = -2$

$c(2) = (2)^3 - (2) - 2 = 4$

Range $= \{-26, -8, -2, 4\}$

65. $f(x) = 2x - 3$

$f(c) = 2c - 3$

$5 = 2c - 3$

$8 = 2c$

$4 = c$

The value of c is 4. The corresponding ordered pair is (4, 5).

67. $f(x) = 1 - 2x$

$f(c) = 1 - 2c$

$-7 = 1 - 2c$

$-8 = -2c$

$4 = c$

The value of c is 4. The corresponding ordered pair is (4, −7).

69. $f(x) = \dfrac{2}{3}x - 2$

$f(c) = \dfrac{2}{3}c - 2$

$0 = \dfrac{2}{3}c - 2$

$2 = \dfrac{2}{3}c$

$3 = c$

The value of c is 3. The corresponding ordered pair is (3, 0).

Objective 3.2.2 Exercises

71. Because the points A and C lie on the line and point B does not, the plotted points A and C belong to the function.

73. The point (4, −1) does not fall on the graphed curve. Therefore, the ordered pair (4, −1) does not belong to the function.

75. $f(x) = 3x - 4$

$f(-1) = 3(-1) - 4 = -7$

Yes, the ordered pair (−1, −7) belongs to the function.

77. Evaluate the function for the given values of x.

x	$y = f(x) = x - 3$	(x, y)
−1	$f(-1) = (-1) - 3 = -4$	(−1, −4)
0	$f(0) = (0) - 3 = -3$	(0, −3)
1	$f(1) = (1) - 3 = -2$	(1, −2)
2	$f(2) = (2) - 3 = -1$	(2, −1)
3	$f(3) = (3) - 3 = 0$	(3, 0)
4	$f(4) = (4) - 3 = 1$	(4, 1)
5	$f(5) = (5) - 3 = 2$	(5, 2)

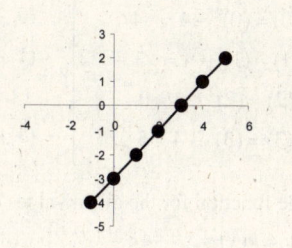

Chapter 3: Linear Functions And Inequalities In Two Variables

79. Evaluate the function for the given values of x.

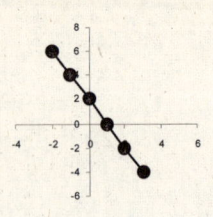

x	$y = g(x) = -2x + 2$	(x, y)
-2	$g(-2) = -2(-2) + 2 = 6$	$(-2, 6)$
-1	$g(-1) = -2(-1) + 2 = 4$	$(-1, 4)$
0	$g(0) = -2(0) + 2 = 2$	$(0, 2)$
1	$g(1) = -2(1) + 2 = 0$	$(1, 0)$
2	$g(2) = -2(2) + 2 = -2$	$(2, -2)$
3	$g(3) = -2(3) + 2 = -4$	$(3, -4)$

81. Evaluate the function for the given values of x.

x	$y = h(x) = \frac{1}{2}x - 1$	(x, y)
-6	$h(-6) = \frac{1}{2}(-6) - 1 = -4$	$(-6, -4)$
-4	$h(-4) = \frac{1}{2}(-4) - 1 = -3$	$(-4, -3)$
-2	$h(-2) = \frac{1}{2}(-2) - 1 = -2$	$(-2, -2)$
0	$h(0) = \frac{1}{2}(0) - 1 = -1$	$(0, -1)$
2	$h(2) = \frac{1}{2}(2) - 1 = 0$	$(2, 0)$
4	$h(4) = \frac{1}{2}(4) - 1 = 1$	$(4, 1)$
6	$h(6) = \frac{1}{2}(6) - 1 = 2$	$(6, 2)$

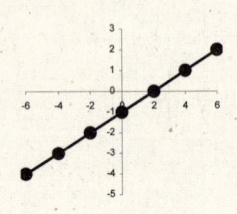

83. Evaluate the function for the given values of x.

x	$y = f(x) = x^2 - 4$	(x, y)
-3	$f(-3) = (-3)^2 - 4 = 5$	$(-3, 5)$
-2	$f(-2) = (-2)^2 - 4 = 0$	$(-2, 0)$
-1	$f(-1) = (-1)^2 - 4 = -3$	$(-1, -3)$
0	$f(0) = (0)^2 - 4 = -4$	$(0, -4)$
1	$f(1) = (1)^2 - 4 = -3$	$(1, -3)$
2	$f(2) = (2)^2 - 4 = 0$	$(2, 0)$
3	$f(3) = (3)^2 - 4 = 5$	$(3, -1)$

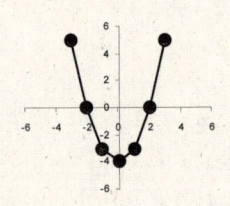

85. Evaluate the function for the given values of x.

x	$y = h(x) = x^2 - 4x$	(x, y)
-1	$h(-1) = (-1)^2 - 4(-1) = 5$	$(-1, 5)$
0	$h(0) = (0)^2 - 4(0) = 0$	$(0, 0)$
1	$h(1) = (1)^2 - 4(1) = -3$	$(1, -3)$
2	$h(2) = (2)^2 - 4(2) = -4$	$(2, -4)$
3	$h(3) = (3)^2 - 4(3) = -3$	$(3, -3)$
4	$h(4) = (4)^2 - 4(4) = 0$	$(4, 0)$
5	$h(5) = (5)^2 - 4(5) = 5$	$(5, 5)$

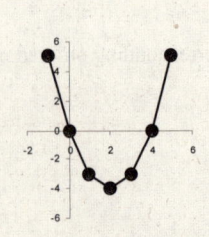

87. Evaluate the function for the given values of x.

x	$y = g(x) = x^2 + 2x - 3$	(x, y)
-4	$g(-4) = (-4)^2 + 2(-4) - 3 = 5$	$(-4, 5)$
-3	$g(-3) = (-3)^2 + 2(-3) - 3 = 0$	$(-3, 0)$
-2	$g(-2) = (-2)^2 + 2(-2) - 3 = -3$	$(-2, -3)$
-1	$g(-1) = (-1)^2 + 2(-1) - 3 = -4$	$(-1, -4)$
0	$g(0) = (0)^2 + 2(0) - 3 = -3$	$(0, -3)$
1	$g(1) = (1)^2 + 2(1) - 3 = 0$	$(1, 0)$
2	$g(2) = (2)^2 + 2(2) - 3 = 5$	$(2, 5)$

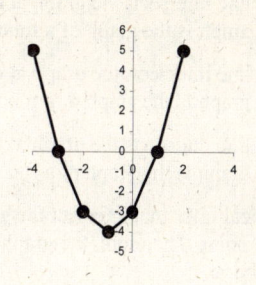

89. Evaluate the function for the given values of x.

x	$y = f(x) = |x| - 3$	(x, y)
-6	$f(-6) = |-6| - 3 = 3$	$(-6, 3)$
-4	$f(-4) = |-4| - 3 = 1$	$(-4, 1)$
-2	$f(-2) = |-2| - 3 = -1$	$(-2, -1)$
0	$f(0) = |0| - 3 = -3$	$(0, -3)$
2	$f(2) = |2| - 3 = -1$	$(2, -1)$
4	$f(4) = |4| - 3 = 1$	$(4, 1)$
6	$f(6) = |6| - 3 = 3$	$(6, 3)$

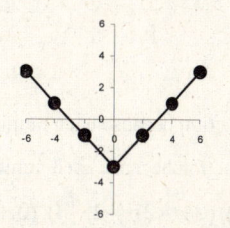

91. Evaluate the function for the given values of x.

x	$y = h(x) = -2|x| + 5$	(x, y)
-3	$h(-3) = -2|-3| + 5 = -1$	$(-3, -1)$
-2	$h(-2) = -2|-2| + 5 = 1$	$(-2, 1)$
-1	$h(-1) = -2|-1| + 5 = 3$	$(-1, 3)$
0	$h(0) = -2|0| + 5 = 5$	$(0, 5)$
1	$h(1) = -2|1| + 5 = 3$	$(1, 3)$
2	$h(2) = -2|2| + 5 = 1$	$(2, 1)$
3	$h(3) = -2|3| + 5 = -1$	$(3, -1)$

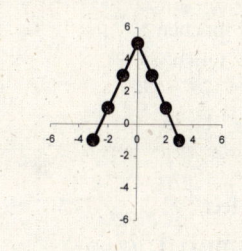

93. Evaluate the function for the given values of x.

x	$y = f(x) = |x - 2| - 3$	(x, y)
-6	$f(-6) = |-6 - 2| - 3 = 5$	$(-6, 5)$
-4	$f(-4) = |-4 - 2| - 3 = 3$	$(-4, 3)$
0	$f(0) = |0 - 2| - 3 = -1$	$(0, -1)$
2	$f(2) = |2 - 2| - 3 = -3$	$(2, -3)$
4	$f(4) = |4 - 2| - 3 = -1$	$(4, -1)$
5	$f(5) = |5 - 2| - 3 = 0$	$(5, 0)$
6	$f(6) = |6 - 2| - 3 = 1$	$(6, 1)$

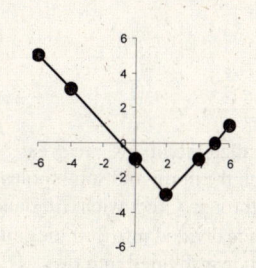

Objective 3.2.3 Exercises

95. Every vertical line intersects the graph at most once. Therefore, the graph is the graph of a function.

97. Every vertical line intersects the graph at most once. Therefore, the graph is the graph of a function.

99. Every vertical line intersects the graph at most once. Therefore, the graph is the graph of a function.

101. There are vertical lines that intersect the graph at more than one point. Therefore, the graph is not the graph of a function.

103. Every vertical line intersects the graph at most once. Therefore, the graph is the graph of a function.

105. a. Yes
 b. Yes
 c. Yes
 d. Yes

111. a. Yes, F is a function because each number in the domain is paired with exactly one number in the range.
 b. Yes, G is a function because each number in the domain is paired with exactly one number in the range.
 c. $F' = \{(-6, -3), (-4, -2), (-2, -1), (0, 0), (2, 1), (4, 2), (6, 3)\}$. Yes, F' is a function.
 d. $G' = \{(9, -3), (4, -2), (1, -1), (0, 0), (1, 1), (4, 2), (9, 3)\}$. No, G' is not a function because the x-value 9 is with two different values, as are the x-values 4 and 1.
 e. No, if the ordered pairs of a function are reversed, the resulting set is not necessarily a function.

113. Find the value of the function
 $s = f(v) = 0.017v^2$ when $v = 60$.
 $s = 0.017v^2$
 $s = 0.017(60)^2$
 $s = 61.2$
 A car will skid 61.2 feet.

115. $\{(-2, -8), (-1, 1), (0, 0), (1, 1), (2, 8)\}$
 Yes, the set defines a function because each member of the domain is assigned to exactly one member of the range.

117. a. 20 ft/s
 b. 28 ft/s

119. a. 60°F
 b. 52°F

121. Longitude lines, also called meridians, are great circles that pass through the north and south poles. The circle that passes through Greenwich, England, has longitude 0°. Earth is divided into 360 longitude lines. However, longitude is divided into two hemispheres: 180° west of Greenwich, England, and 180° east of Greenwich, England. Latitude lines are circles that are parallel to the equator.

Applying Concepts 3.2

107. $f(x) = |x - 2|$
 $f(c) = |c - 2|$
 $4 = |c - 2|$
 $c - 2 = 4 \qquad c - 2 = -4$
 $c = 6 \qquad c = -2$
 The value of c is either 6 or -2.

109. $f(x) = 4x + 7$
 $f(-2 + h) - f(-2) = 4(-2 + h) + 7 - [4(-2) + 7]$
 $= -8 + 4h + 7 - [-8 + 7]$
 $= -8 + 4h + 7 - (-1)$
 $= 4h - 1 + 1$
 $= 4h$

The circle of the equator has latitude 0°. The north pole has latitude 90° north; the south pole has latitude 90° south. A nautical mile is the distance that makes up 1° of the longitude at the equator. This is approximately 6080 ft. The time zones are approximately the distance that makes up 15° of longitude. This is a result of Earth's rotation. Because Earth revolves on its axis once in 24 hours, it revolves $\frac{1}{24} \cdot 360° = 15°$ each hour.

Section 3.3

Concept Review 3.3

1. Never true
 The equation of a linear function is a first-degree equation. This equation has a variable in the denominator, so it is not a first-degree equation.

3. Always true

5. Sometimes true
 The only time that this occurs is when the line passes through the point (0, 0).

Chapter 3: Linear Functions And Inequalities In Two Variables 59

Objective 3.3.1 Exercises

1. To graph a linear function by plotting the points, find three ordered-pair solutions of the equation. Plot these ordered pairs in a rectangular coordinate system. Draw a straight line through the points.

3. a. When $x = -1$, $y = 2(-1) - 5 = \underline{-7}$. A point on the graph is $(\underline{-1}, \underline{-8})$.

   b. When $x = 0$, $y = 2(0) - 5 = \underline{-5}$. A point on the graph is $(\underline{0}, \underline{-5})$.

   c. When $x = 1$, $y = 2(1) - 5 = \underline{-3}$. A point on the graph is $(\underline{1}, \underline{-3})$.

5.

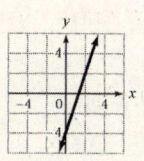

7.

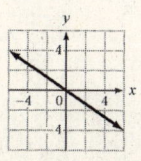

9.

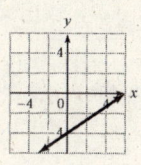

11.

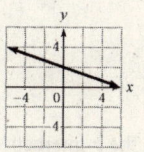

13.

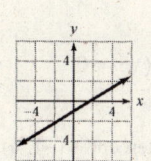

Objective 3.3.2 Exercises

15. $2x + y = -3$
    $y = -2x - 3$

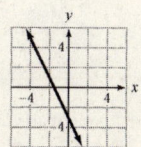

17. $x - 4y = 8$
    $-4y = -x + 8$
    $y = \dfrac{1}{4}x - 2$

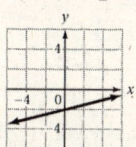

19. $y = 1$

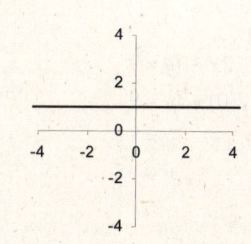

21. $x = 4$

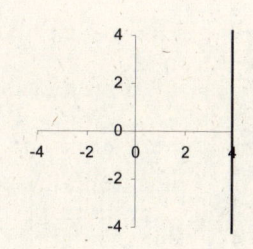

23. $2x - 3y = 12$
    $-3y = -2x + 12$
    $y = \dfrac{2}{3}x - 4$

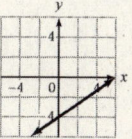

25. $3x - 2y = 8$
    $-2y = -3x + 8$
    $y = \dfrac{3}{2}x - 4$

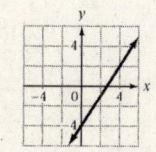

60 Chapter 3: Linear Functions And Inequalities In Two Variables

27. x-intercept: y-intercept
$$x - 2y = -4 \qquad x - 2y = -4$$
$$x - 2(0) = -4 \qquad 0 - 2y = -4$$
$$x = -4 \qquad -2y = -4$$
$$y = 2$$
$$(-4, 0) \qquad (0, 2)$$

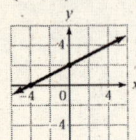

29. x-intercept: y-intercept:
$$2x - 3y = 9 \qquad 2x - 3y = 9$$
$$2x - 3(0) = 9 \qquad 2(0) - 3y = 9$$
$$2x = 9 \qquad -3y = 9$$
$$x = \frac{9}{2} \qquad y = -3$$
$$\left(\frac{9}{2}, 0\right) \qquad (0, -3)$$

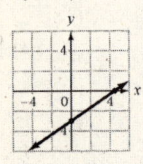

31. x-intercept: y-intercept:
$$2x - y = 4 \qquad 2x - y = 4$$
$$2x - 0 = 4 \qquad 2(0) - y = 4$$
$$2x = 4 \qquad -y = 4$$
$$x = 2 \qquad y = -4$$
$$(2, 0) \qquad (0, -4)$$

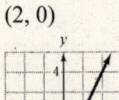

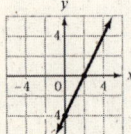

33. x-intercept: y-intercept:
$$3x + 2y = 5 \qquad 3x + 2y = 5$$
$$3x + 2(0) = 5 \qquad 3(0) + 2y = 5$$
$$3x = 5 \qquad 2y = 5$$
$$x = \frac{5}{3} \qquad y = \frac{5}{2}$$
$$\left(\frac{5}{3}, 0\right) \qquad \left(0, \frac{5}{2}\right)$$

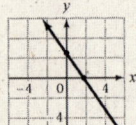

35. x-intercept: y-intercept:
$$3x + 2y = 4 \qquad 3x + 2y = 4$$
$$3x + 2(0) = 4 \qquad 3(0) + 2y = 4$$
$$3x = 4 \qquad 2y = 4$$
$$x = \frac{4}{3} \qquad y = 2$$
$$\left(\frac{4}{3}, 0\right) \qquad (0, 2)$$

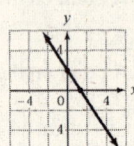

37. x-intercept: y-intercept:
$$3x - 5y = 9 \qquad 3x - 5y = 9$$
$$3x - 5(0) = 9 \qquad 3(0) - 5y = 9$$
$$3x = 9 \qquad -5y = 9$$
$$x = 3 \qquad y = -\frac{9}{5}$$
$$(3, 0) \qquad \left(0, -\frac{9}{5}\right)$$

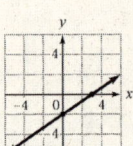

39. Let $A < 0$, $B > 0$, and $C < 0$. To find the y-intercept, let $x = 0$.
$$Ax + By = C$$
$$A(0) + By = C$$
$$By = C$$
$$y = \frac{C}{B}$$
The value $\frac{C}{B}$ is negative; therefore, the y-intercept is below the x-axis.

41. Because the x-intercept is $(-3, 0)$, $f(-3) = 0$. Therefore, the zero of f is -3.

43. From the graph, we see that the x-intercept is $(-1, 0)$. The x-coordinate of the x-intercept is -1, which is the zero of the function.

45. From the graph, we see that the x-intercept is $(3, 0)$. The x-coordinate of the x-intercept is 3, which is the zero of the function.

47. $$f(x) = 2x - 6$$
$$0 = 2x - 6$$
$$6 = 2x$$
$$3 = x$$
The zero is 3.

Chapter 3: Linear Functions And Inequalities In Two Variables

49. $g(x) = -2x + 5$
$0 = -2x + 5$
$-5 = -2x$
$\frac{5}{2} = x$
The zero is $\frac{5}{2}$.

51. $y(x) = \frac{3}{4}x - 6$
$0 = \frac{3}{4}x - 6$
$6 = \frac{3}{4}x$
$8 = x$
The zero is 8.

53. $f(t) = \frac{t}{2} - 3$
$0 = \frac{t}{2} - 3$
$3 = \frac{t}{2}$
$6 = t$
The zero is 6.

55. $g(x) = 3x + 8$
$0 = 3x + 8$
$-8 = 3x$
$-\frac{8}{3} = x$
The zero is $-\frac{8}{3}$.

57. $f(x) = -\frac{3}{4}x + 2$
$0 = -\frac{3}{4}x + 2$
$-2 = -\frac{3}{4}x$
$\frac{8}{3} = x$
The zero is $\frac{8}{3}$.

59.
   a. Find the value of V for $t = 0$: $V = 105(\underline{0}) + 1500 = \underline{1500}$. This means a point on the graph of the equation $V = 105t + 1500$ is $(0, \underline{1500})$.

   b. Find the value of V for $t = 8$: $V = 105(\underline{8}) + 1500 = \underline{2340}$. This means a point on the graph of the equation $V = 105t + 1500$ is $(8, \underline{1500})$.

   c. The graph of $V = 105t + 1500$ for $0 \le t \le 8$ is a straight line joining $(0, \underline{1500})$ and $(8, \underline{2340})$.

61. The roller coaster travels 940 ft in 5 s.

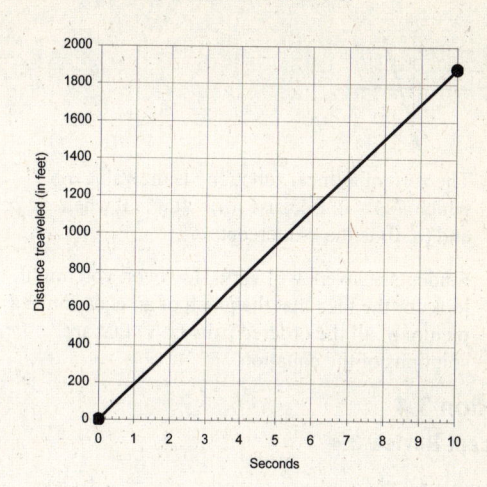

63. The realtor will earn $4000 for selling $60,000 worth of property.

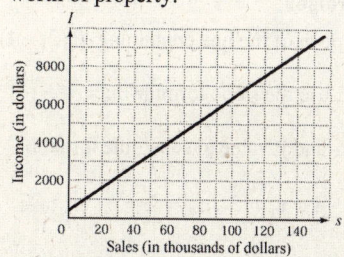

65. The caterer will charge $614 for 120 hot appetizers.

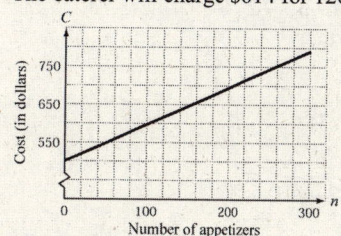

Applying Concepts 3.3

67. $s(p) = 0.80p$
$s(p) = 0.80(200)$
$s(p) = 160$
The sale price is $160.

69.

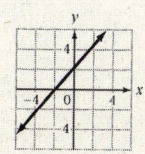

71.

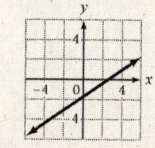

73.

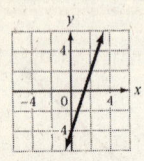

75. The zero of a linear function f is the value a for which $f(a) = 0$. Because $f(a) = 0$, $y = 0$ when $x = a$, and $(a, 0)$ is the x-intercept.

77. Students answers will vary. However, you might look for the idea that the graph of an equation is a picture of all the ordered pairs (x, y) that are solutions of the equation.

Section 3.4

Concept Review 3.4

1. Always true

3. Never true
   The slope of a vertical line is undefined.

Objective 3.4.1 Exercises

1. To find the slope m of the line containing $P_1(2, -4)$ and $P_2(3, -5)$, use the slope formula $m = \dfrac{y_2 - y_1}{x_2 - x_1}$.
   Identify each x- and y-value to substitute into the slope formula.
   $y_2 = \underline{-5}$, $y_1 = \underline{-4}$, $x_2 = \underline{3}$, $x_1 = \underline{2}$

3. $P_1(1, 3), P_2(3, 1)$
   $m = \dfrac{y_2 - y_1}{x_2 - x_1} = \dfrac{1 - 3}{3 - 1} = \dfrac{-2}{2} = -1$
   The slope is -1.

5. $P_1(-1, 4), P_2(2, 5)$
   $m = \dfrac{y_2 - y_1}{x_2 - x_1} = \dfrac{5 - 4}{2 - (-1)} = \dfrac{1}{3}$
   The slope is $\dfrac{1}{3}$.

7. $P_1(-1, 3), P_2(-4, 5)$
   $m = \dfrac{y_2 - y_1}{x_2 - x_1} = \dfrac{5 - 3}{-4 - (-1)} = \dfrac{2}{-3} = -\dfrac{2}{3}$
   The slope is $-\dfrac{2}{3}$.

9. $P_1(0, 3), P_2(4, 0)$
   $m = \dfrac{y_2 - y_1}{x_2 - x_1} = \dfrac{0 - 3}{4 - 0} = \dfrac{-3}{4} = -\dfrac{3}{4}$
   The slope is $-\dfrac{3}{4}$.

11. $P_1(2, 4), P_2(2, -2)$
    $m = \dfrac{y_2 - y_1}{x_2 - x_1} = \dfrac{-2 - 4}{2 - 2} = \dfrac{-6}{0}$
    The slope is undefined.

13. $P_1(2, 5), P_2(-3, -2)$
    $m = \dfrac{y_2 - y_1}{x_2 - x_1} = \dfrac{-2 - 5}{-3 - 2} = \dfrac{-7}{-5} = \dfrac{7}{5}$
    The slope is $\dfrac{7}{5}$.

15. $P_1(2, 3), P_2(-1, 3)$
    $m = \dfrac{y_2 - y_1}{x_2 - x_1} = \dfrac{3 - 3}{-1 - 2} = \dfrac{0}{-3} = 0$
    The line has zero slope.

17. $P_1(0, 4), P_2(-2, 5)$
    $m = \dfrac{y_2 - y_1}{x_2 - x_1} = \dfrac{5 - 4}{-2 - 0} = \dfrac{1}{-2} = -\dfrac{1}{2}$
    The slope is $-\dfrac{1}{2}$.

19. $P_1(-3, -1), P_2(-3, 4)$
    $m = \dfrac{y_2 - y_1}{x_2 - x_1} = \dfrac{4 - (-1)}{-3 - (-3)} = \dfrac{5}{0}$
    The slope is undefined.

21. $m = \dfrac{240 - 80}{6 - 2} = \dfrac{160}{4} = 40$
    The slope is the average speed of the motorist in miles per hour.

23. $m = \dfrac{13 - 6}{40 - 180} = \dfrac{7}{-140} = -0.05$
    For each mile the car is driven, approximately 0.05 gallons of fuel is used.

25. Strategy
    Find lines that have a slope that match the rates of the runners.

    Solution
    Lois has the highest rate so the line with the steepest slope represents Lois. Line A has the steepest slope so it represents Lois's distance. The line which represents Tanya's distance must have a slope of $\dfrac{6}{1}$.
    Line B goes through the points (1, 6) and (0, 0) so its slope is $\dfrac{6}{1}$. So line B represents Tanya's distance.
    In one hour the difference between Lois's and Tanya's distances is 3. Line C goes through the point (1, 3) so it represents the distance between Lois and Tanya.

Chapter 3: Linear Functions And Inequalities In Two Variables   **63**

27. a. The slope for a ramp that is 6 inches high and 5 feet long is $\dfrac{6 \text{ inches}}{60 \text{ inches}} = \dfrac{1}{10}$ or 0.1.

    $\dfrac{1}{12} = 0.0833$. Since $0.1 > 0.0833$, the ramp does not meet the ANSI requirements.

    b. The slope for a ramp that is 12 inches high and 170 inches long is $\dfrac{12}{170}$ or 0.07. Since $0.07 < 0.08$, the ramp does meet the ANSI requirements.

29. The slope of the line with equation $y = 5x - 3$ is <u>5</u>, and its y-intercept is <u>(0, -3)</u>.

Objective 3.4.2 Exercises

31.

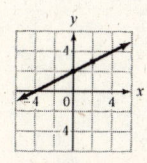

33.

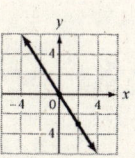

35.

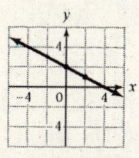

37. $x - 3y = 3$
    $-3y = -x + 3$
    $y = \dfrac{1}{3}x - 1$

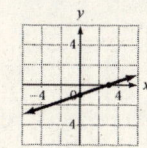

39. $4x + y = 2$
    $y = -4x + 2$

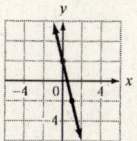

41. Let $A > 0$, $B < 0$, and $C > 0$. Rewrite the equation in slope-intercept form.
    $Ax + By = C$
    $By = -Ax + C$
    $y = -\dfrac{A}{B}x + \dfrac{C}{B}$

    The value $\dfrac{C}{B}$ is negative; therefore, the y-intercept lies below the x-axis. The value $-\dfrac{A}{B}$ is positive; therefore, the graph slants upward to the right.

43.

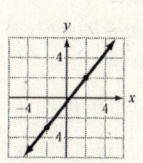

45.

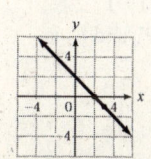

Applying Concepts 3.4

47. increases by 2

49. increases by $\dfrac{1}{2}$

51. $P_1 = (3, 2)$
    $P_2 = (4, 6)$
    $P_3 = (5, k)$

    P_1 to $P_2 : m = \dfrac{6-2}{4-3} = 4$

    The slope from P_1 to P_3 and that from P_2 to P_3 must also be 4.
    Set the slope from P_1 to P_3 equal to 4.

    $\dfrac{2-k}{3-5} = 4$

    $\dfrac{2-k}{-2} = 4$

    $2 - k = -8$
    $k = 10$

    This checks out against P_2 to P_3, so $k = 10$.

Copyright © Houghton Mifflin Company. All rights reserved.

53. The graph for i has a negative slope and y-intercept of 4. i and D;
The graph for ii has negative slope and y-intercept of –4. ii and C;
The graph for iii has zero slope and y-intercept of 2. iii and B;
The graph of iv has a negative slope and y-intercept of 0. iv and F;
The graph of v has a positive slope and y-intercept of 4. v and E;
The graph of vi has a negative slope and y-intercept of –2. vi and A

55. Three given points lie on the same line if the slope between each pair of points is the same.

  a. Let $P_1 = (2, 5)$, $P_2 = (-1, -1)$, and $P_3 = (3, 7)$. Then the slope of the line between P_1 and P_2 is $m = \dfrac{-1-5}{-1-2} = 2$, the slope of the line between P_2 and P_3 is $m = \dfrac{7-(-1)}{3-(-1)} = 2$, and the slope of the line between P_1 and P_3 is $m = \dfrac{7-5}{3-2} = 2$.
  Yes, the points lie on the same line.

  b. Let $P_1 = (-1, 5)$, $P_2 = (0, 3)$, and $P_3 = (-3, 4)$. Then the slope of the line between P_1 and P_2 is $m = \dfrac{3-5}{0-(-1)} = -2$, the slope of the line between P_2 and P_3 is $m = \dfrac{4-3}{-3-0} = -\dfrac{1}{3}$, and the slope of the line between P_1 and P_3 is $m = \dfrac{4-5}{-3-(-1)} = \dfrac{1}{2}$.
  No, the points do not lie on the same line.

Section 3.5

Concept Review 3.5

1. Always true

3. Never true
The point-slope formula is given by $y - y_1 = m(x - x_1)$.

5. Never true
The line represented by the equation $y = 2x - \dfrac{1}{2}$ has slope 2 and y-intercept $-\dfrac{1}{2}$.

Objective 3.5.1 Exercises

1. In the equation of the line that has slope $-\dfrac{6}{5}$ and y-intercept (0, 2), m is $-\dfrac{6}{5}$ and b is $\underline{2}$. The equation of the line is $y = -\dfrac{6}{5}x + 2$.

3. $m = 2, b = 5$
$y = mx + b$
$y = 2x + 5$
The equation of the line is $y = 2x + 5$.

5. $m = \dfrac{1}{2}$, $(x_1, y_1) = (2, 3)$
$y - y_1 = m(x - x_1)$
$y - 3 = \dfrac{1}{2}(x - 2)$
$y - 3 = \dfrac{1}{2}x - 1$
$y = \dfrac{1}{2}x + 2$
The equation of the line is $y = \dfrac{1}{2}x + 2$.

7. $m = -\dfrac{5}{3}$, $(x_1, y_1) = (3, 0)$
$y - y_1 = m(x - x_1)$
$y - 0 = -\dfrac{5}{3}(x - 3)$
$y = -\dfrac{5}{3}x + 5$
The equation of the line is $y = -\dfrac{5}{3}x + 5$.

9. $m = -3$, $(x_1, y_1) = (-1, 7)$
$y - y_1 = m(x - x_1)$
$y - 7 = -3[x - (-1)]$
$y - 7 = -3(x + 1)$
$y - 7 = -3x - 3$
$y = -3x + 4$
The equation of the line is $y = -3x + 4$.

11. $m = \dfrac{1}{2}$, $(x_1, y_1) = (0, 0)$
$y - y_1 = m(x - x_1)$
$y - 0 = \dfrac{1}{2}(x - 0)$
$y = \dfrac{1}{2}x$
The equation of the line is $y = \dfrac{1}{2}x$.

13. $m = 3$, $(x_1, y_1) = (2, -3)$
$y - y_1 = m(x - x_1)$
$y - (-3) = 3(x - 2)$
$y + 3 = 3x - 6$
$y = 3x - 9$
The equation of the line is $y = 3x - 9$.

15. $m = -\dfrac{2}{3}, (x_1, y_1) = (3, 5)$

$y - y_1 = m(x - x_1)$

$y - 5 = -\dfrac{2}{3}(x - 3)$

$y - 5 = -\dfrac{2}{3}x + 2$

$y = -\dfrac{2}{3}x + 7$

The equation of the line is $y = -\dfrac{2}{3}x + 7$.

17. $m = -1, (x_1, y_1) = (0, -3)$

$y - y_1 = m(x - x_1)$

$y - (-3) = -1(x - 0)$

$y + 3 = -x + 0$

$y = -x - 3$

The equation of the line is $y = -x - 3$.

19. The slope is undefined; $(x_1, y_1) = (3, -4)$.

The line is a vertical line. All points on the line have an abscissa of 3. The equation of the line is $x = 3$.

21. $m = 0, (x_1, y_1) = (-2, -3)$

$y - y_1 = m(x - x_1)$

$y - (-3) = 0[x - (-2)]$

$y + 3 = 0$

$y = -3$

The equation of the line is $y = -3$.

23. $m = -2, (x_1, y_1) = (4, -5)$

$y - y_1 = m(x - x_1)$

$y - (-5) = -2(x - 4)$

$y + 5 = -2x + 8$

$y = -2x + 3$

The equation of the line is $y = -2x + 3$.

25. The slope is undefined; $(x_1, y_1) = (-5, -1)$.

The line is a vertical line. All points on the line have an abscissa of -5. The equation of the line is $x = -5$.

Objective 3.5.2 Exercises

27. The line slants downward to the right, so the value of m must be negative. The line appears to cross the y-axis at $(0, 2)$, so the value of b cannot be 6.

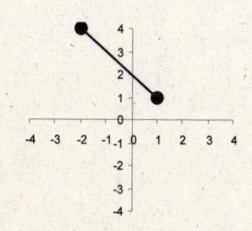

29. $P_1(0, 2), P_2(3, 5)$

$m = \dfrac{y_2 - y_1}{x_2 - x_1} = \dfrac{5 - 2}{3 - 0} = \dfrac{3}{3} = 1$

$y - y_1 = m(x - x_1)$

$y - 2 = 1(x - 0)$

$y - 2 = x$

$y = x + 2$

The equation of the line is $y = x + 2$.

31. $P_1(0, -3), P_2(-4, 5)$

$m = \dfrac{y_2 - y_1}{x_2 - x_1} = \dfrac{5 - (-3)}{-4 - 0} = \dfrac{8}{-4} = -2$

$y - y_1 = m(x - x_1)$

$y - (-3) = -2(x - 0)$

$y + 3 = -2x$

$y = -2x - 3$

The equation of the line is $y = -2x - 3$.

33. $P_1(-1, 3), P_2(2, 4)$

$m = \dfrac{y_2 - y_1}{x_2 - x_1} = \dfrac{4 - 3}{2 - (-1)} = \dfrac{1}{3}$

$y - y_1 = m(x - x_1)$

$y - 3 = \dfrac{1}{3}[x - (-1)]$

$y - 3 = \dfrac{1}{3}(x + 1)$

$y - 3 = \dfrac{1}{3}x + \dfrac{1}{3}$

$y = \dfrac{1}{3}x + \dfrac{10}{3}$

The equation of the line is $y = \dfrac{1}{3}x + \dfrac{10}{3}$.

35. $P_1(0, 3), P_2(2, 0)$

$m = \dfrac{y_2 - y_1}{x_2 - x_1} = \dfrac{0 - 3}{2 - 0} = \dfrac{-3}{2} = -\dfrac{3}{2}$

$y - y_1 = m(x - x_1)$

$y - 3 = -\dfrac{3}{2}(x - 0)$

$y - 3 = -\dfrac{3}{2}x$

$y = -\dfrac{3}{2}x + 3$

The equation of the line is $y = -\dfrac{3}{2}x + 3$.

66 Chapter 3: Linear Functions And Inequalities In Two Variables

37. $P_1(-2, -3), P_2(-1, -2)$

$m = \dfrac{y_2 - y_1}{x_2 - x_1} = \dfrac{-2-(-3)}{-1-(-2)} = \dfrac{1}{1} = 1$

$y - y_1 = m(x - x_1)$

$y - (-3) = 1[x - (-2)]$

$y + 3 = x + 2$

$y = x - 1$

The equation of the line is $y = x - 1$.

39. $P_1(2, 3), P_2(5, 5)$

$m = \dfrac{y_2 - y_1}{x_2 - x_1} = \dfrac{5-3}{5-2} = \dfrac{2}{3}$

$y - y_1 = m(x - x_1)$

$y - 3 = \dfrac{2}{3}(x - 2)$

$y - 3 = \dfrac{2}{3}x - \dfrac{4}{3}$

$y = \dfrac{2}{3}x + \dfrac{5}{3}$

The equation of the line is $y = \dfrac{2}{3}x + \dfrac{5}{3}$.

41. $P_1(2, 0), P_2(0, -1)$

$m = \dfrac{y_2 - y_1}{x_2 - x_1} = \dfrac{-1-0}{0-2} = \dfrac{-1}{-2} = \dfrac{1}{2}$

$y - y_1 = m(x - x_1)$

$y - 0 = \dfrac{1}{2}(x - 2)$

$y = \dfrac{1}{2}x - 1$

The equation of the line is $y = \dfrac{1}{2}x - 1$.

43. $P_1(3, -4), P_2(-2, -4)$

$m = \dfrac{y_2 - y_1}{x_2 - x_1} = \dfrac{-4-(-4)}{-2-3} = \dfrac{0}{-5} = 0$

$y - y_1 = m(x - x_1)$

$y - (-4) = 0(x - 3)$

$y + 4 = 0$

$y = -4$

The equation of the line is $y = -4$.

45. $P_1(0, 0), P_2(4, 3)$

$m = \dfrac{y_2 - y_1}{x_2 - x_1} = \dfrac{3-0}{4-0} = \dfrac{3}{4}$

$y - y_1 = m(x - x_1)$

$y - 0 = \dfrac{3}{4}(x - 0)$

$y = \dfrac{3}{4}x$

The equation of the line is $y = \dfrac{3}{4}x$.

47. $P_1(-2, 5), P_2(-2, -5)$

$m = \dfrac{y_2 - y_1}{x_2 - x_1} = \dfrac{-5-5}{-2-(-2)} = \dfrac{-10}{0}$

The slope is undefined. The line is a vertical line. All points on the line have an abscissa of –2. The equation of the line is $x = -2$.

49. $P_1(2, 1), P_2(-2, -3)$

$m = \dfrac{y_2 - y_1}{x_2 - x_1} = \dfrac{-3-1}{-2-2} = \dfrac{-4}{-4} = 1$

$y - y_1 = m(x - x_1)$

$y - 1 = 1(x - 2)$

$y - 1 = x - 2$

$y = x - 1$

The equation of the line is $y = x - 1$.

51. $P_1(0, 3), P_2(3, 0)$

$m = \dfrac{y_2 - y_1}{x_2 - x_1} = \dfrac{0-3}{3-0} = \dfrac{-3}{3} = -1$

$y - y_1 = m(x - x_1)$

$y - 3 = -1(x - 0)$

$y - 3 = -x$

$y = -x + 3$

The equation of the line is $y = -x + 3$.

Objective 3.5.3 Exercises

53. Then $m = \underline{1000}$ and an ordered pair that is a solution of the equation is $(0, \underline{500})$.

55. Strategy
Use the slope-intercept form to determine the equation of the line.

Solution
y-intercept = 0
$m = 1200$
$y = mx + b$
$y = 1200x + 0$
The equation that represents the ascent of the plane is $y = 1200x$ where x stands for the number of minutes after take-off. To find the height of the plane when $x = 11$
$y = 1200(11)$
$y = 13,200$
The plane will be 13,200 feet in the air after 11 minutes.

Copyright © Houghton Mifflin Company. All rights reserved.

57. **Strategy**
Use the slope-intercept form to find the equation.

Solution
The y-intercept is 4.95 because it is the charge for 0 minutes.
$b = 4.95$
$m = 0.59$
$y = mx + b$
$y = 0.59x + 4.95$
The equation for the monthly cost of the phone is $y = 0.59x + 4.95$.
To find the cost for 13 minutes of use,
$y = 0.59(13) + 4.95$
$y = 12.62$
It costs \$12.62 to use the cellular phone for 13 minutes.

59. **Strategy**
Use the point-slope formula.

Solution
$(x_1, y_1) = (2, 126), (x_2, y_2) = (3, 189)$
$m = \dfrac{189 - 126}{3 - 2} = 63$
$y - y_1 = m(x - x_1)$
$y - y_1 = 63(x - x_1)$
$y - 126 = 63(x - 2)$
$y - 126 = 63x - 126$
$y = 63x$
The equation to approximate the number of calories in a hamburger is $y = 63x$ where x represents the ounces in the hamburger. To predict the number of calories in a 5-ounce serving of hamburger,
$y = 63 \cdot 5$
$y = 315$
A 5-ounce serving of hamburger will contain 315 calories.

61. **Strategy**
Use the point-slope formula.

Solution
$(x_1, y_1) = (1927, 33.5), (x_2, y_2) = (1997, 3.3)$
$m = \dfrac{y_2 - y_1}{x_2 - x_1} = \dfrac{3.3 - 33.5}{1997 - 1927} = \dfrac{-30.2}{70} = -0.431$
$y - y_1 = m(x - x_1)$
$y - 33.5 = -0.431(x - 1927)$
$y - 33.5 = -0.431x + 830.5$
$y = -0.431x + 864$

The equation to predict how long a flight took is $y = -0.431x + 864$ where x represents the year. To predict how long a flight between the two cities would have taken in 1967,
$y = -0.431(1967) + 864$
$= -847.8 + 864 = 16.2$
It would have taken 16.2 hours in 1967 for a plane to cross the Atlantic.

63. **Strategy**
Use the slope-intercept form of the equation.

Solution
The y-intercept is 16 gallons because 0 miles have been driven at the beginning of the trip. The slope is -0.032.
$y = mx + b$
$y = -0.032x + 16$
The equation to find the number of gallons in the tank is $y = -0.032x + 16$ where x represents the number of miles traveled. To predict the number of gallons in the tank after 150 miles,
$y = -0.032(150) + 16$
$y = 11.2$
11.2 gallons will be in the tank when 150 miles have been driven.

Applying Concepts 3.5

65. To find the x-intercept, set y to 0.
$0 = mx + b$
$-b = mx$
$-\dfrac{b}{m} = x$
The x-intercept is $\left(-\dfrac{b}{m}, 0\right)$.

67. Find the equation of the line. The two points are $(1, 3)$ and $(-1, 5)$.
$m = \dfrac{5 - 3}{-1 - 1} = \dfrac{2}{-2} = -1$
$y - 3 = -1(x - 1)$
$y - 3 = -x + 1$
$y = -x + 4$
The function is $f(x) = -x + 4$
$f(4) = -4 + 4 = 0$

69. If m is a given constant, changing b causes the graph of the line to move up or down.

71. The slope of any line parallel to the y-axis is undefined. In order to use the point-slope formula, we must be able to substitute the slope of the line for m. In other words, to use the point-slope formula, the slope of the line must be defined. Therefore, we cannot use the point-slope formula to find the equation of a line with undefined slope – that is, one that is parallel to the y-axis.

Section 3.6

Concept Review 3.6

1. Sometimes true
   The only time perpendicular lines have the same y-intercept is when the point of intersection of the two lines is on the y-axis.

3. Never true
   The product of the slopes of two perpendicular lines is equal to -1.
   $\left(\dfrac{3}{2}\right)\left(-\dfrac{3}{2}\right) = -\dfrac{9}{4}$; the lines are not perpendicular.

5. Always true

7. Always true

Objective 3.6.1 Exercises

1. We can determine whether two lines are parallel by looking at their slopes. If the slopes are equal, then the lines are parallel.

3. The slope is -5. Parallel lines have the same slope.

5. The slope is $-\dfrac{1}{4}$, the negative reciprocal of 4.

7. $-\dfrac{1}{3}$

9. $x = -2$ is a vertical line.
   $y = 3$ is a horizontal line.
   The lines are perpendicular.

11. $x = -3$ is a vertical line.
    $y = \dfrac{1}{3}$ is a horizontal line.
    The lines are not parallel.

13. $y = \dfrac{2}{3}x - 4, m_1 = \dfrac{2}{3}$
    $y = -\dfrac{3}{2}x - 4, m_2 = -\dfrac{3}{2}$
    $m_1 \neq m_2$
    The lines are not parallel.

15. $y = \dfrac{4}{3}x - 2, m_1 = \dfrac{4}{3}$
    $y = -\dfrac{3}{4}x + 2, m_2 = -\dfrac{3}{4}$
    $m_1 \cdot m_2 = \dfrac{4}{3}\left(-\dfrac{3}{4}\right) = -1$
    The lines are perpendicular.

17. $2x + 3y = 2$
    $3y = -2x + 2$
    $y = -\dfrac{2}{3}x + \dfrac{2}{3}, m_1 = -\dfrac{2}{3}$
    $2x + 3y = -4$
    $3y = -2x - 4$
    $y = -\dfrac{2}{3}x - \dfrac{4}{3}, m_2 = -\dfrac{2}{3}$
    $m_1 = m_2 = -\dfrac{2}{3}$
    The lines are parallel.

19. $x - 4y = 2$
    $-4y = -x + 2$
    $y = \dfrac{1}{4}x - \dfrac{1}{2}, m_1 = \dfrac{1}{4}$
    $4x + y = 8$
    $y = -4x + 8, m_2 = -4$
    $m_1 \cdot m_2 = \dfrac{1}{4}(-4) = -1$
    The lines are perpendicular.

21. $m_1 = \dfrac{6-2}{1-3} = \dfrac{4}{-2} = -2$
    $m_2 = \dfrac{-1-3}{-1-(-1)} = \dfrac{-4}{0}$
    $m_1 \neq m_2$
    The lines are not parallel.

23. $m_1 = \dfrac{-1-2}{4-(-3)} = \dfrac{-3}{7} = -\dfrac{3}{7}$
    $m_2 = \dfrac{-4-3}{-2-1} = \dfrac{-7}{-3} = \dfrac{7}{3}$
    $m_1 \cdot m_2 = -\dfrac{3}{7}\left(\dfrac{7}{3}\right) = -1$
    The lines are perpendicular.

25. $m_1 = \dfrac{2-0}{0-(-5)} = \dfrac{2}{5}$
    $m_2 = \dfrac{-1-1}{0-5} = \dfrac{-2}{-5} = \dfrac{2}{5}$
    $m_1 = m_2 = \dfrac{2}{5}$
    The lines are parallel.

27. $2x - 3y = 2$
$-3y = -2x + 2$
$y = \frac{2}{3}x - \frac{2}{3}$
$m = \frac{2}{3}$
$y - y_1 = m(x - x_1)$
$y - (-4) = \frac{2}{3}[x - (-2)]$
$y + 4 = \frac{2}{3}(x + 2)$
$y + 4 = \frac{2}{3}x + \frac{4}{3}$
$y = \frac{2}{3}x - \frac{8}{3}$

The equation of the line is $y = \frac{2}{3}x - \frac{8}{3}$.

29. $y = -3x + 4$
$m_1 = -3$
$m_1 \cdot m_2 = -1$
$-3 \cdot m_2 = -1$
$m_2 = \frac{1}{3}$
$y - y_1 = m(x - x_1)$
$y - 1 = \frac{1}{3}(x - 4)$
$y - 1 = \frac{1}{3}x - \frac{4}{3}$
$y = \frac{1}{3}x - \frac{1}{3}$

The equation of the line is $y = \frac{1}{3}x - \frac{1}{3}$.

31. $3x - 5y = 2$
$-5y = -3x + 2$
$y = \frac{3}{5}x - \frac{2}{5}$
$m_1 = \frac{3}{5}$
$m_1 \cdot m_2 = -1$
$\frac{3}{5} \cdot m_2 = -1$
$m_2 = -\frac{5}{3}$
$y - y_1 = m(x - x_1)$
$y - (-3) = -\frac{5}{3}[x - (-1)]$
$y + 3 = -\frac{5}{3}(x + 1)$
$y + 3 = -\frac{5}{3}x - \frac{5}{3}$
$y = -\frac{5}{3}x - \frac{14}{3}$

The equation of the line is $y = -\frac{5}{3}x - \frac{14}{3}$.

33. Let $A > 0$ and $B > 0$. Rewrite the equation in slope-intercept form.
$Ax + By = C$
$By = -Ax + C$
$y = -\frac{A}{B}x + \frac{C}{B}$

The value of the slope, $-\frac{A}{B}$, is negative, so the graph slants downward to the right. Therefore, a line parallel to l also slants downward to the right.

Applying Concepts 3.6

35. Write the equations of the lines in slope-intercept form.

(1) $A_1x + B_1y = C_1$
$B_1y = C_1 - A_1x$
$y = \frac{C_1}{B_1} - \frac{A_1}{B_1}x$

(2) $A_2x + B_2y = C_2$
$B_2y = C_2 - A_2x$
$y = \frac{C_2}{B_2} - \frac{A_2}{B_2}x$

The slopes of the two lines must be the same for the lines to be parallel, so $\frac{A_1}{B_1} = \frac{A_2}{B_2}$.

Chapter 3: Linear Functions And Inequalities In Two Variables

37. Strategy
To find the equation of the line on the initial path

- Find the slope of the line of the string
- Find the slope of the line on the initial path which is perpendicular to the line of the string.
- Use the point-slope formula to find the equation of the line of the initial path.

Solution
Slope of the string,
$m_1 = \dfrac{3-0}{6-0} = \dfrac{1}{2}$

Slope of the line on the initial path,
$m_1 \cdot m_2 = -1$
$\dfrac{1}{2} \cdot m_2 = -1$
$m_2 = -2$

Equation of the line,
$y - y_1 = m(x - x_1)$
$y - 3 = -2[x - 6]$
$y - 3 = -2x + 12$
$y = -2x + 15$

The equation of the line is $y = -2x + 15$.

39. Strategy
Find a line perpendicular to either of the given lines. The line must form a triangle with the other two lines, so the line cannot go through the point (6, −1).

Solution
Start with the line $y = -\dfrac{1}{2}x + 2$.

The slope of a line perpendicular to this line is 2, and the line has the form $y = 2x + b$. Since the line cannot go through the point (6, −1), the value of b cannot be −13. So one possible solution is any equation of the form $y = 2x + b$, where $b \neq -13$.

Now consider the line $y = \dfrac{2}{3}x + 5$. The slope of a line perpendicular to this line is $-\dfrac{3}{2}$, and the line has the form $y = -\dfrac{3}{2}x + c$. Since the line cannot go through the point (6, −1), the value of c cannot be 8. So the other possible solution is any equation of the form $y = -\dfrac{3}{2}x + c$, where $c \neq 8$.

Section 3.7
Concept Review 3.7

1. Never true
The solution of a linear inequality is a half-plane.

3. Always true

5. Always true

Objective 3.7.1 Exercises

1. A half-plane is the set of points on one side of a line in the plane.

3. $0 > 2(0) - 7$
$0 > -7$
Yes

5. $0 \leq -\dfrac{2}{3}(0) - 8$
$0 \not\leq -8$
No

7. $y \leq \dfrac{3}{2}x - 3$

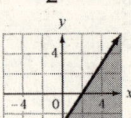

9. $y < \dfrac{4}{5}x - 2$

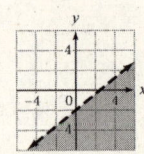

11. $y < -\dfrac{1}{3}x + 2$

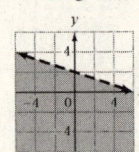

13. $x + 3y < 4$
$3y < -x + 4$
$y < -\dfrac{1}{3}x + \dfrac{4}{3}$

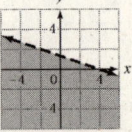

15. $2x + 3y \geq 6$
$3y \geq -2x + 6$
$y \geq -\dfrac{2}{3}x + 2$

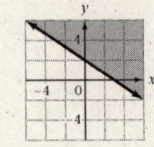

17. $-x + 2y > -8$
$2y > x - 8$
$y > \frac{1}{2}x - 4$

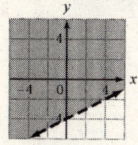

19. $y - 4 < 0$
$y < 4$

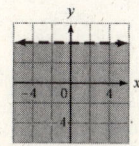

21. $6x + 5y < 15$
$5y < -6x + 15$
$y < -\frac{6}{5}x + 3$

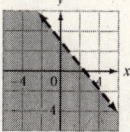

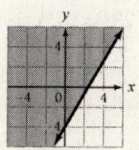

23. $-5x + 3y \geq -12$
$3y \geq 5x - 12$
$y \geq \frac{5}{3}x - 4$

25. If $(0, 0)$ is a point on the graph of the inequality, then
$Ax + By > C$
$A(0) + B(0) > C$
$0 + 0 > C$
$0 > C$
which implies that C is negative.

Applying Concepts 3.7

27. The inequality $y < 3x - 1$ is not a function because, given a value of x, there is more than one value of y. For example, both $(3, 2)$ and $(3, -1)$ are ordered pairs that satisfy the inequality, and this contradicts the definition of function because there are two ordered pairs with the same first coordinate and different second coordinates.

29. There are no points whose coordinates satisfy both $y \leq x - 1$ and $y \geq x + 2$. The solution set of $y \leq x - 1$ is all points on and below the line $y = x - 1$. The solution set of $y \geq x + 2$ is all points on and above the line $y = x + 2$. Because the lines $y = x - 1$ and $y = x + 2$ are parallel lines, and $y = x + 2$ is above the line $y = x - 1$, there are no points that lie both below $y = x - 1$ and above $y = x + 2$.

CHAPTER 3 REVIEW EXERCISES

1. $y = \frac{x}{x-2}$
$y = \frac{4}{4-2}$
$y = 2$
The ordered pair is $(4, 2)$.

2. $x_m = \frac{x_1 + x_2}{2} = \frac{-2 + 3}{2} = \frac{1}{2}$
$y_m = \frac{y_1 + y_2}{2} = \frac{4 + 5}{2} = \frac{9}{2}$
Length $= \sqrt{(x_2 - x_1)^2 + (y_2 - y_1)^2}$
$= \sqrt{[3 - (-2)]^2 + (5 - 4)^2}$
$= \sqrt{5^2 + 1^2}$
$= \sqrt{26}$
The midpoint is $\left(\frac{1}{2}, \frac{9}{2}\right)$ and the length is $\sqrt{26}$.

3. $y = x^2 - 2$
Ordered pairs: $(-3, 7)$
$(-2, 2)$
$(-1, -1)$
$(0, -2)$
$(1, -1)$
$(2, 2)$
$(3, 7)$

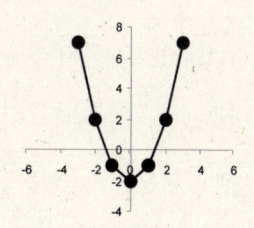

4.

5. $P(x) = 3x + 4$
$P(-2) = 3(-2) + 4 = -2$

$P(a) = 3(a) + 4$
$P(a) = 3a + 4$

6. Domain = {–1, 0, 1, 2, 5}
Range = {0, 2, 3}
Yes, because each x-coordinate is paired with exactly one y-value, the relation is a function.

7.
$f(x) = x^2 - 2$
$f(-2) = (-2)^2 - 2 = 2$
$f(-1) = (-1)^2 - 2 = -1$
$f(0) = 0^2 - 2 = -2$
$f(1) = 1^2 - 2 = -1$
$f(2) = 2^2 - 2 = 2$
Range = {–2, –1, 2}

8. $f(c) = 3c - 1$
$5 = 3c - 1$
$6 = 3c$
$2 = c$
An ordered pair of the function is (2, 5).

9. To find the x-intercept, let $y = 0$.
$4x - 6(0) = 12$
$4x = 12$
$x = 3$
The x-intercept is (3, 0).
To find the y-intercept, let $x = 0$.
$4(0) - 6y = 12$
$-6y = 12$
$y = -2$
The y-intercept is (0, –2).

10.

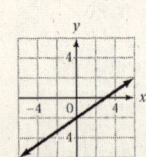

11.

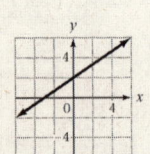

12. $m = \dfrac{y_2 - y_1}{x_2 - x_1}$
$m = \dfrac{2 - (-2)}{-1 - 3} = \dfrac{4}{-4} = -1$

13. x-intercept y-intercept
$3x + 2y = -4$ $3x + 2y = -4$
$3x + 2(0) = -4$ $3(0) + 2y = -4$
$3x = -4$ $2y = -4$
$x = -\dfrac{4}{3}$ $y = -2$

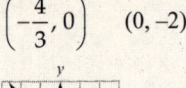

$\left(-\dfrac{4}{3}, 0\right)$ $(0, -2)$

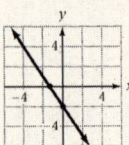

14.

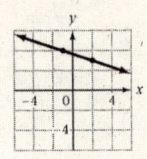

15. Use the point-slope form to find the equation of the line.
$y - y_1 = m(x - x_1)$
$y - 4 = \dfrac{5}{2}[x - (-3)]$
$y - 4 = \dfrac{5}{2}(x + 3)$
$y - 4 = \dfrac{5}{2}x + \dfrac{15}{2}$
$y = \dfrac{5}{2}x + \dfrac{23}{2}$

The equation of the line is $y = \dfrac{5}{2}x + \dfrac{23}{2}$.

16. $P_1(-2, 4), P_2(4, -3)$
$m = \dfrac{y_2 - y_1}{x_2 - x_1} = \dfrac{-3 - 4}{4 - (-2)} = \dfrac{-7}{6} = -\dfrac{7}{6}$
$y - y_1 = m(x - x_1)$
$y - 4 = -\dfrac{7}{6}[x - (-2)]$
$y - 4 = -\dfrac{7}{6}(x + 2)$
$y - 4 = -\dfrac{7}{6}x - \dfrac{7}{3}$
$y = -\dfrac{7}{6}x + \dfrac{5}{3}$

The equation of the line is $y = -\dfrac{7}{6}x + \dfrac{5}{3}$.

17.
$y = -3x + 4$
$m = -3$
$y - y_1 = m(x - x_1)$
$y - (-2) = -3(x - 3)$
$y + 2 = -3x + 9$
$y = -3x + 7$
The equation of the line is $y = -3x + 7$.

18.
$2x - 3y = 4$
$-3y = -2x + 4$
$y = \dfrac{2}{3}x - \dfrac{4}{3}$
$m = \dfrac{2}{3}$
$y - y_1 = m(x - x_1)$
$y - (-4) = \dfrac{2}{3}[x - (-2)]$
$y + 4 = \dfrac{2}{3}(x + 2)$
$y + 4 = \dfrac{2}{3}x + \dfrac{4}{3}$
$y = \dfrac{2}{3}x - \dfrac{8}{3}$
The equation of the line is $y = \dfrac{2}{3}x - \dfrac{8}{3}$.

19.
$y = -\dfrac{2}{3}x + 6$
$m_1 = -\dfrac{2}{3}$
$m_1 \cdot m_2 = -1$
$-\dfrac{2}{3}m_2 = -1$
$m_2 = \dfrac{3}{2}$
$y - y_1 = m(x - x_1)$
$y - 5 = \dfrac{3}{2}(x - 2)$
$y - 5 = \dfrac{3}{2}x - 3$
$y = \dfrac{3}{2}x + 2$
The equation of the line is $y = \dfrac{3}{2}x + 2$.

20.
$4x - 2y = 7$
$-2y = -4x + 7$
$y = 2x - \dfrac{7}{2}$
$m_1 = 2$
$m_1 \cdot m_2 = -1$
$2m_2 = -1$
$m_2 = -\dfrac{1}{2}$
$y - y_1 = m(x - x_1)$
$y - (-1) = -\dfrac{1}{2}[x - (-3)]$
$y + 1 = -\dfrac{1}{2}(x + 3)$
$y + 1 = -\dfrac{1}{2}x - \dfrac{3}{2}$
$y = -\dfrac{1}{2}x - \dfrac{5}{2}$
The equation of the line is $y = -\dfrac{1}{2}x - \dfrac{5}{2}$.

21.
$f(x) = -3x + 12$
$0 = -3x + 12$
$3x = 12$
$x = 4$
The zero of f is 4.

22. Since no vertical line passes through the graph at more than one point, this is the graph of a function.

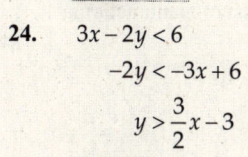

23. $y \geq 2x - 3$

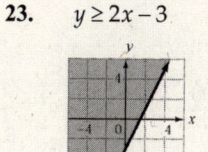

24.
$3x - 2y < 6$
$-2y < -3x + 6$
$y > \dfrac{3}{2}x - 3$

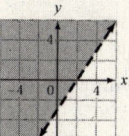

25.

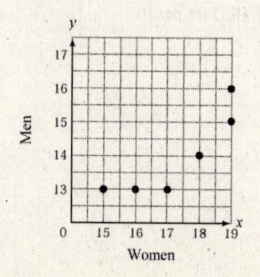

74 Chapter 3: Linear Functions And Inequalities In Two Variables

26. a. average rate of change $= \dfrac{28.4 - 14.1}{2000 - 1980}$

$= \dfrac{14.3}{20}$

$= 0.715$

The average annual rate of change was 715,000 people.

b. For 1970 to 1980,

average rate of change $= \dfrac{14.1 - 9.6}{1980 - 1970}$

$= \dfrac{4.5}{10}$

$= 0.45$

For 1980 to 1990,

average rate of change $= \dfrac{19.8 - 14.1}{1990 - 1980}$

$= \dfrac{5.7}{10}$

$= 0.57$

450,000 < 570,000

The average annual rate of change from 1970 to 1980 was 450,000 people, which was less than the average annual rate of change from 1980 to 1990.

27.

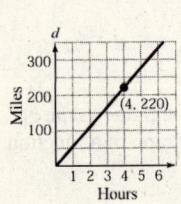

In 4 h, the car has traveled 220 miles.

28. $m = \dfrac{y_2 - y_1}{x_2 - x_1} = \dfrac{12{,}000 - 6000}{500 - 200} = \dfrac{6000}{300} = 20$

The slope is 20. The slope represents the cost per calculator manufactured. The cost of manufacturing one calculator is $20.

29. The y-intercept is (0, 25,000).
The slope is 80.
$y = mx + b$
$y = 80x + 25{,}000$
The linear function is $y = 80x + 25{,}000$.
Predict the cost of building a house with 2000 ft².
$y = 80(2000) + 25{,}000$
$y = 185{,}000$
The house will cost $185,000 to build.

CHAPTER 3 TEST

1. $f(x) = -x^2 + 2$
Ordered pairs: $(-2, -2)$
$(-1, 1)$
$(0, 2)$
$(1, 1)$
$(2, -2)$

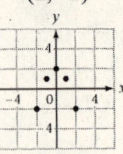

2. $y = 2x + 6$
$y = 2(-3) + 6$
$y = -6 + 6$
$y = 0$
The ordered-pair solution is $(-3, 0)$.

3.

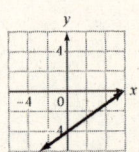

4. $2x + 3y = -3$
$3y = -2x - 3$
$y = -\dfrac{2}{3}x - 1$

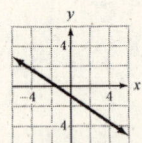

5. The equation of the vertical line that contains $(-2, 3)$ is $x = -2$.

6. $x_m = \dfrac{x_1 + x_2}{2} = \dfrac{4 + (-5)}{2} = -\dfrac{1}{2}$

$y_m = \dfrac{y_1 + y_2}{2} = \dfrac{2 + 8}{2} = 5$

Length $= \sqrt{(x_2 - x_1)^2 + (y_2 - y_1)^2}$

$= \sqrt{(-5 - 4)^2 + (8 - 2)^2}$

$= \sqrt{81 + 36}$

$= \sqrt{117}$

The midpoint is $\left(-\dfrac{1}{2}, 5\right)$ and the length is $\sqrt{117}$.

7. $P_1(-2, 3), P_2(4, 2)$

$m = \dfrac{y_2 - y_1}{x_2 - x_1} = \dfrac{2 - 3}{4 - (-2)} = -\dfrac{1}{6}$

The slope of the line is $-\dfrac{1}{6}$.

8. $P(x) = 3x^2 - 2x + 1$
   $P(2) = 3(2)^2 - 2(2) + 1$
   $P(2) = 9$

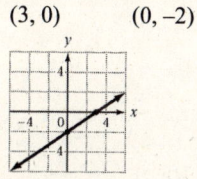

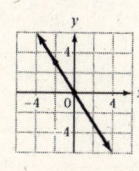

9. x-intercept: y-intercept:
   $2x - 3y = 6$ $2x - 3y = 6$
   $2x - 3(0) = 6$ $2(0) - 3y = 6$
   $2x = 6$ $-3y = 6$
   $x = 3$ $y = -2$
   (3, 0)   (0, -2)

10.

11. $m = \frac{2}{5}, (x_1, y_1) = (-5, 2)$
    $y - 2 = \frac{2}{5}[x - (-5)]$
    $y - 2 = \frac{2}{5}(x + 5)$
    $y - 2 = \frac{2}{5}x + 2$
    $y = \frac{2}{5}x + 4$

    The equation of the line is $y = \frac{2}{5}x + 4$.

12. $f(c) = 5c - 2$
    $3 = 5c - 2$
    $5 = 5c$
    $1 = c$
    One ordered pair of the function is (3, 1).

13. $P_1(3, -4), P_2(-2, 3)$
    $m = \frac{y_2 - y_1}{x_2 - x_1} = \frac{3 - (-4)}{-2 - 3} = \frac{3 + 4}{-5} = -\frac{7}{5}$
    $y - y_1 = m(x - x_1)$
    $y - (-4) = -\frac{7}{5}(x - 3)$
    $y + 4 = -\frac{7}{5}x + \frac{21}{5}$
    $y = -\frac{7}{5}x + \frac{1}{5}$

    The equation of the line is $y = -\frac{7}{5}x + \frac{1}{5}$.

14. $s(t) = \frac{4}{3}t - 8$
    $0 = \frac{4}{3}t - 8$
    $8 = \frac{4}{3}t$
    $\frac{3}{4} \cdot 8 = \frac{3}{4} \cdot \frac{4}{3}t$
    $6 = t$
    The zero of s is 6.

15. Domain = {-4, -2, 0, 3}
    Range = {0, 2, 5}
    Yes, the relation is a function.

16. $y = -\frac{3}{2}x - 6$
    $m = -\frac{3}{2}$
    $y - y_1 = m(x - x_1)$
    $y - 2 = -\frac{3}{2}(x - 1)$
    $y - 2 = -\frac{3}{2}x + \frac{3}{2}$
    $y = -\frac{3}{2}x + \frac{7}{2}$

    The equation of the line is $y = -\frac{3}{2}x + \frac{7}{2}$.

17. $y = -\frac{1}{2}x - 3$
    $m_1 = -\frac{1}{2}$
    $m_1 \cdot m_2 = -1$
    $-\frac{1}{2}m_2 = -1$
    $m_2 = 2$
    $y - y_1 = m(x - x_1)$
    $y - (-3) = 2[x - (-2)]$
    $y + 3 = 2(x + 2)$
    $y + 3 = 2x + 4$
    $y = 2x + 1$
    The equation of the line is $y = 2x + 1$.

18. $3x - 4y > 8$
    $-4y > -3x + 8$
    $y < \frac{3}{4}x - 2$

Chapter 3: Linear Functions And Inequalities In Two Variables

19. No, the graph is not a function because there are vertical lines through the graph that pass through it at more than one point.

20. Dependent variable: number of students: (y)
    Independent variable: tuition cost (x)
    $$m = \frac{\text{change in } y}{\text{change in } x} = \frac{-6}{20} = -\frac{3}{10}$$
    $P_1(250, 100)$
    Use the point-slope formula to find the equation.
    $$y - y_1 = m(x - x_1)$$
    $$y - 100 = -\frac{3}{10}(x - 250)$$
    $$y - 100 = -\frac{3}{10}x + 75$$
    $$y = -\frac{3}{10}x + 175$$
    The equation that predicts the number of students for a certain tuition is $y = -\frac{3}{10}x + 175$. Predict the number of students when the tuition is $300.
    $$y = -\frac{3}{10}x + 175$$
    $$y = -\frac{3}{10}(300) + 175$$
    $$y = 85$$
    When the tuition is $300, 85 students will enroll.

21. **Strategy**
    Use two points on the graph to find the slope of the line.

    Solution
    $(x_1, y_1) = (3, 40{,}000), (x_2, y_2) = (12, 10{,}000)$
    $$m = \frac{y_2 - y_1}{x_2 - x_1} = \frac{10{,}000 - 40{,}000}{12 - 3} = -\frac{10{,}000}{3}$$
    The value of the house decreases by $3333.33 per year.

22.

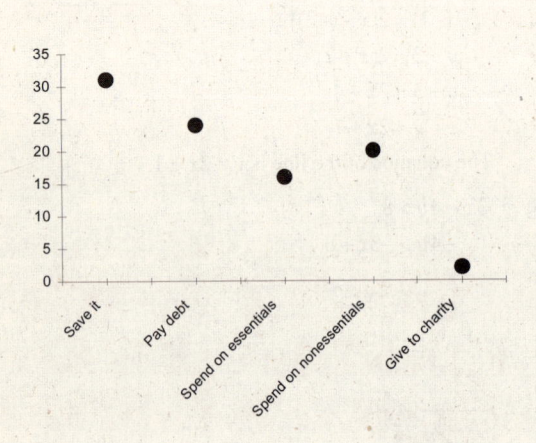

CUMULATIVE REVIEW EXERCISES

1. The Commutative Property of Multiplication

2. $$3 - \frac{x}{2} = \frac{3}{4}$$
   $$4\left(3 - \frac{x}{2}\right) = \frac{3}{4}(4)$$
   $$12 - 2x = 3$$
   $$12 - 2x - 12 = 3 - 12$$
   $$-2x = -9$$
   $$x = \frac{9}{2}$$
   The solution is $\frac{9}{2}$.

3. $$2[y - 2(3 - y) + 4] = 4 - 3y$$
   $$2(y - 6 + 2y + 4) = 4 - 3y$$
   $$2(3y - 2) = 4 - 3y$$
   $$6y - 4 = 4 - 3y$$
   $$9y - 4 = 4$$
   $$9y = 8$$
   $$y = \frac{8}{9}$$
   The solution is $\frac{8}{9}$.

4. $$\frac{1 - 3x}{2} + \frac{7x - 2}{6} = \frac{4x + 2}{9}$$
   $$18\left(\frac{1 - 3x}{2} + \frac{7x - 2}{6}\right) = 18\left(\frac{4x + 2}{9}\right)$$
   $$9(1 - 3x) + 3(7x - 2) = 2(4x + 2)$$
   $$9 - 27x + 21x - 6 = 8x + 4$$
   $$-6x + 3 = 8x + 4$$
   $$-14x = 1$$
   $$x = -\frac{1}{14}$$
   The solution is $-\frac{1}{14}$.

5. $x - 3 < -4$ or $\quad 2x + 2 > 3$
   $\quad x < -1 \qquad\qquad 2x > 1$
   $\{x \mid x < -1\} \qquad\quad x > \frac{1}{2}$
   $\qquad\qquad\qquad\quad \left\{x \mid x > \frac{1}{2}\right\}$
   $\{x \mid x < -1\} \cup \left\{x \mid x > \frac{1}{2}\right\} = \left\{x \mid x < -1 \text{ or } x > \frac{1}{2}\right\}$

6. $8 - |2x - 1| = 4$
$-|2x - 1| = -4$
$|2x - 1| = 4$
$2x - 1 = 4 \quad 2x - 1 = -4$
$2x = 5 \quad\quad 2x = -3$
$x = \dfrac{5}{2} \quad\quad x = -\dfrac{3}{2}$

The solutions are $\dfrac{5}{2}$ and $-\dfrac{3}{2}$.

7. $|3x - 5| < 5$
$-5 < 3x - 5 < 5$
$-5 + 5 < 3x - 5 + 5 < 5 + 5$
$0 < 3x < 10$
$\dfrac{1}{3}(0) < \dfrac{1}{3}(3x) < 10\left(\dfrac{1}{3}\right)$
$0 < x < \dfrac{10}{3}$
$\left\{x \mid 0 < x < \dfrac{10}{3}\right\}$

8. $4 - 2(4 - 5)^3 + 2 = 4 - 2(-1)^3 + 2$
$= 4 + 2 + 2$
$= 8$

9. $(a - b)^2 \div ab$ for $a = 4$ and $b = -2$
$[4 - (-2)]^2 \div 4(-2) = 6^2 \div 4(-2)$
$= 36 \div 4(-2)$
$= 9 \cdot (-2)$
$= -18$

10. $\{x \mid x < -2\} \cup \{x \mid x > 0\}$

$\longleftarrow\!\!+\!\!+\!\!+\!\!)\!\!+\!\!+\!\!(\!\!+\!\!+\!\!+\!\!+\!\!+\!\!\longrightarrow$
$-5\ -4\ -3\ -2\ -1\ \ 0\ \ 1\ \ 2\ \ 3\ \ 4\ \ 5$

11. Solve each inequality.
$3x - 1 < 4 \quad\quad x - 2 > 2$
$3x < 5 \quad\quad\quad x > 4$
$x < \dfrac{5}{3}$

The solution is $\left\{x \mid x < \dfrac{5}{3} \text{ and } x > 4\right\}$, and there is no such value, so the solution is the null set.

12. $P(x) = x^2 + 5$
$P(-3) = (-3)^2 + 5$
$P(-3) = 14$

13. $y = -\dfrac{5}{4}x + 3$
$y = -\dfrac{5}{4}(-8) + 3$
$y = 10 + 3$
$y = 13$

The ordered-pair solution is $(-8, 13)$.

14. $P_1(-1, 3), P_2(3, -4)$
$m = \dfrac{y_2 - y_1}{x_2 - x_1} = \dfrac{-4 - 3}{3 - (-1)} = \dfrac{-7}{4} = -\dfrac{7}{4}$

15. $m = \dfrac{3}{2}, (x_1, y_1) = (-1, 5)$
$y - y_1 = m(x - x_1)$
$y - 5 = \dfrac{3}{2}[x - (-1)]$
$y - 5 = \dfrac{3}{2}(x + 1)$
$y - 5 = \dfrac{3}{2}x + \dfrac{3}{2}$
$y = \dfrac{3}{2}x + \dfrac{13}{2}$

The equation of the line is $y = \dfrac{3}{2}x + \dfrac{13}{2}$.

16. $(x_1, y_1) = (4, -2), (x_2, y_2) = (0, 3)$
$m = \dfrac{y_2 - y_1}{x_2 - x_1} = \dfrac{3 - (-2)}{0 - 4} = \dfrac{3 + 2}{-4} = -\dfrac{5}{4}$
$y - y_1 = m(x - x_1)$
$y - (-2) = -\dfrac{5}{4}(x - 4)$
$y + 2 = -\dfrac{5}{4}x + 5$
$y = -\dfrac{5}{4}x + 3$

The equation of the line is $y = -\dfrac{5}{4}x + 3$.

17. $y = -\dfrac{3}{2}x + 2, m = -\dfrac{3}{2}$
$y - y_1 = m(x - x_1)$
$y - 4 = -\dfrac{3}{2}(x - 2)$
$y - 4 = -\dfrac{3}{2}x + 3$
$y = -\dfrac{3}{2}x + 7$

The equation of the line is $y = -\dfrac{3}{2}x + 7$.

Copyright © Houghton Mifflin Company. All rights reserved.

18. $3x - 2y = 5$
$-2y = -3x + 5$
$y = \frac{3}{2}x - \frac{5}{2}$
$m_1 = \frac{3}{2}$
$m_1 \cdot m_2 = -1$
$\frac{3}{2}m_2 = -1$
$m_2 = -\frac{2}{3}$
$y - y_1 = m(x - x_1)$
$y - 0 = -\frac{2}{3}(x - 4)$
$y = -\frac{2}{3}x + \frac{8}{3}$

The equation of the line is $y = -\frac{2}{3}x + \frac{8}{3}$.

19. x-intercept:
$3x - 5y = 15$
$3x - 5(0) = 15$
$3x = 15$
$x = 5$
$(5, 0)$

y-intercept:
$3x - 5y = 15$
$3(0) - 5y = 15$
$-5y = 15$
$y = -3$
$(0, -3)$

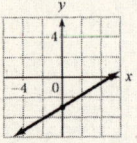

20.

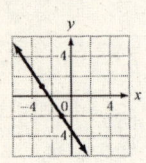

21. $3x - 2y \geq 6$
$-2y \geq -3x + 6$
$y \leq \frac{3}{2}x - 3$

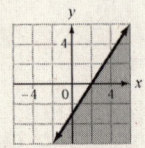

22. Strategy
- Number of nickels: $3x$
  Number of quarters: x

Coin	Number	Value	Total Value
Nickels	$3x$	5	$3x(5)$
Quarters	x	25	$25x$

- The sum of the total values of each denomination of coin equals the total value of all the coins (160).
  $3(x)(5) + 25x = 160$

Solution
$15x + 25x = 160$
$40x = 160$
$x = 4$
$3x = 12$

There are 12 nickels in the purse.

23. Strategy
- Rate of first plane: x
  Rate of second plane: $2x$

	Rate	Time	Distance
First plane	x	3	$3x$
Second plane	$2x$	3	$3(2x)$

- The two planes travel a total distance of 1800 miles.
  $3x + 3(2x) = 1800$

Solution
$3x + 3(2x) = 1800$
$3x + 6x = 1800$
$9x = 1800$
$x = 200$
$2x = 400$

The rate of the first plane is 200 mph and the rate of the second plane is 400 mph.

24. Strategy
 - Pounds of coffee costing $3.00: x
 - Pounds of coffee costing $8.00: $80 - x$

	Amount	Cost	Value
$3.00 coffee	x	3.00	$3x$
$8.00 coffee	$80 - x$	8.00	$8(80 - x)$
$5.00 mixture	80	5.00	$5(80)$

 - The sum of the values of each part of the mixture equals the value of the mixture.
$3x + 8(80 - x) = 5(80)$

Solution
$$3x + 640 - 8x = 400$$
$$-5x + 640 = 400$$
$$-5x = -240$$
$$x = 48$$
$$80 - x = 32$$

The mixture consists of 48 lb of $3 coffee and 32 lb of $8 coffee.

25. Strategy
To write the equation

 - Use points on the graph to find the slope of the line.

 - Locate the y-intercept of the line on the graph.

 - Use the slope-intercept form of an equation to write the equation of the line.

Solution
$(x_1, y_1) = (0, 15,000), (x_2, y_2) = (6, 0)$
The y-intercept is $(0, 15,000)$.
$$m = \frac{y_2 - y_1}{x_2 - x_1} = \frac{0 - 15,000}{6 - 0} = -2500$$
$$y = mx + b$$
$$y = -2500x + 15,000$$
The value of the truck decreases by $2500 each year.

CHAPTER 4: SYSTEMS OF EQUATIONS AND INEQUALITIES

CHAPTER 4 PREP TEST

1. $6x + 5y$ [1.3.3]

2. $3x + 2y - z$ [1.3.2]
   $3(-1) + 2(4) - (-2)$
   $= -3 + 8 + 2$
   $= 7$

3. $3x - 2(-2) = 4$ [2.1.1]
   $3x + 4 = 4$
   $3x = 0$
   $x = 0$

4. $3x + 4(-2x - 5) = -5$ [2.1.2]
   $3x - 8x - 20 = -5$
   $-5x = 15$
   $x = -3$

5. $0.45x + 0.06(-x + 4000) = 630$ [2.1.2]
   $0.45x - 0.06x + 240 = 630$
   $0.39x = 390$
   $x = 1000$

6.

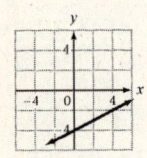

7. $3x - 2y = 6$ [3.3.2]
   $-2y = -3x + 6$
   $y = \frac{3}{2}x - 3$

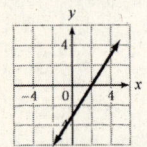

8.

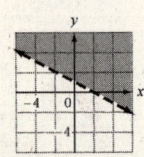

Section 4.1

Concept Review 4.1

1. Always true

3. Sometimes true
   A system of equations with two unknowns has one solution, an infinite number of solutions, or no solution.

5. Always true

Objective 4.1.1 Exercises

1. $3x - 2y = 2$
   $3(0) - 2(-1) = 2$
   $2 = 2$

   $x + 2y = 6$
   $0 + 2(-1) = 6$
   $-2 \neq 6$
   No

3. $x + y = -8$
   $-3 + -5 = -8$
   $-8 = -8$

   $2x + 5y = -31$
   $2(-3) + 5(-5) = -31$
   $-6 - 25 = -31$
   $-31 = -31$
   Yes

5. Independent

7. Inconsistent

9. The solution is (−2, 3).

11.

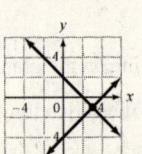

    The solution is (3, −1).

13.

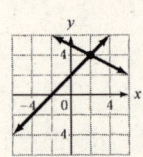

    The solution is (2, 4).

15.

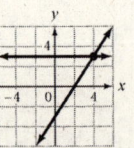

    The solution is (4, 3).

17.

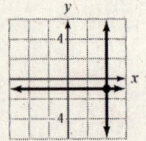

    The solution is (4, −1).

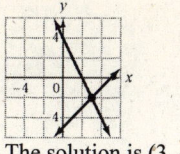

19. The solution is (3, −2).

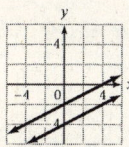

21. The lines are parallel and therefore do not intersect. The system of equations has no solution. The system is inconsistent.

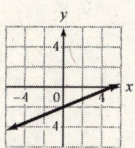

23. The system of equations is dependent. The solutions are the ordered pairs $\left(x, \dfrac{2}{5}x - 2\right)$.

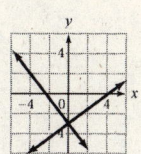

25. The solution is (0, −3).

27. Inconsistent

Objective 4.1.2 Exercises

29. When the substitution results in a false equation, such as 0 = 9, with no variable, the system of equations is inconsistent.

31. To solve the system by substitution, substitute 3 for x in equation (1): $y = 2(3) - 8 = \underline{-2}$. The solution of the system is ($\underline{3}$, $\underline{-2}$).

33. (1) $3x - 2y = 4$
(2) $x = 2$
Substitute the value of x into equation (1).
$3x - 2y = 4$
$3(2) - 2y = 4$
$6 - 2y = 4$
$-2y = -2$
$y = 1$
The solution is (2, 1).

35. (1) $y = 2x - 1$
(2) $x + 2y = 3$
Substitute $2x - 1$ for y in equation (2).
$x + 2y = 3$
$x + 2(2x - 1) = 3$
$x + 4x - 2 = 3$
$5x - 2 = 3$
$5x = 5$
$x = 1$
Substitute into equation (1).
$y = 2x - 1$
$y = 2(1) - 1$
$y = 2 - 1$
$y = 1$
The solution is (1, 1).

37. (1) $4x - 3y = 5$
(2) $y = 2x - 3$
Substitute $2x - 3$ for y in equation (1).
$4x - 3y = 5$
$4x - 3(2x - 3) = 5$
$4x - 6x + 9 = 5$
$-2x + 9 = 5$
$-2x = -4$
$x = 2$
Substitute into equation (2).
$y = 2x - 3$
$y = 2(2) - 3$
$y = 4 - 3$
$y = 1$
The solution is (2, 1).

39. (1) $x = 2y + 4$
(2) $4x + 3y = -17$
Substitute $2y + 4$ for x in equation (2).
$4x + 3y = -17$
$4(2y + 4) + 3y = -17$
$8y + 16 + 3y = -17$
$11y + 16 = -17$
$11y = -33$
$y = -3$
Substitute into equation (1).
$x = 2y + 4$
$x = 2(-3) + 4$
$x = -6 + 4$
$x = -2$
The solution is (−2, −3).

Chapter 4: Systems Of Equations And Inequalities

41. (1) $5x + 4y = -1$
(2) $\quad\quad y = 2 - 2x$
Substitute $2 - 2x$ for y in equation (1).
$5x + 4y = -1$
$5x + 4(2 - 2x) = -1$
$5x + 8 - 8x = -1$
$-3x + 8 = -1$
$-3x = -9$
$x = 3$
Substitute into equation (2).
$y = 2 - 2x$
$y = 2 - 2(3)$
$y = 2 - 6$
$y = -4$
The solution is $(3, -4)$.

43. (1) $7x - 3y = 3$
(2) $\quad\quad x = 2y + 2$
Substitute $2y + 2$ for x in equation (1).
$7x - 3y = 3$
$7(2y + 2) - 3y = 3$
$14y + 14 - 3y = 3$
$11y + 14 = 3$
$11y = -11$
$y = -1$
Substitute into equation (2).
$x = 2y + 2$
$x = 2(-1) + 2$
$x = -2 + 2$
$x = 0$
The solution is $(0, -1)$.

45. (1) $2x + 2y = 7$
(2) $\quad\quad y = 4x + 1$
Substitute $4x + 1$ for y in equation (1).
$2x + 2y = 7$
$2x + 2(4x + 1) = 7$
$2x + 8x + 2 = 7$
$10x + 2 = 7$
$10x = 5$
$x = \dfrac{1}{2}$
Substitute into equation (2).
$y = 4x + 1$
$y = 4\left(\dfrac{1}{2}\right) + 1$
$y = 2 + 1$
$y = 3$
The solution is $\left(\dfrac{1}{2}, 3\right)$.

47. (1) $3x + y = 5$
(2) $2x + 3y = 8$
Solve equation (1) for y.
$3x + y = 5$
$y = -3x + 5$
Substitute into equation (2).
$2x + 3y = 8$
$2x + 3(-3x + 5) = 8$
$2x - 9x + 15 = 8$
$-7x + 15 = 8$
$-7x = -7$
$x = 1$
Substitute into equation (1).
$3x + y = 5$
$3(1) + y = 5$
$3 + y = 5$
$y = 2$
The solution is $(1, 2)$.

49. (1) $x + 3y = 5$
(2) $2x + 3y = 4$
Solve equation (1) for x.
$x + 3y = 5$
$x = -3y + 5$
Substitute into equation (2).
$2x + 3y = 4$
$2(-3y + 5) + 3y = 4$
$-6y + 10 + 3y = 4$
$-3y + 10 = 4$
$-3y = -6$
$y = 2$
Substitute into equation (1).
$x + 3y = 5$
$x + 3(2) = 5$
$x + 6 = 5$
$x = -1$
The solution is $(-1, 2)$.

51. (1) $3x + 4y = 14$
(2) $2x + y = 1$
Solve equation (2) for y.
$2x + y = 1$
$y = -2x + 1$
Substitute into equation (1).
$3x + 4y = 14$
$3x + 4(-2x + 1) = 14$
$3x - 8x + 4 = 14$
$-5x + 4 = 14$
$-5x = 10$
$x = -2$
Substitute into equation (2).
$2x + y = 1$
$2(-2) + y = 1$
$-4 + y = 1$
$y = 5$
The solution is $(-2, 5)$.

53. (1) $3x + 5y = 0$
(2) $x - 4y = 0$
Solve equation (2) for x.
$x - 4y = 0$
$x = 4y$
Substitute into equation (1).
$3x + 5y = 0$
$3(4y) + 5y = 0$
$12y + 5y = 0$
$17y = 0$
$y = 0$
Substitute into equation (2).
$x - 4y = 0$
$x - 4(0) = 0$
$x = 0$
The solution is $(0, 0)$.

55. (1) $5x - 3y = -2$
(2) $-x + 2y = -8$
Solve equation (2) for x.
$-x + 2y = -8$
$-x = -2y - 8$
$x = 2y + 8$
Substitute $2y + 8$ for x in equation (1).
$5x - 3y = -2$
$5(2y + 8) - 3y = -2$
$10y + 40 - 3y = -2$
$7y + 40 = -2$
$7y = -42$
$y = -6$

Substitute into equation (2).
$-x + 2y = -8$
$-x + 2(-6) = -8$
$-x - 12 = -8$
$-x = 4$
$x = -4$
The solution is $(-4, -6)$.

57. (1) $x + 3y = 4$
(2) $x = 5 - 3y$
Substitute $5 - 3y$ for x in equation (1).
$x + 3y = 4$
$(5 - 3y) + 3y = 4$
$5 - 3y + 3y = 4$
$5 \neq 4$
Inconsistent

59. (1) $2x - 4y = 16$
(2) $-x + 2y = -8$
Solve equation (2) for x.
$-x + 2y = -8$
$-x = -2y - 8$
$x = 2y + 8$
Substitute $2y + 8$ for x in equation (1).
$2x - 4y = 16$
$2(2y + 8) - 4y = 16$
$4y + 16 - 4y = 16$
$16 = 16$
The system of equations is dependent.
Solve equation (2) for y.
$-x + 2y = -8$
$2y = x - 8$
$y = \frac{1}{2}x - 4$
The solution is the ordered pairs $\left(x, \frac{1}{2}x - 4\right)$

61. (1) $3x - y = 10$
(2) $6x - 2y = 5$
Solve equation (1) for y.
$3x - y = 10$
$-y = -3x + 10$
$y = 3x - 10$
Substitute $3x - 10$ for y in equation (1).
$6x - 2y = 5$
$6x - 2(3x - 10) = 5$
$6x - 6x + 20 = 5$
$20 \neq 5$
Inconsistent

84 Chapter 4: Systems Of Equations And Inequalities

63. (1) $y = 3x + 2$
    (2) $y = 2x + 3$
    Substitute $2x + 3$ for y in equation (1).
    $y = 3x + 2$
    $2x + 3 = 3x + 2$
    $2x = 3x - 1$
    $-x = -1$
    $x = 1$
    Substitute into equation (2).
    $y = 2x + 3$
    $y = 2(1) + 3$
    $y = 2 + 3$
    $y = 5$
    The solution is (1, 5).

65. (1) $x = 2y + 1$
    (2) $x = 3y - 1$
    Substitute $3y - 1$ for x in equation (1).
    $x = 2y + 1$
    $3y - 1 = 2y + 1$
    $3y = 2y + 2$
    $y = 2$
    Substitute into equation (2).
    $x = 3y - 1$
    $x = 3(2) - 1$
    $x = 6 - 1$
    $x = 5$
    The solution is (5, 2).

67. (1) $y = 5x - 1$
    (2) $y = 5 - x$
    Substitute $5 - x$ for y in equation (1).
    $y = 5x - 1$
    $5 - x = 5x - 1$
    $-x = 5x - 6$
    $-6x = -6$
    $x = 1$
    Substitute into equation (2).
    $y = 5 - x$
    $y = 5 - 1$
    $y = 4$
    The solution is (1, 4).

69. Dependent

Applying Concepts 4.1

71. Inconsistent equations have the same slope but different y-intercepts. Solve the two equations for y, and then set the slopes equal to each other.
    (1) $2x - 2y = 5$
    $-2y = -2x + 5$
    $y = x - \dfrac{5}{2}$

    (2) $kx - 2y = 3$
    $-2y = -kx + 3$
    $y = \dfrac{k}{2}x - \dfrac{3}{2}$
    $\dfrac{k}{2} = 1$
    $k = 2$
    The value of k is 2.

73. Inconsistent equations have the same slope but different y-intercepts. Solve the two equations for y, and then set the slopes equal to each other.
    (1) $x = 6y + 6$
    $x - 6 = 6y$
    $\dfrac{1}{6}x - 1 = y$

    (2) $kx - 3y = 6$
    $-3y = -kx + 6$
    $y = \dfrac{k}{3}x - 2$
    $\dfrac{k}{3} = \dfrac{1}{6}$
    $k = \dfrac{1}{2}$
    The value of k is $\dfrac{1}{2}$.

75. Strategy
    Solve a system of equations using x to represent one number and y to represent the second number.

    Solution
    (1) $x + y = 44$
    (2) $x - 8 = y$
    Solve equation (1) for y.
    $x + y = 44$
    $y = -x + 44$
    Substitute $-x + 44$ for y in equation (2).
    $x - 8 = y$
    $x - 8 = -x + 44$
    $2x = 52$
    $x = 26$
    Substitute into equation (1).
    $x + y = 44$
    $26 + y = 44$
    $y = 18$
    The numbers are 26 and 18.

77. **Strategy**
Solve a system of equations using x to represent one number and y to represent the second number.

Solution
(1) $x + y = 19$
(2) $2x - 5 = y$

Solve equation (1) for y.
$x + y = 19$
$y = -x + 19$

Substitute $-x + 19$ for y in equation (2).
$2x - 5 = y$
$2x - 5 = -x + 19$
$3x = 24$
$x = 8$

Substitute into equation (1).
$x + y = 19$
$8 + y = 19$
$y = 11$

The numbers are 8 and 11.

79.
$\dfrac{2}{a} + \dfrac{3}{b} = 4$

$2\left(\dfrac{1}{a}\right) + 3\left(\dfrac{1}{b}\right) = 4$

(1) $2x + 3y = 4$

$\dfrac{4}{a} + \dfrac{1}{b} = 3$

$4\left(\dfrac{1}{a}\right) + \dfrac{1}{b} = 3$

(2) $4x + y = 3$

Solve equation (1) for y.
$2x + 3y = 4$
$3y = -2x + 4$
$y = \left(-\dfrac{2}{3}\right)x + \dfrac{4}{3}$

Substitute $-\dfrac{2}{3}x + \dfrac{4}{3}$ for y in equation (2).
$4x + y = 3$
$4x + \left(-\dfrac{2}{3}x + \dfrac{4}{3}\right) = 3$
$\dfrac{10}{3}x + \dfrac{4}{3} = 3$
$10x + 4 = 9$
$10x = 5$
$x = \dfrac{1}{2}$

Substitute into equation (1).
$2x + 3y = 4$
$2\left(\dfrac{1}{2}\right) + 3y = 4$
$1 + 3y = 4$
$3y = 3$
$y = 1$

Replace x by $\dfrac{1}{a}$.
$x = \dfrac{1}{2}$
$\dfrac{1}{a} = \dfrac{1}{2}$
$2a\left(\dfrac{1}{a}\right) = 2a\left(\dfrac{1}{2}\right)$
$2 = a$

Replace y by $\dfrac{1}{b}$.
$y = 1$
$\dfrac{1}{b} = 1$
$b\left(\dfrac{1}{b}\right) = b(1)$
$1 = b$

The solution is (2, 1).

81. $\dfrac{1}{a}+\dfrac{3}{b}=2$

$\dfrac{1}{a}+3\left(\dfrac{1}{b}\right)=2$

(1) $\quad x+3y=2$

$\dfrac{4}{a}-\dfrac{1}{b}=3$

$4\left(\dfrac{1}{a}\right)-\dfrac{1}{b}=3$

(2) $\quad 4x-y=3$

Solve equation (1) for y.
$x+3y=2$
$3y=-x+2$
$y=-\dfrac{1}{3}x+\dfrac{2}{3}$

Substitute $-\dfrac{1}{3}x+\dfrac{2}{3}$ for y in equation (2).

$4x-y=3$

$4x-\left(-\dfrac{1}{3}x+\dfrac{2}{3}\right)=3$

$4x+\dfrac{1}{3}x-\dfrac{2}{3}=3$

$\dfrac{13}{3}x=\dfrac{11}{3}$

$x=\dfrac{11}{13}$

Substitute into equation (1).
$x+3y=2$

$\dfrac{11}{13}+3y=2$

$3y=\dfrac{15}{13}$

$y=\dfrac{15}{39}=\dfrac{5}{13}$

Replace x by $\dfrac{1}{a}$.

$x=\dfrac{11}{13}$

$\dfrac{1}{a}=\dfrac{11}{13}$

$13a\left(\dfrac{1}{a}\right)=13a\left(\dfrac{11}{13}\right)$

$13=11a$

$\dfrac{13}{11}=a$

Replace y by $\dfrac{1}{b}$.

$y=\dfrac{5}{13}$

$\dfrac{1}{b}=\dfrac{5}{13}$

$13b\left(\dfrac{1}{b}\right)=13b\left(\dfrac{5}{13}\right)$

$13=5b$

$\dfrac{13}{5}=b$

The solution is $\left(\dfrac{13}{11},\dfrac{13}{5}\right)$.

83. $y=-\dfrac{1}{2}x+2$

$y=2x-1$

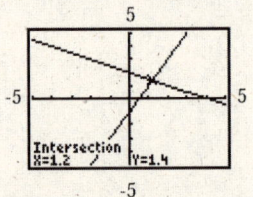

The solution is (1.20, 1.40).

85. $y=\sqrt{2}x-1$

$y=-\sqrt{3}x+1$

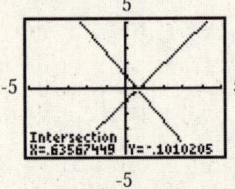

The solution is (0.64, −0.10).

87. a. $y_1 = -0.0036x + 9.559$
$= -0.0036(2010) + 9.559 = 2.323$
$y_2 = 0.0419x - 82.156$
$= 0.0419(2010) - 82.516 = 1.703$

Since $2.323 > 1.703$ the population of Orlando will not exceed the population of Pittsburgh in the year 2010.

b. $y_1 = y_2$
$-0.0036x + 9.559 = 0.0419x - 82.156$
$91.715 = 0.0455x$
$2015.7143 = x$

In 2016 the population of Orlando will first exceed the population of Pittsburgh.

c. The slope of y_1 is $-0.0036(1,000,000) = 3600$. The slope indicates that the population of Pittsburgh is decreasing at the rate of 3600 people per year.

d. The slope of y_2 is $0.0419(1,000,000) = 41,900$. The slope indicates that the population of Orlando is increasing at the rate of 41,900 people per year.

Section 4.2

Concept Review 4.2

1. Always true

3. Sometimes true
The solution of a system of three equations in three variables may be a point, a line, or a plane, or the system may not have a solution.

5. Never true
The system is inconsistent and has no solutions.

Objective 4.2.1 Exercises

1. The solution to a system of equations is an ordered pair or a set of ordered pairs. 5 is not a solution to the system. It is the x-coordinate of the solution. To find the solution, substitute 5 for x into one of the equations in the system to find the y-coordinate of the solution.

3. (1) $x - y = 5$
(2) $x + y = 7$
Eliminate y. Add the equations.
$2x = 12$
$x = 6$
Replace x in equation (1).
$x - y = 5$
$6 - y = 5$
$-y = -1$
$y = 1$
The solution is (6, 1).

5. (1) $3x + y = 4$
(2) $x + y = 2$
Eliminate y.
$3x + y = 4$
$-1(x + y) = -1(2)$

$3x + y = 4$
$-x - y = -2$
Add the equations.
$2x = 2$
$x = 1$
Replace x in equation (2).
$x + y = 2$
$1 + y = 2$
$y = 1$
The solution is (1, 1).

7. (1) $3x + y = 7$
(2) $x + 2y = 4$
Eliminate y.
$-2(3x + y) = -2(7)$
$x + 2y = 4$

$-6x - 2y = -14$
$x + 2y = 4$
Add the equations.
$-5x = -10$
$x = 2$
Replace x in equation (2).
$x + 2y = 4$
$2 + 2y = 4$
$2y = 2$
$y = 1$
The solution is (2, 1).

9. (1) $3x - y = 4$
(2) $6x - 2y = 8$
Eliminate y.
$-2(3x - y) = -2(4)$
$6x - 2y = 8$

$-6x + 2y = -8$
$6x - 2y = 8$
Add the equations.
$0 = 0$
This is a true equation. The equations are dependent. The solutions are the ordered pairs $(x, 3x - 4)$.

Chapter 4: Systems Of Equations And Inequalities

11. (1) $2x + 5y = 9$
(2) $4x - 7y = -16$
Eliminate x.
$-2(2x + 5y) = -2(9)$
$4x - 7y = -16$

$-4x - 10y = -18$
$4x - 7y = -16$
Add the equations.
$-17y = -34$
$y = 2$
Replace y in equation (1).
$2x + 5y = 9$
$2x + 5(2) = 9$
$2x + 10 = 9$
$2x = -1$
$x = -\dfrac{1}{2}$

The solution is $\left(-\dfrac{1}{2}, 2\right)$.

13. (1) $4x - 6y = 5$
(2) $2x - 3y = 7$
Eliminate y.
$4x - 6y = 5$
$-2(2x - 3y) = -2(7)$

$4x - 6y = 5$
$-4x + 6y = -14$
Add the equations.
$0 = -9$
This is not a true equation. The system of equations is inconsistent and therefore has no solution.

15. (1) $3x - 5y = 7$
(2) $x - 2y = 3$
Eliminate x.
$3x - 5y = 7$
$-3(x - 2y) = -3(3)$

$3x - 5y = 7$
$-3x + 6y = -9$
Add the equations.
$y = -2$
Replace y in equation (2).
$x - 2y = 3$
$x - 2(-2) = 3$
$x + 4 = 3$
$x = -1$
The solution is $(-1, -2)$.

17. (1) $3x + 2y = 16$
(2) $2x - 3y = -11$
Eliminate y.
$3(3x + 2y) = 3(16)$
$2(2x - 3y) = 2(-11)$

$9x + 6y = 48$
$4x - 6y = -22$
Add the equations.
$13x = 26$
$x = 2$
Replace x in equation (1).
$3x + 2y = 16$
$3(2) + 2y = 16$
$6 + 2y = 16$
$2y = 10$
$y = 5$
The solution is $(2, 5)$.

19. (1) $4x + 4y = 5$
(2) $2x - 8y = -5$
Eliminate y.
$2(4x + 4y) = 2(5)$
$2x - 8y = -5$

$8x + 8y = 10$
$2x - 8y = -5$
Add the equations.
$10x = 5$
$x = \dfrac{1}{2}$
Replace x in equation (1).
$4x + 4y = 5$
$4\left(\dfrac{1}{2}\right) + 4y = 5$
$2 + 4y = 5$
$4y = 3$
$y = \dfrac{3}{4}$
The solution is $\left(\dfrac{1}{2}, \dfrac{3}{4}\right)$.

21. (1) $5x + 4y = 0$
(2) $3x + 7y = 0$
Eliminate x.
$-3(5x + 4y) = -3(0)$
$5(3x + 7y) = 5(0)$

$-15x - 12y = 0$
$15x + 35y = 0$
Add the equations.
$23y = 0$
$y = 0$
Replace y in equation (1).
$5x + 4y = 0$
$5x + 4(0) = 0$
$5x = 0$
$x = 0$
The solution is $(0, 0)$.

23. (1) $3x - 6y = 6$
(2) $9x - 3y = 8$
Eliminate y.
$3x - 6y = 6$
$-2(9x - 3y) = -2(8)$

$3x - 6y = 6$
$-18x + 6y = -16$
Add the equations.
$-15x = -10$
$x = \dfrac{2}{3}$
Replace x in the equation (1).
$3x - 6y = 6$
$3\left(\dfrac{2}{3}\right) - 6y = 6$
$2 - 6y = 6$
$-6y = 4$
$y = -\dfrac{2}{3}$
The solution is $\left(\dfrac{2}{3}, -\dfrac{2}{3}\right)$.

25. (1) $5x + 2y = 2x + 1$
(2) $2x - 3y = 3x + 2$
Write the equations in the form $Ax + By = C$.
$5x + 2y = 2x + 1$
$3x + 2y = 1$

$2x - 3y = 3x + 2$
$-x - 3y = 2$

Solve the system.
$3x + 2y = 1$
$-x - 3y = 2$
Eliminate x.
$3x + 2y = 1$
$3(-x - 3y) = 3(2)$

$3x + 2y = 1$
$-3x - 9y = 6$
Add the equations.
$-7y = 7$
$y = -1$
Replace y in the equation $-x - 3y = 2$.
$-x - 3y = 2$
$-x - 3(-1) = 2$
$-x + 3 = 2$
$-x = -1$
$x = 1$
The solution is $(1, -1)$.

27. (1) $\dfrac{2}{3}x - \dfrac{1}{2}y = 3$
(2) $\dfrac{1}{3}x - \dfrac{1}{4}y = \dfrac{3}{2}$
Clear the fractions.
$6\left(\dfrac{2}{3}x - \dfrac{1}{2}y\right) = 6(3)$
$12\left(\dfrac{1}{3}x - \dfrac{1}{4}y\right) = 12\left(\dfrac{3}{2}\right)$

$4x - 3y = 18$
$4x - 3y = 18$
Eliminate x.
$-1(4x - 3y) = -1(18)$
$4x - 3y = 18$

$-4x + 3y = -18$
$4x - 3y = 18$
Add the equations.
$0 = 0$
This is a true equation. The equations are dependent.
The solutions are the ordered pairs $\left(x, \dfrac{4}{3}x - 6\right)$.

Chapter 4: Systems Of Equations And Inequalities

29. (1) $\dfrac{2}{5}x - \dfrac{1}{3}y = 1$

(2) $\dfrac{3}{5}x + \dfrac{2}{3}y = 5$

Clear the fractions.

$15\left(\dfrac{2}{5}x - \dfrac{1}{3}y\right) = 15(1)$

$15\left(\dfrac{3}{5}x + \dfrac{2}{3}y\right) = 15(5)$

$6x - 5y = 15$
$9x + 10y = 75$

Eliminate y.

$2(6x - 5y) = 2(15)$
$9x + 10y = 75$

$12x - 10y = 30$
$9x + 10y = 75$

Add the equations.
$21x = 105$
$x = 5$

Replace x in equation (1).

$\dfrac{2}{5}x - \dfrac{1}{3}y = 1$

$\dfrac{2}{5}(5) - \dfrac{1}{3}y = 1$

$2 - \dfrac{1}{3}y = 1$

$-\dfrac{1}{3}y = -1$

$y = 3$

The solution is $(5, 3)$.

31. (1) $\dfrac{3}{4}x + \dfrac{2}{5}y = -\dfrac{3}{20}$

(2) $\dfrac{3}{2}x - \dfrac{1}{4}y = \dfrac{3}{4}$

Clear the fractions.

$20\left(\dfrac{3}{4}x + \dfrac{2}{5}y\right) = 20\left(-\dfrac{3}{20}\right)$

$4\left(\dfrac{3}{2}x - \dfrac{1}{4}y\right) = 4\left(\dfrac{3}{4}\right)$

$15x + 8y = -3$
$6x - y = 3$

Eliminate y.

$15x + 8y = -3$
$8(6x - y) = 8(3)$

$15x + 8y = -3$
$48x - 8y = 24$

Add the equations.
$63x = 21$

$x = \dfrac{1}{3}$

Replace x in equation (2).

$\dfrac{3}{2}x - \dfrac{1}{4}y = \dfrac{3}{4}$

$\dfrac{3}{2}\left(\dfrac{1}{3}\right) - \dfrac{1}{4}y = \dfrac{3}{4}$

$\dfrac{1}{2} - \dfrac{1}{4}y = \dfrac{3}{4}$

$-\dfrac{1}{4}y = \dfrac{1}{4}$

$y = -1$

The solution is $\left(\dfrac{1}{3}, -1\right)$.

33. (1) $4x - 5y = 3y + 4$

(2) $2x + 3y = 2x + 1$

Write the equations in the form $Ax + By = C$.

$4x - 5y = 3y + 4$
$4x - 8y = 4$

$2x + 3y = 2x + 1$
$3y = 1$

Solve the system.
$4x - 8y = 4$
$3y = 1$

Solve the equation $3y = 1$ for y.
$3y = 1$

$y = \dfrac{1}{3}$

Replace y in the equation $4x - 8y = 4$.
$4x - 8y = 4$

$4x - 8\left(\dfrac{1}{3}\right) = 4$

$4x - \dfrac{8}{3} = 4$

$4x = \dfrac{20}{3}$

$x = \dfrac{5}{3}$

The solution is $\left(\dfrac{5}{3}, \dfrac{1}{3}\right)$.

35. (1) $2x + 5y = 5x + 1$
(2) $3x - 2y = 3y + 3$
Write the equations in the form $Ax + By = C$.
$2x + 5y = 5x + 1$
$-3x + 5y = 1$

$3x - 2y = 3y + 3$
$3x - 5y = 3$
Solve the system.
$-3x + 5y = 1$
$3x - 5y = 3$
Add the equations.
$0 = 4$
This is not a true equation. The system of equations is inconsistent and therefore has no solution.

37. If, after adding the equations, the result is a true equation, such as $3 = 3$, with no variable, the system of equations is dependent.

39. Dependent

Objective 4.2.2 Exercises

41. The solution of an independent system of linear equations in three variables is an ordered triple of the form (x, y, z).

43. The first and second equations, or the second and third equations.

45. (1) $x + 3y + z = 6$
(2) $3x + y - z = -2$
(3) $2x + 2y - z = 1$
Eliminate z. Add equations (1) and (2).
$x + 3y + z = 6$
$3x + y - z = -2$

$4x + 4y = 4$

Multiply both sides of the equation by $\frac{1}{4}$.
(4) $x + y = 1$
Add equations (1) and (3).
$x + 3y + z = 6$
$2x + 2y - z = 1$

(5) $3x + 5y = 7$
Multiply equation (4) by -3 and add to equation (5).
$-3(x + y) = -3(1)$
$3x + 5y = 7$

$-3x - 3y = -3$
$3x + 5y = 7$

$2y = 4$
$y = 2$

Replace y by 2 in equation (4).
$x + y = 1$
$x + 2 = 1$
$x = -1$
Replace x by -1 and y by 2 in equation (1).
$x + 3y + z = 6$
$-1 + 3(2) + z = 6$
$-1 + 6 + z = 6$
$5 + z = 6$
$z = 1$
The solution is $(-1, 2, 1)$.

47. (1) $x - 2y + z = 6$
(2) $x + 3y + z = 16$
(3) $3x - y - z = 12$
Eliminate z. Add equations (1) and (3).
$x - 2y + z = 6$
$3x - y - z = 12$

(4) $4x - 3y = 18$
Add equations (2) and (3).
$x + 3y + z = 16$
$3x - y - z = 12$

$4x + 2y = 28$

Multiply both sides of the equation by $\frac{1}{2}$.
(5) $2x + y = 14$
Multiply equation (5) by 3 and add to equation (4).
$3(2x + y) = 3(14)$
$4x - 3y = 18$

$6x + 3y = 42$
$4x - 3y = 18$

$10x = 60$
$x = 6$
Replace x by 6 in equation (5).
$2x + y = 14$
$2(6) + y = 14$
$12 + y = 14$
$y = 2$
Replace x by 6 and y by 2 in equation (1).
$x - 2y + z = 6$
$6 - 2(2) + z = 6$
$6 - 4 + z = 6$
$2 + z = 6$
$z = 4$
The solution is $(6, 2, 4)$.

Chapter 4: Systems Of Equations And Inequalities

49.
(1) $2y + z = 7$
(2) $2x - z = 3$
(3) $x - y = 3$

Eliminate z. Add equations (1) and (2).
$2y + z = 7$
$2x - z = 3$

$2x + 2y = 10$

Multiply both sides of the equation by $\frac{1}{2}$.

(4) $x + y = 5$

Add equations (3) and (4).
$x - y = 3$
$x + y = 5$

$2x = 8$
$x = 4$

Replace x by 4 in equation (4).
$x + y = 5$
$4 + y = 5$
$y = 1$

Replace y by 1 in equation (1).
$2y + z = 7$
$2(1) + z = 7$
$2 + z = 7$
$z = 5$

The solution is (4, 1, 5).

51.
(1) $2x + y - 3z = 7$
(2) $x - 2y + 3z = 1$
(3) $3x + 4y - 3z = 13$

Eliminate z. Add equations (1) and (2).
$2x + y - 3z = 7$
$x - 2y + 3z = 1$

(4) $3x - y = 8$

Add equations (2) and (3).
$x - 2y + 3z = 1$
$3x + 4y - 3z = 13$

$4x + 2y = 14$

Multiply each side of the equation by $\frac{1}{2}$.

(5) $2x + y = 7$

Add equations (4) and (5).
$3x - y = 8$
$2x + y = 7$

$5x = 15$
$x = 3$

Replace x by 3 in equation (5).
$2x + y = 7$
$2(3) + y = 7$
$6 + y = 7$
$y = 1$

Replace x by 3 and y by 1 in equation (1).
$2x + y - 3z = 7$
$2(3) + 1 - 3z = 7$
$6 + 1 - 3z = 7$
$7 - 3z = 7$
$-3z = 0$
$z = 0$

The solution is (3, 1, 0).

53.
(1) $3x + 4z = 5$
(2) $2y + 3z = 2$
(3) $2x - 5y = 8$

Eliminate z. Multiply equation (1) by -3 and equation (2) by 4.
Then add the equations.
$-3(3x + 4z) = -3(5)$
$4(2y + 3z) = 4(2)$

$-9x - 12z = -15$
$8y + 12z = 8$

(4) $-9x + 8y = -7$

Multiply equation (3) by 9 and equation (4) by 2.
Then add the equations.
$9(2x - 5y) = 9(8)$
$2(-9x + 8y) = 2(-7)$

$18x - 45y = 72$
$-18x + 16y = -14$

$-29y = 58$
$y = -2$

Replace y by -2 in equation (3).
$2x - 5y = 8$
$2x - 5(-2) = 8$
$2x + 10 = 8$
$2x = -2$
$x = -1$

Replace x by -1 in equation (1).
$3x + 4z = 5$
$3(-1) + 4z = 5$
$-3 + 4z = 5$
$4z = 8$
$z = 2$

The solution is $(-1, -2, 2)$.

55. (1) $x - 3y + 2z = 1$
    (2) $x - 2y + 3z = 5$
    (3) $2x - 6y + 4z = 3$

Eliminate x. Multiply equation (1) by -1 and add to equation (2).
$-1(x - 3y + 2z) = -1(1)$
$x - 2y + 3z = 5$

$-x + 3y - 2z = -1$
$x - 2y + 3z = 5$

    (4) $y + z = 4$

Multiply equation (1) by -2 and add to equation (3).
$-2(x - 3y + 2z) = -2(1)$
$2x - 6y + 4z = 3$

$-2x + 6y - 4z = -2$
$2x - 6y + 4z = 3$

$0 = 1$

This is not a true equation. The system of equations is inconsistent and therefore has no solution.

57. (1) $3x - y - 2z = 11$
    (2) $2x + y - 2z = 11$
    (3) $x + 3y - z = 8$

Eliminate z. Multiply equation (1) by -1 and add to equation (2).
$-1(3x - y - 2z) = -1(11)$
$2x + y - 2z = 11$

$-3x + y + 2z = -11$
$2x + y - 2z = 11$

    (4) $-x + 2y = 0$

Multiply equation (3) by -2 and add to equation (1).
$3x - y - 2z = 11$
$-2(x + 3y - z) = -2(8)$

$3x - y - 2z = 11$
$-2x - 6y + 2z = -16$

    (5) $x - 7y = -5$

Add equations (4) and (5).
$-x + 2y = 0$
$x - 7y = -5$

$-5y = -5$
$y = 1$

Replace y by 1 in equation (4).
$-x + 2y = 0$
$-x + 2(1) = 0$
$-x + 2 = 0$
$-x = -2$
$x = 2$

Replace x by 2 and y by 1 in equation (3).
$x + 3y - z = 8$
$2 + 3(1) - z = 8$
$2 + 3 - z = 8$
$5 - z = 8$
$-z = 3$
$z = -3$

The solution is $(2, 1, -3)$.

59.
(1) $4x + 5y + z = 6$
(2) $2x - y + 2z = 11$
(3) $x + 2y + 2z = 6$

Eliminate z. Multiply equation (1) by –2 and add to equation (2).
$-2(4x + 5y + z) = -2(6)$
$2x - y + 2z = 11$

$-8x - 10y - 2z = -12$
$2x - y + 2z = 11$

(4) $-6x - 11y = -1$

Multiply equation (2) by –1 and add to equation (3).
$-1(2x - y + 2z) = -1(11)$
$x + 2y + 2z = 6$

$-2x + y - 2z = -11$
$x + 2y + 2z = 6$

(5) $-x + 3y = -5$

Multiply equation (5) by –6 and add to equation (4).
$-6x - 11y = -1$
$-6(-x + 3y) = -6(-5)$

$-6x - 11y = -1$
$6x - 18y = 30$

$-29y = 29$
$y = -1$

Replace y by –1 in equation (5).
$-x + 3y = -5$
$-x + 3(-1) = -5$
$-x - 3 = -5$
$-x = -2$
$x = 2$

Replace x by 2 and y by –1 in equation (1).
$4x + 5y + z = 6$
$4(2) + 5(-1) + z = 6$
$8 - 5 + z = 6$
$3 + z = 6$
$z = 3$

The solution is (2, –1, 3).

61.
(1) $3x + 2y - 3z = 8$
(2) $2x + 3y + 2z = 10$
(3) $x + y - z = 2$

Eliminate z. Multiply equation (1) by 2 and equation (2) by 3.
Then add the equations.
$2(3x + 2y - 3z) = 2(8)$
$3(2x + 3y + 2z) = 3(10)$

$6x + 4y - 6z = 16$
$6x + 9y + 6z = 30$

(4) $12x + 13y = 46$

Multiply equation (3) by 2 and add to equation (2).
$2x + 3y + 2z = 10$
$2(x + y - z) = 2(2)$

$2x + 3y + 2z = 10$
$2x + 2y - 2z = 4$

(5) $4x + 5y = 14$

Multiply equation (5) by –3 and add to equation (4).
$12x + 13y = 46$
$-3(4x + 5y) = -3(14)$

$12x + 13y = 46$
$-12x - 15y = -42$

$-2y = 4$
$y = -2$

Replace y by –2 in equation (5).
$4x + 5y = 14$
$4x + 5(-2) = 14$
$4x - 10 = 14$
$4x = 24$
$x = 6$

Replace x by 6 and y by –2 in equation (3).
$x + y - z = 2$
$6 + (-2) - z = 2$
$4 - z = 2$
$-z = -2$
$z = 2$

The solution is (6, –2, 2).

63. (1) $3x - 3y + 4z = 6$
    (2) $4x - 5y + 2z = 10$
    (3) $x - 2y + 3z = 4$

Eliminate x. Multiply equation (3) by -3 and add to equation (1).
$3x - 3y + 4z = 6$
$-3(x - 2y + 3z) = -3(4)$

$3x - 3y + 4z = 6$
$-3x + 6y - 9z = -12$

    (4) $3y - 5z = -6$

Multiply equation (3) by -4 and add to equation (2).
$4x - 5y + 2z = 10$
$-4(x - 2y + 3z) = -4(4)$

$4x - 5y + 2z = 10$
$-4x + 8y - 12z = -16$

    (5) $3y - 10z = -6$

Multiply equation (4) by -1 and add to equation (5).
$-1(3y - 5z) = -1(-6)$
$3y - 10z = -6$

$-3y + 5z = 6$
$3y - 10z = -6$

$-5z = 0$
$z = 0$

Replace z by 0 in equation (4).
$3y - 5z = -6$
$3y - 5(0) = -6$
$3y = -6$
$y = -2$

Replace y by -2 and z by 0 in equation (3).
$x - 2y + 3z = 4$
$x - 2(-2) + 3(0) = 4$
$x + 4 = 4$
$x = 0$

The solution is $(0, -2, 0)$.

65. (1) $2x + 2y + 3z = 13$
    (2) $-3x + 4y - z = 5$
    (3) $5x - 3y + z = 2$

Eliminate z. Multiply equation (2) by 3 and add to equation (1).
$2x + 2y + 3z = 13$
$3(-3x + 4y - z) = 3(5)$

$2x + 2y + 3z = 13$
$-9x + 12y - 3z = 15$

$-7x + 14y = 28$

Multiply each side of the equation by $\frac{1}{7}$.

    (4) $-x + 2y = 4$

Add equations (2) and (3).
$-3x + 4y - z = 5$
$5x - 3y + z = 2$

    (5) $2x + y = 7$

Multiply equation (4) by 2 and add to equation (5).
$2(-x + 2y) = 2(4)$
$2x + y = 7$

$-2x + 4y = 8$
$2x + y = 7$

$5y = 15$
$y = 3$

Replace y by 3 in equation (5).
$2x + y = 7$
$2x + 3 = 7$
$2x = 4$
$x = 2$

Replace x by 2 and y by 3 in equation (3).
$5x - 3y + z = 2$
$5(2) - 3(3) + z = 2$
$10 - 9 + z = 2$
$1 + z = 2$
$z = 1$

The solution is $(2, 3, 1)$.

67. (1) $5x + 3y - z = 5$
(2) $3x - 2y + 4z = 13$
(3) $4x + 3y + 5z = 22$

Eliminate z. Multiply equation (1) by 4 and add to equation (2).
$4(5x + 3y - z) = 4(5)$
$3x - 2y + 4z = 13$

$20x + 12y - 4z = 20$
$3x - 2y + 4z = 13$

(4) $23x + 10y = 33$

Multiply equation (1) by 5 and add to equation (3).
$5(5x + 3y - z) = 5(5)$
$4x + 3y + 5z = 22$

$25x + 15y - 5z = 25$
$4x + 3y + 5z = 22$

(5) $29x + 18y = 47$

Multiply equation (4) by -18 and equation (5) by 10. Then add the equations.
$-18(23x + 10y) = -18(33)$
$10(29x + 18y) = 10(47)$

$-414x - 180y = -594$
$290x + 180y = 470$

$-124x = -124$
$x = 1$

Replace x by 1 in equation (4).
$23x + 10y = 33$
$23(1) + 10y = 33$
$23 + 10y = 33$
$10y = 10$
$y = 1$

Replace x by 1 and y by 1 in equation (1).
$5x + 3y - z = 5$
$5(1) + 3(1) - z = 5$
$5 + 3 - z = 5$
$8 - z = 5$
$-z = -3$
$z = 3$

The solution is (1, 1, 3).

Applying Concepts 4.2

69. (1) $0.4x - 0.9y = -0.1$
(2) $0.3x + 0.2y = 0.8$

Multiply both sides of each equation by 10.
(1) $4x - 9y = -1$
(2) $3x + 2y = 8$

Eliminate y.
$2(4x - 9y) = 2(-1)$
$9(3x + 2y) = 9(8)$

$8x - 18y = -2$
$27x + 18y = 72$

Add the equations.
$35x = 70$
$x = 2$

Replace x in equation (1).
$4x - 9y = -1$
$4(2) - 9y = -1$
$8 - 9y = -1$
$-9y = -9$
$y = 1$

The solution is (2, 1).

71. (1) $2.25x + 1.5y = 3$
(2) $1.75x + 2.25y = 1.25$

Multiply both sides of each equation by 100.
(1) $225x + 150y = 300$
(2) $175x + 225y = 125$

Eliminate y.
$3(225x + 150y) = 3(300)$
$-2(175x + 225y) = -2(125)$

$675x + 450y = 900$
$-350x - 450y = -250$

Add the equations.
$325x = 650$
$x = 2$

Replace x in equation (1).
$225x + 150y = 300$
$225(2) + 150y = 300$
$450 + 150y = 300$
$150y = -150$
$y = -1$

The solution is (2, −1).

73. (1) $1.6x - 0.9y + 0.3z = 2.9$
(2) $1.6x + 0.5y - 0.1z = 3.3$
(3) $0.8x - 0.7y + 0.1z = 1.5$

Multiply both sides of each equation by 10.
(1) $16x - 9y + 3z = 29$
(2) $16x + 5x - z = 33$
(3) $8x - 7y + z = 15$

Eliminate z. Add equations (2) and (3).
$16x + 5y - z = 33$
$8x - 7y + z = 15$

(4) $24x - 2y = 48$

Multiply equation (2) by 3 and add to equation (1).
$3(16x + 5y - z) = 3(33)$
$16x - 9y + 3z = 29$

$48x + 15y - 3z = 99$
$16x - 9y + 3z = 29$

(5) $64x + 6y = 128$

Multiply equation (4) by 3 and add to equation (5).
$3(24x - 2y) = 3(48)$
$64x + 6y = 128$

$72x - 6y = 144$
$64x + 6y = 128$

$136x = 272$
$x = 2$

Replace x by 2 in equation (4).
$24x - 2y = 48$
$24(2) - 2y = 48$
$48 - 2y = 48$
$-2y = 0$
$y = 0$

Replace x by 2 and y by 0 in equation (1).
$16x - 9y + 3z = 29$
$16(2) - 9(0) + 3z = 29$
$32 - 0 + 3z = 29$
$3z = -3$
$z = -1$

The solution is $(2, 0, -1)$.

75. Strategy
Substitute 3 for x, -2 for y, and 4 for z in the equations. Solve for A, B, and C.

Solution
$Ax + 3y + 2z = 8$
$A(3) + 3(-2) + 2(4) = 8$
$3A - 6 + 8 = 8$
$3A + 2 = 8$
$3A = 6$
$A = 2$

$2x + By - 3z = -12$
$2(3) + B(-2) - 3(4) = -12$
$6 - 2B - 12 = -12$
$-2B - 6 = -12$
$-2B = -6$
$B = 3$

$3x - 2y + Cz = 1$
$3(3) - 2(-2) + C(4) = 1$
$9 + 4 + 4C = 1$
$4C + 13 = 1$
$4C = -12$
$C = -3$

The value of A is 2. The value of B is 3. The value of C is -3.

Chapter 4: Systems Of Equations And Inequalities

77. Strategy
Solve a system of equations using x to represent the number of nickels, y to represent the number of dimes, and z to represent the number of quarters.

Solution
(1) $\quad x + y + z = 30$
(2) $\quad 5x + 10y + 25z = 325$

Eliminate x by multiplying equation (1) by (-5) and adding equation (2).
$-5x - 5y - 5z = -150$
$5x + 10y + 25z = 325$
(3) $\overline{\quad 5y + 20z = 175}$

Solve equation (3) for y in terms of z.
(3) $5y + 20z = 175$
$\quad 5y = -20z + 175$
$\quad y = -4z + 35$

Replace y by $-4z + 35$ in equation (1) and solve for x in terms of z.
$\quad x + y + z = 30$
$x + (-4z + 35) + z = 30$
$\quad x - 4z + 35 + z = 30$
$\quad x - 3z + 35 = 30$
$\quad x = 3z - 5$

The number of nickels is $3z - 5$, the number of dimes is $-4z + 35$, and the number of quarters is z, where $z = 2, 3, 4, 5, 6, 7,$ or 8. All other values of z make the number of nickels or dimes negative, which is not possible.

79.
(1) $\quad \dfrac{1}{x} + \dfrac{2}{y} = 3$

(2) $\quad \dfrac{1}{x} - \dfrac{3}{y} = -2$

Clear the fractions.
$xy\left(\dfrac{1}{x} + \dfrac{2}{y}\right) = xy \cdot 3$

$xy\left(\dfrac{1}{x} - \dfrac{3}{y}\right) = xy \cdot (-2)$

$\quad y + 2x = 3xy$
$\quad y - 3x = -2xy$

Eliminate y.
$y + 2x = 3xy$
$-y + 3x = 2xy$
$\quad 5x = 5xy$
$\quad y = 1$

Substitute y into equation (2).
$\dfrac{1}{x} - \dfrac{3}{y} = -2$

$\dfrac{1}{x} - \dfrac{3}{1} = -2$

$\dfrac{1}{x} = 1$

$x = 1$

The solution is $(1, 1)$.

81.
(1) $\quad \dfrac{3}{x} - \dfrac{5}{y} = -\dfrac{3}{2}$

(2) $\quad \dfrac{1}{x} - \dfrac{2}{y} = -\dfrac{2}{3}$

Clear fractions.
$2xy\left(\dfrac{3}{x} - \dfrac{5}{y}\right) = 2xy\left(-\dfrac{3}{2}\right)$

$3xy\left(\dfrac{1}{x} - \dfrac{2}{y}\right) = 3xy\left(-\dfrac{2}{3}\right)$

$\quad 6y - 10x = -3xy$
$\quad 3y - 6x = -2xy$

Eliminate y.
$\quad 6y - 10x = -3xy$
$-6y + 12x = 4xy$
$\quad 2x = xy$
$\quad y = 2$

Substitute y into equation (2).
$\dfrac{1}{x} - \dfrac{2}{y} = -\dfrac{2}{3}$

$\dfrac{1}{x} - \dfrac{2}{2} = -\dfrac{2}{3}$

$\dfrac{1}{x} = \dfrac{1}{3}$

$x = 3$

The solution is $(3, 2)$.

83. a. The graph of $x = 3$ in an xyz-coordinate system is a plane parallel to the yz plane at $x = 3$.

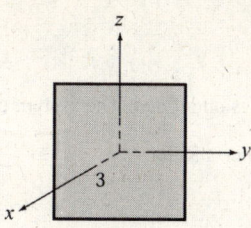

   b. The graph of $y = 4$ in an xyz-coordinate system is a plane parallel to the xz-plane at $y = 4$.

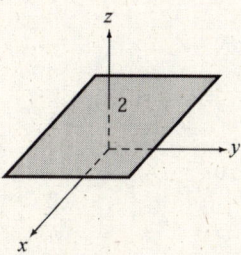

   c. The graph of $z = 2$ in an xyz-coordinate system is a plane at $z = 2$.

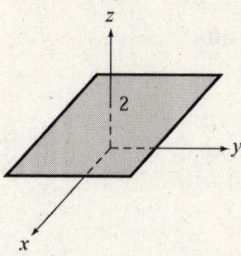

   d. The graph of $y = x$ in an xyz-coordinate system is a vertical plane perpendicular to the xy-plane and 45° from the xz- and yz-planes.

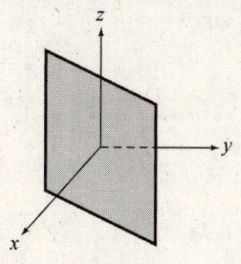

Section 4.3
Concept Review 4.3
1. Always true

3. Sometimes true
   A square matrix has the same number of rows and columns. A matrix may have different numbers of rows and columns.

5. Always true

Objective 4.3.1 Exercises
1. In matrix B, element b_{21} is $\underline{5}$.

3. Because matrix B has the same number of rows and columns, it is called a $\underline{\text{square}}$ matrix.

5. The determinant of B is written $\begin{vmatrix} 1 & -1 \\ 5 & 8 \end{vmatrix}$. The value of this determinant is $(1)(\underline{\,8\,}) - 5(\underline{-1}) = \underline{13}$.

7. $\begin{vmatrix} 2 & -1 \\ 3 & 4 \end{vmatrix} = 2(4) - 3(-1) = 8 + 3 = 11$

9. $\begin{vmatrix} 6 & -2 \\ -3 & 4 \end{vmatrix} = 6(4) - (-3)(-2) = 24 - 6 = 18$

11. $\begin{vmatrix} 3 & 6 \\ 2 & 4 \end{vmatrix} = 3(4) - 2(6) = 12 - 12 = 0$

13. $\begin{vmatrix} 1 & -1 & 2 \\ 3 & 2 & 1 \\ 1 & 0 & 4 \end{vmatrix} = 1\begin{vmatrix} 2 & 1 \\ 0 & 4 \end{vmatrix} + 1\begin{vmatrix} 3 & 1 \\ 1 & 4 \end{vmatrix} + 2\begin{vmatrix} 3 & 2 \\ 1 & 0 \end{vmatrix}$
    $= 1(8 - 0) + 1(12 - 1) + 2(0 - 2)$
    $= 8 + 11 - 4 = 15$

15. $\begin{vmatrix} 3 & -1 & 2 \\ 0 & 1 & 2 \\ 3 & 2 & -2 \end{vmatrix} = 3\begin{vmatrix} 1 & 2 \\ 2 & -2 \end{vmatrix} + 1\begin{vmatrix} 0 & 2 \\ 3 & -2 \end{vmatrix} + 2\begin{vmatrix} 0 & 1 \\ 3 & 2 \end{vmatrix}$
    $= 3(-2 - 4) + 1(0 - 6) + 2(0 - 3)$
    $= 3(-6) + 1(-6) + 2(-3)$
    $= -18 - 6 - 6 = -30$

17. $\begin{vmatrix} 4 & 2 & 6 \\ -2 & 1 & 1 \\ 2 & 1 & 3 \end{vmatrix} = 4\begin{vmatrix} 1 & 1 \\ 1 & 3 \end{vmatrix} - 2\begin{vmatrix} -2 & 1 \\ 2 & 3 \end{vmatrix} + 6\begin{vmatrix} -2 & 1 \\ 2 & 1 \end{vmatrix}$
    $= 4(3 - 1) - 2(-6 - 2) + 6(-2 - 2)$
    $= 4(2) - 2(-8) + 6(-4)$
    $= 8 + 16 - 24 = 0$

Objective 4.3.2 Exercises
19. a. The coefficient determinant for the system is $D = \begin{vmatrix} 2 & -3 \\ -5 & 4 \end{vmatrix}$.

    b. The value of the coefficient determinant is $D = \underline{-7}$.

100 Chapter 4: Systems Of Equations And Inequalities

21. a. The numerator determinant for y is
$$D_y = \begin{vmatrix} 2 & 5 \\ -5 & -2 \end{vmatrix}.$$
    b. The value of the numerator determinant for y is $D_y = \underline{21}$.

23. $2x - 5y = 26$
$5x + 3y = 3$
$$D = \begin{vmatrix} 2 & -5 \\ 5 & 3 \end{vmatrix} = 31, \quad D_x = \begin{vmatrix} 26 & -5 \\ 3 & 3 \end{vmatrix} = 93,$$
$$D_y = \begin{vmatrix} 2 & 26 \\ 5 & 3 \end{vmatrix} = -124$$
$$x = \frac{D_x}{D} = \frac{93}{31} = 3 \qquad y = \frac{D_y}{D} = \frac{-124}{31} = -4$$
The solution is $(3, -4)$.

25. $x - 4y = 8$
$3x + 7y = 5$
$$D = \begin{vmatrix} 1 & -4 \\ 3 & 7 \end{vmatrix} = 19, \quad D_x = \begin{vmatrix} 8 & -4 \\ 5 & 7 \end{vmatrix} = 76,$$
$$D_y = \begin{vmatrix} 1 & 8 \\ 3 & 5 \end{vmatrix} = -19$$
$$x = \frac{D_x}{D} = \frac{76}{19} = 4 \qquad y = \frac{D_y}{D} = \frac{-19}{19} = -1$$
The solution is $(4, -1)$.

27. $2x + 3y = 4$
$6x - 12y = -5$
$$D = \begin{vmatrix} 2 & 3 \\ 6 & -12 \end{vmatrix} = -42, \quad D_x = \begin{vmatrix} 4 & 3 \\ -5 & -12 \end{vmatrix} = -33,$$
$$D_y = \begin{vmatrix} 2 & 4 \\ 6 & -5 \end{vmatrix} = -34$$
$$x = \frac{D_x}{D} = \frac{-33}{-42} = \frac{11}{14} \qquad y = \frac{D_y}{D} = \frac{-34}{-42} = \frac{17}{21}$$
The solution is $\left(\frac{11}{14}, \frac{17}{21}\right)$.

29. $2x + 5y = 6$
$6x - 2y = 1$
$$D = \begin{vmatrix} 2 & 5 \\ 6 & -2 \end{vmatrix} = -34, \quad D_x = \begin{vmatrix} 6 & 5 \\ 1 & -2 \end{vmatrix} = -17,$$
$$D_y = \begin{vmatrix} 2 & 6 \\ 6 & 1 \end{vmatrix} = -34$$
$$x = \frac{D_x}{D} = \frac{-17}{-34} = \frac{1}{2} \qquad y = \frac{D_y}{D} = \frac{-34}{-34} = 1$$
The solution is $\left(\frac{1}{2}, 1\right)$.

31. $-2x + 3y = 7$
$4x - 6y = 9$
$$D = \begin{vmatrix} -2 & 3 \\ 4 & -6 \end{vmatrix} = 0$$
Since $D = 0$, $\frac{D_x}{D}$ is undefined. The system cannot be solved by Cramer's Rule.

33. $2x - 5y = -2$
$3x - 7y = -3$
$$D = \begin{vmatrix} 2 & -5 \\ 3 & -7 \end{vmatrix} = 1, \quad D_x = \begin{vmatrix} -2 & -5 \\ -3 & -7 \end{vmatrix} = -1,$$
$$D_y = \begin{vmatrix} 2 & -2 \\ 3 & -3 \end{vmatrix} = 0$$
$$x = \frac{D_x}{D} = \frac{-1}{1} = -1 \qquad y = \frac{D_y}{D} = \frac{0}{1} = 0$$
The solution is $(-1, 0)$.

35. $2x - y + 3z = 9$
$x + 4y + 4z = 5$
$3x + 2y + 2z = 5$
$$D = \begin{vmatrix} 2 & -1 & 3 \\ 1 & 4 & 4 \\ 3 & 2 & 2 \end{vmatrix} = -40,$$
$$D_x = \begin{vmatrix} 9 & -1 & 3 \\ 5 & 4 & 4 \\ 5 & 2 & 2 \end{vmatrix} = -40,$$
$$D_y = \begin{vmatrix} 2 & 9 & 3 \\ 1 & 5 & 4 \\ 3 & 5 & 2 \end{vmatrix} = 40,$$
$$D_z = \begin{vmatrix} 2 & -1 & 9 \\ 1 & 4 & 5 \\ 3 & 2 & 5 \end{vmatrix} = -80$$
$$x = \frac{D_x}{D} = \frac{-40}{-40} = 1 \qquad y = \frac{D_y}{D} = \frac{40}{-40} = -1$$
$$z = \frac{D_z}{D} = \frac{-80}{-40} = 2$$
The solution is $(1, -1, 2)$.

37.
$3x - y + z = 11$
$x + 4y - 2z = -12$
$2x + 2y - z = -3$

$D = \begin{vmatrix} 3 & -1 & 1 \\ 1 & 4 & -2 \\ 2 & 2 & -1 \end{vmatrix} = -3,$

$D_x = \begin{vmatrix} 11 & -1 & 1 \\ -12 & 4 & -2 \\ -3 & 2 & -1 \end{vmatrix} = -6,$

$D_y = \begin{vmatrix} 3 & 11 & 1 \\ 1 & -12 & -2 \\ 2 & -3 & -1 \end{vmatrix} = 6,$

$D_z = \begin{vmatrix} 3 & -1 & 11 \\ 1 & 4 & -12 \\ 2 & 2 & -3 \end{vmatrix} = -9$

$x = \dfrac{D_x}{D} = \dfrac{-6}{-3} = 2 \qquad y = \dfrac{D_y}{D} = \dfrac{6}{-3} = -2$

$z = \dfrac{D_z}{D} = \dfrac{-9}{-3} = 3$

The solution is $(2, -2, 3)$.

39.
$4x - 2y + 6z = 1$
$3x + 4y + 2z = 1$
$2x - y + 3z = 2$

$D = \begin{vmatrix} 4 & -2 & 6 \\ 3 & 4 & 2 \\ 2 & -1 & 3 \end{vmatrix} = 0$

Since $D = 0$, $\dfrac{D_x}{D}$ is undefined. The system cannot be solved by Cramer's Rule.

41.
$5x - 4y + 2z = 4$
$3x - 5y + 3z = -4$
$3x + y - 5z = 12$

$D = \begin{vmatrix} 5 & -4 & 2 \\ 3 & -5 & 3 \\ 3 & 1 & -5 \end{vmatrix} = 50,$

$D_x = \begin{vmatrix} 4 & -4 & 2 \\ -4 & -5 & 3 \\ 12 & 1 & -5 \end{vmatrix} = 136,$

$D_y = \begin{vmatrix} 5 & 4 & 2 \\ 3 & -4 & 3 \\ 3 & 12 & -5 \end{vmatrix} = 112,$

$D_z = \begin{vmatrix} 5 & -4 & 4 \\ 3 & -5 & -4 \\ 3 & 1 & 12 \end{vmatrix} = -16$

$x = \dfrac{D_x}{D} = \dfrac{136}{50} = \dfrac{68}{25} \qquad y = \dfrac{D_y}{D} = \dfrac{112}{50} = \dfrac{56}{25}$

$z = \dfrac{D_z}{D} = \dfrac{-16}{50} = -\dfrac{8}{25}$

The solution is $\left(\dfrac{68}{25}, \dfrac{56}{25}, -\dfrac{8}{25}\right)$.

Objective 4.3.3 Exercises

43. The elements of the main diagonal of the augmented matrix are 1, –3, and 2.

45. The constant for the third equation of the system of equations associated with the augmented matrix is 9

47.
$\begin{bmatrix} -3 & 6 & -9 & | & -6 \\ 2 & -3 & 1 & | & 1 \\ 3 & -1 & 2 & | & 9 \end{bmatrix}$

49. The equation that corresponds to the first row is $x - 2y + 3z = 2$. The equation that corresponds to the second row is $y - 5z = -3$. The equation that corresponds to the third row is $z = 1$.

51.
$\begin{bmatrix} 2 & -4 & | & 1 \\ 3 & -7 & | & -1 \end{bmatrix}$

$\dfrac{1}{2}R_1 \to \begin{bmatrix} 1 & -2 & | & \dfrac{1}{2} \\ 3 & -7 & | & -1 \end{bmatrix}$

$-3R_1 + R_2 \to \begin{bmatrix} 1 & -2 & | & \dfrac{1}{2} \\ 0 & -1 & | & -\dfrac{5}{2} \end{bmatrix}$

$-1R_2 \to \begin{bmatrix} 1 & -2 & | & \dfrac{1}{2} \\ 0 & 1 & | & \dfrac{5}{2} \end{bmatrix}$

53.
$\begin{bmatrix} 5 & -2 & | & 3 \\ -7 & 3 & | & 1 \end{bmatrix}$

$\begin{bmatrix} 5 & -2 & | & 3 \\ -7 & 3 & | & 1 \end{bmatrix}$

$\dfrac{1}{5}R_1 \to \begin{bmatrix} 1 & -\dfrac{2}{5} & | & \dfrac{3}{5} \\ -7 & 3 & | & 1 \end{bmatrix}$

$7R_1 + R_2 \to \begin{bmatrix} 1 & -\dfrac{2}{5} & | & \dfrac{3}{5} \\ 0 & \dfrac{1}{5} & | & \dfrac{26}{5} \end{bmatrix}$

$5R_2 \to \begin{bmatrix} 1 & -\dfrac{2}{5} & | & \dfrac{3}{5} \\ 0 & 1 & | & 26 \end{bmatrix}$

55.
$$\begin{bmatrix} 1 & 4 & 1 & | & -2 \\ 3 & 11 & -1 & | & 2 \\ 2 & 3 & 1 & | & 4 \end{bmatrix}$$

$\begin{matrix} -3R_1 + R_2 \to \\ -2R_1 + R_3 \to \end{matrix} \begin{bmatrix} 1 & 4 & 1 & | & -2 \\ 0 & -1 & -4 & | & 8 \\ 0 & -5 & -1 & | & 8 \end{bmatrix}$

$-1R_2 \to \begin{bmatrix} 1 & 4 & 1 & | & -2 \\ 0 & 1 & 4 & | & -8 \\ 0 & -5 & -1 & | & 8 \end{bmatrix}$

$5R_2 + R_3 \to \begin{bmatrix} 1 & 4 & 1 & | & -2 \\ 0 & 1 & 4 & | & -8 \\ 0 & 0 & 19 & | & -32 \end{bmatrix}$

$\frac{1}{19}R_3 \to \begin{bmatrix} 1 & 4 & 1 & | & -2 \\ 0 & 1 & 4 & | & -8 \\ 0 & 0 & 1 & | & -\frac{32}{19} \end{bmatrix}$

57.
$$\begin{bmatrix} -2 & 6 & -1 & | & 3 \\ 1 & -2 & 2 & | & 1 \\ 3 & -6 & 7 & | & 6 \end{bmatrix}$$

$-\frac{1}{2}R_1 \to \begin{bmatrix} 1 & -3 & \frac{1}{2} & | & -\frac{3}{2} \\ 1 & -2 & 2 & | & 1 \\ 3 & -6 & 7 & | & 6 \end{bmatrix}$

$\begin{matrix} -1R_1 + R_2 \to \\ -3R_1 + R_3 \to \end{matrix} \begin{bmatrix} 1 & -3 & \frac{1}{2} & | & -\frac{3}{2} \\ 0 & 1 & \frac{3}{2} & | & \frac{5}{2} \\ 0 & 3 & \frac{11}{2} & | & \frac{21}{2} \end{bmatrix}$

$-3R_2 + R_3 \to \begin{bmatrix} 1 & -3 & \frac{1}{2} & | & -\frac{3}{2} \\ 0 & 1 & \frac{3}{2} & | & \frac{5}{2} \\ 0 & 0 & 1 & | & 3 \end{bmatrix}$

59. Writing the augmented matrix as a system of equations,
$x - y + 3z = -2$
$y - z = 1$
$z = 3$
Using substitution,
$y - 3 = 1$
$y = 4$
$x - 4 + 3(3) = -2$
$x = -7$
The solution is $(-7, 4, 3)$.

61. $3x + y = 6$
$2x - y = -1$
$$\begin{bmatrix} 3 & 1 & | & 6 \\ 2 & -1 & | & -1 \end{bmatrix}$$

$\frac{1}{3}R_1 \to \begin{bmatrix} 1 & \frac{1}{3} & | & 2 \\ 2 & -1 & | & -1 \end{bmatrix}$

$-2R_1 + R_2 \to \begin{bmatrix} 1 & \frac{1}{3} & | & 2 \\ 0 & -\frac{5}{3} & | & -5 \end{bmatrix}$

$-\frac{3}{5}R_2 \to \begin{bmatrix} 1 & \frac{1}{3} & | & 2 \\ 0 & 1 & | & 3 \end{bmatrix}$

$x + \left(\frac{1}{3}\right)y = 2$
$y = 3$
$x + \left(\frac{1}{3}\right)(3) = 2$
$x + 1 = 2$
$x = 1$
The solution is $(1, 3)$.

63. $x - 3y = 8$
$3x - y = 0$
$$\begin{bmatrix} 1 & -3 & | & 8 \\ 3 & -1 & | & 0 \end{bmatrix}$$

$-3R_1 + R_2 \to \begin{bmatrix} 1 & -3 & | & 8 \\ 0 & 8 & | & -24 \end{bmatrix}$

$\frac{1}{8}R_2 \to \begin{bmatrix} 1 & 3 & | & 8 \\ 0 & 1 & | & -3 \end{bmatrix}$

$x - 3y = 8$
$y = -3$
$x - 3(-3) = 8$
$x + 9 = 8$
$x = -1$
The solution is $(-1, -3)$.

65. $y = 4x - 10$
$2y = 5x - 11$

$4x - y = 10$
$5x - 2y = 11$

$\begin{bmatrix} 4 & -1 & | & 10 \\ 5 & -2 & | & 11 \end{bmatrix}$

$\frac{1}{4}R_1 \to \begin{bmatrix} 1 & -\frac{1}{4} & | & \frac{5}{2} \\ 5 & -2 & | & 11 \end{bmatrix}$

$-5R_1 + R_2 \to \begin{bmatrix} 1 & -\frac{1}{4} & | & \frac{5}{2} \\ 0 & -\frac{3}{4} & | & -\frac{3}{2} \end{bmatrix}$

$-\frac{4}{3}R_2 \to \begin{bmatrix} 1 & -\frac{1}{4} & | & \frac{5}{2} \\ 0 & 1 & | & 2 \end{bmatrix}$

$x - \left(\frac{1}{4}\right)y = \frac{5}{2}$
$y = 2$

$x - \left(\frac{1}{4}\right)(2) = \frac{5}{2}$
$x - \frac{1}{2} = \frac{5}{2}$
$x = 3$

The solution is (3, 2).

67. $2x - y = -4$
$y = 2x - 8$

$2x - y = -4$
$2x - y = 8$

$\begin{bmatrix} 2 & -1 & | & -4 \\ 2 & -1 & | & 8 \end{bmatrix}$

$\frac{1}{2}R_1 \to \begin{bmatrix} 1 & -\frac{1}{2} & | & -2 \\ 2 & -1 & | & 8 \end{bmatrix}$

$-2R_1 + R_2 \to \begin{bmatrix} 1 & -\frac{1}{2} & | & -2 \\ 0 & 0 & | & 12 \end{bmatrix}$

$x - \frac{1}{2}y = -2$
$0 = 12$

This is not a true equation. The system of equations is inconsistent.

69. $4x - 3y = -14$
$3x + 4y = 2$

$\begin{bmatrix} 4 & -3 & | & -14 \\ 3 & 4 & | & 2 \end{bmatrix}$

$\frac{1}{4}R_1 \to \begin{bmatrix} 1 & -\frac{3}{4} & | & -\frac{7}{2} \\ 3 & 4 & | & 2 \end{bmatrix}$

$-3R_1 + R_2 \to \begin{bmatrix} 1 & -\frac{3}{4} & | & -\frac{7}{2} \\ 0 & \frac{25}{4} & | & \frac{25}{2} \end{bmatrix}$

$\frac{4}{25}R_2 \to \begin{bmatrix} 1 & -\frac{3}{4} & | & -\frac{7}{2} \\ 0 & 1 & | & 2 \end{bmatrix}$

$x - \left(\frac{3}{4}\right)y = -\frac{7}{2}$
$y = 2$

$x - \left(\frac{3}{4}\right)(2) = -\frac{7}{2}$
$x - \frac{3}{2} = -\frac{7}{2}$
$x = -2$

The solution is (–2, 2).

71. $5x + 4y + 3z = -9$
$x - 2y + 2z = -6$
$x - y - z = 3$

$$\begin{bmatrix} 5 & 4 & 3 & | & -9 \\ 1 & -2 & 2 & | & -6 \\ 1 & -1 & -1 & | & 3 \end{bmatrix}$$

$R_1 \leftrightarrow R_2 \begin{bmatrix} 1 & -2 & 2 & | & -6 \\ 5 & 4 & 3 & | & -9 \\ 1 & -1 & -1 & | & 3 \end{bmatrix}$

$\begin{matrix} -5R_1 + R_2 \to \\ -1R_1 + R_3 \to \end{matrix} \begin{bmatrix} 1 & -2 & 2 & | & -6 \\ 0 & 14 & -7 & | & 21 \\ 0 & 1 & -3 & | & 9 \end{bmatrix}$

$\frac{1}{14}R_2 \to \begin{bmatrix} 1 & -2 & 2 & | & -6 \\ 0 & 1 & -\frac{1}{2} & | & \frac{3}{2} \\ 0 & 1 & -3 & | & 9 \end{bmatrix}$

$-1R_2 + R_3 \to \begin{bmatrix} 1 & -2 & 2 & | & -6 \\ 0 & 1 & -\frac{1}{2} & | & \frac{3}{2} \\ 0 & 0 & -\frac{5}{2} & | & \frac{15}{2} \end{bmatrix}$

$-\frac{2}{5}R_3 \to \begin{bmatrix} 1 & -2 & 2 & | & -6 \\ 0 & 1 & -\frac{1}{2} & | & \frac{3}{2} \\ 0 & 0 & 1 & | & -3 \end{bmatrix}$

$x - 2y + 2z = -6$
$y - \left(\frac{1}{2}\right)z = \frac{3}{2}$
$z = -3$

$y - \left(\frac{1}{2}\right)(-3) = \frac{3}{2}$
$y + \frac{3}{2} = \frac{3}{2}$
$y = 0$

$x - 2(0) + 2(-3) = -6$
$x - 6 = -6$
$x = 0$

The solution is $(0, 0, -3)$.

73. $5x - 5y + 2z = 8$
$2x + 3y - z = 0$
$x + 2y - z = 0$

$$\begin{bmatrix} 5 & -5 & 2 & | & 8 \\ 2 & 3 & -1 & | & 0 \\ 1 & 2 & -1 & | & 0 \end{bmatrix}$$

$R_1 \leftrightarrow R_3 \begin{bmatrix} 1 & 2 & -1 & | & 0 \\ 2 & 3 & -1 & | & 0 \\ 5 & -5 & 2 & | & 8 \end{bmatrix}$

$\begin{matrix} -2R_1 + R_2 \to \\ -5R_1 + R_3 \to \end{matrix} \begin{bmatrix} 1 & 2 & -1 & | & 0 \\ 0 & -1 & 1 & | & 0 \\ 0 & -15 & 7 & | & 8 \end{bmatrix}$

$-1R_2 \to \begin{bmatrix} 1 & 2 & -1 & | & 0 \\ 0 & 1 & -1 & | & 0 \\ 0 & -15 & 7 & | & 8 \end{bmatrix}$

$15R_2 + R_3 \to \begin{bmatrix} 1 & 2 & -1 & | & 0 \\ 0 & 1 & -1 & | & 0 \\ 0 & 0 & -8 & | & 8 \end{bmatrix}$

$-\frac{1}{8}R_3 \to \begin{bmatrix} 1 & 2 & -1 & | & 0 \\ 0 & 1 & -1 & | & 0 \\ 0 & 0 & 1 & | & -1 \end{bmatrix}$

$x + 2y - z = 0$
$y - z = 0$
$z = -1$

$y - (-1) = 0$
$y + 1 = 0$
$y = -1$

$x + 2(-1) - (-1) = 0$
$x - 2 + 1 = 0$
$x - 1 = 0$
$x = 1$

The solution is $(1, -1, -1)$.

75.
$$2x + 3y + z = 5$$
$$3x + 3y + 3z = 10$$
$$4x + 6y + 2z = 5$$

$$\begin{bmatrix} 2 & 1 & 3 & | & 5 \\ 3 & 3 & 3 & | & 10 \\ 4 & 6 & 2 & | & 5 \end{bmatrix}$$

$$\tfrac{1}{2}R_1 \to \begin{bmatrix} 1 & \tfrac{1}{2} & \tfrac{3}{2} & | & \tfrac{5}{2} \\ 3 & 3 & 3 & | & 10 \\ 4 & 6 & 2 & | & 5 \end{bmatrix}$$

$$\begin{aligned} -3R_1 + R_2 &\to \\ -4R_1 + R_2 &\to \end{aligned} \begin{bmatrix} 1 & \tfrac{1}{2} & \tfrac{3}{2} & | & \tfrac{5}{2} \\ 0 & \tfrac{3}{2} & \tfrac{3}{2} & | & \tfrac{5}{2} \\ 0 & 0 & 0 & | & -5 \end{bmatrix}$$

$$x + \left(\tfrac{3}{2}\right)y + \left(\tfrac{1}{2}\right)z = \tfrac{5}{2}$$
$$-\left(\tfrac{3}{2}\right)y + \left(\tfrac{3}{2}\right)z = \tfrac{5}{2}$$
$$0 = -5$$

This is not a true equation. The system of equations is inconsistent.

77.
$$3x + 2y + 3z = 2$$
$$6x - 2y + z = 1$$
$$3x + 4y + 2z = 3$$

$$\begin{bmatrix} 3 & 2 & 3 & | & 2 \\ 6 & -2 & 1 & | & 1 \\ 3 & 4 & 2 & | & 3 \end{bmatrix}$$

$$\tfrac{1}{3}R_1 \to \begin{bmatrix} 1 & \tfrac{2}{3} & 1 & | & \tfrac{2}{3} \\ 6 & -2 & 1 & | & 1 \\ 3 & 4 & 2 & | & 3 \end{bmatrix}$$

$$\begin{aligned} -6R_1 + R_2 &\to \\ -3R_1 + R_3 &\to \end{aligned} \begin{bmatrix} 1 & \tfrac{2}{3} & 1 & | & \tfrac{2}{3} \\ 0 & -6 & -5 & | & -3 \\ 0 & 2 & -1 & | & 1 \end{bmatrix}$$

$$-\tfrac{1}{6}R_2 \to \begin{bmatrix} 1 & \tfrac{2}{3} & 1 & | & \tfrac{2}{3} \\ 0 & 1 & \tfrac{5}{6} & | & \tfrac{1}{2} \\ 0 & 2 & -1 & | & 1 \end{bmatrix}$$

$$-2R_2 + R_3 \to \begin{bmatrix} 1 & \tfrac{2}{3} & 1 & | & \tfrac{2}{3} \\ 0 & 1 & \tfrac{5}{6} & | & \tfrac{1}{2} \\ 0 & 0 & -\tfrac{8}{3} & | & 0 \end{bmatrix}$$

$$-\tfrac{3}{8}R_3 \to \begin{bmatrix} 1 & \tfrac{2}{3} & 1 & | & \tfrac{2}{3} \\ 0 & 1 & \tfrac{5}{6} & | & \tfrac{1}{2} \\ 0 & 0 & 1 & | & 0 \end{bmatrix}$$

$$x + \left(\tfrac{2}{3}\right)y + z = \tfrac{2}{3}$$
$$y + \left(\tfrac{5}{6}\right)z = \tfrac{1}{2}$$
$$z = 0$$

$$y + \left(\tfrac{5}{6}\right)(0) = \tfrac{1}{2}$$
$$y = \tfrac{1}{2}$$

$$x + \left(\tfrac{2}{3}\right)\left(\tfrac{1}{2}\right) + 0 = \tfrac{2}{3}$$
$$x + \tfrac{1}{3} = \tfrac{2}{3}$$
$$x = \tfrac{1}{3}$$

The solution is $\left(\tfrac{1}{3}, \tfrac{1}{2}, 0\right)$.

79.
$$5x - 5y - 5z = 2$$
$$5x + 5y - 5z = 6$$
$$10x + 10y + 5z = 3$$

$$\begin{bmatrix} 5 & -5 & -5 & | & 2 \\ 5 & 5 & -5 & | & 6 \\ 10 & 10 & 5 & | & 3 \end{bmatrix}$$

$\frac{1}{5}R_1 \to \begin{bmatrix} 1 & -1 & -1 & | & \frac{2}{5} \\ 5 & 5 & -5 & | & 6 \\ 10 & 10 & 5 & | & 3 \end{bmatrix}$

$\begin{matrix} -5R_1 + R_2 \to \\ -10R_1 + R_3 \to \end{matrix} \begin{bmatrix} 1 & -1 & -1 & | & \frac{2}{5} \\ 0 & 10 & 0 & | & 4 \\ 0 & 20 & 15 & | & -1 \end{bmatrix}$

$\frac{1}{10}R_2 \to \begin{bmatrix} 1 & -1 & -1 & | & \frac{2}{5} \\ 0 & 1 & 0 & | & \frac{2}{5} \\ 0 & 20 & 15 & | & -1 \end{bmatrix}$

$-20R_2 + R_3 \to \begin{bmatrix} 1 & -1 & -1 & | & \frac{2}{5} \\ 0 & 1 & 0 & | & \frac{2}{5} \\ 0 & 0 & 15 & | & -9 \end{bmatrix}$

$\frac{1}{15}R_3 \to \begin{bmatrix} 1 & -1 & -1 & | & \frac{2}{5} \\ 0 & 1 & 0 & | & \frac{2}{5} \\ 0 & 0 & 1 & | & -\frac{3}{5} \end{bmatrix}$

$$x - y - z = \frac{2}{5}$$
$$y = \frac{2}{5}$$
$$z = -\frac{3}{5}$$

$$x - \frac{2}{5} - \left(-\frac{3}{5}\right) = \frac{2}{5}$$
$$x - \frac{2}{5} + \frac{3}{5} = \frac{2}{5}$$
$$x + \frac{1}{5} = \frac{2}{5}$$
$$x = \frac{1}{5}$$

The solution is $\left(\frac{1}{5}, \frac{2}{5}, -\frac{3}{5}\right)$.

81.
$$4x + 4y - 3z = 3$$
$$8x + 2y + 3z = 0$$
$$4x - 4y + 6z = -3$$

$$\begin{bmatrix} 4 & 4 & -3 & | & 3 \\ 8 & 2 & 3 & | & 0 \\ 4 & -4 & 6 & | & -3 \end{bmatrix}$$

$\frac{1}{4}R_1 \to \begin{bmatrix} 1 & 1 & -\frac{3}{4} & | & \frac{3}{4} \\ 8 & 2 & 3 & | & 0 \\ 4 & -4 & 6 & | & -3 \end{bmatrix}$

$\begin{matrix} -8R_1 + R_2 \to \\ -4R_1 + R_3 \to \end{matrix} \begin{bmatrix} 1 & 1 & -\frac{3}{4} & | & \frac{3}{4} \\ 0 & -6 & 9 & | & -6 \\ 0 & -8 & 9 & | & -6 \end{bmatrix}$

$-\frac{1}{6}R_2 \to \begin{bmatrix} 1 & 1 & -\frac{3}{4} & | & \frac{3}{4} \\ 0 & 1 & -\frac{3}{2} & | & 1 \\ 0 & -8 & 9 & | & -6 \end{bmatrix}$

$8R_2 + R_3 \to \begin{bmatrix} 1 & 1 & -\frac{3}{4} & | & \frac{3}{4} \\ 0 & 1 & -\frac{3}{2} & | & 1 \\ 0 & 0 & -3 & | & 2 \end{bmatrix}$

$-\frac{1}{3}R_3 \to \begin{bmatrix} 1 & 1 & -\frac{3}{4} & | & \frac{3}{4} \\ 0 & 1 & -\frac{3}{2} & | & 1 \\ 0 & 0 & 1 & | & -\frac{2}{3} \end{bmatrix}$

$$x + y - \left(\frac{3}{4}\right)z = \frac{3}{4}$$
$$y - \left(\frac{3}{2}\right)z = 1$$
$$z = -\frac{2}{3}$$

$$y - \left(\frac{3}{2}\right)\left(-\frac{2}{3}\right) = 1$$
$$y + 1 = 1$$
$$y = 0$$

$$x + 0 - \left(\frac{3}{4}\right)\left(-\frac{2}{3}\right) = \frac{3}{4}$$
$$x + \frac{1}{2} = \frac{3}{4}$$
$$x = \frac{1}{4}$$

The solution is $\left(\frac{1}{4}, 0, -\frac{2}{3}\right)$.

Applying Concepts 4.3

83.
$$\begin{vmatrix} 1 & 0 & 2 \\ 4 & 3 & -1 \\ 0 & 2 & x \end{vmatrix} = -24$$

Expand the minors of the first row.
$$1(3x+2) - 0 + 2(8-0) = -24$$
$$3x + 2 + 16 = -24$$
$$3x + 18 = -24$$
$$3x = -42$$
$$x = -14$$

The solution is −14.

85. If all the elements in one row or one column of a 2×2 matrix are zeros, the value of the determinant of the matrix is 0.
For example,
$$\begin{vmatrix} a_1 & 0 \\ b_1 & 0 \end{vmatrix} = a_1(0) - (0)b_1 = 0 - 0 = 0$$

87. a.
$$\begin{vmatrix} x & x & a \\ y & y & b \\ z & z & c \end{vmatrix}$$

Expand by minors of the first row.
$$x(cy - bz) - x(cy - bz) + a(yz - yz)$$
$$= cxy - bxz - cxy + bxz + ayz - ayz$$
$$= 0$$

b. If two columns of a determinant contain identical elements, the value of the determinant is 0.

89. $A = \dfrac{1}{2}\left\{ \begin{vmatrix} x_1 & x_2 \\ y_1 & y_2 \end{vmatrix} + \begin{vmatrix} x_2 & x_3 \\ y_2 & y_3 \end{vmatrix} + \begin{vmatrix} x_3 & x_4 \\ y_3 & y_4 \end{vmatrix} + \ldots + \begin{vmatrix} x_n & x_1 \\ y_n & y_1 \end{vmatrix} \right\}$

$A = \dfrac{1}{2}\left\{ \begin{vmatrix} 9 & 26 \\ -3 & 6 \end{vmatrix} + \begin{vmatrix} 26 & 18 \\ 6 & 21 \end{vmatrix} + \begin{vmatrix} 18 & 16 \\ 21 & 10 \end{vmatrix} + \begin{vmatrix} 16 & 1 \\ 10 & 11 \end{vmatrix} + \begin{vmatrix} 1 & 9 \\ 11 & -3 \end{vmatrix} \right\}$

$A = \dfrac{1}{2}(132 + 438 - 156 + 166 - 102) = 239$

The area of the polygon is 239 ft^2.

91. The last row of this matrix indicates $0x + 0y + 0z = -3$. This equation is not true for any values of x, y, and z. Therefore, the system of equations has no solution.

Section 4.4

Concept Review 4.4

1. Never true
The rate up the river is $(y - x)$ mph.

3. Always true

Objective 4.4.1 Exercises

1.

	Rate, r	•	Time, t	=	Distance, d
With current	$b + c$	•	3	=	$3(b + c)$
Against current	$B - c$	•	5	=	$5(b - c)$

3. n is less than m

5. Strategy
Rate of the plane in calm air: p
Rate of the wind: w

	Rate	Time	Distance
With wind	$p + w$	2	$2(p + w)$
Against wind	$p - w$	2	$2(p - w)$

• The distance traveled with the wind is 320 mi.
The distance traveled against the wind is 280 mi.
$2(p + w) = 320$
$2(p - w) = 280$

Solution
$2(p + w) = 320$
$2(p - w) = 280$

$\dfrac{1}{2} \cdot 2(p + w) = \dfrac{1}{2} \cdot 320$
$\dfrac{1}{2} \cdot 2(p - w) = \dfrac{1}{2} \cdot 280$

$p + w = 160$
$p - w = 140$

$2p = 300$
$p = 150$

$p + w = 160$
$150 + w = 160$
$w = 10$

The rate of the plane in calm air is 150 mph.
The rate of the wind is 10 mph.

Chapter 4: Systems Of Equations And Inequalities

7. Strategy
Rate of the cabin cruiser in calm water: x
Rate of the current: y

	Rate	Time	Distance
With current	$x+y$	3	$3(x+y)$
Against current	$x-y$	4	$4(x-y)$

- The distance traveled with the current is 48 mi.
- The distance traveled against the current is 48 mi.

$3(x+y) = 48$
$4(x-y) = 48$

Solution
$3(x+y) = 48$
$4(x-y) = 48$

$\frac{1}{3} \cdot 3(x+y) = \frac{1}{3} \cdot 48$
$\frac{1}{4} \cdot 4(x-y) = \frac{1}{4} \cdot 48$

$x + y = 16$
$x - y = 12$

$2x = 28$
$x = 14$

$x + y = 16$
$14 + y = 16$
$y = 2$

The rate of the cabin cruiser in calm water is 14 mph. The rate of the current is 2 mph.

9. Strategy
Rate of the plane in calm air: x
Rate of the wind: y

	Rate	Time	Distance
With wind	$x+y$	2.5	$2.5(x+y)$
Against wind	$x-y$	3	$3(x-y)$

- The distance traveled with the wind is 450 mi.
- The distance traveled against the wind is 450 mi.

$2.5(x+y) = 450$
$3(x-y) = 450$

Solution
$2.5(x+y) = 450$
$3(x-y) = 450$

$\frac{1}{2.5} \cdot 2.5(x+y) = \frac{1}{2.5} \cdot 450$
$\frac{1}{3} \cdot 3(x-y) = \frac{1}{3} \cdot 450$

$x + y = 180$
$x - y = 150$

$2x = 330$
$x = 165$

$x + y = 180$
$165 + y = 180$
$y = 15$

The rate of the plane in calm air is 165 mph.
The rate of the wind is 15 mph.

11. Strategy
The rate of the boat in calm water: x
The rate of the current: y

	Rate	Time	Distance
With current	$x+y$	4	$4(x+y)$
Against current	$x-y$	4	$4(x-y)$

- The distance traveled with the current is 88 km.
- The distance traveled against the current is 64 km.

$4(x+y) = 88$
$4(x-y) = 64$

Solution
$4(x+y) = 88$
$4(x-y) = 64$

$\frac{1}{4} \cdot 4(x+y) = \frac{1}{4} \cdot 88$
$\frac{1}{4} \cdot 4(x-y) = \frac{1}{4} \cdot 64$

$x + y = 22$
$x - y = 16$

$2x = 38$
$x = 19$

$x + y = 22$
$19 + y = 22$
$y = 3$

The rate of the boat in calm water is 19 km/h. The rate of the current is 3 km/h.

Chapter 4: Systems Of Equations And Inequalities

13. Strategy
Rate of the plane in calm air: x
Rate of the wind: y

	Rate	Time	Distance
With wind	$x+y$	3	$3(x+y)$
Against wind	$x-y$	4	$4(x-y)$

• The distance traveled with the wind is 360 mi.
• The distance traveled against the wind is 360 mi.

$3(x+y) = 360$
$4(x-y) = 360$

Solution
$3(x+y) = 360$
$4(x-y) = 360$

$\frac{1}{3} \cdot 3(x+y) = \frac{1}{3} \cdot 360$

$\frac{1}{4} \cdot 4(x-y) = \frac{1}{4} \cdot 360$

$x + y = 120$
$x - y = 90$

$2x = 210$
$x = 105$

$x + y = 120$
$105 + y = 120$
$y = 15$

The rate of the plane in calm air is 105 mph. The rate of the wind is 15 mph.

15. Strategy
Rate of the boat in calm water: x
Rate of the current: y

	Rate	Time	Distance
With current	$x+y$	3	$3(x+y)$
Against current	$x-y$	3.6	$3.6(x-y)$

• The distance traveled with the current is 54 mi.
• The distance traveled against the current is 54 mi.

$3(x+y) = 54$
$3.6(x-7) = 54$

Solution
$3(x+y) = 54$
$3.6(x-y) = 54$

$\frac{1}{3} \cdot 3(x+y) = \frac{1}{3} \cdot 54$

$\frac{1}{3.6} \cdot 3.6(x-y) = \frac{1}{3.6} \cdot 54$

$x + y = 18$
$x - y = 15$

$2x = 33$
$x = 16.5$

$x + y = 18$
$16.5 + y = 18$
$y = 1.5$

The rate of the boat in calm water is 16.5 mph.
The rate of the current is 1.5 mph.

Objective 4.4.2 Exercises

17.

First Project	Amount	•	Unit cost	=	Value
Hardwood flooring	200	•	h	=	$200h$
Wall-to-wall carpet	300	•	w	=	$300w$

Second Project	Amount	•	Unit cost	=	Value
Hardwood flooring	350	•	h	=	$350h$
Wall-to-wall carpet	100	•	w	=	$100w$

19. The cost per pound of the dark roast coffee is greater than the cost per pound of the light roast coffee.

Chapter 4: Systems Of Equations And Inequalities

21. Strategy
Cost of redwood: x
Cost of pine: y
First purchase:

	Amount	Cost	Total Value
Redwood	50	x	$50x$
Pine	90	y	$90y$

Second Purchase:

	Amount	Cost	Total Value
Redwood	200	x	$200x$
Pine	100	y	$100y$

- The first purchase cost $31.20.
- The second purchase cost $78.

$$50x + 90y = 31.20$$
$$200x + 100y = 78$$

Solution
$$50x + 90y = 31.20$$
$$200x + 100y = 78$$

$$-4(50x + 90y) = -4(31.20)$$
$$200x + 100y = 78$$

$$-200x - 360y = -124.80$$
$$200x + 100y = 78$$

$$-260y = -46.80$$
$$y = 0.18$$

$$50x + 90y = 31.20$$
$$50x + 90(0.18) = 31.20$$
$$50x + 16.20 = 31.20$$
$$50x = 15$$
$$x = 0.30$$

The cost of the pine is $.18 per foot.
The cost of the redwood is $.30 per foot.

23. Strategy
Cost per unit of electricity: x
Cost per unit of gas: y
First month:

	Amount	Rate	Total Value
Electricity	400	x	$400x$
Gas	120	y	$120y$

Second month:

	Amount	Rate	Total Value
Electricity	350	x	$350x$
Gas	200	y	$200y$

- The total cost for the first month was $147.20.
- The total cost for the second month was $144.

$$400x + 120y = 147.20$$
$$350x + 200y = 144$$

Solution
$$400x + 120y = 147.20$$
$$350x + 200y = 144$$
$$-5(400x + 120y) = -5(147.20)$$
$$3(350x + 200y) = 3(144)$$
$$-2000x - 600y = -736$$
$$1050x + 600y = 432$$
$$-950x = -304$$
$$x = 0.32$$
$$400x + 120y = 147.20$$
$$400(0.32) + 120y = 147.20$$
$$128 + 120y = 147.20$$
$$120y = 19.2$$
$$y = 0.16$$

The cost per unit of gas is $.16.

25. Strategy
Number of quarters in the bank: q
Number of dimes in the bank: d
Coins in the bank now:

Coin	Number	Value	Total Value
Quarter	q	25	$25q$
Dime	d	10	$10d$

Coins in the bank if the quarters were dimes and the dimes were quarters:

Coin	Number	Value	Total Value
Quarter	d	25	$25d$
Dime	q	10	$10q$

- The value of the quarters and dimes in the bank is $6.90.
- The value of the quarters and dimes in the bank would be $7.80.

$$25q + 10d = 690$$
$$10q + 25d = 780$$

Solution
$$25q + 10d = 690$$
$$10q + 25d = 780$$

$$5(25q + 10d) = 5(690)$$
$$-2(10q + 25d) = -2(780)$$

$$125q + 50d = 3450$$
$$-20q - 50d = -1560$$

$$105q = 1890$$
$$q = 18$$

There are 18 quarters in the bank.

27. Strategy

Number of mountain bikes to be manufactured: t
Number of trail bikes to be manufactured: s
Cost of materials:

Type of Bicycle	Number	Cost	Total Cost
Mountain	t	70	$70t$
Trail	s	50	$50s$

Cost of labor:

Type of Bicycle	Number	Cost	Total Cost
Mountain	t	80	$80t$
Trail	s	40	$40s$

- The company has budgeted $2500 for material.
- The company has budgeted $2600 for labor.

$70t + 50s = 2500$
$80t + 40s = 2600$

Solution
$70t + 50s = 2500$
$80t + 40s = 2600$

$4(70t + 50s) = 4(2500)$
$-5(80t + 40s) = -5(2600)$

$280t + 200s = 10{,}000$
$-400t - 200s = -13{,}000$

$-120t = -3000$
$t = 25$

The company plans to manufacture 25 mountain bikes during the week.

29. Strategy

Amount of the first powder to be used: x
Amount of the second powder to be used: y
Vitamin B_1:

	Amount	Percent	Quantity
1st powder	x	0.25	$0.25x$
2nd powder	y	0.15	$0.15y$

Vitamin B_2:

	Amount	Percent	Quantity
1st powder	x	0.15	$0.15x$
2nd powder	y	0.20	$0.20y$

The mixture contains 117.5 mg of vitamin B_1.
The mixture contains 120 mg of vitamin B_2.

$0.25x + 0.15y = 117.5$
$0.15x + 0.20y = 120$

Solution
$0.25x + 0.15y = 117.5$
$0.15x + 0.20y = 120$

$-4(0.25x + 0.15y) = -4(117.5)$
$3(0.15x + 0.20y) = 3(120)$

$-1.0x - 0.60y = -470$
$0.45x + 0.60y = 360$

$-0.55x = -110$
$x = 200$

$0.25x + 0.15y = 117.5$
$0.25(200) + 0.15y = 117.5$
$50 + 0.15y = 117.5$
$0.15y = 67.5$
$y = 450$

The pharmacist should use 200 mg of the first powder and 450 mg of the second powder.

31. Strategy

Cost of a Model II computer: x
Cost of a Model VI computer: y
Cost of a Model IX computer: z

First shipment:

Computer	Number	Cost	Total Cost
Model II	4	x	$4x$
Model VI	6	y	$6y$
Model IX	10	z	$10z$

Second shipment:

Computer	Number	Cost	Total Cost
Model II	8	x	$8x$
Model VI	3	y	$3y$
Model IX	5	z	$5z$

Third Shipment:

Computer	Number	Cost	Total Cost
Model II	2	x	$2x$
Model VI	9	y	$9y$
Model IX	5	z	$5z$

- The bill for the first shipment was $114,000.
- The bill for the second shipment was $72,000.
- The bill for the third shipment was $81,000.

$4x + 6y + 10z = 114{,}000$
$8x + 3y + 5z = 72{,}000$
$2x + 9y + 5z = 81{,}000$

Solution
$$4x + 6y + 10z = 114{,}000$$
$$8x + 3y + 5z = 72{,}000$$
$$2x + 9y + 5z = 81{,}000$$

$$4x + 6y + 10z = 114{,}000$$
$$-2(8x + 3y + 5z) = -2(72{,}000)$$

$$4x + 6y + 10z = 114{,}000$$
$$-16x - 6y - 10z = -144{,}000$$

$$-12x = -30{,}000$$
$$x = 2500$$

$$4x + 6y + 10z = 114{,}000$$
$$-2(2x + 9y + 5z) = -2(81{,}000)$$

$$4x + 6y + 10z = 114{,}000$$
$$-4x - 18y - 10z = 162{,}000$$

$$-12y = -48{,}000$$
$$y = 4000$$

$$4x + 6y + 10z = 114{,}000$$
$$4(2500) + 6(4000) + 10z = 114{,}000$$
$$10{,}000 + 24{,}000 + 10z = 114{,}000$$
$$34{,}000 + 10z = 114{,}000$$
$$10z = 80{,}000$$
$$z = 8000$$

The manufacturer charges $2500 for a Model II, $4000 for a Model VI, and $8000 for a Model IX computer.

33. Strategy
Amount earning 9% interest: x
Amount earning 7% interest: y
Amount earning 5% interest: z

Interest Rate	Amount deposited	Amount earned
9%	x	$0.09x$
7%	y	$0.07y$
5%	z	$0.05z$

- The total amount deposited is $18,000.
- The total interest earned is $1340.

$$x + y + z = 18{,}000$$
$$x = 2z$$
$$0.09x + 0.07y + 0.05z = 1340$$

Solution
$$x + y + z = 18{,}000$$
$$x = 2z$$
$$0.09x + 0.07y + 0.05z = 1340$$

$$x + y + z = 18{,}000$$
$$2z + y + z = 18{,}000$$
$$3z + y = 18{,}000$$

$$0.09x + 0.07y + 0.05z = 1340$$
$$0.09(2z) + 0.07y + 0.05z = 1340$$
$$0.18z + 0.07y + 0.05z = 1340$$
$$0.23z + 0.07y = 1340$$

$$0.23z + 0.07y = 1340$$
$$-0.07(3z + y) = -0.07(18{,}000)$$

$$0.23z + 0.07y = 1340$$
$$-0.21z - 0.07y = -1260$$
$$0.02z = 80$$
$$z = 4000$$

$$x = 2z$$
$$x = 2(4000)$$
$$x = 8000$$

$$x + y + z = 18{,}000$$
$$8000 + y + 4000 = 18{,}000$$
$$12{,}000 + y = 18{,}000$$
$$y = 6000$$

The amounts deposited in each account are $8000 at 9% interest, $6000 at 7% interest, and $4000 at 5% interest.

35. Strategy

The sum of d_1 and d_3 is equal to the length of the rod, which is 15 in.

$w_1 = 5, w_2 = 1,$ and $w_3 = 3$ so the equation that ensures the mobile will balance is $5d_1 = d_2 + 3d_3$.

$d_1 + d_3 = 15$
$5d_1 = d_2 + 3d_3$
$d_3 = 3d_2$

Solution
$d_1 + d_3 = 15$
$5d_1 = d_2 + 3d_3$
$d_3 = 3d_2$

$5d_1 = d_2 + 3d_3$
$5d_1 = d_2 + 3(3d_2)$
$5d_1 = d_2 + 9d_2$
$5d_1 = 10d_2$
$5d_1 - 10d_2 = 0$

$d_1 + d_3 = 15$
$d_1 + 3d_2 = 15$

$5d_1 - 10d_2 = 0$
$-5(d_1 + 3d_2) = -5(15)$

$5d_1 - 10d_2 = 0$
$-5d_1 - 15d_2 = -75$
$-25d_2 = -75$
$d_2 = 3$

$d_3 = 3d_2$
$d_3 = 3(3)$
$d_3 = 9$

$d_1 + d_3 = 15$
$d_1 + 9 = 15$
$d_1 = 6$

The distances are
$d_1 = 6$ in., $d_2 = 3$ in., and $d_3 = 9$ in.

37. Strategy

Number of nickels in the bank: n
Number of dimes in the bank: d
Number of quarters in the bank: q
Coins in the bank:

Coin	Number	Value	Total Value
Nickel	n	5	$5n$
Dime	d	10	$10d$
Quarter	q	25	$25q$

The value of the nickels, dimes, and quarters in the bank is 200 cents.
$n + d + q = 19$
$5n + 10d + 25q = 200$
$n = 2d$

Solution
$n + d + q = 19$
$5n + 10d + 25q = 200$
$n = 2d$

$n + d + q = 19$
$2d + d + q = 19$
$3d + q = 19$

$5n + 10d + 25q = 200$
$5(2d) + 10d + 25q = 200$
$10d + 10d + 25q = 200$
$20d + 25q = 200$

$-25(3d + q) = -25(19)$
$20d + 25q = 200$

$-75d - 25q = -475$
$20d + 25q = 200$

$-55d = -275$
$d = 5$

$n = 2d$
$n = 10$

$n + d + q = 19$
$10 + 5 + q = 19$
$15 + q = 19$
$q = 4$

There are 10 nickels, 5 dimes, and 4 quarters in the bank.

39. Strategy

Amount invested at 9%: x
Amount invested at 12%: y
Amount invested at 8%: z

Interest Rate	Amount deposited	Amount earned
9%	x	$0.09x$
12%	y	$0.12y$
8%	z	$0.08z$

The total amount invested is $33,000.
$$x + y + z = 33{,}000$$
$$0.09x + 0.12y + 0.08z = 3290$$
$$y = x + z - 5000$$

Solution
$$x + y + z = 33{,}000$$
$$x + (x + z - 5000) + z = 33{,}000$$
$$2x + 2z - 5000 = 33{,}000$$
$$2x + 2z = 38{,}000$$

$$0.09x + 0.12y + 0.08z = 3290$$
$$0.09x + 0.12(x + z - 5000) + 0.08z = 3290$$
$$0.09x + 0.12x + 0.12z - 600 + 0.08z = 3290$$
$$0.21x + 0.2z - 600 = 3290$$
$$0.21x + 0.2z = 3890$$

$$2x + 2z = 38{,}000$$
$$-10(0.21x + 0.2z) = -10(3890)$$

$$2x + 2z = 38{,}000$$
$$-2.1x - 2z = -38{,}900$$
$$-0.1x = -900$$
$$x = 9000$$

$$2x + 2z = 38{,}000$$
$$2(9000) + 2z = 38{,}000$$
$$18{,}000 + 2z = 38{,}000$$
$$2z = 20{,}000$$
$$z = 10{,}000$$

$$y = x + z - 5000$$
$$y = 9000 + 10{,}000 - 5000$$
$$y = 14{,}000$$

The amounts invested are $9000 at 9% interest, $14,000 at 12% interest, and $10,000 at 8% interest.

Applying Concepts 4.4

41. Strategy

Write and solve a system of equations using x and y to represent the measures of the two angles.

Solution
$$x + y = 90$$
$$y = 8x + 9$$

$$x + (8x + 9) = 90$$
$$9x + 9 = 90$$
$$9x = 81$$
$$x = 9$$

$$x + y = 90$$
$$9 + y = 90$$
$$y = 81$$

The measures of the two angles are 9 and 81 degrees.

43. Strategy

Number of nickels in the bank: n
Number of dimes in the bank: d
Number of quarters in the bank: q
Coins in the bank now:

Coin	Number	Value	Total Value
Nickel	n	5	$5n$
Dime	d	10	$10d$
Quarter	q	25	$25q$

Coins in the bank if nickels were dimes and dimes were nickels:

Coin	Number	Value	Total Value
Nickel	d	5	$5d$
Dime	n	10	$10n$
Quarter	q	25	$25q$

Coins in the bank if quarters were dimes and dimes were quarters:

Coin	Number	Value	Total Value
Nickel	n	5	$5n$
Dime	q	10	$10q$
Quarter	d	25	$25d$

- The value of the nickels, dimes, and quarters in the bank is 350 cents.
- The value of the nickels, dimes, and quarters would be 425 cents if the nickels were dimes and the dimes were nickels.
- The value of the nickels, dimes, and quarters would be 425 cents if the dimes were quarters and the quarters were dimes.

Solution
$$5n + 10d + 25q = 350$$
$$10n + 5d + 25q = 425$$
$$5n + 25d + 10q = 425$$

$$5n + 10d + 25q = 350$$
$$-(5n + 25d + 10q) = -(425)$$

$$5n + 10d + 25q = 350$$
$$-5n - 25d - 10q = -425$$

$$-15d + 15q = -75$$

$$-2(5n + 10d + 25q) = -2(350)$$
$$10n + 5d + 25q = 425$$

$$-10n - 20d - 50q = -700$$
$$10n + 5d + 25q = 425$$

$$-15d - 25q = -275$$

$$-15d - 25q = -275$$
$$-(-15d + 15q) = -(-75)$$

$$-15d - 25q = -275$$
$$15d - 15q = 75$$
$$-40q = -200$$
$$q = 5$$

$$-15d + 15q = -75$$
$$-15d + 15(5) = -75$$
$$-15d + 75 = -75$$
$$-15d = -150$$
$$d = 10$$

$$5n + 10d + 25q = 350$$
$$5n + 10(10) + 25(5) = 350$$
$$5n + 100 + 125 = 350$$
$$5n + 225 = 350$$
$$5n = 125$$
$$n = 25$$

There are 25 nickels, 10 dimes, and 5 quarters in the bank.

Section 4.5
Concept Review 4.5

1. Sometimes true
   The solution set of a system of inequalities can be a portion of the plane or the empty set.

3. Always true

Objective 4.5.1 Exercises

1. Check the ordered pair (5, 1) in the system of inequalities

$2(5) - 1$	4
$10 - 1$	4
9	4

   $9 > 4$
   The ordered pair (5, 1) is not a solution.

   Check the ordered pair (−3, −5) in the system of inequalities.

$2(-3) - (-5)$	4		$-3 - 3(-5)$	6
$-6 + 5$	4		$-3 + 15$	6
-1	4		12	6

   $-1 < 4$ and $12 \geq 6$
   The ordered pair (−3, −5) is a solution.

3. a. Points in the solution set of the inequality $y \leq 2x + 2$ are located in regions <u>A</u> and <u>D</u>.

   b. Points in the solution set of the inequality $y \geq -x + 5$ are located in regions <u>A</u> and <u>B</u>.

   c. Points in the solution set of the system of inequalities $\begin{cases} y \leq 2x + 2 \\ y \geq -x + 5 \end{cases}$ are located in region <u>A</u>.

5. Solve each inequality for y.
   $$x - y \geq 3 \qquad x + y \leq 5$$
   $$-y \geq -x + 3 \qquad y \leq -x + 5$$
   $$y \leq x - 3$$

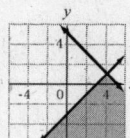

7. $y > 3x - 3 \qquad 2x + y \geq 2$
   $$y \geq -2x + 2$$

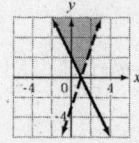

9. Solve each inequality for y.
   $$2x + y \geq -2 \qquad 6x + 3y \leq 6$$
   $$y \geq -2x - 2 \qquad 3y \leq -6x + 6$$
   $$y \leq -2x + 2$$

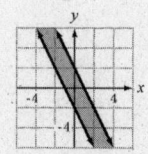

Chapter 4: Systems Of Equations And Inequalities

11. Solve each inequality for y.
    $3x - 2y < 6 \qquad y \leq 3$
    $-2y < -3x + 6$
    $y > \dfrac{3}{2}x - 3$

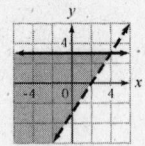

13. Solve the inequality for y.
    $y > 2x - 6 \qquad x + y < 0$
    $\qquad\qquad\qquad y < -x$

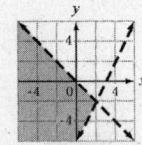

15. Solve each inequality for the variable.
    $x + 1 \geq 0 \qquad y - 3 \leq 0$
    $x \geq -1 \qquad\quad y \leq 3$

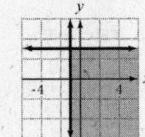

17. Solve each inequality for y.
    $2x + y \geq 4 \qquad 3x - 2y < 6$
    $y \geq -2x + 4 \qquad -2y < -3x + 6$
    $\qquad\qquad\qquad\quad y > \dfrac{3}{2}x - 3$

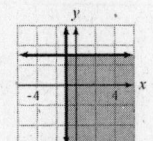

19. Solve each inequality for y.
    $x - 2y \leq 6 \qquad 2x + 3y \leq 6$
    $-2y \leq -x + 6 \qquad 3y \leq -2x + 6$
    $y \geq \dfrac{1}{2}x - 3 \qquad y \leq -\dfrac{2}{3}x + 2$

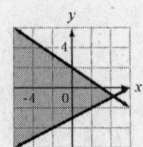

21. Solve each inequality for y.
    $x - 2y \leq 4 \qquad 3x + 2y \leq 8$
    $-2y \leq -x + 4 \qquad 2y \leq -3x + 8$
    $y \geq \dfrac{1}{2}x - 2 \qquad y \leq -\dfrac{3}{2}x + 4$
    $y \geq \dfrac{1}{2}x - 2$
    $y \leq -\dfrac{3}{2}x + 4$

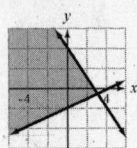

23. Points below the line $x + y = b$

25. Region between the parallel lines $x + y = a$ and $x + y = b$

Applying Concepts 4.5

27. Solve each inequality for y.
    $2x + 3y \leq 15 \qquad 3x - y \leq 6$
    $3y \leq -2x + 15 \qquad -y \leq -3x + 6$
    $y \leq -\dfrac{2}{3}x + 5 \qquad y \geq 3x - 6$
    $y \leq -\dfrac{2}{3}x + 5$
    $y \geq 3x - 6$
    $y \geq 0$

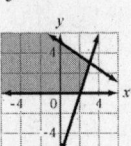

29. Solve each inequality for y.
    $x - y \leq 5 \qquad 2x - y \geq 6 \qquad y \geq 0$
    $-y \leq -x + 5 \qquad -y \geq -2x + 6$
    $y \geq x - 5 \qquad\quad y \leq 2x - 6$

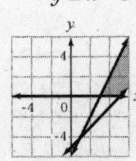

31. Solve each inequality for y.
    $2x - y \leq 4 \qquad 3x + y < 1 \qquad y \leq 0$
    $-y \leq -2x + 4 \qquad y < 1 - 3x$
    $y \geq 2x - 4$

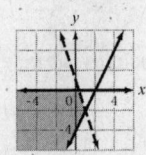

CHAPTER 4 REVIEW EXERCISES

1. (1) $2x - 6y = 15$
  (2) $x = 3y + 8$
Substitute $3y + 8$ for x in equation (1).
$2(3y + 8) - 6y = 15$
$6y + 16 - 6y = 15$
$16 = 15$
This is not a true equation. The lines are parallel and the system is inconsistent.

2. (1) $3x + 12y = 18$
  (2) $x + 4y = 6$
Solve equation (2) for x.
$x + 4y = 6$
$x = -4y + 6$
Substitute into equation (1).
$3x + 12y = 18$
$3(-4y + 6) + 12y = 18$
$-12y + 18 + 12y = 18$
$18 = 18$
This is a true equation. The equations are dependent.
The solutions are the ordered pairs $\left(x, -\dfrac{1}{4}x + \dfrac{3}{2}\right)$.

3. (1) $3x + 2y = 2$
  (2) $x + y = 3$
Eliminate y. Multiply equation (2) by -2 and add to equation (1).
$3x + 2y = 2$
$-2(x + y) = 3(-2)$

$3x + 2y = 2$
$-2x - 2y = -6$
Add the equations.
$x = -4$
Replace x in equation (2).
$x + y = 3$
$-4 + y = 3$
$y = 7$
The solution is $(-4, 7)$.

4. (1) $5x - 15y = 30$
  (2) $x - 3y = 6$
Eliminate x. Multiply equation (2) by -5 and add to equation (1).
$5x - 15y = 30$
$-5(x - 3y) = 6(-5)$

$5x - 15y = 30$
$-5x + 15y = -30$
Add the equations.
$0 = 0$
This is a true equation. The equations are dependent.
The solutions are the ordered pairs $\left(x, \dfrac{1}{3}x - 2\right)$.

5. (1) $3x + y = 13$
  (2) $2y + 3z = 5$
  (3) $x + 2z = 11$
Eliminate y. Multiply equation (1) by -2 and add to equation (2).
$-2(3x + y) = 13(-2)$
$2y + 3z = 5$

$-6x - 2y = -26$
$2y + 3z = 5$

(4) $-6x + 3z = -21$
Multiply equation (3) by 6 and add to equation (4).
$6(x + 2z) = 6(11)$
$-6x + 3z = -21$

$6x + 12z = 66$
$-6x + 3z = -21$

$15z = 45$
$z = 3$
Replace z by 3 in equation (3).
$x + 2z = 11$
$x + 2(3) = 11$
$x + 6 = 11$
$x = 5$
Replace x by 5 in equation (1).
$3x + y = 13$
$3(5) + y = 13$
$15 + y = 13$
$y = -2$
The solution is $(5, -2, 3)$.

118 Chapter 4: Systems Of Equations And Inequalities

6. (1) $3x - 4y - 2z = 17$
(2) $4x - 3y + 5z = 5$
(3) $5x - 5y + 3z = 14$

Eliminate z. Multiply equation (1) by 5 and equation (2) by 2 and add the new equations.
$5(3x - 4y - 2z) = 17(5)$
$2(4x - 3y + 5z) = 5(2)$

$15x - 20y - 10z = 85$
$8x - 6y + 10z = 10$

(4) $23x - 26y = 95$

Multiply equation (1) by 3 and equation (3) by 2 and add the new equations.
$3(3x - 4y - 2z) = (17)3$
$2(5x - 5y + 3z) = (14)2$

$9x - 12y - 6z = 51$
$10x - 10y + 6z = 28$

(5) $19x - 22y = 79$

Multiply equation (4) by −11 and equation (5) by 13 and add the new equations.
$-11(23x - 26y) = 95(-11)$
$13(19x - 22y) = 79(13)$

$-253x + 286y = -1045$
$247x - 286y = 1027$

$-6x = -18$
$x = 3$

Substitute x for 3 in equation (4).
$23x - 26y = 95$
$23(3) - 26y = 95$
$69 - 26y = 95$
$-26y = 26$
$y = -1$

Substitute x by 3 and y by −1 in equation (1).
$3x - 4y - 2z = 17$
$3(3) - 4(-1) - 2z = 17$
$9 + 4 - 2z = 17$
$-2z = 4$
$z = -2$

The solution is $(3, -1, -2)$.

7. $\begin{vmatrix} 6 & 1 \\ 2 & 5 \end{vmatrix} = 6(5) - 2(1) = 30 - 2 = 28$

8. $\begin{vmatrix} 1 & 5 & -2 \\ -2 & 1 & 4 \\ 4 & 3 & -8 \end{vmatrix} = 1\begin{vmatrix} 1 & 4 \\ 3 & -8 \end{vmatrix} - 5\begin{vmatrix} -2 & 4 \\ 4 & -8 \end{vmatrix} - 2\begin{vmatrix} -2 & 1 \\ 4 & 3 \end{vmatrix}$
$= 1(-8 - 12) - 5(16 - 16) - 2(-6 - 4)$
$= 1(-20) - 5(0) - 2(-10)$
$= -20 - 0 + 20$
$= 0$

9. $2x - y = 7$
$3x + 2y = 7$

$D = \begin{vmatrix} 2 & -1 \\ 3 & 2 \end{vmatrix} = 7$

$D_x = \begin{vmatrix} 7 & -1 \\ 7 & 2 \end{vmatrix} = 21$

$D_y = \begin{vmatrix} 2 & 7 \\ 3 & 7 \end{vmatrix} = -7$

$x = \dfrac{D_x}{D} = \dfrac{21}{7} = 3$

$y = \dfrac{D_y}{D} = \dfrac{-7}{7} = -1$

The solution is $(3, -1)$.

10. $3x - 4y = 10$
$2x + 5y = 15$

$D = \begin{vmatrix} 3 & -4 \\ 2 & 5 \end{vmatrix} = 23$

$D_x = \begin{vmatrix} 10 & -4 \\ 15 & 5 \end{vmatrix} = 110$

$D_y = \begin{vmatrix} 3 & 10 \\ 2 & 15 \end{vmatrix} = 25$

$x = \dfrac{D_x}{D} = \dfrac{110}{23}$

$y = \dfrac{D_y}{D} = \dfrac{25}{23}$

The solution is $\left(\dfrac{110}{23}, \dfrac{25}{23}\right)$.

11. $x + y + z = 0$
$x + 2y + 3z = 5$
$2x + y + 2z = 3$

$D = \begin{vmatrix} 1 & 1 & 1 \\ 1 & 2 & 3 \\ 2 & 1 & 2 \end{vmatrix} = 2$

$D_x = \begin{vmatrix} 0 & 1 & 1 \\ 5 & 2 & 3 \\ 3 & 1 & 2 \end{vmatrix} = -2$

$D_y = \begin{vmatrix} 1 & 0 & 1 \\ 1 & 5 & 3 \\ 2 & 3 & 2 \end{vmatrix} = -6$

$D_z = \begin{vmatrix} 1 & 1 & 0 \\ 1 & 2 & 5 \\ 2 & 1 & 3 \end{vmatrix} = 8$

$x = \dfrac{D_x}{D} = \dfrac{-2}{2} = -1$

$y = \dfrac{D_y}{D} = \dfrac{-6}{2} = -3$

$z = \dfrac{D_z}{D} = \dfrac{8}{2} = 4$

The solution is $(-1, -3, 4)$.

12. $x + 3y + z = 6$
$2x + y - z = 12$
$x + 2y - z = 13$

$D = \begin{vmatrix} 1 & 3 & 1 \\ 2 & 1 & -1 \\ 1 & 2 & -1 \end{vmatrix} = 7$

$D_x = \begin{vmatrix} 6 & 3 & 1 \\ 12 & 1 & -1 \\ 13 & 2 & -1 \end{vmatrix} = 14$

$D_y = \begin{vmatrix} 1 & 6 & 1 \\ 2 & 12 & -1 \\ 1 & 13 & -1 \end{vmatrix} = 21$

$D_z = \begin{vmatrix} 1 & 3 & 6 \\ 2 & 1 & 12 \\ 1 & 2 & 13 \end{vmatrix} = -35$

$x = \dfrac{D_x}{D} = \dfrac{14}{7} = 2$

$y = \dfrac{D_y}{D} = \dfrac{21}{7} = 3$

$z = \dfrac{D_z}{D} = \dfrac{-35}{7} = -5$

The solution is $(2, 3, -5)$.

13. (1) $x - 2y + z = 7$
(2) $3x - z = -1$
(3) $3y + z = 1$

Eliminate z. Add equations (2) and (3).
$3x - z = -1$
$3y + z = 1$

(4) $3x + 3y = 0$

Multiply equation (1) by -1 and add to equation (3).
$-(x - 2y + z) = -7$
$3y + z = 1$

$-x + 2y - z = -7$
$3y + z = 1$

(5) $-x + 5y = -6$

Multiply equation (4) by $\dfrac{1}{3}$ and add to equation (5).

$\dfrac{1}{3}(3x + 3y) = \dfrac{1}{3}(0)$
$-x + 5y = -6$

$x + y = 0$
$-x + 5y = -6$

$6y = -6$
$y = -1$

Replace y in equation (5).
$-x + 5y = -6$
$-x + 5(-1) = -6$
$-x - 5 = -6$
$-x = -1$
$x = 1$

Replace x in equation (2).
$3x - z = -1$
$3(1) - z = -1$
$3 - z = -1$
$-z = -4$
$z = 4$

The solution is $(1, -1, 4)$.

14. $3x - 2y = 2$
$-2x + 3y = 1$

$D = \begin{vmatrix} 3 & -2 \\ -2 & 3 \end{vmatrix} = 5$

$D_x = \begin{vmatrix} 2 & -2 \\ 1 & 3 \end{vmatrix} = 8$

$D_y = \begin{vmatrix} 3 & 2 \\ -2 & 1 \end{vmatrix} = 7$

$x = \dfrac{D_x}{D} = \dfrac{8}{5}$

$y = \dfrac{D_y}{D} = \dfrac{7}{5}$

The solution is $\left(\dfrac{8}{5}, \dfrac{7}{5}\right)$.

15. $2x - 2y - 6z = 1$
$4x + 2y + 3z = 1$
$2x - 3y - 3z = 3$

$\begin{bmatrix} 2 & -2 & -6 & | & 1 \\ 4 & 2 & 3 & | & 1 \\ 2 & -3 & -3 & | & 3 \end{bmatrix}$

$\dfrac{1}{2}R_1 \rightarrow \begin{bmatrix} 1 & -1 & -3 & | & \dfrac{1}{2} \\ 4 & 2 & 3 & | & 1 \\ 2 & -3 & -3 & | & 3 \end{bmatrix}$

$\begin{matrix} -4R_1 + R_2 \rightarrow \\ -2R_1 + R_3 \rightarrow \end{matrix} \begin{bmatrix} 1 & -1 & -3 & | & \dfrac{1}{2} \\ 0 & 6 & 15 & | & -1 \\ 0 & -1 & 3 & | & 2 \end{bmatrix}$

$\dfrac{1}{6}R_2 \rightarrow \begin{bmatrix} 1 & -1 & -3 & | & \dfrac{1}{2} \\ 0 & 1 & \dfrac{5}{2} & | & -\dfrac{1}{6} \\ 0 & -1 & 3 & | & 2 \end{bmatrix}$

$R_2 + R_3 \rightarrow \begin{bmatrix} 1 & -1 & -3 & | & \dfrac{1}{2} \\ 0 & 1 & \dfrac{5}{2} & | & -\dfrac{1}{6} \\ 0 & 0 & \dfrac{11}{2} & | & \dfrac{11}{6} \end{bmatrix}$

$\dfrac{2}{11}R_3 \rightarrow \begin{bmatrix} 1 & -1 & -3 & | & \dfrac{1}{2} \\ 0 & 1 & \dfrac{5}{2} & | & -\dfrac{1}{6} \\ 0 & 0 & 1 & | & \dfrac{1}{3} \end{bmatrix}$

$x - y - 3z = \dfrac{1}{2}$

$y + \left(\dfrac{5}{2}\right)z = -\dfrac{1}{6}$

$z = \dfrac{1}{3}$

$y + \left(\dfrac{5}{2}\right)z = -\dfrac{1}{6}$

$y + \left(\dfrac{5}{2}\right)\left(\dfrac{1}{3}\right) = -\dfrac{1}{6}$

$y + \dfrac{5}{6} = -\dfrac{1}{6}$

$y = -1$

$x - y - 3z = \dfrac{1}{2}$

$x - (-1) - 3\left(\dfrac{1}{3}\right) = \dfrac{1}{2}$

$x + 1 - 1 = \dfrac{1}{2}$

$x = \dfrac{1}{2}$

The solution is $\left(\dfrac{1}{2}, -1, \dfrac{1}{3}\right)$.

16. $\begin{vmatrix} 3 & -2 & 5 \\ 4 & 6 & 3 \\ 1 & 2 & 1 \end{vmatrix} = 3\begin{vmatrix} 6 & 3 \\ 2 & 1 \end{vmatrix} + 2\begin{vmatrix} 4 & 3 \\ 1 & 1 \end{vmatrix} + 5\begin{vmatrix} 4 & 6 \\ 1 & 2 \end{vmatrix}$

$= 3(6 - 6) + 2(4 - 3) + 5(8 - 6)$
$= 3(0) + 2(1) + 5(2)$
$= 0 + 2 + 10 = 12$

17. $4x - 3y = 17$
$3x - 2y = 12$

$D = \begin{vmatrix} 4 & -3 \\ 3 & -2 \end{vmatrix} = 1$

$D_x = \begin{vmatrix} 17 & -3 \\ 12 & -2 \end{vmatrix} = 2$

$D_y = \begin{vmatrix} 4 & 17 \\ 3 & 12 \end{vmatrix} = -3$

$x = \dfrac{D_x}{D} = \dfrac{2}{1} = 2$

$y = \dfrac{D_y}{D} = \dfrac{-3}{1} = -3$

The solution is $(2, -3)$.

18. $3x + 2y - z = -1$
$x + 2y + 3z = -1$
$3x + 4y + 6z = 0$

$$\begin{bmatrix} 3 & 2 & -1 & | & -1 \\ 1 & 2 & 3 & | & -1 \\ 3 & 4 & 6 & | & 0 \end{bmatrix}$$

$R_1 \leftrightarrow R_2 \begin{bmatrix} 1 & 2 & 3 & | & -1 \\ 3 & 2 & -1 & | & -1 \\ 3 & 4 & 6 & | & 0 \end{bmatrix}$

$\begin{matrix} -3R_1 + R_2 \to \\ -3R_1 + R_3 \to \end{matrix} \begin{bmatrix} 1 & 2 & 3 & | & -1 \\ 0 & 1 & \frac{5}{2} & | & -\frac{1}{2} \\ 0 & -2 & -3 & | & 3 \end{bmatrix}$

$2R_2 + R_3 \to \begin{bmatrix} 1 & 2 & 3 & | & -1 \\ 0 & 1 & \frac{5}{2} & | & -\frac{1}{2} \\ 0 & 0 & 2 & | & 2 \end{bmatrix}$

$\frac{1}{2}R_3 \to \begin{bmatrix} 1 & 2 & 3 & | & -1 \\ 0 & 1 & \frac{5}{2} & | & -\frac{1}{2} \\ 0 & 0 & 1 & | & 1 \end{bmatrix}$

$x + 2y + 3z = -1$
$y + \frac{5}{2}z = -\frac{1}{2}$
$z = 1$

$y + \left(\frac{5}{2}\right)z = -\frac{1}{2}$
$y + \frac{5}{2}(1) = -\frac{1}{2}$
$y + \frac{5}{2} = -\frac{1}{2}$
$y = -3$

$x + 2y + 3z = -1$
$x + 2(-3) + 3(1) = -1$
$x - 6 + 3 = -1$
$x - 3 = -1$
$x = 2$

The solution is $(2, -3, 1)$.

19.

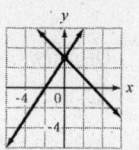

The solution is $(0, 3)$.

20.

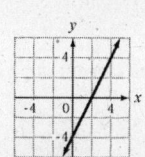

The two equations represent the same line. The solutions are the ordered pairs $(x, 2x - 4)$.

21. Solve each inequality for y.

$\quad x + 3y < 6 \qquad\qquad 2x - y > 4$
$\quad\quad 3y < -x + 6 \qquad\quad -y > -2x + 4$
$\quad\quad\quad y < -\frac{1}{3}x + 2 \qquad\quad y < 2x - 4$

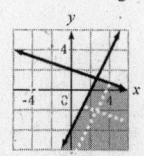

22. Solve each inequality for y.

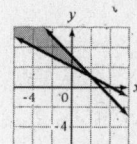

$\quad 2x + 4y \geq 8 \qquad\qquad x + y \leq 3$
$\quad\quad 4y \geq -2x + 8 \qquad\quad y \leq -x + 3$
$\quad\quad\quad y \geq -\frac{1}{2}x + 2$

Chapter 4: Systems Of Equations And Inequalities

23. Strategy
Rate of the cabin cruiser in calm water: x
Rate of the current: y

	Rate	Time	Distance
With current	$x+y$	3	$3(x+y)$
Against current	$x-y$	5	$5(x-y)$

- The distance traveled with the current is 60 mi.
- The distance traveled against the current is 60 mi.

$3(x+y) = 60$
$5(x-y) = 60$

Solution
$3(x+y) = 60$
$5(x-y) = 60$

$\frac{1}{3} \cdot 3(x+y) = \frac{1}{3}(60)$
$\frac{1}{5} \cdot 5(x-y) = \frac{1}{5}(60)$

$x + y = 20$
$x - y = 12$

$2x = 32$
$x = 16$

$x + y = 20$
$16 + y = 20$
$y = 4$

The rate of the cabin cruiser in calm water is 16 mph. The rate of the current is 4 mph.

24. Strategy
Rate of the plane in calm air: p
Rate of the wind: w

	Rate	Time	Distance
With wind	$p+w$	3	$3(p+w)$
Against wind	$p-w$	4	$4(p-w)$

- The distance traveled with the wind is 600 mi.
- The distance traveled against the wind is 600 mi.

$3(p+w) = 600$
$4(p-w) = 600$

Solution
$3(p+w) = 600$
$4(p-w) = 600$

$\frac{1}{3} \cdot 3(p+w) = \frac{1}{3}(600)$
$\frac{1}{4} \cdot 4(p-w) = \frac{1}{4}(600)$

$p + w = 200$
$p - w = 150$

$2p = 350$
$p = 175$

$p + w = 200$
$175 + w = 200$
$w = 25$

The rate of the plane in calm air is 175 mph.
The rate of the wind is 25 mph.

25. Strategy
Number of children's tickets sold Friday: x
Number of adult's tickets sold Friday: y

Friday:

	Number	Value	Total Value
Children	x	5	$5x$
Adults	y	8	$8y$

Saturday:

	Number	Value	Total Value
Children	$3x$	5	$5(3x)$
Adults	$\frac{1}{2}y$	8	$8(\frac{1}{2}y)$

- The total receipts for Friday were $2500.
- The total receipts for Saturday were $2500.

$5x + 8y = 2500$
$5(3x) + 8\left(\frac{1}{2}y\right) = 2500$

Solution
$5x + 8y = 2500$
$15x + 4y = 2500$

$5x + 8y = 2500$
$-2(15x + 4y) = -2(2500)$

$5x + 8y = 2500$
$-30x - 8y = -5000$

$-25x = -2500$
$x = 100$

The number of children attending on Friday was 100.

26. Strategy
Amount invested at 8%: x
Amount invested at 6%: y
Amount invested at 4%: z

- The total invested was $25,000.
- The three accounts earn total annual interest of $1520.

$$x + y + z = 25{,}000$$
$$x = 2y$$
$$0.08x + 0.06y + 0.04z = 1520$$

Solution
$$x + y + z = 25{,}000$$
$$x = 2y$$
$$0.08x + 0.06y + 0.04z = 1520$$

$$x + y + z = 25{,}000$$
$$x = 2y$$
$$100(0.08x + 0.06y + 0.04z) = 100(1520)$$
$$x + y + z = 25{,}000$$
$$x = 2y$$
$$8x + 6y + 4z = 152{,}000$$

$$2y + y + z = 25{,}000$$
$$8(2y) + 6y + 4z = 152{,}000$$

$$3y + z = 25{,}000$$
$$22y + 4z = 152{,}000$$

$$-4(3y + z) = -4(25{,}000)$$
$$22y + 4z = 152{,}000$$

$$-12y - 4z = -100{,}000$$
$$22y + 4z = 152{,}000$$

$$10y = 52{,}000$$
$$y = 5200$$
$$x = 2y$$
$$x = 2(5200) = 10{,}400$$
$$x + y + z = 25{,}000$$
$$10{,}400 + 5200 + z = 25{,}000$$
$$z = 9400$$

There was $10,400 invested at 8%, $5200 invested at 6%, and $9400 invested at 4%.

CHAPTER 4 TEST

1. (1) $3x + 2y = 4$
(2) $x = 2y - 1$

Substitute $2y - 1$ for x in equation (1).
$$3(2y - 1) + 2y = 4$$
$$6y - 3 + 2y = 4$$
$$8y = 7$$
$$y = \frac{7}{8}$$

Substitute into equation (2).
$$x = 2y - 1$$
$$x = 2\left(\frac{7}{8}\right) - 1$$
$$x = \frac{7}{4} - 1 = \frac{3}{4}$$

The solution is $\left(\frac{3}{4}, \frac{7}{8}\right)$.

2. (1) $5x + 2y = -23$
(2) $2x + y = -10$

Solve equation (2) for y.
$$2x + y = -10$$
$$y = -2x - 10$$

Substitute $-2x - 10$ for y in equation (1).
$$5x + 2y = -23$$
$$5x + 2(-2x - 10) = -23$$
$$5x - 4x - 20 = -23$$
$$x - 20 = -23$$
$$x = -3$$

Substitute into equation (2).
$$2x + y = -10$$
$$2(-3) + y = -10$$
$$-6 + y = -10$$
$$y = -4$$

The solution is $(-3, -4)$.

3. (1) $y = 3x - 7$
(2) $y = -2x + 3$

Substitute equation (2) into equation (1).
$$-2x + 3 = 3x - 7$$
$$-5x + 3 = -7$$
$$-5x = -10$$
$$x = 2$$

Substitute into equation (1).
$$y = 3x - 7$$
$$y = 3(2) - 7 = 6 - 7 = -1$$

The solution is $(2, -1)$.

4. $3x + 4y = -2$
$2x + 5y = 1$

$\begin{bmatrix} 3 & 4 & | & -2 \\ 2 & 5 & | & 1 \end{bmatrix}$

$-1R_2 + R_1 \to \begin{bmatrix} 1 & -1 & | & -3 \\ 2 & 5 & | & 1 \end{bmatrix}$

$-2R_1 + R_2 \to \begin{bmatrix} 1 & -1 & | & -3 \\ 0 & 7 & | & 7 \end{bmatrix}$

$\frac{1}{7}R_2 \to \begin{bmatrix} 1 & -1 & | & -3 \\ 0 & 1 & | & 1 \end{bmatrix}$

$x - y = -3$
$y = 1$
$x - 1 = -3$
$x = -2$

The solution is $(-2, 1)$.

5. (1) $4x - 6y = 5$
(2) $6x - 9y = 4$

Multiply equation (1) by -3. Multiply equation (2) by 2. Add the new equations.
$-3(4x - 6y) = -3(5)$
$2(6x - 9y) = 2(4)$

$-12x + 18y = -15$
$12x - 18y = 8$

$0 = -7$

This is not a true equation. The system of equations is inconsistent and therefore has no solution.

6. (1) $3x - y = 2x + y - 1$
(2) $5x + 2y = y + 6$

Write the equation in the form $Ax + By = C$.
(3) $x - 2y = -1$
(4) $5x + y = 6$

Multiply equation (4) by 2 and add to equation (3).
$x - 2y = -1$
$2(5x + y) = 2(6)$

$x - 2y = -1$
$10x + 2y = 12$

$11x = 11$
$x = 1$

Substitute into equation (4).
$5x + y = 6$
$5(1) + y = 6$
$y = 1$

The solution is $(1, 1)$.

7. (1) $2x + 4y - z = 3$
(2) $x + 2y + z = 5$
(3) $4x + 8y - 2z = 7$

Eliminate z. Add equations (1) and (2).
$2x + 4y - z = 3$
$x + 2y + z = 5$

(4) $3x + 6y = 8$

Multiply equation (2) by 2 and add to equation (3).
$2(x + 2y + z) = 2(5)$
$4x + 8y - 2z = 7$

$2x + 4y + 2z = 10$
$4x + 8y - 2z = 7$

(5) $6x + 12y = 17$

Multiply equation (4) by -2 and add to equation (5).
$-2(3x + 6y) = -2(8)$
$6x + 12y = 17$

$-6x - 12y = -16$
$6x + 12y = 17$

$0 = 1$

This is not a true equation. The system of equations is inconsistent and therefore has no solution.

8. $x - y - z = 5$
$2x + z = 2$
$3y - 2z = 1$

$$\begin{bmatrix} 1 & -1 & -1 & | & 5 \\ 2 & 0 & 1 & | & 2 \\ 0 & 3 & -2 & | & 1 \end{bmatrix}$$

$-2R_1 + R_2 \to \begin{bmatrix} 1 & -1 & -1 & | & 5 \\ 0 & 2 & 3 & | & -8 \\ 0 & 3 & -2 & | & 1 \end{bmatrix}$

$R_2 \leftrightarrow R_3 \begin{bmatrix} 1 & -1 & -1 & | & 5 \\ 0 & 3 & -2 & | & 1 \\ 0 & 2 & 3 & | & -8 \end{bmatrix}$

$-1R_3 + R_2 \to \begin{bmatrix} 1 & -1 & -1 & | & 5 \\ 0 & 1 & -5 & | & 9 \\ 0 & 2 & 3 & | & -8 \end{bmatrix}$

$-2R_2 + R_3 \to \begin{bmatrix} 1 & -1 & -1 & | & 5 \\ 0 & 1 & -5 & | & 9 \\ 0 & 0 & 13 & | & -26 \end{bmatrix}$

$\frac{1}{13}R_3 \to \begin{bmatrix} 1 & -1 & -1 & | & 5 \\ 0 & 1 & -5 & | & 9 \\ 0 & 0 & 1 & | & -2 \end{bmatrix}$

$x - y - z = 5$
$y - 5z = 9$
$z = -2$

$y - 5z = 9$
$y - 5(-2) = 9$
$y + 10 = 9$
$y = -1$

$x - y - z = 5$
$x - (-1) - (-2) = 5$
$x + 1 + 2 = 5$
$x + 3 = 5$
$x = 2$

The solution is $(2, -1, -2)$.

9. $\begin{vmatrix} 3 & -1 \\ -2 & 4 \end{vmatrix} = 3(4) - (-2)(-1) = 12 - 2 = 10$

10. $\begin{vmatrix} 1 & -2 & 3 \\ 3 & 1 & 1 \\ 2 & -1 & -2 \end{vmatrix} = 1\begin{vmatrix} 1 & 1 \\ -1 & -2 \end{vmatrix} - (-2)\begin{vmatrix} 3 & 1 \\ 2 & -2 \end{vmatrix} + 3\begin{vmatrix} 3 & 1 \\ 2 & -1 \end{vmatrix}$

$= 1(-2 - (-1)) + 2(-6 - 2) + 3(-3 - 2)$
$= 1(-2 + 1) + 2(-8) + 3(-5)$
$= -1 - 16 - 15$
$= -32$

11. $x - y = 3$
$2x + y = -4$

$D = \begin{vmatrix} 1 & -1 \\ 2 & 1 \end{vmatrix} = 3$

$D_x = \begin{vmatrix} 3 & -1 \\ -4 & 1 \end{vmatrix} = -1$

$D_y = \begin{vmatrix} 1 & 3 \\ 2 & -4 \end{vmatrix} = -10$

$x = \dfrac{D_x}{D} = -\dfrac{1}{3}$

$y = \dfrac{D_y}{D} = -\dfrac{10}{3}$

The solution is $\left(-\dfrac{1}{3}, -\dfrac{10}{3}\right)$.

12. $x - y + z = 2$
$2x - y - z = 1$
$x + 2y - 3z = -4$

$D = \begin{vmatrix} 1 & -1 & 1 \\ 2 & -1 & -1 \\ 1 & 2 & -3 \end{vmatrix} = 5$

$D_x = \begin{vmatrix} 2 & -1 & 1 \\ 1 & -1 & -1 \\ -4 & 2 & -3 \end{vmatrix} = 1$

$D_y = \begin{vmatrix} 1 & 2 & 1 \\ 2 & 1 & -1 \\ 1 & -4 & -3 \end{vmatrix} = -6$

$D_z = \begin{vmatrix} 1 & -1 & 2 \\ 2 & -1 & 1 \\ 1 & 2 & -4 \end{vmatrix} = 3$

$x = \dfrac{D_x}{D} = \dfrac{1}{5}$

$y = \dfrac{D_y}{D} = -\dfrac{6}{5}$

$z = \dfrac{D_z}{D} = \dfrac{3}{5}$

The solution is $\left(\dfrac{1}{5}, -\dfrac{6}{5}, \dfrac{3}{5}\right)$.

126   Chapter 4: Systems Of Equations And Inequalities

13. $3x + 2y + 2z = 2$
$x - 2y - z = 1$
$2x - 3y - 3z = -3$

$D = \begin{vmatrix} 3 & 2 & 2 \\ 1 & -2 & -1 \\ 2 & -3 & -3 \end{vmatrix} = 13$

$D_x = \begin{vmatrix} 2 & 2 & 2 \\ 1 & -2 & -1 \\ -3 & -3 & -3 \end{vmatrix} = 0$

$D_y = \begin{vmatrix} 3 & 2 & 2 \\ 1 & 1 & -1 \\ 2 & -3 & 3 \end{vmatrix} = -26$

$D_z = \begin{vmatrix} 3 & 2 & 2 \\ 1 & -2 & 1 \\ 2 & -3 & -3 \end{vmatrix} = 39$

$\dfrac{D_x}{D} = \dfrac{0}{13} = 0$

$\dfrac{D_y}{D} = \dfrac{-26}{13} = -2$

$\dfrac{D_z}{D} = \dfrac{39}{13} = 3$

The solution is $(0, -2, 3)$.

14. The solution is $(3, 4)$.

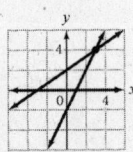

15. The solution is $(-5, 0)$.

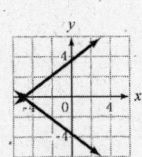

16. Solve each inequality for y.
$2x - y < 3$ $4x + 3y < 11$
$-y < -2x + 3$ $3y < -4x + 11$
$y > 2x - 3$ $y < -\dfrac{4}{3}x + \dfrac{11}{3}$

$y > 2x - 3$
$y < -\dfrac{4}{3}x + \dfrac{11}{3}$

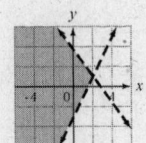

17. Solve each inequality for y.
$x + y > 2$ $2x - y < -1$
$y > -x + 2$ $-y < -2x - 1$
                       $y > 1 + 2x$

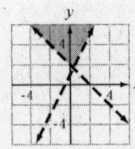

18. $(-0.14, 2.43)$

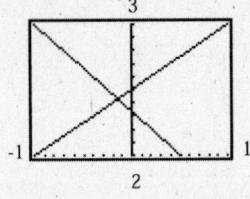

19. Strategy
 • Rate of the plane in calm air: x
   Rate of the wind: y

	Rate	Time	Distance
With wind	$x + y$	2	$2(x + y)$
Against wind	$x - y$	2.8	$2.8(x - y)$

 • The distance traveled with the wind is 350 mi.
 • The distance traveled against the wind is 350 mi.

$2(x + y) = 350$
$2.8(x - y) = 350$

Solution
$2(x + y) = 350$
$2.8(x - y) = 350$

$\dfrac{1}{2} \cdot 2(x + y) = \dfrac{1}{2} \cdot 350$
$\dfrac{1}{2.8} \cdot 2.8(x - y) = \dfrac{1}{2.8} \cdot 350$

$x + y = 175$
$x - y = 125$

$2x = 300$
$x = 150$

$x + y = 175$
$150 + y = 175$
$y = 25$

The rate of the plane in calm air is 150 mph. The rate of the wind is 25 mph.

20. Strategy

Cost per yard of cotton: x
Cost per yard of wool: y
First purchase:

	Amount	Cost	Total Value
Cotton	60	x	$60x$
Wool	90	y	$90y$

Second purchase:

	Amount	Cost	Total Value
Cotton	80	x	$80x$
Wool	20	y	$20y$

- The total cost of the first purchase was $1800.
- The total cost of the second purchase was $1000.

$60x + 90y = 1800$
$80x + 20y = 1000$

Solution
$-4(60x + 90y) = -4(1800)$
$3(80x + 20y) = 3(1000)$

$-240x - 360y = -7200$
$240x + 60y = 3000$

$-300y = -4200$
$y = 14$

$60x + 90(14) = 1800$
$60x + 1260 = 1800$
$60x = 540$
$x = 9$

The cost per yard of cotton is $9.00. The cost per yard of wool is $14.00.

CUMULATIVE REVIEW EXERCISES

1.
$$\frac{3}{2}x - \frac{3}{8} + \frac{1}{4}x = \frac{7}{12}x - \frac{5}{6}$$
$$24\left(\frac{3}{2}x - \frac{3}{8} + \frac{1}{4}x\right) = 24\left(\frac{7}{12}x - \frac{5}{6}\right)$$
$$36x - 9 + 6x = 14x - 20$$
$$42x - 9 = 14x - 20$$
$$28x - 9 = -20$$
$$28x = -11$$
$$x = -\frac{11}{28}$$

The solution is $-\frac{11}{28}$.

2.
$(x_1, y_1) = (2, -1), (x_2, y_2) = (3, 4)$

$m = \dfrac{y_2 - y_1}{x_2 - x_1} = \dfrac{4 - (-1)}{3 - 2} = \dfrac{5}{1} = 5$

$y - y_1 = m(x - x_1)$
$y - (-1) = 5(x - 2)$
$y + 1 = 5x - 10$
$y = 5x - 11$

The equation of the line is $y = 5x - 11$.

3. $3[x - 2(5 - 2x) - 4x] + 6 = 3(x - 10 + 4x - 4x) + 6$
$= 3(x - 10) + 6$
$= 3x - 30 + 6$
$= 3x - 24$

4. $a = 4, b = 8, c = -2$
$a + bc \div 2 = 4 + 8(-2) \div 2$
$= 4 - 16 \div 2$
$= 4 - 8$
$= -4$

5. $2x - 3 < 9$ or $5x - 1 < 4$
Solve each inequality.
$2x - 3 < 9$ or $5x - 1 < 4$
$2x < 12$ $\qquad$ $5x < 5$
$x < 6$ $\qquad$ $x < 1$
is the same as $\{x | x < 6\} \cup \{x | x < 1\}$, which is $\{x | x < 6\}$.

6. $|x - 2| - 4 < 2$
$|x - 2| < 6$

$-6 < x - 2 < 6$
$-6 + 2 < x - 2 + 2 < 6 + 2$
$-4 < x < 8$
$\{x | -4 < x < 8\}$

7. $|2x - 3| > 5$
Solve each inequality.
$2x - 3 < -5$ or $2x - 3 > 5$
$2x < -2$ $\qquad$ $2x > 8$
$x < -1$ $\qquad$ $x > 4$
This is the set $\{x | x < -1\} \cup \{x | x > 4\}$ or $\{x | x < -1 \text{ or } x > 4\}$.

8. $f(x) = 3x^3 - 2x^2 + 1$
$f(-3) = 3(-3)^3 - 2(-3)^2 + 1$
$f(-3) = 3(-27) - 2(9) + 1$
$f(-3) = -98$

Chapter 4: Systems Of Equations And Inequalities

9. The range is the set of numbers found by substituting in the set of numbers in the domain.
$f(-2) = 3(-2)^2 - 2(-2) = 16$
$f(-1) = 3(-1)^2 - 2(-1) = 5$
$f(0) = 3(0)^2 - 2(0) = 0$
$f(1) = 3(1)^2 - 2(1) = 1$
$f(2) = 3(2)^2 - 2(2) = 8$
The range is {0, 1, 5, 8, 16}.

10. $F(x) = x^2 - 3$
$F(2) = 2^2 - 3 = 1$

11. $f(x) = 3x - 4$
$f(2+h) = 3(2+h) - 4$
$= 6 + 3h - 4$
$= 2 + 3h$

$f(2) = 3(2) - 4 = 2$
$f(2+h) - f(2) = 2 + 3h - 2 = 3h$

12. $\{x | x \leq 2\} \cap \{x | x > -3\}$

    ⟵+—+—(+—+—+—+—]—+—+—+⟶
    -5 -4 -3 -2 -1 0 1 2 3 4 5

13. Slope $= -\frac{2}{3}$, Point $= (-2, 3)$
$y - 3 = -\frac{2}{3}[x - (-2)]$
$y - 3 = -\frac{2}{3}(x + 2)$
$y = -\frac{2}{3}x - \frac{4}{3} + 3$
$y = -\frac{2}{3}x + \frac{5}{3}$

14. The slope of the line $2x - 3y = 7$ is found by rearranging the equation as follows:
$-3y = 7 - 2x$
$y = -\frac{7}{3} + \frac{2}{3}x$
Slope $= \frac{2}{3}$

The perpendicular line has slope $= -\frac{3}{2}$.
The line is found using
$y - 2 = -\frac{3}{2}[x - (-1)]$
$y = -\frac{3}{2}(x + 1) + 2$
$y = -\frac{3}{2}x - \frac{3}{2} + 2$
$y = -\frac{3}{2}x + \frac{1}{2}$

15. The distance between points is
$\sqrt{(x_2 - x_1)^2 + (y_2 - y_1)^2}$.
$d = \sqrt{[2 - (-4)]^2 + (0 - 2)^2}$
$d = \sqrt{6^2 + (-2)^2}$
$d = \sqrt{36 + 4}$
$d = \sqrt{40}$
$d = 2\sqrt{10}$

16. The midpoint is found using $\left(\frac{x_1 + x_2}{2}, \frac{y_1 + y_2}{2}\right)$.
Midpoint $= \left(\frac{-4+3}{2}, \frac{3+5}{2}\right)$
$= \left(-\frac{1}{2}, 4\right)$

17. $2x - 5y = 10$
$-5y = -2x + 10$
$y = \frac{2}{5}x - 2$
The y-intercept is -2.
The slope is $\frac{2}{5}$.

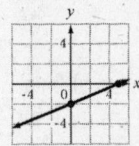

18. $3x - 4y \geq 8$
$-4y \geq -3x + 8$
$y \leq \frac{3}{4}x - 2$
The y-intercept is -2. The slope is $\frac{3}{4}$.

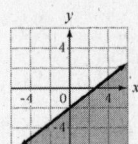

19. (1) $3x - 2y = 7$
    (2) $y = 2x - 1$
Solve by the substitution method.
$3x - 2(2x - 1) = 7$
$3x - 4x + 2 = 7$
$-x + 2 = 7$
$-x = 5$
$x = -5$
Substitute -5 for x in equation (2).
$y = 2x - 1$
$y = 2(-5) - 1 = -10 - 1 = -11$
The solution is $(-5, -11)$.

20.
(1) $3x + 2z = 1$
(2) $2y - z = 1$
(3) $x + 2y = 1$

Multiply equation (2) by –1 and add to equation (3).
$-1(2y - z) = -1(1)$
$x + 2y = 1$

$-2y + z = -1$
$x + 2y = 1$

(4) $x + z = 0$

Multiply equation (4) by –2 and add to equation (1).
$(-2)(x + z) = -2(0)$
$3x + 2z = 1$

$-2x - 2z = 0$
$3x + 2z = 1$

$x = 1$

Substitute 1 for x in equation (4).
$x + z = 0$
$1 + z = 0$
$z = -1$

Substitute 1 for x in equation (3).
$x + 2y = 1$
$1 + 2y = 1$
$2y = 0$
$y = 0$

The solution is (1, 0, –1).

21.
$\begin{vmatrix} 2 & -5 & 1 \\ 3 & 1 & 2 \\ 6 & -1 & 4 \end{vmatrix} = 2\begin{vmatrix} 1 & 2 \\ -1 & 4 \end{vmatrix} - 3\begin{vmatrix} -5 & 1 \\ -1 & 4 \end{vmatrix} + 6\begin{vmatrix} -5 & 1 \\ 1 & 2 \end{vmatrix}$

$= 2(4 + 2) - 3(-20 + 1) + 6(-10 - 1)$
$= 2(6) - 3(-19) + 6(-11)$
$= 12 + 57 - 66$
$= 3$

22.

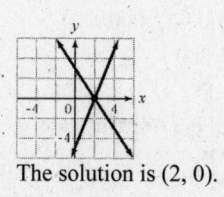

The solution is (2, 0).

23.
$D = \begin{vmatrix} 4 & -3 \\ 3 & -2 \end{vmatrix} = 4(-2) - 3(-3) = 1$

$D_x = \begin{vmatrix} 17 & -3 \\ 12 & -2 \end{vmatrix} = 17(-2) - 12(-3) = 2$

$D_y = \begin{vmatrix} 4 & 17 \\ 3 & 12 \end{vmatrix} = 4(12) - 3(17) = -3$

$x = \dfrac{D_x}{D} = \dfrac{2}{1} = 2$

$y = \dfrac{D_y}{D} = \dfrac{-3}{1} = -3$

The solution is (2, –3).

24. Solve each inequality for y.

$3x - 2y \geq 4$ $\quad x + y < 3$
$-2y \geq -3x + 4$ $\quad y < -x + 3$
$y \leq \dfrac{3}{2}x - 2$

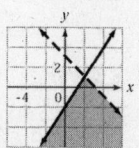

25. Strategy

The unknown number of quarters: x
The unknown number of dimes: $3x$
The unknown number of nickels:
$40 - (x + 3x)$

	Number	Value	Total Value
Quarters	x	25	$25x$
Dimes	$3x$	10	$10(3x)$
Nickels	$40 - 4x$	5	$5(40 - 4x)$

The sum of the total values of the denominations is $4.10 (410 cents). $25x + 10(3x) + 5(40 - 4x) = 410$

Solution
$25x + 10(3x) + 5(40 - 4x) = 410$
$25x + 30x + 200 - 20x = 410$
$35x + 200 = 410$
$35x = 210$
$x = 6$

$40 - 4x = 40 - 4(6) = 40 - 24 = 16$
There are 16 nickels in the purse.

26. Strategy
The unknown amount of pure water: x

	Amount	Percent	Quantity
Water	x	0	$0x$
4%	100	0.04	$100 \cdot (0.04)$
2.5%	$100+x$	0.025	$(100+x) \cdot (0.025)$

The sum of the quantities before mixing equals the quantity after mixing.
$0 \cdot x + 100(0.04) = (100+x)0.025$

Solution
$$0 \cdot x + 100(0.04) = (100+x)0.025$$
$$0 + 4 = 2.5 + 0.025x$$
$$1.5 = 0.025x$$
$$60 = x$$
The amount of water that should be added is 60 ml.

27. Strategy
Rate of the plane in calm air: x
Rate of the wind: y

	Rate	Time	Distance
With wind	$x+y$	2	$2(x+y)$
Against wind	$x-y$	3	$3(x-y)$

- The distance traveled with the wind is 150 mi.
- The distance traveled against the wind is 150 mi.

$2(x+y) = 150$
$3(x-y) = 150$

Solution
$2(x+y) = 150$
$3(x-y) = 150$

$$\frac{1}{2} \cdot 2(x+y) = \frac{1}{2} \cdot 150$$
$$\frac{1}{3} \cdot 3(x-y) = \frac{1}{3} \cdot 150$$

$x + y = 75$
$x - y = 50$

$2x = 125$
$x = 62.5$

$x + y = 75$
$62.5 + y = 75$
$y = 12.5$

The rate of the wind is 12.5 mph.

28. Strategy
Cost per pound of hamburger: x
Cost per pound of steak: y
First purchase:

	Amount	Cost	Value
Hamburger	100	x	$100x$
Steak	50	y	$50y$

Second purchase:

	Amount	Cost	Value
Hamburger	150	x	$150x$
Steak	100	y	$100y$

- The total cost of the first purchase is $490.
- The total cost of the second purchase is $860.

$100x + 50y = 490$
$150x + 100y = 860$

Solution
$100x + 50y = 490$
$150x + 100y = 860$

$3(100x + 50y) = 3(490)$
$-2(150x + 100y) = -2(860)$

$300x + 150y = 1470$
$-300x - 200y = -1720$

$-50y = -250$
$y = 5$

The cost per pound of steak is $5.

29. Strategy
Let M be the number of ohms, T the tolerance, and r the given amount of resistance. Find the tolerance and solve $|M - r| \leq T$ for M.

Solution
$T = 0.15 \cdot 12{,}000 = 1800$ ohms

$$|M - 12{,}000| \leq 1800$$
$$-1800 \leq M - 12{,}000 \leq 1800$$
$$-1800 + 12{,}000 \leq M - 12{,}000 + 12{,}000 \leq 1800 + 12{,}000$$
$$10{,}200 \leq M \leq 13{,}800$$

The lower and upper limits of the resistance are 10,200 ohms and 13,800 ohms.

30.
The slope of the line is
$$\frac{5000 - 1000}{100 - 0} = \frac{4000}{100} = 40$$
The slope represents the marginal income or the income generated per number of sales. The account executive earns $40 for each $1000 of sales.

CHAPTER 5: POLYNOMIALS AND EXPONENTS

CHAPTER 5 PREP TEST

1. $-4(3y) = -12y$ [1.3.3]
2. $(-2)^3 = -8$ [1.2.3]
3. $-4a - 8b + 7a = 3a - 8b$ [1.3.3]
4. $3x - 2[y - 4(x+1) + 5]$ [1.3.3]
   $= 3x - 2[y - 4x - 4 + 5]$
   $= 3x - 2[y - 4x + 1]$
   $= 3x - 2y + 8x - 2$
   $= 11x - 2y - 2$
5. $-(x - y) = -x + y$ [1.3.3]
6. $40 = 2 \cdot 2 \cdot 2 \cdot 5$ [1.2.2]
7. 4 [1.2.2]
8. $x^3 - 2x^2 + x + 5$ [1.3.2]
   $= (-2)^3 - 2(-2)^2 + (-2) + 5$
   $= -8 - 2(4) - 2 + 5$
   $= -8 - 8 + 3$
   $= -13$
9. $3x + 1 = 0$ [2.1.1]
   $3x = -1$
   $x = -\dfrac{1}{3}$

Section 5.1

Concept Review 5.1

1. Never true
   $2^{-4} = \dfrac{1}{2^4} = \dfrac{1}{16}$

3. Never true
   $(2+3)^{-1} = 5^{-1} = \dfrac{1}{5}$

5. Never true
   There is no rule to add two numbers with the same base.

Objective 5.1.1 Exercises

1. Of the two expressions $x^4 + y^5$ and x^4y^5, the one which is a monomial is x^4y^5. The degree of this monomial is _9_.

3. a. Yes (Rule for Simplifying Powers of Products)
   b. No
   c. No
   d. Yes (Rule for Simplifying Powers of Products)

5. $(ab^3)(a^3b) = a^4b^4$
7. $(9xy^2)(-2x^2y^2) = -18x^3y^4$
9. $(x^4y^4)^4 = x^8y^{16}$
11. $(-3x^2y^3)^4 = (-3)^4 x^8 y^{12} = 81x^8 y^{12}$
13. $(3^3 a^5 b^3)^2 = 3^6 a^{10} b^6 = 729 a^{10} b^6$
15. $(x^2y^2)(xy^3)^3 = (x^2y^2)(x^3y^9) = x^5 y^{11}$
17. $[(3x)^3]^2 = (3x)^6 = 3^6 x^6 = 729 x^6$
19. $[(ab)^3]^6 = (ab)^{18} = a^{18}b^{18}$
21. $[(2xy)^3]^4 = (2xy)^{12} = 2^{12} x^{12} y^{12} = 4096 x^{12} y^{12}$
23. $[(2a^4b^3)^3]^2 = (2a^4b^3)^6 = 2^6 a^{24} b^{18} = 64 a^{24} b^{18}$
25. $x^n \cdot x^{n+1} = x^{n+n+1} = x^{2n+1}$
27. $y^{3n} \cdot y^{3n-2} = y^{3n+3n-2} = y^{6n-2}$
29. $(a^n)^{3n} = a^{3n^2}$
31. $(x^{3n})^5 = x^{15n}$
33. $(2xy)(-3x^2yz)(x^2y^3z^3) = -6x^5y^5z^4$
35. $(3b^5)(2ab^2)(-2ab^2c^2) = -12a^2b^9c^2$
37. $(-2x^2y^3z)(3x^2yz^4) = -6x^4y^4z^5$
39. $(-3ab^3)^3(-2^2a^2b)^2 = [(-3)^3 a^3 b^9][(-2^2)^2 a^4 b^2]$
    $= (-27a^3b^9)(16a^4b^2)$
    $= -432a^7b^{11}$
41. $(-2ab^2)(-3a^4b^5)^3 = (-2ab^2)[(-3)^3 a^{12} b^{15}]$
    $= (-2ab^2)(-27a^{12}b^{15})$
    $= 54a^{13}b^{17}$

Objective 5.1.2 Exercises

43. If a variable has a negative exponent, write the expression in the denominator of a fraction whose numerator is 1 and change the sign of the exponent. For example, $x^{-3} = \dfrac{1}{x^3}$. If the expression is in the denominator of a fraction, write it in the numerator of the fraction and change the sign of the exponent. For example, $\dfrac{1}{x^{-5}} = \dfrac{x^5}{1} = x^5$.

45.
$$\left(\frac{a^{-3}}{a^5}\right)^2 = (a^{-8})^2$$
$$= a^{-16}$$
$$= \frac{1}{a^{16}}$$

47. $\dfrac{a^8}{a^5} = a^{8-5} = a^3$

49. $\dfrac{a^7 b}{a^2 b^4} = a^{7-2} b^{1-4} = a^5 b^{-3} = \dfrac{a^5}{b^3}$

51. $\dfrac{1}{3^{-5}} = 3^5 = 243$

53. $\dfrac{1}{y^{-3}} = y^3$

55. $\dfrac{a^3}{4b^{-2}} = \dfrac{a^3 b^2}{4}$

57. $x^{-3} \cdot x^{-5} = x^{-8} = \dfrac{1}{x^8}$

59. $(5x^2)^{-3} = 5^{-3} x^{-6} = \dfrac{1}{5^3 x^6} = \dfrac{1}{125 x^6}$

61. $\dfrac{x^4}{x^{-5}} = x^9$

63. $a^{-5} \cdot a^7 = a^2$

65. $(x^3 y^5)^{-2} = x^{-6} y^{-10} = \dfrac{1}{x^6 y^{10}}$

67.
$$(3a)^{-3}(9a^{-1})^{-2} = (3a)^{-3}(3^2 a^{-1})^{-2}$$
$$= (3^{-3} a^{-3})(3^{-4} a^2)$$
$$= 3^{-7} a^{-1}$$
$$= \dfrac{1}{3^7 a}$$
$$= \dfrac{1}{2187 a}$$

69.
$$(x^{-1} y^2)^{-3}(x^2 y^{-4})^{-3} = (x^3 y^{-6})(x^{-6} y^{12})$$
$$= x^{-3} y^6$$
$$= \dfrac{y^6}{x^3}$$

71. $\left(\dfrac{x^2 y^{-1}}{xy}\right)^{-4} = \left(\dfrac{x}{y^2}\right)^{-4} = \dfrac{x^{-4}}{y^{-8}} = \dfrac{y^8}{x^4}$

73. $\dfrac{a^2 b^3 c^7}{a^6 b c^5} = \dfrac{b^2 c^2}{a^4}$

75. $\dfrac{(-3a^2 b^3)^2}{(-2ab^4)^3} = \dfrac{(-3)^2 a^4 b^6}{(-2)^3 a^3 b^{12}} = \dfrac{9 a^4 b^6}{-8 a^3 b^{12}} = -\dfrac{9a}{8b^6}$

77. $\left(\dfrac{a^{-2} b}{a^3 b^{-4}}\right)^2 = \left(\dfrac{b^5}{a^5}\right)^2 = \dfrac{b^{10}}{a^{10}}$

79. $\dfrac{y^{2n}}{-y^{8n}} = -\dfrac{1}{y^{8n-2n}} = -\dfrac{1}{y^{6n}}$

81.
$$\dfrac{x^{2n-1} y^{n-3}}{x^{n+4} y^{n+3}} = x^{2n-1-(n+4)} y^{n-3-(n+3)}$$
$$= x^{2n-1-n-4} y^{n-3-n-3}$$
$$= x^{n-5} y^{-6}$$
$$= \dfrac{x^{n-5}}{y^6}$$

83. $\dfrac{(3x^{-2} y)^{-2}}{(4xy^{-2})^{-1}} = \dfrac{3^{-2} x^4 y^{-2}}{4^{-1} x^{-1} y^2} = \dfrac{4x^5}{9y^4}$

85.
$$\left(\dfrac{9ab^{-2}}{8a^{-2}b}\right)^{-2} \left(\dfrac{3a^{-2}b}{2a^2 b^{-2}}\right)^3 = \left(\dfrac{9a^3 b^{-3}}{8}\right)^{-2} \left(\dfrac{3a^{-4} b^3}{2}\right)^3$$
$$= \dfrac{9^{-2} a^{-6} b^6}{8^{-2}} \cdot \dfrac{3^3 a^{-12} b^9}{2^3}$$
$$= \dfrac{8^2 \cdot 3^3 a^{-18} b^{15}}{9^2 \cdot 2^3}$$
$$= \dfrac{64 \cdot 27 b^{15}}{81 \cdot 8 a^{18}}$$
$$= \dfrac{8 b^{15}}{3 a^{18}}$$

87. $[(x^{-2} y^{-1})^2]^{-3} = (x^{-4} y^{-2})^{-3} = x^{12} y^6$

89. $\left[\left(\dfrac{a^2}{b}\right)^{-1}\right]^2 = \left(\dfrac{a^{-2}}{b^{-1}}\right)^2 = \left(\dfrac{b}{a^2}\right)^2 = \dfrac{b^2}{a^4}$

91.

a. False. $\dfrac{a^n}{b^m}$ cannot be simplified any further.

b. True

Objective 5.1.3 Exercises

93. To write the number 0.00000078 in scientific notation, move the decimal point _7_ places to the right. This means the exponent on 10 is _7_.

95. $0.00000005 = 5 \times 10^{-8}$

97. $4,300,000 = 4.3 \times 10^{6}$

99. $9,800,000,000 = 9.8 \times 10^{9}$

101. $6.2 \times 10^{-12} = 0.0000000000062$

103. $6.34 \times 10^{5} = 634,000$

105. $4.35 \times 10^{9} = 4,350,000,000$

107. $(8.9 \times 10^{-5})(3.2 \times 10^{-6}) = (8.9)(3.2) \times 10^{-5+(-6)}$
$= 28.48 \times 10^{-11}$
$= 2.848 \times 10^{-10}$

109. $(480,000)(0.0000000096)$
$= (4.8 \times 10^{5})(9.6 \times 10^{-9})$
$= (4.8)(9.6) \times 10^{5+(-9)}$
$= 46.08 \times 10^{-4}$
$= 4.608 \times 10^{-3}$

111. $\dfrac{2.7 \times 10^{4}}{3 \times 10^{-6}} = 0.9 \times 10^{4-(-6)}$
$= 0.9 \times 10^{10}$
$= 9 \times 10^{9}$

113. $\dfrac{4800}{0.00000024} = \dfrac{4.8 \times 10^{3}}{2.4 \times 10^{-7}}$
$= 2 \times 10^{3-(-7)}$
$= 2 \times 10^{10}$

115. $\dfrac{0.000000346}{0.0000005} = \dfrac{3.46 \times 10^{-7}}{5 \times 10^{-7}}$
$= 0.692 \times 10^{-7-(-7)}$
$= 0.692 \times 10^{0} = 6.92 \times 10^{-1}$

117. $\dfrac{(6.9 \times 10^{27})(8.2 \times 10^{-13})}{4.1 \times 10^{15}}$
$= \dfrac{(6.9)(8.2) \times 10^{27+(-13)-15}}{4.1}$
$= 13.8 \times 10^{-1}$
$= 1.38$

119. $\dfrac{(720)(0.0000000039)}{(26,000,000,000)(0.018)}$
$= \dfrac{7.2 \times 10^{2} \times 3.9 \times 10^{-9}}{2.6 \times 10^{10} \times 1.8 \times 10^{-2}}$
$= \dfrac{(7.2)(3.9) \times 10^{2+(-9)-10-(-2)}}{(2.6)(1.8)}$
$= 6 \times 10^{-15}$

121. Since $n < 0$ and $m < 0$, the sum $n + m < 0$. The expression $(a \times 10^{n})(b \times 10^{m})$ is less than one.

Objective 5.1.4 Exercises

123. The speed of light is 3×10^{5} km/s. Find how many kilometers light travels in five hours:
$d = (3 \times 10^{5})(1.8 \times 10^{4}) = \underline{5.4} \times 10^{\underline{9}}$ km

125. Strategy
To find the number of arithmetic operations:

Find the reciprocal of 2×10^{-9}, which is the number of operations performed in one second.

Write the number of seconds in one minute (60) in scientific notation.

Multiply the number of arithmetic operations per second by the number of seconds in one minute.

Solution
$\dfrac{1}{2 \times 10^{-9}} = \dfrac{1}{2} \times 10^{9}$
$60 = 6 \times 10$
$\left(\dfrac{1}{2} \times 10^{9}\right)(6 \times 10) = \dfrac{1}{2} \times 6 \times 10^{10}$
$= 3 \times 10^{10}$

The computer can perform 3×10^{10} operations in one minute.

127. Strategy
To find the distance traveled:
Write the number of seconds in one day in scientific notation.
Use the equation $d = rt$, where r is the speed of light and t is the number of seconds in one day.

Solution
$r = 3 \times 10^{8}$
$24 \cdot 60 \cdot 60 = 86,400 = 8.64 \times 10^{4}$
$d = rt$
$d = (3 \times 10^{8})(8.64 \times 10^{4})$
$d = 3 \times 8.64 \times 10^{12}$
$d = 25.92 \times 10^{12}$
$d = 2.592 \times 10^{13}$

Light travels 2.592×10^{13} m in one day.

129. Strategy
To find the number of times heavier the proton is, divide the mass of the proton by the mass of the electron.

Solution
$$\frac{1.673 \times 10^{-27}}{9.109 \times 10^{-31}} = 0.183664508 \times 10^4$$
$$= 1.83664508 \times 10^3$$
The proton is 1.83664508×10^3 times heavier than the electron.

131. Strategy
To find the rate, divide the number of miles by the time.

Solution
$$\frac{119 \times 10^6}{11} = 1.0\overline{81} \times 10^7 \text{ mi/min}$$
The signals traveled at $1.0\overline{81} \times 10^7$ mi/min.

133. Strategy
To find the weight of one seed:
Write the number of seeds per ounce in scientific notation.
Find the reciprocal of the number of seeds per ounce, which is the number of ounces per seed.

Solution
$31,000,000 = 3.1 \times 10^7$
$$\frac{1}{3.1 \times 10^7} = 0.32258065 \times 10^{-7}$$
$$= 3.2258065 \times 10^{-8}$$
The weight of one orchid seed is 3.2258065×10^{-8} oz.

135. Strategy
To find the time, use the equation $d = rt$, where r is the speed of the satellite and d is the distance to Saturn.

Solution
$$d = rt$$
$$8.86 \times 10^8 = (1 \times 10^5)t$$
$$\frac{8.86 \times 10^8}{1 \times 10^5} = t$$
$$8.86 \times 10^3 = t$$
It will take the satellite 8.86×10^3 h to reach Saturn.

137. Strategy
To find the volume, use the formula $V = \frac{4}{3}\pi r^3$, where $r = 1.5 \times 10^{-4}$ mm.

Solution
$$V = \frac{4}{3}\pi r^3$$
$$= \frac{4}{3}\pi(1.5 \times 10^{-4})^3$$
$$= \frac{4}{3}\pi(1.5)^3 \times 10^{-12}$$
$$= 14.1371669 \times 10^{-12}$$
$$= 1.41371669 \times 10^{-11}$$
The volume of the cell is $1.41371669 \times 10^{-11}$ mm^3.

139. Strategy
To find the time:
Write the speed of the space ship in scientific notation.
Use the equation $d = rt$, where r is the speed of the space ship and d is the distance across the galaxy.

Solution
$25,000 = 2.5 \times 10^4$
$$d = rt$$
$$5.6 \times 10^{10} = (2.5 \times 10^4)t$$
$$\frac{5.6 \times 10^{19}}{2.5 \times 10^4} = t$$
$$2.24 \times 10^{15} = t$$
The space ship travels across the galaxy in 2.24×10^{15} h.

Applying Concepts 5.1

141.
$$\frac{3}{x} + 1$$
No, the expression is not a polynomial because there is a variable in the denominator.

143.
$$\sqrt{5}x + 2$$
Yes, the expression is a polynomial.

145.
$$x + \sqrt{3}$$
Yes, the expression is a polynomial.

147.
$$(6x^3 + kx^2 - 2x - 1) - (4x^3 - 3x^2 + 1) = 2x^3 - x^2 - 2x - 2$$
$$(6x^3 + kx^2 - 2x - 1) + (-4x^3 + 3x^2 - 1) = 2x^3 - x^2 - 2x - 2$$
$$2x^3 + (k+3)x^2 - 2x - 2 = 2x^3 - x^2 - 2x - 2$$
$$k + 3 = -1$$
$$k = -4$$

149. **Strategy**
To find the perimeter, replace the variables a, b, and c in the equation $P = a + b + c$ by the given values and solve for P.

Solution
$P = a + b + c$
$P = 4x^n + 3x^n + 3x^n$
$P = 10x^n$
The perimeter is $10x^n$.

151. **Strategy**
To find the area, replace the variables b and h in the equation $A = \frac{1}{2}bh$ by the given values and solve for A.

Solution
$A = \frac{1}{2}bh$
$A = \frac{1}{2}(8xy)(5xy)$
$A = 4xy(5xy)$
$A = 20x^2y^2$
The area is $20x^2y^2$.

153.
$\frac{5x^3}{y^{-6}} + \left(\frac{x^{-1}}{y^2}\right)^{-3} = 5x^3y^6 + (x^{-1}y^{-2})^{-3}$
$= 5x^3y^6 + x^3y^6$
$= 6x^3y^6$

155.
$\left(\frac{2m^3n^{-2}}{4m^4n}\right)^{-2} \div \left(\frac{mn^5}{m^{-1}n^3}\right)^3$
$= \left(\frac{m^{-1}n^{-3}}{2}\right)^{-2} \div (m^2n^2)^3$
$= \frac{m^2n^6}{2^{-2}} \div m^6n^6$
$= 2^2 m^2n^6 \div m^6n^6$
$= \frac{4m^2n^6}{m^6n^6}$
$= \frac{4}{m^4}$

157. No, let $a = -1$, and $b = 1$. Then $a < b$, but $a^{-1} = \frac{1}{-1} = -1 < b^{-1} = \frac{1}{1} = 1$ because $-1 < 1$.

Section 5.2

Concept Review 5.2

1. Always true

3. Always true

5. Sometimes true
The sum of $-x^2 + 2x - 3$ and $x^2 + 5x - 10$ is the binomial $7x - 13$.

Objective 5.2.1 Exercises

1. The degree of a polynomial in one variable is the greatest of the degrees of any of its terms.
$x^3 - 2x^2 + 4$ and $7 + 5x - 6x^3$ are third-degree polynomials.

3. The degree of P is $\underline{2}$. The coefficient -3 is called the <u>leading</u> coefficient. The term 10 is called the <u>constant</u> term.

5. Polynomial: (a) –1 (b) 8 (c) 2

7. Not a polynomial.

9. Not a polynomial.

11. Polynomial: (a) 3 (b) π (c) 5

13. Polynomial: (a) –5 (b) 2 (c) 3

15. Polynomial: (a) 14 (b) 14 (c) 0

17. $P(3) = 3(3)^2 - 2(3) - 8$
$P(3) = 13$

19. $R(2) = 2(2)^3 - 3(2)^2 + 4(2) - 2$
$R(2) = 10$

21. $f(-1) = (-1)^4 - 2(-1)^2 - 10$
$f(-1) = -11$

23.

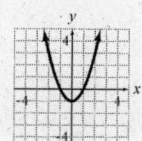

25.

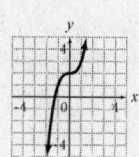

27.

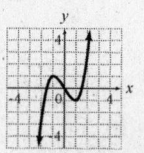

29. $L(s) = 0.641s^2$
$L(6) = 0.641(6)^2$
$L(6) = 23.076$
The length of the wave is 23.1 m.

31. $T(n) = n^2 - n$
    $T(8) = (8)^2 - 8$
    $T(8) = 56$
    The league must schedule 56 games.

33. $M(r) = 6.14r^2 + 6.14r + 2.094$
    $M(6) = 6.14(6)^2 + 6.14(6) + 2.094$
    $M(6) = 259.974$
    260 in^3 of meringue are needed.

35. Because this is a cubic function, its graph is like the one shown in C.

Objective 5.2.2 Exercises

37. The additive inverse of $8x^2 + 3x - 5$ is
    $-(8x^2 + 3x - 5) = \underline{-8x^2 - 3x + 5}$.

39. $5x^2 + 2x - 7$
    $x^2 - 8x + 12$
    $\overline{6x^2 - 6x + 5}$

41. $x^2 - 3x + 8$
    $-2x^2 + 3x - 7$
    $\overline{-x^2 + 1}$

43. $(3y^2 - 7y) + (2y^2 - 8y + 2)$
    $= (3y^2 + 2y^2) + (-7y - 8y) + 2$
    $= 5y^2 - 15y + 2$

45. $(2a^2 - 3a - 7) - (-5a^2 - 2a - 9)$
    $= (2a^2 + 5a^2) + (-3a + 2a) + (-7 + 9)$
    $= 7a^2 - a + 2$

47. $P(x) + R(x) = (3x^3 - 4x^2 - x + 1) + (2x^3 + 5x - 8)$
    $= 5x^3 - 4x^2 + 4x - 7$

49. $P(x) + R(x) = (x^{2n} + 7x^n - 3) + (-x^{2n} + 2x^n + 8)$
    $= 9x^n + 5$

51. $S(x) = P(x) + R(x)$
    $= (3x^4 - 3x^3 - x^2) + (3x^3 - 7x^2 + 2x)$
    $= 3x^4 - 8x^2 + 2x$

53. $D(x) = P(x) - R(x)$
    $= (x^2 + 2x + 1) - (2x^3 - 3x^2 + 2x - 7)$
    $= (x^2 + 2x + 1) + (-2x^3 + 3x^2 - 2x + 7)$
    $= -2x^3 + 4x^2 + 8$

55. $Q(x) - P(x) = (dx^2 - ex + f) - (-ax^2 - bx + c)$
    $= (dx^2 - ex + f) + (ax^2 + bx - c)$
    $= dx^2 + ax^2 - ex + bx + f - c$
    $= (d + a)x^2 + (-e + b)x + (f - c)$

   a. Because a and d are both positive, the coefficient of the x^2-term, $d + a$, is positive.

   b. Because $b > e$, the coefficient of the x-term, $-e + b$, is positive.

   c. Because $c > f$, the coefficient of the constant term, $f - c$, is negative.

Applying Concepts 5.2

57. $(2x^3 + 3x^2 + kx + 5) - (x^3 + x^2 - 5x - 2) = x^3 + 2x^2 + 3x + 7$
    $(2x^3 - x^3) + (3x^2 - x^2) + (kx + 5x) + (5 + 2) = x^3 + 2x^2 + 3x + 7$
    $x^3 + 2x^2 + (k + 5)x + 7 = x^3 + 2x^2 + 3x + 7$
    $(k + 5)x = 3x$
    $k + 5 = 3$
    $k = -2$

59. The degree of $P(x) + Q(x)$ will be 4.
    Example:
    $P(x) = 2x^3 + 4x^2 - 3x + 9$
    $Q(x) = x^4 - 5x + 1$
    $P(x) + Q(x) = x^4 + 2x^3 + 4x^2 - 8x + 10$,
    a fourth-degree polynomial.

61. Strategy
    Determine the midpoint of the beam, x, and substitute x into $D(x)$ to determine the deflection.

    Solution
    Length of beam = 10 ft
    Midpoint of beam = 5
    $D(x) = 0.005x^4 - 0.1x^3 + 0.5x^2$
    $D(5) = 0.005(5)^4 - 0.1(5)^3 + 0.5(5)^2$
    $D(5) = 3.125$
    The maximum deflection of the beam is 3.125 in.

63.
$P(x) = 4x^4 - 3x^2 + 6x + c$
$-3 = 4(-1)^4 - 3(-1)^2 + 6(-1) + c$
$-3 = 4 - 3 - 6 + c$
$-3 = -5 + c$
$2 = c$
The value of c is 2.

65.

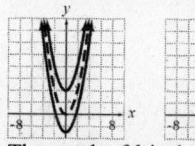

The graph of k is the graph of f moved 2 units down.

67. The graphs of $f(x) = x^2$, $g(x) = (x-3)^2$, and $h(x) = x^2 - 3$ are all graphs of the parabola whose equation is $f(x) = x^2$. The graph of g is shifted 3 units to the right; the graph of h is shifted three units down. See accompanying diagram.

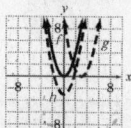

Section 5.3
Concept Review 5.3
1. Always true

3. Always true

5. Always true

Objective 5.3.1 Exercises
1. The Distributive Property is used to multiply expressions when one or both of the expressions have more than one term.

3. Use the Distributive Property to multiply each term of $(y - 10)$ by $-3y^3$.
$-3y^3(y^2 - 10) = \underline{-3y^3}(y^2) - (\underline{-3y^3})(10)$
$\qquad\qquad\qquad = \underline{-3y^5} + \underline{30y^3}$

5. $2x(x-3) = 2x^2 - 6x$

7. $3x^2(2x^2 - x) = 6x^4 - 3x^3$

9. $3xy(2x - 3y) = 6x^2y - 9xy^2$

11. $x^n(x+1) = x^{n+1} + x^n$

13. $x^n(x^n + y^n) = x^{2n} + x^n y^n$

15. $2b + 4b(2-b) = 2b + 8b - 4b^2 = -4b^2 + 10b$

17. $-2a^2(3a^2 - 2a + 3) = -6a^4 + 4a^3 - 6a^2$

19. $3b(3b^4 - 3b^2 + 8) = 9b^5 - 9b^3 + 24b$

21. $-5x^2(4 - 3x + 3x^2 + 4x^3)$
$= -20x^2 + 15x^3 - 15x^4 - 20x^5$
$= -20x^5 - 15x^4 + 15x^3 - 20x^2$

23. $-2x^2y(x^2 - 3xy + 2y^2) = -2x^4y + 6x^3y^2 - 4x^2y^3$

25. $x^n(x^{2n} + x^n + x) = x^{3n} + x^{2n} + x^{n+1}$

27. $a^{n+1}(a^n - 3a + 2) = a^{2n+1} - 3a^{n+2} + 2a^{n+1}$

29. $2y^2 - y[3 - 2(y - 4) - y] = 2y^2 - y[3 - 2y + 8 - y]$
$\qquad\qquad\qquad\qquad\qquad = 2y^2 - y[11 - 3y]$
$\qquad\qquad\qquad\qquad\qquad = 2y^2 - 11y + 3y^2$
$\qquad\qquad\qquad\qquad\qquad = 5y^2 - 11y$

31. $2y - 3[y - 2y(y-3) + 4y]$
$= 2y - 3[y - 2y^2 + 6y + 4y]$
$= 2y - 3[11y - 2y^2]$
$= 2y - 33y + 6y^2$
$= 6y^2 - 31y$

33. $2x - x(5x - 3) = 2x - 5x^2 + 3x$
$\qquad\qquad\qquad = -5x^2 + 5x$
c and e

Objective 5.3.2 Exercises
35. For the product $(3x - 2)(x + 5)$:
The First terms are $\underline{3x}$ and $\underline{x}$.
The Outer terms are $\underline{3x}$ and $\underline{5}$.
The Inner terms are $\underline{-2}$ and $\underline{x}$.
The Last terms are $\underline{-2}$ and $\underline{5}$.

37. $(5x - 7)(3x - 8) = 15x^2 - 40x - 21x + 56$
$\qquad\qquad\qquad\quad = 15x^2 - 61x + 56$

39. $(7x - 3y)(2x - 9y) = 14x^2 - 63xy - 6xy + 27y^2$
$\qquad\qquad\qquad\qquad = 14x^2 - 69xy + 27y^2$

41. $(3a - 5b)(a + 7b) = 3a^2 + 21ab - 5ab - 35b^2$
$\qquad\qquad\qquad\qquad = 3a^2 + 16ab - 35b^2$

43. $(5x + 9y)(3x + 2y) = 15x^2 + 10xy + 27xy + 18y^2$
$\qquad\qquad\qquad\qquad = 15x^2 + 37xy + 18y^2$

45. $(5x - 9y)(6x - 5y) = 30x^2 - 25xy - 54xy + 45y^2$
$\qquad\qquad\qquad\qquad = 30x^2 - 79xy + 45y^2$

47. $(xy - 5)(2xy + 7) = 2x^2y^2 + 7xy - 10xy - 35$
$\qquad\qquad\qquad\quad = 2x^2y^2 - 3xy - 35$

49. $(x^2 - 4)(x^2 - 6) = x^4 - 6x^2 - 4x^2 + 24$
$\qquad\qquad\qquad\quad = x^4 - 10x^2 + 24$

138 Chapter 5: Polynomials And Exponents

51. $(x^2 - 2y^2)(x^2 + 4y^2)$
$= x^4 + 4x^2y^2 - 2x^2y^2 - 8y^4$
$= x^4 + 2x^2y^2 - 8y^4$

53. $(x^n - 4)(x^n - 5) = x^{2n} - 5x^n - 4x^n + 20$
$= x^{2n} - 9x^n + 20$

55. $(5b^n - 1)(2b^n + 4) = 10b^{2n} + 20b^n - 2b^n - 4$
$= 10b^{2n} + 18b^n - 4$

57. $(3x^n + b^n)(x^n + 2b^n)$
$= 3x^{2n} + 6x^n b^n + x^n b^n + 2b^{2n}$
$= 3x^{2n} + 7x^n b^n + 2b^{2n}$

59. $\quad\quad x^2 + 5x - 8$
$\underline{\times \quad\quad\quad x + 3}$
$\quad\quad 3x^2 + 15x - 24$
$\underline{x^3 + 5x^2 - 8x \quad\quad\quad}$
$x^3 + 8x^2 + 7x - 24$

61. $\quad\quad a^3 - 3a^2 \quad\quad + 7$
$\underline{\times \quad\quad\quad\quad\quad\quad a + 2}$
$\quad\quad 2a^3 - 6a^2 \quad\quad + 14$
$\underline{a^4 - 3a^3 \quad\quad + 7a \quad\quad}$
$a^4 - a^3 - 6a^2 + 7a + 14$

63. $\quad\quad 2a^2 - 5ab - 3b^2$
$\underline{\times \quad\quad\quad\quad\quad 3a + b}$
$\quad\quad 2a^2b - 5ab^2 - 3b^3$
$\underline{6a^3 - 15a^2b - 9ab^2 \quad\quad\quad}$
$6a^3 - 13a^2b - 14ab^2 - 3b^3$

65. $\quad\quad 3b^2 - 3b + 6$
$\underline{\times \quad\quad\quad 2b^2 \quad - 3}$
$\quad\quad\quad\quad - 9b^2 + 9b - 18$
$\underline{6b^4 - 6b^3 + 12b^2 \quad\quad\quad}$
$6b^4 - 6b^3 + 3b^2 + 9b - 18$

67. $\quad\quad 3a^4 \quad\quad - 3a^2 + 2a - 5$
$\underline{\times \quad\quad\quad\quad\quad\quad\quad\quad 2a - 5}$
$\quad\quad -15a^4 \quad + 15a^2 - 10a + 25$
$\underline{6a^5 \quad\quad - 6a^3 + 4a^2 - 10a \quad\quad\quad}$
$6a^5 - 15a^4 - 6a^3 + 19a^2 - 20a + 25$

69. $\quad\quad\quad\quad x^2 - 3x + 1$
$\underline{\times \quad\quad\quad\quad x^2 - 2x + 7}$
$\quad\quad\quad\quad 7x^2 - 21x + 7$
$\quad\quad -2x^3 + 6x^2 - 2x$
$\underline{x^4 - 3x^3 + \quad x^2 \quad\quad\quad}$
$x^4 - 5x^3 + 14x^2 - 23x + 7$

71. $(b - 3)(3b - 2)(b - 1)$
$= (3b^2 - 2b - 9b + 6)(b - 1)$
$= (3b^2 - 11b + 6)(b - 1)$
$\quad\quad\quad = \quad\quad 3b^2 - 11b + 6$
$\quad\quad\quad\underline{\times \quad\quad\quad\quad\quad b - 1}$
$\quad\quad\quad\quad\quad -3b^2 + 11b - 6$
$\quad\quad\quad\underline{3b^3 - 11b^2 + 6b \quad\quad\quad}$
$\quad\quad\quad 3b^3 - 14b^2 + 17b - 6$

73. $\quad\quad\quad x^{2n} - 3x^n y^n - y^{2n}$
$\underline{\times \quad\quad\quad\quad\quad\quad x^n - y^n}$
$\quad\quad - x^{2n} y^n + 3x^n y^{2n} + y^{3n}$
$\underline{x^{3n} - 3x^{2n} y^n - x^n y^{2n} \quad\quad\quad}$
$x^{3n} - 4x^{2n} y^n + 2x^n y^{2n} + y^{3n}$

75. No. The constant term of the product is the product of two negative numbers, which gives a positive product.

Objective 5.3.3 Exercises

77. The product $(a + b)^2 = a^2 + 2ab + b^2$, so
$(4x + 1)^2 = (\underline{4x})^2 + 2(\underline{4x})(\underline{1}) + (\underline{1})^2 = \underline{16x^2} + \underline{8x} + \underline{1}$.

79. $(b - 7)(b + 7) = b^2 - 49$

81. $(b - 11)(b + 11) = b^2 - 121$

83. $(5x - 4y)^2 = 25x^2 - 40xy + 16y^2$

85. $(x^2 + y^2)^2 = x^4 + 2x^2 y^2 + y^4$

87. $(2a - 3b)(2a + 3b) = 4a^2 - 9b^2$

89. $(x^2 + 1)(x^2 - 1) = x^4 - 1$

91. $(2x^n + y^n)^2 = 4x^{2n} + 4x^n y^n + y^{2n}$

93. $(5a - 9b)(5a + 9b) = 25a^2 - 81b^2$

95. $(2x^n - 5)(2x^n + 5) = 4x^{2n} - 25$

97. $(6 - x)(6 + x) = 36 - x^2$

99. $(3a - 4b)^2 = 9a^2 - 24ab + 16b^2$

101. $(3x^n + 2)^2 = 9x^{2n} + 12x^n + 4$

103. $(x^n + 3)(x^n - 3) = x^{2n} - 9$

105. $(x^n - 1)^2 = x^{2n} - 2x^n + 1$

107. $(2x^n + 5y^n)^2 = 4x^{2n} + 20x^n y^n + 25y^{2n}$

109. For $(ax + b)(ax - b) = a^2 x^2 - b^2$, where $a > 0$ and $b > 0$, the coefficient of the x-term is zero.

111. For $(ax+b)^2 = a^2x^2 + 2abx + b^2$, where $a < 0$ and $b < 0$, the coefficient of the x-term is positive.

Objective 5.3.4 Exercises

113. a. We can use the properties of rectangles to show that both unlabeled sides of the figure have length x. The unlabeled vertical side has length $4x-3x = x$. The unlabeled horizontal side has length $x+3-3 = x$.

b. Divide the figure into two rectangles by extending the unlabeled horizontal side. The area of the upper rectangle is $3x(\underline{3}) = \underline{9x}$. The area of the lower rectangle is $(x+3)(\underline{x}) = \underline{x^2 + 3x}$. The area of the whole figure is $\underline{9x} + \underline{x^2 + 3x} = \underline{x^2 + 12x}$.

115. Strategy
To find the area, replace the variables b and h in the equation $A = \frac{1}{2}bh$ by the given values and solve for A.

Solution
$A = \frac{1}{2}bh$
$A = \frac{1}{2}(x+2)(2x-3)$
$A = \left(\frac{1}{2}x + 1\right)(2x - 3)$
$A = x^2 - \frac{3}{2}x + 2x - 3$
$A = x^2 + \frac{x}{2} - 3$
The area is $\left(x^2 + \frac{x}{2} - 3\right)$ ft^2.

117. Strategy
To find the area, subtract the four small rectangles from the large rectangle.
Large rectangle:
Length = $L_1 = x + 8$
Width = $W_1 = x + 4$
Small rectangles:
Length = $L_2 = 2$
Width = $W_2 = 2$

Solution
A = Area of the large rectangle $- 4$(area of small rectangle)
$A = (L_1 \cdot W_1) - 4(L_2 \cdot W_2)$
$A = (x+8)(x+4) - 4(2)(2)$
$A = x^2 + 4x + 8x + 32 - 16$
$A = x^2 + 12x + 16$
The area is $(x^2 + 12x + 16)$ ft^2.

119. Strategy
Length of the box: $18 - 2x$
Width of the box: $18 - 2x$
Height of the box: x
To find the volume, replace the variables L, W, and H in the equation $V = LWH$ and solve for V.

Solution
$V = LWH$
$V = (18 - 2x)(18 - 2x)x$
$V = (324 - 36x - 36x + 4x^2)x$
$V = (324 - 72x + 4x^2)x$
$V = 324x - 72x^2 + 4x^3$
The volume is $(4x^3 - 72x^2 + 324x)$ in^3.

121. Strategy
To find the volume, replace the variables, L, W, and H in the equation $V = LWH$ by the given values and solve for V.

Solution
$V = L \cdot W \cdot H$
$V = (2x+3)(x-5)(x)$
$V = (2x^2 - 7x - 15)(x)$
$V = 2x^3 - 7x^2 - 15x$
The volume is $(2x^3 - 7x^2 - 15x)$ cm^3.

123. Strategy
To find the volume, add the volume of the small rectangular solid to the volume of the large rectangular solid.
Large rectangular solid:
Length = $L_1 = 3x + 4$
Width = $W_1 = x + 6$
Height = $H_1 = x$
Small rectangular solid:
Length = $L_2 = x + 4$
Width = $W_2 = x + 6$
Height = $H_2 = x$

Solution
$V = (L_1 \cdot W_1 \cdot H_1) + (L_2 \cdot W_2 \cdot H_2)$
$V = (3x+4)(x+6)(x) + (x+4)(x+6)(x)$
$V = (3x^2 + 22x + 24)(x) + (x^2 + 10x + 24)(x)$
$V = 3x^3 + 22x^2 + 24x + x^3 + 10x^2 + 24x$
$V = 4x^3 + 32x^2 + 48x$
The volume is $(4x^3 + 32x^2 + 48x)$ cm^3.

Applying Concepts 5.3

125. $\dfrac{(3x-5)^6}{(3x-5)^4} = (3x-5)^{6-4}$
$= (3x-5)^2$
$= (3x-5)(3x-5)$
$= 9x^2 - 30x + 25$

127. $(x+2y)^2 + (x+2y)(x-2y) = (x+2y)(x+2y) + (x+2y)(x-2y)$
$= (x^2 + 4xy + 4y^2) + (x^2 - 4y^2)$
$= 2x^2 + 4xy$

129. $2x^2(3x^3 + 4x - 1) - 5x^2(x^2 - 3) = (6x^5 + 8x^3 - 2x^2) - (5x^4 - 15x^2)$
$= (6x^5 + 8x^3 - 2x^2) + (-5x^4 + 15x^2)$
$= 6x^5 - 5x^4 + 8x^3 + 13x^2$

131. $(3x-2y)^2 - (2x-3y)^2 = (3x-2y)(3x-2y) - (2x-3y)(2x-3y)$
$= (9x^2 - 12xy + 4y^2) - (4x^2 - 12xy + 9y^2)$
$= (9x^2 - 12xy + 4y^2) + (-4x^2 + 12xy - 9y^2)$
$= 5x^2 - 5y^2$

133. $[x^2(2y-1)]^2 = x^{2 \cdot 2}(2y-1)^2$
$= x^4(4y^2 - 4y + 1)$
$= 4x^4y^2 - 4x^4y + x^4$

135. $(kx-7)(kx+2) = k^2x^2 + 5x - 14$
$k^2x^2 + 2kx - 7kx - 14 = k^2x^2 + 5x - 14$
$k^2x^2 - 5kx - 14 = k^2x^2 + 5x - 14$
$-5kx = 5x$
$-5k = 5$
$k = -1$

137. $\dfrac{a^m}{a^n}, m = n+2$
$\dfrac{a^{n+2}}{a^n} = a^{(n+2)-n} = a^2$

139. $(x+7)(2x-3) = 2x^2 - 3x + 14x - 21$
$= 2x^2 + 11x - 21$

141. $(6x^2 + 12xy - 2y^2) - (5x - y)(x + 3y)$
$= (6x^2 + 12xy - 2y^2) - (5x^2 + 15xy - xy - 3y^2)$
$= (6x^2 + 12xy - 2y^2) - (5x^2 + 14xy - 3y^2)$
$= (6x^2 + 12xy - 2y^2) + (-5x^2 - 14xy + 3y^2)$
$= x^2 - 2xy + y^2$

143. Find the value of n.
$3(2n-1) = 5(n-1)$
$6n - 3 = 5n - 5$
$n - 3 = -5$
$n = -2$
Substitute the value of n into the expression.
$(-2n^3)^2 = [-2(-2)^3]^2 = [-2(-8)]^2 = 16^2 = 256$

Section 5.4
Concept Review 5.4

1. Sometimes true
For example, 1 and x are monomials but $\dfrac{1}{x}$ is not a monomial.

3. Always true

Chapter 5: Polynomials And Exponents 141

Objective 5.4.1 Exercises

1. $$\frac{9y^2+6y}{3y} = \frac{9y^2}{3y} + \frac{6y}{3y} = 3y+2$$

3. $$\frac{4a-8}{4} = \frac{4a}{4} - \frac{8}{4} = a-2$$

5. $$\frac{6w^2+4w}{2w} = \frac{6w^2}{2w} + \frac{4w}{2w} = 3w+2$$

7. $$\frac{3t^3-9t^2+12t}{3t} = \frac{3t^3}{3t} - \frac{9t^2}{3t} + \frac{12t}{3t} = t^2-3t+4$$

9. $$\frac{16x^3-24x^2+48x}{-8x^2} = \frac{16x^3}{-8x^2} - \frac{24x^2}{-8x^2} + \frac{48x}{-8x^2} = -2x+3-\frac{6}{x}$$

11. $$\frac{8v^3-6v^2+12v}{4v} = \frac{8v^3}{4v} - \frac{6v^2}{4v} + \frac{12v}{4v} = 2v^2 - \frac{3}{2}v + 3$$

13. $$\frac{-2xy^2+4x^2y}{2xy} = \frac{-2xy^2}{2xy} + \frac{4x^2y}{2xy} = -y+2x$$

15. $$\frac{12x^2y^2-16x^2y+20xy^2}{4x^2y} = \frac{12x^2y^2}{4x^2y} - \frac{16x^2y}{4x^2y} + \frac{20xy^2}{4x^2y} = 3y-4+\frac{5y}{x}$$

17. $$\frac{-18x^3y^3-12x^2y^2+9xy}{3x^2y^2} = \frac{-18x^3y^3}{3x^2y^2} - \frac{12x^2y^2}{3x^2y^2} + \frac{9xy}{3x^2y^2} = -6xy-4+\frac{3}{xy}$$

19. $$\frac{6n^2m-12nm^2-9mn}{-3m^2n^2} = \frac{6n^2m}{-3m^2n^2} - \frac{12nm^2}{-3m^2n^2} - \frac{9mn}{-3m^2n^2} = -\frac{2}{m}+\frac{4}{n}+\frac{3}{mn}$$

21. The related multiplication equation is $12x^2+6x = 3x(4x+2)$.

Objective 5.4.2 Exercises

23. The degree of the quotient of two polynomials is equal to the degree of the dividend minus the degree of the divisor.

25.
$$\begin{array}{r} x+8 \\ x-5\overline{)x^2+3x-40} \\ \underline{x^2-5x} \\ 8x-40 \\ \underline{8x-40} \\ 0 \end{array}$$

$(x^2+3x-40) \div (x-5) = x+8$

27.
$$\begin{array}{r} x^2 \\ x-3\overline{)x^3-3x^2+0x+2} \\ \underline{x^3-3x^2} \\ 2 \end{array}$$

$(x^3-3x^2+2) \div (x-3) = x^2 + \frac{2}{x-3}$

29.
$$\begin{array}{r} 3x+5 \\ 2x+1\overline{)6x^2+13x+8} \\ \underline{6x^2+3x} \\ 10x+8 \\ \underline{10x+5} \\ 3 \end{array}$$

$(6x^2+13x+8) \div (2x+1) = 3x+5+\frac{3}{2x+1}$

31.
$$\begin{array}{r} 5x+7 \\ 2x-1\overline{)10x^2+9x-5} \\ \underline{10x^2-5x} \\ 14x-5 \\ \underline{14x-7} \\ 2 \end{array}$$

$(10x^2+9x-5) \div (2x-1) = 5x+7+\frac{2}{2x-1}$

Chapter 5: Polynomials And Exponents

33.

$$\begin{array}{r} 4x^2+6x+9 \\ 2x-3{\overline{\smash{\big)}\,8x^3+0x^2+0x-9}} \\ \underline{8x^3-12x^2} \\ 12x^2+0x \\ \underline{12x^2-18x} \\ 18x-9 \\ \underline{18x-27} \\ 18 \end{array}$$

$(8x^3-9)\div(2x-3)=4x^2+6x+9+\dfrac{18}{2x-3}$

35.

$$\begin{array}{r} 3x^2+1 \\ 2x^2-5{\overline{\smash{\big)}\,6x^4+0x^3-13x^2+0x-4}} \\ \underline{6x^4-15x^2} \\ 2x^2+0x-4 \\ \underline{2x^2-5} \\ 1 \end{array}$$

$(6x^4-13x^2-4)\div(2x^2-5)=3x^2+1+\dfrac{1}{2x^2-5}$

37.

$$\begin{array}{r} x^2-3x-10 \\ 3x+1{\overline{\smash{\big)}\,3x^3-8x^2-33x-10}} \\ \underline{3x^3+x^2} \\ -9x^2-33x \\ \underline{-9x^2-3x} \\ -30x-10 \\ \underline{-30x-10} \\ 0 \end{array}$$

$\dfrac{3x^3-8x^2-33x-10}{3x+1}=x^2-3x-10$

39.

$$\begin{array}{r} -x^2+2x-1 \\ x-3{\overline{\smash{\big)}\,-x^3+5x^2-7x+4}} \\ \underline{-x^3+3x^2} \\ 2x^2-7x \\ \underline{2x^2-6x} \\ -x+4 \\ \underline{-x+3} \\ 1 \end{array}$$

$\dfrac{4-7x+5x^2-x^3}{x-3}=-x^2+2x-1+\dfrac{1}{x-3}$

41.

$$\begin{array}{r} 2x^3-3x^2+x-4 \\ x-5{\overline{\smash{\big)}\,2x^4-13x^3+16x^2-9x+20}} \\ \underline{2x^4-10x^3} \\ -3x^3+16x^2 \\ \underline{-3x^3+15x^2} \\ x^2-9x \\ \underline{x^2-5x} \\ -4x+20 \\ \underline{-4x+20} \\ 0 \end{array}$$

$\dfrac{16x^2-13x^3+2x^4-9x+20}{x-5}=2x^3-3x^2+x-4$

43.

$$\begin{array}{r} x-4 \\ x^2+1{\overline{\smash{\big)}\,x^3-4x^2+2x-1}} \\ \underline{x^3+x} \\ -4x^2+x-1 \\ \underline{-4x^2-4} \\ x+3 \end{array}$$

$\dfrac{x^3-4x^2+2x-1}{x^2+1}=x-4+\dfrac{x+3}{x^2+1}$

45.

$$\begin{array}{r} 2x-3 \\ x^2-1{\overline{\smash{\big)}\,2x^3-3x^2-x+4}} \\ \underline{2x^3-2x} \\ -3x^2+x+4 \\ \underline{-3x^2+3} \\ x+1 \end{array}$$

$\dfrac{2x^3-x+4-3x^2}{x^2-1}=2x-3+\dfrac{x+1}{x^2-1}$

$\phantom{\dfrac{2x^3-x+4-3x^2}{x^2-1}}=2x-3+\dfrac{1}{x-1}$

47.

$$\begin{array}{r} 3x+1 \\ 2x^2-3{\overline{\smash{\big)}\,6x^3+2x^2+x+4}} \\ \underline{6x^3-9x} \\ 2x^2+10x+4 \\ \underline{2x^2-3} \\ 10x+7 \end{array}$$

$\dfrac{6x^3+2x^2+x+4}{2x^2-3}=3x+1+\dfrac{10x+7}{2x^2-3}$

49. False. When a tenth-degree polynomial is divided by a second-degree polynomial, the quotient is an eighth-degree polynomial.

Objective 5.4.3 Exercises

51. Synthetic division is a shorter method of dividing a polynomial by a <u>binomial</u> of the form $\underline{x-a}$.

53.

$$\begin{array}{r|rrr} -1 & 2 & -6 & -8 \\ & & -2 & 8 \\ \hline & 2 & -8 & 0 \end{array}$$

$(2x^2 - 6x - 8) \div (x+1) = 2x - 8$

55.

$$\begin{array}{r|rrr} 1 & 3 & 0 & -4 \\ & & 3 & 3 \\ \hline & 3 & 3 & -1 \end{array}$$

$(3x^2 - 4) \div (x-1) = 3x + 3 - \dfrac{1}{x-1}$

57.

$$\begin{array}{r|rrr} -4 & 1 & 0 & -9 \\ & & -4 & 16 \\ \hline & 1 & -4 & 7 \end{array}$$

$(x^2 - 9) \div (x+4) = x - 4 + \dfrac{7}{x+4}$

59.

$$\begin{array}{r|rrr} -2 & 1 & 0 & 12 \\ & & -2 & 4 \\ \hline & 1 & -2 & 16 \end{array}$$

$(2x^2 + 24) \div (2x+4) = x - 2 + \dfrac{16}{x+2}$

61.

$$\begin{array}{r|rrrr} -1 & 2 & -1 & 6 & 9 \\ & & -2 & 3 & -9 \\ \hline & 2 & -3 & 9 & 0 \end{array}$$

$(2x^3 - x^2 + 6x + 9) \div (x+1) = 2x^2 - 3x + 9$

63.

$$\begin{array}{r|rrrr} 3 & 1 & -6 & 11 & -6 \\ & & 3 & -9 & 6 \\ \hline & 1 & -3 & 2 & 0 \end{array}$$

$(x^3 - 6x^2 + 11x - 6) \div (x-3) = x^2 - 3x + 2$

65.

$$\begin{array}{r|rrrr} -2 & 1 & -3 & 6 & -9 \\ & & -2 & 10 & -32 \\ \hline & 1 & -5 & 16 & -41 \end{array}$$

$(6x - 3x^2 + x^3 - 9) \div (x+2)$
$= x^2 - 5x + 16 - \dfrac{41}{x+2}$

67.

$$\begin{array}{r|rrrr} -1 & 1 & 0 & 1 & -2 \\ & & -1 & 1 & -2 \\ \hline & 1 & -1 & 2 & -4 \end{array}$$

$(x^3 + x - 2) \div (x+1) = x^2 - x + 2 - \dfrac{4}{x+1}$

69.

$$\begin{array}{r|rrrr} 2 & 4 & 0 & -1 & -18 \\ & & 8 & 16 & 30 \\ \hline & 4 & 8 & 15 & 12 \end{array}$$

$(18 + x - 4x^3) \div (2-x) = 4x^2 + 8x + 15 + \dfrac{12}{x-2}$

71.

$$\begin{array}{r|rrrrr} 5 & 2 & -13 & 16 & -9 & 20 \\ & & 10 & -15 & 5 & -20 \\ \hline & 2 & -3 & 1 & -4 & 0 \end{array}$$

$\dfrac{16x^2 - 13x^3 + 2x^4 - 9x + 20}{x-5} = 2x^3 - 3x^2 + x - 4$

73.

$$\begin{array}{r|rrrrr} 2 & 3 & -4 & 8 & -5 & -5 \\ & & 6 & 4 & 24 & 38 \\ \hline & 3 & 2 & 12 & 19 & 33 \end{array}$$

$\dfrac{5 + 5x - 8x^2 + 4x^3 - 3x^4}{2-x}$
$= 3x^3 + 2x^2 + 12x + 19 + \dfrac{33}{x-2}$

75.

$$\begin{array}{r|rrrrr} -1 & 3 & 3 & -1 & 3 & 2 \\ & & -3 & 0 & 1 & -4 \\ \hline & 3 & 0 & -1 & 4 & -2 \end{array}$$

$\dfrac{3x^4 + 3x^3 - x^2 + 3x + 2}{x+1}$
$= 3x^3 - x + 4 - \dfrac{2}{x+1}$

77.

$$\begin{array}{r|rrrrr} 3 & 2 & 0 & -1 & 0 & 2 \\ & & 6 & 18 & 51 & 153 \\ \hline & 2 & 6 & 17 & 51 & 155 \end{array}$$

$\dfrac{2x^4 - x^2 + 2}{x-3}$
$= 2x^3 + 6x^2 + 17x + 51 + \dfrac{155}{x-3}$

79.

–5	1	0	0	125
		–5	25	–125
	1	–5	25	0

$$\frac{x^3+125}{x+5}=x^2-5x+25$$

81. Yes, you know the quotient and the divisor; therefore, you can multiply these to determine the dividend, $P(x)$.

Objective 5.4.4 Exercises

83.

2	3	–2	1	–9
		6	8	18
	1	–7	9	9

The value of $P(2)=9$.

85.

3	2	–3	–1
		6	9
	2	3	8

$P(3)=8$

87.

4	1	–2	3	–1
		4	8	44
	1	2	11	43

$R(4)=43$

89.

–2	2	–4	3	–1
		–4	16	–38
	2	–8	19	–39

$P(-2)=-39$

91.

2	1	3	–2	4	–9
		2	10	16	40
	1	5	8	20	31

$Q(2)=31$

93.

–3	2	–1	0	–2	–5
		–6	21	–63	195
	2	–7	21	–65	190

$F(-3)=178$

95.

5	1	0	0	–3
		5	25	125
	1	5	25	122

$P(5)=122$

97.

–3	4	0	–3	0	5
		–12	36	–99	297
	4	–12	33	–99	302

$R(-3)=302$

99.

2	1	0	–4	–2	5	–2
		2	4	0	–4	2
	1	2	0	–2	1	0

$Q(2)=0$

Applying Concepts 5.4

101.

$$\begin{array}{r} x-y \\ 3x+2y\overline{)3x^2-xy-2y^2} \\ \underline{3x^2+2xy} \\ -3xy-2y^2 \\ \underline{-3xy-2y^2} \\ 0 \end{array}$$

$$\frac{3x^2-xy-2y^2}{3x+2y}=x-y$$

103.

$$\begin{array}{r} a^2+ab+b^2 \\ a-b\overline{)a^3+0a^2b+0ab^2-b^3} \\ \underline{a^3-a^2b} \\ a^2b+0ab^2 \\ \underline{a^2b-ab^2} \\ ab^2-b^3 \\ \underline{ab^2-b^3} \\ 0 \end{array}$$

$$\frac{a^3-b^3}{a-b}=a^2+ab+b^2$$

105.

$$\require{enclose}\begin{array}{r}x^4-x^3y+x^2y^2-xy^3+y^4\\x+y\enclose{longdiv}{x^5+0x^4y+0x^3y^2+0x^2y^3+0xy^4+y^5}\end{array}$$

$$\underline{x^5+x^4y}$$
$$-x^4y+0x^3y^2$$
$$\underline{-x^4y-x^3y^2}$$
$$x^3y^2+0x^2y^3$$
$$\underline{x^3y^2+x^2y^3}$$
$$-x^2y^3+0xy^4$$
$$\underline{-x^2y^3-xy^4}$$
$$xy^4+y^5$$
$$\underline{xy^4+y^5}$$
$$0$$

$$\frac{x^5+y^5}{x+y}=x^4-x^3y+x^2y^2-xy^3+y^4$$

107.

$$\begin{array}{r|rrrr}3 & 1 & -3 & -1 & k \\ & & 3 & 0 & -3 \\ \hline & 1 & 0 & -1 & k-3\end{array}$$

$k-3=0$
$k=3$
The remainder is zero when k equals 3.

109.

$$\begin{array}{r|rrr}3 & 1 & k & -6 \\ & & 3 & 9+3k \\ \hline & 1 & 3+k & 3+3k\end{array}$$

$3+3k=0$
$3k=-3$
$k=-1$
The remainder is zero when k equals -1.

111. Note: (Quotient)(Divisor)+Remainder = Dividend
(Quotient)(Divisor)=Dividend − Remainder

$$\text{Divisor}=\frac{\text{Dividend}-\text{Remainder}}{\text{Quotient}}$$

Therefore, $\text{Divisor}=\dfrac{(x^2+x+2)-14}{x+4}$

$$=\frac{x^2+x-12}{x+4}$$

$$\require{enclose}\begin{array}{r}x-3\\x+4\enclose{longdiv}{x^2+x-12}\end{array}$$
$$\underline{x^2+4x}$$
$$-3x-12$$
$$\underline{-3x-12}$$
$$0$$

The polynomial is $x-3$.
Check: $(x+4)(x-3)+14=x^2+x-12+14$
$=x^2+x+2$

Section 5.5

Concept Review 5.5

1. Always true

3. Sometimes true
There are some trinomials that are nonfactorable over the integers.

Objective 5.5.1 Exercises

1. The GCF of two monomials is the product of all common factors, each with its smallest exponent. The GCF of $7x^2y$ and $2xy^3$ is xy.

3. Of the two expressions $15x^2-10x$ and $5x(3x-2)$, $5x(3x-2)$ is written in factored form.

5. The GCF of $6a^2$ and $15a$ is $3a$.
$6a^2-15a=3a(2a-5)$

7. The GCF of $4x^3$ and $3x^2$ is x^2.
$4x^3-3x^2=x^2(4x-3)$

9. There is no common factor.
$3a^2-10b^3$ is nonfactorable over the integers.

11. The GCF of x^5, x^3, and x is x.
$x^5-x^3-x=x(x^4-x^2-1)$

13. The GCF of $16x^2$, $12x$ and 24 is 4.
$16x^2-12x+24=4(4x^2-3x+6)$

15. The GCF of $5b^2, 10b^3$, and $25b^4$ is $5b^2$.
$5b^2 - 10b^3 + 25b^4 = 5b^2(1 - 2b + 5b^2)$

17. The GCF of x^{2n} and x^n is x^n.
$x^{2n} - x^n = x^n(x^n - 1)$

19. The GCF of x^{3n} and x^{2n} is x^{2n}.
$x^{3n} - x^{2n} = x^{2n}(x^n - 1)$

21. The GCF of a^{2n+2} and a^2 is a^2.
$a^{2n+2} + a^2 = a^2(a^{2n} + 1)$

23. The GCF of $12x^2y^2, 18x^3y$, and $24x^2y$ is $6x^2y$.
$12x^2y^2 - 18x^3y + 24x^2y = 6x^2y(2y - 3x + 4)$

25. The GCF of $16a^2b^4, 4a^2b^2$, and $24a^3b^2$ is $4a^2b^2$.
$-16a^2b^4 - 4a^2b^2 + 24a^3b^2$
$= 4a^2b^2(-4b^2 - 1 + 6a)$

27. The GCF of y^{2n+2}, y^{n+2}, and y^2 is y^2.
$y^{2n+2} + y^{n+2} - y^2$
$= y^2(y^{2n} + y^n - 1)$

29. No, because $a < b < c < d$, the polynomial is written in descending order.

31. Because $b = 2a$, $c = 3a$, and $d = 4a$, then $x^a + x^b + x^c + x^d$ in factored form is
$x^a + x^b + x^c + x^d = x^a + x^{2a} + x^{3a} + x^{4a}$
$= x^a(1 + x^a + x^{2a} + x^{3a})$

Objective 5.5.2 Exercises

33. $x(a+2) - 2(a+2) = (a+2)(x-2)$

35. $a(x-2) - b(2-x) = a(x-2) + b(x-2)$
$= (x-2)(a+b)$

37. $x^2 + 3x + 2x + 6 = x(x+3) + 2(x+3)$
$= (x+3)(x+2)$

39. $xy + 4y - 2x - 8 = y(x+4) - 2(x+4)$
$= (x+4)(y-2)$

41. $ax + bx - ay - by = x(a+b) - y(a+b)$
$= (a+b)(x-y)$

43. $x^2y - 3x^2 - 2y + 6 = x^2(y-3) - 2(y-3)$
$= (y-3)(x^2-2)$

45. $6 + 2y + 3x^2 + x^2y = 2(3+y) + x^2(3+y)$
$= (3+y)(2+x^2)$

47. $2ax^2 + bx^2 - 4ay - 2by = x^2(2a+b) - 2y(2a+b)$
$= (2a+b)(x^2-2y)$

49. $x^n y - 5x^n + y - 5 = x^n(y-5) + (y-5)$
$= (y-5)(x^n+1)$

51. $x^3 + x^2 + 2x + 2 = x^2(x+1) + 2(x+1)$
$= (x+1)(x^2+2)$

53. $2x^3 - x^2 + 4x - 2 = x^2(2x-1) + 2(2x-1)$
$= (2x-1)(x^2+2)$

55.
I. $xy + 6y + 3x - 18 = (xy + 6y) + (3x - 18) = y(x+6) + 3(x-6)$

II. $xy - 6y - 3x - 18 = (xy - 6y) + (-3x - 18) = y(x-6) - 3(x+6)$

III. $xy + 6y - 3x - 18 = (xy + 6y) + (-3x - 18)$
$= y(x+6) - 3(x+6)$
$= (x+6)(y-3)$

The one expression that can be factored by grouping is III.

Objective 5.5.3 Exercises

57. A quadratic trinomial is a trinomial of the form $ax^2 + bx + c$, where a and b are nonzero coefficients and x is a nonzero constant. To factor a quadratic trinomial means to express the trinomial as a product of two binomials.

59. a. The numbers -3 and 7 have a product of -21 and a sum of 4.

b. The numbers -4 and -5 have a product of 20 and a sum of -9.

61. $x^2 - 8x + 15 = (x-5)(x-3)$

63. $a^2 + 12a + 11 = (a+11)(a+1)$

65. $b^2 + 2b - 35 = (b+7)(b-5)$

67. $y^2 - 16y + 39 = (y-3)(y-13)$

69. $b^2 + 4b - 32 = (b+8)(b-4)$

71. $a^2 - 15a + 56 = (a-7)(a-8)$

73. $y^2 + 13y + 12 = (y+12)(y+1)$

75. $x^2 + 4x - 5 = (x+5)(x-1)$

77. $a^2 + 11ab + 30b^2 = (a+6b)(a+5b)$

79. $x^2 - 14xy + 24y^2 = (x-12y)(x-2y)$

81. $y^2 + 2xy - 63x^2 = (y+9x)(y-7x)$

83. $x^2 - 35x - 36 = (x-36)(x+1)$

85. $a^2 + 13a + 36 = (a+9)(a+4)$

87. There are no binomial factors whose product $x^2 - 7x - 12$. The trinomial is nonfactorable over the integers.

89. For $x^2 + bx + c = (x-n)(x-m)$, let b and c be nonzero and n and m be positive.

 a. Because $c = (-n)(-m)$, the value of c must be positive.

 b. Because $bx = -mx - nx$, so $b = -m - n$, the value of b must be negative.

 c. Because b is negative and c is positive, b is less than c.

Objective 5.5.4 Exercises

91. To factor $ax^2 + bx + c$ by grouping when b is negative and c is positive, both factors of ac should be negative.

93. To factor $ax^2 + bx + c$ by grouping when b is positive and c is positive, both factors of ac should be positive.

95. $2x^2 - 11x - 40 = (2x+5)(x-8)$

97. $4y^2 - 15y + 9 = (4y-3)(y-3)$

99. $2a^2 + 13a + 6 = (2a+1)(a+6)$

101. There are no binomial factors whose product is $12y^2 - 13y - 72$. The trinomial is nonfactorable over the integers.

103. $5x^2 + 26x + 5 = (5x+1)(x+5)$

105. $11x^2 - 122x + 11 = (11x-1)(x-11)$

107. $12x^2 - 17x + 5 = (12x-5)(x-1)$

109. $8y^2 - 18y + 9 = (4y-3)(2y-3)$

111. There are no binomial factors whose product is $6a^2 - 5a - 2$. The trinomial is nonfactorable over the integers.

113. There are no binomial factors whose product is $2x^2 + 5x + 12$. The trinomial is nonfactorable over the integers.

115. $6x^2 + 5xy - 21y^2 = (2x-3y)(3x+7y)$

117. $4a^2 + 43ab + 63b^2 = (4a+7b)(a+9b)$

119. $10x^2 - 23xy + 12y^2 = (5x-4y)(2x-3y)$

121. $24 + 13x - 2x^2 = (8-x)(3+2x)$

123. There are no binomial factors whose product is $8 - 13x + 6x^2$. The trinomial is nonfactorable over the integers.

125. $15 - 14a - 8a^2 = (3-4a)(5+2a)$

127. The GCF of $5y^4, 29y^3$, and $20y^2$ is y^2.
$5y^4 - 29y^3 + 20y^2 = y^2(5y^2 - 29y + 20)$
$= y^2(5y-4)(y-5)$

129. The GCF of $4x^3, 10x^2y$ and $24xy^2$ is $2x$.
$4x^3 + 10x^2y - 24xy^2 = 2x(2x^2 + 5xy - 12y^2)$
$= 2x(2x-3y)(x+4y)$

131. The GCF of $100, 5x$, and $5x^2$ is 5.
$100 - 5x - 5x^2 = 5(20 - x - x^2)$
$= 5(5+x)(4-x)$

133. The GCF of $320x, 8x^2$, and $4x^3$ is $4x$.
$320x - 8x^2 - 4x^3 = 4x(80 - 2x - x^2)$
$= 4x(10+x)(8-x)$

135. The GCF of $20x^2, 38x^3$, and $30x^4$ is $2x^2$.
$20x^2 - 38x^3 - 30x^4 = 2x^2(10 - 19x - 15x^2)$
$= 2x^2(5+3x)(2-5x)$

137. The GCF of $a^4b^4, 3a^3b^3$, and $10a^2b^2$ is a^2b^2.
$a^4b^4 - 3a^3b^3 - 10a^2b^2 = a^2b^2(a^2b^2 - 3ab - 10)$
$= a^2b^2(ab-5)(ab+2)$

139. The GCF of $90a^2b^2$, $45ab$, and 10 is 5.
$90a^2b^2 + 45ab + 10 = 5(18a^2b^2 + 9ab + 2)$

141. There is no common factor.
$4x^4 - 45x^2 + 80$ is nonfactorable over the integers.

143. The GCF of $2a^5, 14a^3$, and $20a$ is $2a$.
$2a^5 + 14a^3 + 20a = 2a(a^4 + 7a^2 + 10)$
$= 2a(a^2+5)(a^2+2)$

145. The GCF of $3x^4y^2, 39x^2y^2$, and $120y^2$ is $3y^2$.
$$3x^4y^2 - 39x^2y^2 + 120y^2 = 3y^2(x^4 - 13x^2 + 40)$$
$$= 3y^2(x^2 - 5)(x^2 - 8)$$

147. The GCF of $45a^2b^2, 6ab^2$, and $72b^2$ is $3b^2$.
$$45a^2b^2 + 6ab^2 - 72b^2 = 3b^2(15a^2 + 2a - 24)$$
$$= 3b^2(3a + 4)(5a - 6)$$

149. The GCF of $36x^3y, 24x^2y^2$, and $45xy^3$ is $3xy$.
$$36x^3y + 24x^2y^2 - 45xy^3$$
$$= 3xy(12x^2 + 8xy - 15y^2)$$
$$= 3xy(6x - 5y)(2x + 3y)$$

151. The GCF of $48a^2b^2, 36ab^3$, and $54b^4$ is $6b^2$.
$$48a^2b^2 - 36ab^3 - 54b^4$$
$$= 6b^2(8a^2 - 6ab - 9b^2)$$
$$= 6b^2(2a - 3b)(4a + 3b)$$

153. The GCF of $10x^{2n}, 25x^n$, and 60 is 5.
$$10x^{2n} + 25x^n - 60 = 5(2x^{2n} + 5x^n - 12)$$
$$= 5(2x^n - 3)(x^n + 4)$$

Applying Concepts 5.5

155. $3x^3y - xy^3 - 2x^2y^2 = 3x^3y - 2x^2y^2 - xy^3$
$$= xy(3x^2 - 2xy - y^2)$$
$$= xy(3x + y)(x - y)$$

157. $9b^3 + 3b^5 - 30b = 3b^5 + 9b^3 - 30b$
$$= 3b(b^4 + 3b^2 - 10)$$
$$= 3b(b^2 + 5)(b^2 - 2)$$

159. $x^2 + kx - 6$

Factors of –6	Sum
1, –6	–5
–1, 6	5
2, –3	–1
–2, 3	1

The possible integer values of k are 5, –5, 1, and –1.

161. $2x^2 - kx - 5$
$2x^2 + (-k)x - 5$

Factors of 2	Factors of –5
1, 2	1, –5
	–1, 5

Trial Factors	Middle Term
$(x + 1)(2x - 5)$	$-5x + 2x = -3x$
$(x - 5)(2x + 1)$	$x - 10x = -9x$
$(x - 1)(2x + 5)$	$5x - 2x = 3x$
$(x + 5)(2x - 1)$	$-x + 10x = 9x$

The possible integer values of $-k$ are 3, –3, 9, and –9.
The possible integer values of k are 3, 9, –3, and –9.

163. $2x^2 + kx - 3$

Factors of 2	Factors of –3
1, 2	1, –3
	–1, 3

Trial Factors	Middle Term
$(x + 1)(2x - 3)$	$-3x + 2x = -x$
$(x - 3)(2x + 1)$	$x - 6x = -5x$
$(x - 1)(2x + 3)$	$3x - 2x = x$
$(x + 3)(2x - 1)$	$-x + 6x = 5x$

The possible integer values of k are –1, –5, 1, and 5.

Section 5.6
Concept Review 5.6

1. Never true
   $a^2b^2c^2$ is a monomial.

3. Never true
   The difference of two perfect cubes is always factorable. For example,
   $$8^2 - x^3 = 2^3 - x^3$$
   $$= (2 - x)(4 + 2x + x^2)$$

5. Never true
   $b^3 + 1$ is the sum of two perfect cubes.
   $b^3 + 1^3 = (b + 1)(b^2 - b + 1)$

Objective 5.6.1 Exercises

1. $4; 25x^6; 100x^4y^4$

3. $4z^4$

5. $9a^2b^3$

7. a. The binomial $16x^2 - 1$ is in the form $a^2 - b^2$, where $a = \underline{4x}$ and $b = \underline{1}$.

   b. Use the formula $a^2 - b^2 = (a + b)(a - b)$ to factor $16x^2 - 1$: $16x^2 - 1 = (\underline{4x + 1})(\underline{4x - 1})$.

9. The factors of the difference of two perfect squares are the sum and difference of the square roots of the perfect squares.

11. $x^2 - 16 = x^2 - 4^2$
    $$= (x + 4)(x - 4)$$

13. $4x^2 - 1 = (2x)^2 - 1^2$
    $$= (2x + 1)(2x - 1)$$

15. $b^2 - 2b + 1 = (b - 1)^2$

17. $16x^2 - 40x + 25 = (4x - 5)^2$

19. $x^2y^2 - 100 = (xy)^2 - 10^2$
    $$= (xy + 10)(xy - 10)$$

21. $x^2 + 4$ is nonfactorable over the integers.

23. $x^2 + 6xy + 9y^2 = (x + 3y)^2$

25. $4x^2 - y^2 = (2x)^2 - 4^2 = (2x + y)(2x - y)$

27. $a^{2n} - 1 = (a^n)^2 - 1^2$
$= (a^n + 1)(a^n - 1)$

29. $a^2 + 4a + 4 = (a + 2)^2$

31. $x^2 - 12x + 36 = (x - 6)^2$

33. $16x^2 - 121 = (4x)^2 - 11^2$
$= (4x + 11)(4x - 11)$

35. $1 - 9a^2 = 1^2 - (3a)^2$
$= (1 + 3a)(1 - 3a)$

37. $4a^2 + 4a - 1$ is nonfactorable over the integers.

39. $b^2 + 7b + 14$ is nonfactorable over the integers.

41. $25 - a^2b^2 = 5^2 - (ab)^2$
$= (5 + ab)(5 - ab)$

43. $25a^2 - 40ab + 16b^2 = (5a - 4b)^2$

45. $x^{2n} + 6x^n + 9 = (x^n + 3)^2$

47. $x^{16} - 81 = (x^8)^2 - (9)^2$
$= (x^8 + 9)(x^8 - 9)$
$= (x^8 + 9)((x^4)^2 - 3^2)$
$= (x^8 + 9)(x^4 + 3)(x^4 - 3)$
Expressions b and d are equivalent to $x^{16} - 81$.

Objective 5.6.2 Exercises

49. $8; x^9; 27c^{15}d^{18}$

51. $2x^3$

53. $4a^2b^6$

55. a. The binomial $125x^3 + 8$ is in the form $a^3 + b^3$, where $a = \underline{5x}$ and $b = \underline{2}$.

b. Use the formula
$a^3 + b^3 = (a + b)(a^2 - ab + b^2)$ to factor
$125x^3 + 8$:
$125x^3 + 8 = \underline{(5x + 2)(25x^2 - 10x + 4)}$.

57. $x^3 - 27 = x^3 - 3^3$
$= (x - 3)(x^2 + 3x + 9)$

59. $8x^3 - 1 = (2x)^3 - 1^3$
$= (2x - 1)(4x^2 + 2x + 1)$

61. $x^3 - y^3 = (x - y)(x^2 + xy + y^2)$

63. $m^3 + n^3 = (m + n)(m^2 - mn + n^2)$

65. $64x^3 + 1 = (4x)^3 + 1^3$
$= (4x + 1)(16x^2 - 4x + 1)$

67. $27x^3 - 8y^3 = (3x)^3 - (2y)^3$
$= (3x - 2y)(9x^2 + 6xy + 4y^2)$

69. $x^3y^3 + 64 = (xy)^3 + 4^3$
$= (xy + 4)(x^2y^2 - 4xy + 16)$

71. $16x^3 - y^3$ is nonfactorable over the integers.

73. $8x^3 - 9y^3$ is nonfactorable over the integers.

75. $(a - b)^3 - b^3$
$= [(a - b) - b][(a - b)^2 + b(a - b) + b^2]$
$= (a - 2b)(a^2 - 2ab + b^2 + ab - b^2 + b^2)$
$= (a - 2b)(a^2 - ab + b^2)$

77. $x^{6n} + y^{3n} = (x^{2n})^3 + (y^n)^3$
$= (x^{2n} + y^n)(x^{4n} - x^{2n}y^n + y^{2n})$

Objective 5.6.3 Exercises

79. The trinomial $5x^4y^2 - 17x^2y - 12$ can be written in the quadratic form $5u^2 - 17u - 12$, where $u = \underline{x^2y}$.

81. A polynomial is quadratic in form if it can be written in the form $au^2 + bu + c$, $a \neq 0$.

83. Let $u = xy$.
$x^2y^2 - 8xy + 15 = u^2 - 8u + 15$
$= (u - 3)(u - 5)$
$= (xy - 3)(xy - 5)$

85. Let $u = xy$.
$x^2y^2 - 17xy + 60 = u^2 - 17u + 60$
$= (u - 5)(u - 12)$
$= (xy - 5)(xy - 12)$

87. Let $u = x^2$.
$x^4 - 9x^2 + 18 = u^2 - 9u + 18$
$= (u - 6)(u - 3)$
$= (x^2 - 6)(x^2 - 3)$

89. Let $u = b^2$.
$b^4 - 13b^2 - 90 = u^2 - 13u - 90$
$= (u - 18)(u + 5)$
$= (b^2 - 18)(b^2 + 5)$

91. Let $u = x^2y^2$.
$$x^4y^4 - 8x^2y^2 + 12 = u^2 - 8u + 12$$
$$= (u-6)(u-2)$$
$$= (x^2y^2 - 6)(x^2y^2 - 2)$$

93. Let $u = x^n$.
$$x^{2n} + 3x^n + 2 = u^2 + 3u + 2$$
$$= (u+2)(u+1)$$
$$= (x^n + 2)(x^n + 1)$$

95. Let $u = xy$.
$$3x^2y^2 - 14xy + 15 = 3u^2 - 14u + 15$$
$$= (3u - 5)(u - 3)$$
$$= (3xy - 5)(xy - 3)$$

97. Let $u = ab$.
$$6a^2b^2 - 23ab + 21 = 6u^2 - 23u + 21$$
$$= (2u - 3)(3u - 7)$$
$$= (2ab - 3)(3ab - 7)$$

99. Let $u = x^2$.
$$2x^4 - 13x^2 - 15 = 2u^2 - 13u - 15$$
$$= (2u - 15)(u + 1)$$
$$= (2x^2 - 15)(x^2 + 1)$$

101. Let $u = x^n$.
$$2x^{2n} - 7x^n + 3 = 2u^2 - 7u + 3$$
$$= (2u - 1)(u - 3)$$
$$= (2x^n - 1)(x^n - 3)$$

103. Let $u = a^n$.
$$6a^{2n} + 19a^n + 10 = 6u^2 + 19u + 10$$
$$= (2u + 5)(3u + 2)$$
$$= (2a^n + 5)(3a^n + 2)$$

Objective 5.6.4 Exercises

105. To factor a polynomial with four terms, try factoring by <u>grouping</u>.

107. $12x^2 - 36x + 27 = 3(4x^2 - 12x + 9)$
$$= 3(2x - 3)^2$$

109. $27a^4 - a = a(27a^3 - 1)$
$$= a(3a - 1)(9a^2 + 3a + 1)$$

111. $20x^2 - 5 = 5(4x^2 - 1)$
$$= 5(2x + 1)(2x - 1)$$

113. $y^5 + 6y^4 - 55y^3 = y^3(y^2 + 6y - 55)$
$$= y^3(y + 11)(y - 5)$$

115. $16x^4 - 81 = (4x^2 + 9)(4x^2 - 9)$
$$= (4x^2 + 9)(2x + 3)(2x - 3)$$

117. $16a - 2a^4 = 2a(8 - a^3)$
$$= 2a(2 - a)(4 + 2a + a^2)$$

119. $x^3 + 2x^2 - x - 2 = x^2(x + 2) - 1(x + 2)$
$$= (x + 2)(x^2 - 1)$$
$$= (x + 2)(x + 1)(x - 1)$$

121. $2x^3 + 4x^2 - 3x - 6 = 2x^2(x + 2) - 3(x + 2)$
$$= (x + 2)(2x^2 - 3)$$

123. $x^3 + x^2 - 16x - 16 = x^2(x + 1) - 16(x + 1)$
$$= (x + 1)(x^2 - 16)$$
$$= (x + 1)(x + 4)(x - 4)$$

125. $a^3b^6 - b^3 = b^3(a^3b^3 - 1)$
$$= b^3(ab - 1)(a^2b^2 + ab + 1)$$

127. $x^4 - 2x^3 - 35x^2 = x^2(x^2 - 2x - 35)$
$$= x^2(x - 7)(x + 5)$$

129. $4x^2 + 4x - 1$ is nonfactorable over the integers.

131. $6x^5 + 74x^4 + 24x^3 = 2x^3(3x^2 + 37x + 12)$
$$= 2x^3(3x + 1)(x + 12)$$

133. $16a^4 - b^4 = (4a^2 + b^2)(4a^2 - b^2)$
$$= (4a^2 + b^2)(2a + b)(2a - b)$$

135. $x^4 - 5x^2 - 4$ is nonfactorable over the integers.

137. $3b^5 - 24b^2 = 3b^2(b^3 - 8)$
$$= 3b^2(b - 2)(b^2 + 2b + 4)$$

139. $x^4y^2 - 5x^3y^3 + 6x^2y^4 = x^2y^2(x^2 - 5xy + 6y^2)$
$$= x^2y^2(x - 3y)(x - 2y)$$

141. $16x^3y + 4x^2y^2 - 42xy^3$
$$= 2xy(8x^2 + 2xy - 21y^2)$$
$$= 2xy(4x + 7y)(2x - 3y)$$

143. $x^3 - 2x^2 - x + 2 = x^2(x - 2) - (x - 2)$
$$= (x - 2)(x^2 - 1)$$
$$= (x - 2)(x + 1)(x - 1)$$

145. $8xb - 8x - 4b + 4 = 4(2xb - 2x - b + 1)$
$$= 4[2x(b - 1) - (b - 1)]$$
$$= 4(b - 1)(2x - 1)$$

147. $4x^2y^2 - 4x^2 - 9y^2 + 9$
$$= 4x^2(y^2 - 1) - 9(y^2 - 1)$$
$$= (y^2 - 1)(4x^2 - 9)$$
$$= (y + 1)(y - 1)(2x + 3)(2x - 3)$$

149. $x^5 - 4x^3 - 8x^2 + 32$
$= x^3(x^2 - 4) - 8(x^2 - 4)$
$= (x^2 - 4)(x^3 - 8)$
$= (x + 2)(x - 2)(x - 2)(x^2 + 2x + 4)$
$= (x + 2)(x - 2)^2(x^2 + 2x + 4)$

151. $a^{2n+2} - 6a^{n+2} + 9a^2 = a^2(a^{2n} - 6a^n + 9)$
$= a^2(a^n - 3)^2$

153. $2x^{n+2} - 7x^{n+1} + 3x^n = x^n(2x^2 - 7x + 3)$
$= x^n(2x - 1)(x - 3)$

Applying Concepts 5.6

155. $P(x) = x^n(x - a)(x + b) = x^n(x^2 + bx - ax - ab)$
When this is multiplied out fully, the leading term will have degree x^{n+2}. The degree of the polynomial P is $n + 2$.

157. $4x^2 - kx + 25 = (2x + 5)^2$ or $(2x - 5)^2$
$(2x + 5)(2x + 5) = 4x^2 + 20x + 25$
$(2x - 5)(2x - 5) = 4x^2 - 20x + 25$
The possible values of $-k$ are 20 and -20.
The possible values of k are -20 and 20.

159. $16x^2 + kxy + y^2 = (4x + y)^2$ or $(4x - y)^2$
$(4x + y)(4x + y) = 16x^2 + 8xy + y^2$
$(4x - y)(4x - y) = 16x^2 - 8xy + y^2$
The possible values of k are 8 and -8.

161. $ax^3 + b - bx^3 - a = (ax^3 - bx^3) + (b - a)$
$= x^3(a - b) + [-(a - b)]$
$= x^3(a - b) - (a - b)$
$= (a - b)(x^3 - 1)$
$= (a - b)(x^3 - 1^3)$
$= (a - b)(x - 1)(x^2 + x + 1)$

163. $y^{8n} - 2y^{4n} + 1 = (y^{4n} - 1)^2$
$= [(y^{2n})^2 - 1][(y^{2n})^2 - 1]$
$= (y^{2n} + 1)(y^{2n} - 1)(y^{2n} + 1)(y^{2n} - 1)$
$= (y^{2n} + 1)^2(y^{2n} - 1)^2$
$= (y^{2n} + 1)^2[(y^n)^2 - 1]^2$
$= (y^{2n} + 1)^2[(y^n + 1)(y^n - 1)]^2$
$= (y^{2n} + 1)^2(y^n + 1)^2(y^n - 1)^2$

165. One number is a perfect square less than 63: 1, 4, 9, 16, 25, 36, 49
One number is a prime number less than 63: 2, 3, 5, 7, 11, 13, 17, 19, 23, 29, 31, 37, 41, 43, 47, 53, 59, 61
The product of the numbers is 63: $9 \cdot 7 = 63$
$9 + 7 = 16$
The sum of the numbers is 16.

167. Perfect squares less than 500
1   144
4   169
9   196
16   225
25   256
36   289
49   324
64   361
81   400
100 441
121 484
The palindromic perfect squares less than 500 are 1, 4, 9, 121, and 484.

169. $x^4 + 64$
$= (x^4 + 16x^2 + 64) - 16x^2$
$= (x^2 + 8)(x^2 + 8) - 16x^2$
$= (x^2 + 8)^2 - 16x^2$
$= (x^2 + 8 - 4x)(x^2 + 8 + 4x)$
$= (x^2 - 4x + 8)(x^2 + 4x + 8)$

171. No, a third-degree polynomial cannot have factors $x - 1$, $x + 1$, $x - 3$, and $x + 4$ because the leading term in the product of these factors must be a fourth-degree term.

Section 5.7

Concept Review 5.7

1. Always true

3. Sometimes true
The equation $(x - 2)(x + 3)(x - 1) = 0$ has three solutions.

5. Sometimes true
If $n = 1$, then $n + 1 = 2$, and $n + 3 = 4$, which are even integers.

Objective 5.7.1 Exercises

1. A quadratic equation is an equation of the form $ax^2 + bx + c = 0$, $a \neq 0$. A quadratic equation has a term of degree 2. In a linear equation, the highest exponent on a variable is 1.
$x^2 + 3x - 7 = 0$ is an example of a quadratic equation.
$2x + 4 = 0$ is an example of a linear equation.

3. Of the two quadratic equations $0 = 5x^2 + 2x - 3$ and $x^2 - 4x = 3$, the one that is in standard form is $0 = 5x^2 + 2x - 3$.

5. Let $f(x) = x^2 + x - 15$. To find two values of c for which $f(c) = 5$, solve the equation $x^2 + x - 15 = \underline{5}$.

7. $(y + 4)(y + 6) = 0$
$y + 4 = 0 \quad y + 6 = 0$
$y = -4 \quad y = -6$
The solutions are -4 and -6.

Chapter 5: Polynomials And Exponents

9. $x(x-7) = 0$
$x = 0 \quad x - 7 = 0$
$ x = 7$
The solutions are 0 and 7.

11. $3z(2z + 5) = 0$
$3z = 0 \quad 2z + 5 = 0$
$z = 0 \quad 2z = -5$
$ z = -\dfrac{5}{2}$
The solutions are 0 and $-\dfrac{5}{2}$.

13. $(2x + 3)(x - 7) = 0$
$2x + 3 = 0 \quad x - 7 = 0$
$ 2x = -3 \quad x = 7$
$ x = -\dfrac{3}{2}$
The solutions are $-\dfrac{3}{2}$ and 7.

15. $b^2 - 49 = 0$
$b^2 - 7^2 = 0$
$(b + 7)(b - 7) = 0$
$b + 7 = 0 \quad b - 7 = 0$
$b = -7 \quad b = 7$
The solutions are 7 and –7.

17. $9t^2 - 16 = 0$
$(3t)^2 - 4^2 = 0$
$(3t + 4)(3t - 4) = 0$
$3t + 4 = 0 \quad 3t - 4 = 0$
$3t = -4 \quad 3t = 4$
$t = -\dfrac{4}{3} \quad t = \dfrac{4}{3}$
The solutions are $\dfrac{4}{3}$ and $-\dfrac{4}{3}$.

19. $y^2 + 4y - 5 = 0$
$(y + 5)(y - 1) = 0$
$y + 5 = 0 \quad y - 1 = 0$
$y = -5 \quad y = 1$
The solutions are –5 and 1.

21. $2b^2 - 5b - 12 = 0$
$(2b + 3)(b - 4) = 0$
$2b + 3 = 0 \quad b - 4 = 0$
$2b = -3 \quad b = 4$
$b = -\dfrac{3}{2}$
The solutions are $-\dfrac{3}{2}$ and 4.

23. $x^2 - 9x = 0$
$x(x - 9) = 0$
$x = 0 \quad x - 9 = 0$
$ x = 9$
The solutions are 0 and 9.

25. $3a^2 - 12a = 0$
$3a(a - 4) = 0$
$3a = 0 \quad a - 4 = 0$
$a = 0 \quad a = 4$
The solutions are 0 and 4.

27. $z^2 - 3z = 28$
$z^2 - 3z - 28 = 0$
$(z - 7)(z + 4) = 0$
$z - 7 = 0 \quad z + 4 = 0$
$z = 7 \quad z = -4$
The solutions are 7 and –4.

29. $3t^2 + 13t = 10$
$3t^2 + 13t - 10 = 0$
$(3t - 2)(t + 5) = 0$
$3t - 2 = 0 \quad t + 5 = 0$
$3t = 2 \quad t = -5$
$t = \dfrac{2}{3}$
The solutions are $\dfrac{2}{3}$ and –5.

31. $5b^2 - 17b = -6$
$5b^2 - 17b + 6 = 0$
$(5b - 2)(b - 3) = 0$
$5b - 2 = 0 \quad b - 3 = 0$
$5b = 2 \quad b = 3$
$b = \dfrac{2}{5}$
The solutions are $\dfrac{2}{5}$ and 3.

33. $8x^2 - 10x = 3$
$8x^2 - 10x - 3 = 0$
$(2x - 3)(4x + 1) = 0$
$2x - 3 = 0 \quad 4x + 1 = 0$
$2x = 3 \quad 4x = -1$
$x = \dfrac{3}{2} \quad x = -\dfrac{1}{4}$
The solutions are $\dfrac{3}{2}$ and $-\dfrac{1}{4}$.

35.
$$y(y-2) = 35$$
$$y^2 - 2y = 35$$
$$y^2 - 2y - 35 = 0$$
$$(y-7)(y+5) = 0$$
$$y - 7 = 0 \quad y + 5 = 0$$
$$y = 7 \quad y = -5$$
The solutions are 7 and −5.

37.
$$x(x-12) = -27$$
$$x^2 - 12x = -27$$
$$x^2 - 12x + 27 = 0$$
$$(x-9)(x-3) = 0$$
$$x - 9 = 0 \quad x - 3 = 0$$
$$x = 9 \quad x = 3$$
The solutions are 9 and 3.

39.
$$y(3y-2) = 8$$
$$3y^2 - 2y = 8$$
$$3y^2 - 2y - 8 = 0$$
$$(3y+4)(y-2) = 0$$
$$3y + 4 = 0 \quad y - 2 = 0$$
$$3y = -4 \quad y = 2$$
$$y = -\frac{4}{3}$$
The solutions are $-\frac{4}{3}$ and 2.

41.
$$3a^2 - 4a = 20 - 15a$$
$$3a^2 + 11a = 20$$
$$3a^2 + 11a - 20 = 0$$
$$(3a-4)(a+5) = 0$$
$$3a - 4 = 0 \quad a + 5 = 0$$
$$3a = 4 \quad a = -5$$
$$a = \frac{4}{3}$$
The solutions are $\frac{4}{3}$ and −5.

43.
$$(y+5)(y-7) = -20$$
$$y^2 - 2y - 35 = -20$$
$$y^2 - 2y - 15 = 0$$
$$(y-5)(y+3) = 0$$
$$y - 5 = 0 \quad y + 3 = 0$$
$$y = 5 \quad y = -3$$
The solutions are 5 and −3.

45.
$$(b+5)(b+10) = 6$$
$$b^2 + 15b + 50 = 6$$
$$b^2 + 15b + 44 = 0$$
$$(b+11)(b+4) = 0$$
$$b + 11 = 0 \quad b + 4 = 0$$
$$b = -11 \quad b = -4$$
The solutions are −11 and −4.

47.
$$(t-3)^2 = 1$$
$$t^2 - 6t + 9 = 1$$
$$t^2 - 6t + 8 = 0$$
$$(t-2)(t-4) = 0$$
$$t - 2 = 0 \quad t - 4 = 0$$
$$t = 2 \quad t = 4$$
The solutions are 2 and 4.

49.
$$(3-x)^2 + x^2 = 5$$
$$9 - 6x + x^2 + x^2 = 5$$
$$2x^2 - 6x + 9 = 5$$
$$2x^2 - 6x + 4 = 0$$
$$2(x^2 - 3x + 2) = 0$$
$$2(x-1)(x-2) = 0$$
$$x - 1 = 0 \quad x - 2 = 0$$
$$x = 1 \quad x = 2$$
The solutions are 1 and 2.

51.
$$(a-1)^2 = 3a - 5$$
$$a^2 - 2a + 1 = 3a - 5$$
$$a^2 - 5a + 1 = -5$$
$$a^2 - 5a + 6 = 0$$
$$(a-2)(a-3) = 0$$
$$a - 2 = 0 \quad a - 3 = 0$$
$$a = 2 \quad a = 3$$
The solutions are 2 and 3.

53.
$$x^3 + 4x^2 - x - 4 = 0$$
$$x^2(x+4) - 1(x+4) = 0$$
$$(x+4)(x^2 - 1) = 0$$
$$(x+4)(x+1)(x-1) = 0$$
$$x + 4 = 0 \quad x + 1 = 0 \quad x - 1 = 0$$
$$x = -4 \quad x = -1 \quad x = 1$$
The solutions are −4, −1, and 1.

55.
$$f(c) = 1$$
$$c^2 - 3c + 3 = 1$$
$$c^2 - 3c + 2 = 0$$
$$(c-2)(c-1) = 0$$
$$c - 2 = 0 \quad c - 1 = 0$$
$$c = 2 \quad c = 1$$
The values of c are 1 and 2.

57.
$$f(c) = -4$$
$$2c^2 - c - 5 = -4$$
$$2c^2 - c - 1 = 0$$
$$(2c+1)(c-1) = 0$$
$$2c + 1 = 0 \quad c - 1 = 0$$
$$2c = -1 \quad c = 1$$
$$c = -\frac{1}{2}$$
The values of c are $-\frac{1}{2}$ and 1.

59.
$$f(c) = 2$$
$$4c^2 - 4c + 3 = 2$$
$$4c^2 - 4c + 1 = 0$$
$$(2c-1)(2c-1) = 0$$
$$2c - 1 = 0 \quad 2c - 1 = 0$$
$$2c = 1 \quad 2c = 1$$
$$c = \frac{1}{2} \quad c = \frac{1}{2}$$
The value of c is $\frac{1}{2}$.

61.
$$f(c) = 0$$
$$c^2 + 3c - 4 = 0$$
$$(c-1)(c+4) = 0$$
$$c - 1 = 0 \quad c + 4 = 0$$
$$c = 1 \quad c = -4$$
The zeros of f are -4 and 1.

63.
$$s(c) = 0$$
$$c^2 - 4c - 12 = 0$$
$$(c-6)(c+2) = 0$$
$$c - 6 = 0 \quad c + 2 = 0$$
$$c = 6 \quad c = -2$$
The zeros of s are -2 and 6.

65.
$$f(c) = 0$$
$$2c^2 - 5c + 3 = 0$$
$$(2c-3)(c-1) = 0$$
$$2c - 3 = 0 \quad c - 1 = 0$$
$$2c = 3 \quad c = 1$$
$$c = \frac{3}{2}$$
The zeros of f are 1 and $\frac{3}{2}$.

67.
$$p(c) = 0$$
$$c^2 + 6c + 9 = 0$$
$$(c+3)(c+3) = 0$$
$$c + 3 = 0 \quad c + 3 = 0$$
$$c = -3 \quad c = -3$$
The zeros of p are -3.

Objective 5.7.2 Exercises

69. Let x represent a positive integer. Then the next consecutive positive integer is $x + 1$, and an expression that represents the sum of the squares of the two integers is $x^2 + (x+1)^2$.

71. Strategy
This is an integer problem.
The unknown integer is n.
The sum of the integer and its square is 90.
$n + n^2 = 90$

Solution
$$n + n^2 = 90$$
$$n^2 + n - 90 = 0$$
$$(n+10)(n-9) = 0$$
$$n + 10 = 0 \quad n - 9 = 0$$
$$n = -10 \quad n = 9$$
The integer is -10 or 9.

73. Strategy
This is an integer problem.
The first positive integer is n.
The next consecutive positive integer is $n + 1$.
The sum of the squares of the two consecutive positive integers is equal to 145.
$n^2 + (n+1)^2 = 145$

Solution
$$n^2 + (n+1)^2 = 145$$
$$n^2 + n^2 + 2n + 1 = 145$$
$$2n^2 + 2n - 144 = 0$$
$$2(n^2 + n - 72) = 0$$
$$2(n+9)(n-8) = 0$$
$$n + 9 = 0 \quad n - 8 = 0$$
$$n = -9 \quad n = 8$$
The solutions -9 and -8 are not possible because they are not positive. The integers are 8 and 9.

Chapter 5: Polynomials And Exponents

75. Strategy
This is an integer problem.
The unknown integer is n.
The sum of the cube of the integer and the product of the integer and twelve is equal to seven times the square of the integer.
$x^3 + 12x = 7x^2$

Solution
$$x^3 + 12x = 7x^2$$
$$x^3 - 7x^2 + 12x = 0$$
$$x(x^2 - 7x + 12) = 0$$
$$x(x-3)(x-4) = 0$$
$$x = 0 \quad x - 3 = 0 \quad x - 4 = 0$$
$$ x = 3 \quad x = 4$$
The integer is 0, 3, or 4.

77. Strategy
This is a geometry problem.
The width of the rectangle: x
The length of the rectangle: $2x + 5$
The area of the rectangle is 168 in^2. Use the equation for the area of a rectangle ($A = L \cdot W$).

Solution
$$A = L \cdot W$$
$$168 = (2x+5)x$$
$$168 = 2x^2 + 5x$$
$$0 = 2x^2 + 5x - 168$$
$$0 = (2x+21)(x-8)$$
$$2x + 21 = 0 \quad x - 8 = 0$$
$$2x = -21 \quad x = 8$$
$$x = -\frac{21}{2}$$

Since the width cannot be negative, $-\frac{21}{2}$ cannot be a solution.
$2x + 5 = 2(8) + 5 = 21$
The width is 8 in. The length is 21 in.

79. Strategy
Length of the trough: 6
Width of the trough: $10 - 2x$
Height of the trough: x
The volume of the trough is 72 m^3.
Use the equation for the volume of a rectangular solid ($V = LWH$).

Solution
$$V = LWH$$
$$72 = 6(10-2x)(x)$$
$$72 = (60 - 12x)x$$
$$72 = 60x - 12x^2$$
$$12x^2 - 60x + 72 = 0$$
$$12(x^2 - 5x + 6) = 0$$
$$12(x-3)(x-2) = 0$$
$$x - 3 = 0 \quad x - 2 = 0$$
$$x = 3 \quad x = 2$$
The value of x should be 2 m or 3 m.

81. Strategy
To find the velocity of the rocket, substitute 500 for s in the equation and solve for v.

Solution
$$v^2 = 20s$$
$$v^2 = 20(500)$$
$$v^2 = 10{,}000$$
$$v^2 - 10{,}000 = 0$$
$$(v-100)(v+100) = 0$$
$$v - 100 = 0 \quad v + 100 = 0$$
$$v = 100 \quad v = -100$$

Since the rocket is traveling in the direction it was launched, the velocity is not negative, so −100 is not a solution. The velocity of the rocket is 100 m/s.

83. Strategy
This is a geometry problem.
The height of the triangle: x
The base of the triangle: $3x$
The area of the triangle is 24 cm^2. Use the equation for the area of a triangle $\left(A = \frac{1}{2}bh\right)$.

Solution
$$A = \frac{1}{2}bh$$
$$24 = \frac{1}{2}(3x)x$$
$$24 = \frac{1}{2}(3x^2)$$
$$48 = 3x^2$$
$$0 = 3x^2 - 48$$
$$0 = 3(x^2 - 16)$$
$$0 = 3(x+4)(x-4)$$
$$x + 4 = 0 \quad x - 4 = 0$$
$$x = -4 \quad x = 4$$

Since the height cannot be negative, −4 cannot be a solution.
$3x = 3(4) = 12$
The height is 4 cm. The base is 12 cm.

Chapter 5: Polynomials And Exponents

85. Strategy
To find the time for the object to reach the ground, replace the variables d and v in the equation by their given values and solve for t.

Solution
$$d = vt + 16t^2$$
$$480 = 16t + 16t^2$$
$$0 = 16t^2 + 16t - 480$$
$$0 = 16(t^2 + t - 30)$$
$$0 = t^2 + t - 30$$
$$0 = (t+6)(t-5)$$
$$t + 6 = 0 \quad t - 5 = 0$$
$$t = -6 \quad t = 5$$

Since the time cannot be a negative number, -6 is not a solution. The time is 5 s.

87. Strategy
Increase in length and width: x
Length of the larger rectangle: $6 + x$
Width of the larger rectangle: $3 + x$
Use the equation $A = L \cdot W$.

Solution
Smaller rectangle:
$$A = L \cdot W$$
$$A = 6 \cdot 3 = 18$$
Larger rectangle:
$$A = L \cdot W$$
$$18 + 70 = (6 + x)(3 + x)$$
$$88 = 18 + 9x + x^2$$
$$0 = x^2 + 9x - 70$$
$$0 = (x + 14)(x - 5)$$
$$x + 14 = 0 \quad x - 5 = 0$$
$$x = -14 \quad x = 5$$

Since an increase in length and width cannot be a negative number, -14 is not a solution.
Length $= 6 + x = 6 + 5 = 11$
Width $= 3 + x = 3 + 5 = 8$
The length of the larger rectangle is 11 cm. The width is 8 cm.

Applying Concepts 5.7

89. Strategy
Solve the equation $n(n + 6) = 16$ for n, and then substitute the values of n into the expression $3n^2 + 2n - 1$ and evaluate.

Solution
$$n(n+6) = 16$$
$$n^2 + 6n = 16$$
$$n^2 + 6n - 16 = 0$$
$$(n+8)(n-2) = 0$$
$$n + 8 = 0 \quad n - 2 = 0$$
$$n = -8 \quad n = 2$$

$$\begin{array}{ll} 3n^2 + 2n - 1 & 3n^2 + 2n - 1 \\ = 3(-8)^2 + 2(-8) - 1 & = 3(2)^2 + 2(2) - 1 \\ = 3(64) + 2(-8) - 1 & = 3(4) + 2(2) - 1 \\ = 192 - 16 - 1 & = 12 + 4 - 1 \\ = 175 & = 15 \end{array}$$

$3n^2 + 2n - 1$ is equal to 175 or 15.

91. Strategy
Solve the equation $6a(a - 1) = 36$ for a, and then substitute the values of a into the expression $-a^2 + 5a - 2$ and evaluate.

Solution
$$6a(a-1) = 36$$
$$6a^2 - 6a = 36$$
$$6a^2 - 6a - 36 = 0$$
$$(a+2)(6a-18) = 0$$
$$a + 2 = 0 \quad 6a - 18 = 0$$
$$a = -2 \quad a = 3$$

$$\begin{array}{ll} -a^2 + 5a - 2 & -a^2 + 5a - 2 \\ = -(-2)^2 + 5(-2) - 2 & = -(3)^2 + 5(3) - 2 \\ = -(4) + 5(-2) - 2 & = -(9) + 5(3) - 2 \\ = -4 - 10 - 2 & = -9 + 15 - 2 \\ = -16 & = 4 \end{array}$$

$-a^2 + 5a - 2$ is equal to -16 or 4.

93.
$$f(c) = 1$$
$$c^3 + 3c^2 - 4c - 11 = 1$$
$$c^3 + 3c^2 - 4c - 12 = 0$$
$$c^2(c+3) - 4(c+3) = 0$$
$$(c+3)(c^2 - 4) = 0$$
$$(c+3)(c+2)(c-2) = 0$$
$$c + 3 = 0 \quad c + 2 = 0 \quad c - 2 = 0$$
$$c = -3 \quad c = -2 \quad c = 2$$

The values of c are -3, -2, and 2.

95. Strategy
Width of rectangular piece of cardboard: x
Length of rectangular piece of cardboard: $x + 10$
Width of cardboard box: $x - 4$
Length of cardboard box: $(x + 10) - 4 = x + 6$
Height of cardboard box: 2
The formula for the volume of a rectangular solid is $V = LWH$.

Solution
$V = LWH$
$112 = (x+6)(x-4)(2)$
$56 = (x+6)(x-4)$
$56 = x^2 + 2x - 24$
$0 = x^2 + 2x - 80$
$0 = (x+10)(x-8)$

$x + 10 = 0 \quad x - 8 = 0$
$\quad x = -10 \quad x = 8$

Since width cannot be a negative number, -10 is not a solution.
Length $= x + 10 = 8 + 10 = 18$
The length is 18 in. The width is 8 in.

97.
$(x-3)(x-2)(x+1) = 0$
$(x^2 - 5x + 6)(x+1) = 0$
$(x^3 - 5x^2 + 6x) + (x^2 - 5x + 6) = 0$
$x^3 - 4x^2 + x + 6 = 0$

CHAPTER 5 REVIEW EXERCISES

1. $(3x^2 - 2x - 6) + (-x^2 - 3x + 4) = 2x^2 - 5x - 2$

2. $(5x^2 - 8xy + 2y^2) - (x^2 - 3y^2)$
$= (5x^2 - 8xy + 2y^2) + (-x^2 + 3y^2)$
$= 4x^2 - 8xy + 5y^2$

3. $(5x^2yz^4)(2xy^3z^{-1})(7x^{-2}y^{-2}z^3) = 70xy^2z^6$

4. $(2x^{-1}y^2z^5)^4(-3x^3yz^{-3})$
$= (16x^{-4}y^8z^{20})(-3x^3yz^{-3})$
$= -48x^{-1}y^9z^{17}$
$= -\dfrac{48y^9z^{17}}{x}$

5. $\dfrac{3x^4yz^{-1}}{-12xy^3z^2} = -\dfrac{x^3}{4y^2z^3}$

6. $\dfrac{(2a^4b^{-3}c^2)^3}{(2a^3b^2c^{-1})^4} = \dfrac{8a^{12}b^{-9}c^6}{16a^{12}b^8c^{-4}}$
$= \dfrac{1}{2}a^{12-12}b^{-9-8}c^{6-(-4)}$
$= \dfrac{1}{2}b^{-17}c^{10}$
$= \dfrac{c^{10}}{2b^{17}}$

7. $93{,}000{,}000 = 9.3 \times 10^7$

8. $2.54 \times 10^{-3} = 0.00254$

9. $\dfrac{3 \times 10^{-3}}{15 \times 10^2} = \dfrac{3}{15} \times \dfrac{10^{-3}}{10^2} = 0.2 \times 10^{-5} = 2 \times 10^{-6}$

10. $P(-2) = 2(-2)^3 - (-2) + 7$
$P(-2) = -16 + 2 + 7$
$P(-2) = -7$

11. $y = x^2 + 1$

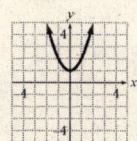

12. a. 3
b. 8
c. 5

13.
$$\begin{array}{r|rrrr} -3 & -2 & 2 & 0 & -4 \\ & & 6 & -24 & 72 \\ \hline & -2 & 8 & -24 & 68 \end{array}$$

$P(-3) = 68$

14. $\dfrac{6x^2y^3 - 18x^3y^2 + 12x^4y}{3x^2y} = \dfrac{6x^2y^3}{3x^2y} - \dfrac{18x^3y^2}{3x^2y} + \dfrac{12x^4y}{3x^2y}$
$= 2y^2 - 6xy + 4x^2$

15.
$$\begin{array}{r} 5x + 4 \\ 3x-2 \overline{\smash{\big)}15x^2 + 2x - 2} \\ \underline{15x^2 - 10x} \\ 12x - 2 \\ \underline{12x - 8} \\ 6 \end{array}$$

$\dfrac{15x^2 + 2x - 2}{3x - 2} = 5x + 4 + \dfrac{6}{3x - 2}$

16.
$$6x+1 \overline{)12x^2-16x-7} \quad \text{quotient: } 2x-3$$
$$\underline{12x^2+2x}$$
$$-18x-7$$
$$\underline{-18x-3}$$
$$-4$$

$$\frac{12x^2-16x-7}{6x+1} = 2x-3-\frac{4}{6x+1}$$

17.

-6	4	27	10	2
		-24	-18	48
	4	3	-8	50

$$\frac{4x^3+27x^2+10x+2}{x+6} = 4x^2+3x-8+\frac{50}{x+6}$$

18.

4	1	0	0	0	-4
		4	16	64	256
	1	4	16	64	252

$$\frac{x^4-4}{x-4} = x^3+4x^2+16x+64+\frac{252}{x-4}$$

19. $4x^2y(3x^3y^2+2xy-7y^3)$
$= 12x^5y^3+8x^3y^2-28x^2y^4$

20. $a^{2n+3}(a^n-5a+2) = a^{3n+3}-5a^{2n+4}+2a^{2n+3}$

21. $5x^2-4x[x-(3x+2)+x]$
$= 5x^2-4x(x-3x-2+x)$
$= 5x^2-4x(-x-2)$
$= 5x^2+4x^2+8x$
$= 9x^2+8x$

22. $(x^{2n}-x)(x^{n+1}-3) = x^{3n+1}-3x^{2n}-x^{n+2}+3x$

23.
$$\begin{array}{r} x^3-3x^2-5x+1 \\ \times \quad\quad\quad x+6 \\ \hline 6x^3-18x^2-30x+6 \\ x^4-3x^3-5x^2+\ x \\ \hline x^4+3x^3-23x^2-29x+6 \end{array}$$

$(x+6)(x^3-3x^2-5x+1)$
$= x^4+3x^3-23x^2-29x+6$

24. $(x-4)(3x+2)(2x-3)$
$= (x-4)(6x^2-5x-6)$
$= 6x^3-5x^2-6x-24x^2+20x+24$
$= 6x^3-29x^2+14x+24$

25. $(5a+2b)(5a-2b) = 25a^2-4b^2$

26. $(4x-3y)^2 = 16x^2-24xy+9y^2$

27. The GCF of $18a^5b^2, 12a^3b^3$, and $30a^2b$ is $6a^2b$.
$18a^5b^2-12a^3b^3+30a^2b$
$= 6a^2b(3a^3b-2ab^2+5)$

28. The GCF of $5x^{n+5}, x^{n+3}$, and $4x^2$ is x^2.
$5x^{n+5}+x^{n+3}+4x^2 = x^2(5x^{n+3}+x^{n+1}+4)$

29. $x(y-3)+4(3-y) = x(y-3)+4[(-1)(y-3)]$
$= x(y-3)-4(y-3)$
$= (y-3)(x-4)$

30. $2ax+4bx-3ay-6by = 2x(a+2b)-3y(a+2b)$
$= (a+2b)(2x-3y)$

31. $x^2+12x+35 = (x+5)(x+7)$

32. $12+x-x^2 = (3+x)(4-x)$

33. $x^2-16x+63 = (x-7)(x-9)$

34. $6x^2-31x+18 = (3x-2)(2x-9)$

35. $24x^2+61x-8 = (8x-1)(3x+8)$

36. Let $u = xy$.
$x^2y^2-9 = u^2-9$
$= (u+3)(u-3)$
$= (xy+3)(xy-3)$

37. $4x^2+12xy+9y^2 = (2x+3y)^2$

38. Let $u = x^n$.
$x^{2n}-12x^n+36 = u^2-12u+36$
$= (u-6)^2$
$= (x^n-6)^2$

39. $36-a^{2n} = 6^2-(a^n)^2 = (6+a^n)(6-a^n)$

40. $64a^3-27b^3 = (4a)^3-(3b)^3$
$= (4a-3b)(16a^2+12ab+9b^2)$

41. $8-y^{3n} = (2)^3-(y^n)^3 = (2-y^n)(4+2y^n+y^{2n})$

42. Let $u = x^2$.
$15x^4+x^2-6 = 15u^2+u-6$
$= (3u+2)(5u-3)$
$= (3x^2+2)(5x^2-3)$

43. Let $u = x^4$.
$$36x^8 - 36x^4 + 5 = 36u^2 - 36u + 5$$
$$= (6u - 5)(6u - 1)$$
$$= (6x^4 - 5)(6x^4 - 1)$$

44. Let $u = x^2y^2$.
$$21x^4y^4 + 23x^2y^2 + 6 = 21u^2 + 23u + 6$$
$$= (7u + 3)(3u + 2)$$
$$= (7x^2y^2 + 3)(3x^2y^2 + 2)$$

45. $3a^6 - 15a^4 - 18a^2 = 3a^2(a^4 - 5a^2 - 6)$
$$= 3a^2(a^2 - 6)(a^2 + 1)$$

46. $x^{4n} - 8x^{2n} + 16 = (x^{2n} - 4)^2$
$$= [(x^n + 2)(x^n - 2)]^2$$
$$= (x^n + 2)^2(x^n - 2)^2$$

47. $3a^4b - 3ab^4 = 3ab(a^3 - b^3)$
$$= 3ab(a - b)(a^2 + ab + b^2)$$

48. $x^3 - x^2 - 6x = 0$
$$x(x^2 - x - 6) = 0$$
$$x(x - 3)(x + 2) = 0$$
$$x = 0 \quad x - 3 = 0 \quad x + 2 = 0$$
$$x = 3 \quad x = -2$$
The solutions are 0, 3, and –2.

49. $6x^2 + 60 = 39x$
$$6x^2 - 39x + 60 = 0$$
$$3(2x^2 - 13x + 20) = 0$$
$$3(2x - 5)(x - 4) = 0$$
$$2x - 5 = 0 \quad x - 4 = 0$$
$$2x = 5 \quad x = 4$$
$$x = \frac{5}{2}$$
The solutions are $\frac{5}{2}$ and 4.

50. $x^3 - 16x = 0$
$$x(x^2 - 16) = 0$$
$$x(x + 4)(x - 4) = 0$$
$$x = 0 \quad x + 4 = 0 \quad x - 4 = 0$$
$$x = -4 \quad x = 4$$
The solutions are 0, –4, and 4.

51. $y^3 + y^2 - 36y - 36 = 0$
$$y^2(y + 1) - 36(y + 1) = 0$$
$$(y + 1)(y^2 - 36) = 0$$
$$(y + 1)(y + 6)(y - 6) = 0$$
$$y + 1 = 0 \quad y + 6 = 0 \quad y - 6 = 0$$
$$y = -1 \quad y = -6 \quad y = 6$$
The solutions are –1, –6, and 6.

52. $f(c) = 0$
$$c^2 - 5c - 6 = 0$$
$$(c - 6)(c + 1) = 0$$
$$c - 6 = 0 \quad c + 1 = 0$$
$$c = 6 \quad c = -1$$
The zeros of f are –1 and 6.

53. Strategy
To find how far Earth is from the Great Galaxy of Andromeda, use the equation $d = rt$, where $r = 5.9 \times 10^{12}$ mph and $t = 2.2 \times 10^6$ years.
$2.2 \times 10^6 \times 24 \times 365 = 1.9272 \times 10^{10}$

Solution
$d = r \cdot t$
$$= (5.9 \times 10^{12})(1.9272 \times 10^{10})$$
$$= 5.9 \times 1.9272 \times 10^{22}$$
$$= 11.37048 \times 10^{22}$$
$$= 1.137048 \times 10^{23}$$
The distance from Earth to the Great Galaxy of Andromeda is 1.137048×10^{23} mi.

54. Strategy
To find how much power is generated by the sun, divide the amount of horsepower that Earth receives by the proportion that this is of the power generated by the sun.

Solution
$$\frac{2.4 \times 10^{14}}{2.2 \times 10^{-7}} = \frac{2.4}{2.2} \times \frac{10^{14}}{10^{-7}} = 1.09 \times 10^{21}$$
The sun generates 1.09×10^{21} horsepower.

55. Strategy
To find the area, replace the variables L and W in the equation $A = LW$ by the given values and solve for A.

Solution
$A = L \cdot W$
$A = (5x + 3)(2x - 7) = 10x^2 - 29x - 21$
The area is $(10x^2 - 29x - 21)$ cm^2.

56. Strategy
This is a geometry problem.
To find the volume, replace the variable s in the equation $V = s^3$ by the given value and solve for V.

Solution
$V = s^3$
$V = (3x-1)^3$
$V = (3x-1)(9x^2 - 6x + 1)$
$V = 27x^3 - 18x^2 + 3x - 9x^2 + 6x - 1$
$V = 27x^3 - 27x^2 + 9x - 1$
The volume is $(27x^3 - 27x^2 + 9x - 1)$ ft^3.

57. Strategy
To find the area, subtract the area of the small square from the area of the large rectangle.
Large rectangle:
$L_1 = 3x - 2$
$W_1 = (x+4) + x$
Small square:
side = x

Solution
A = Area of large rectangle – Area of square
$A = L_1 W_1 - s^2$
$A = (3x-2)[(x+4)+x] - x^2$
$A = (3x-2)(2x+4) - x^2$
$A = 6x^2 + 8x - 8 - x^2$
$A = 5x^2 + 8x - 8$
The area is $(5x^2 + 8x - 8)$ in^2.

58. Strategy
This is an integer problem.
The first even integer: n
The next consecutive even integer: $n + 2$
The sum of the squares of the 2 consecutive even integers is 52.
$n^2 + (n+2)^2 = 52$

Solution
$n^2 + (n+2)^2 = 52$
$n^2 + n^2 + 4n + 4 = 52$
$2n^2 + 4n - 48 = 0$
$2(n^2 + 2n - 24) = 0$
$2(n+6)(n-4) = 0$
$n + 6 = 0 \qquad n - 4 = 0$
$n = -6 \qquad n = 4$
$n + 2 = -6 + 2 = -4 \quad n + 2 = 4 + 2 = 6$
The two integers are –6 and –4, or 4 and 6.

59. Strategy
This is an integer problem.
The unknown integer is n.
The sum of this number and its square is 56.

Solution
$n^2 + n = 56$
$n^2 + n - 56 = 0$
$(n+8)(n-7) = 0$
$n + 8 = 0 \quad n - 7 = 0$
$n = -8 \quad n = 7$
The integer is –8 or 7.

60. Strategy
This is a geometry problem.
Width of the rectangle: x
Length of the rectangle: $2x + 2$
The area of the rectangle is 60 m^2.
Use the equation for the area of a rectangle $(A = LW)$.

Solution
$A = LW$
$60 = (2x+2)(x)$
$60 = 2x^2 + 2x$
$0 = 2x^2 + 2x - 60$
$0 = 2(x^2 + x - 30)$
$0 = 2(x+6)(x-5)$
$x + 6 = 0 \quad x - 5 = 0$
$x = -6 \quad x = 5$
Since the width of a rectangle cannot be negative, –6 cannot be a solution.
$2x + 2 = 2(5) + 2 = 10 + 2 = 12$
The length of the rectangle is 12 m.

CHAPTER 5 TEST

1. $(6x^3 - 7x^2 + 6x - 7) - (4x^3 - 3x^2 + 7)$
$= (6x^3 - 7x^2 + 6x - 7) + (-4x^3 + 3x^2 - 7)$
$= 2x^3 - 4x^2 + 6x - 14$

2. $(-4a^2 b)^3 (-ab^4) = (-64a^6 b^3)(-ab^4) = 64a^7 b^7$

3.
$\dfrac{(2a^{-4}b^2)^3}{4a^{-2}b^{-1}} = \dfrac{8a^{-12}b^6}{4a^{-2}b^{-1}}$
$= 2a^{-12-(-2)} b^{6-(-1)}$
$= 2a^{-10} b^7$
$= \dfrac{2b^7}{a^{10}}$

4. $0.000000501 = 5.01 \times 10^{-7}$

5. Strategy
To find the number of seconds in one week in scientific notation:
Multiply the number of seconds in a minute (60) by the minutes in an hour (60) by the hours in a day (24) by the days in a week (7).
Convert that product to scientific notation.

Solution
$60 \times 60 \times 24 \times 7 = 604,800 = 6.048 \times 10^5$

The number of seconds in a week is 6.048×10^5 s.

6. $(2x^{-3}y)^{-4} = 2^{-4}x^{12}y^{-4} = \dfrac{x^{12}}{2^4 y^4} = \dfrac{x^{12}}{16y^4}$

7. $-5x[3 - 2(2x - 4) - 3x] = -5x[3 - 4x + 8 - 3x]$
$= -5x[-7x + 11]$
$= 35x^2 - 55x$

8. $(3a + 4b)(2a - 7b) = 6a^2 - 13ab - 28b^2$

9.
$$\begin{array}{r} 3t^3 - 4t^2 + 1 \\ \times \quad 2t^2 - 5 \\ \hline -15t^3 + 20t^2 - 5 \\ 6t^5 - 8t^4 \quad\quad + 2t^2 \\ \hline 6t^5 - 8t^4 - 15t^3 + 22t^2 - 5 \end{array}$$

10. $(3z - 5)^2 = 9z^2 - 30z + 25$

11. $\dfrac{25x^2y^2 - 15x^2y + 20xy^2}{5x^2y} = \dfrac{25x^2y^2}{5x^2y} - \dfrac{15x^2y}{5x^2y} + \dfrac{20xy^2}{5x^2y}$
$= 5y - 3 + \dfrac{4y}{x}$

12.
$$\begin{array}{r} 2x^2 + 3x + 5 \\ 2x-3\overline{)4x^3 + 0x^2 + x - 15} \\ \underline{4x^3 - 6x^2} \quad\quad\quad \\ 6x^2 + x \quad\quad \\ \underline{6x^2 - 9x} \quad\quad \\ 10x - 15 \\ \underline{10x - 15} \\ 0 \end{array}$$

$(4x^3 + x - 15) \div (2x - 3) = 2x^2 + 3x + 5$

13.
$$\begin{array}{r|rrrr} 3 & 1 & -5 & 5 & 5 \\ & & 3 & -6 & -3 \\ \hline & 1 & -2 & -1 & 2 \end{array}$$

$(x^3 - 5x^2 + 5x + 5) \div (x - 3) = x^2 - 2x - 1 + \dfrac{2}{x-3}$

14. $P(2) = 3(2)^2 - 8(2) + 1$
$P(2) = 12 - 16 + 1$
$P(2) = -3$

15.
$$\begin{array}{r|rrrr} -2 & -1 & 0 & 4 & -8 \\ & & 2 & -4 & 0 \\ \hline & -1 & 2 & 0 & -8 \end{array}$$

$P(-2) = -8$

16. Let $u = a^2$.
$6a^4 - 13a^2 - 5 = 6u^2 - 13u - 5$
$= (2u - 5)(3u + 1)$
$= (2a^2 - 5)(3a^2 + 1)$

17. $12x^3 + 12x^2 - 45x = 3x(4x^2 + 4x - 15)$
$= 3x(2x - 3)(2x + 5)$

18. $16x^2 - 25 = (4x - 5)(4x + 5)$

19. $16t^2 + 24t + 9 = (4t + 3)^2$

20. $27x^3 - 8 = (3x)^3 - (2)^3 = (3x - 2)(9x^2 + 6x + 4)$

21. $6x^2 - 4x - 3xa + 2a = 2x(3x - 2) - a(3x - 2)$
$= (3x - 2)(2x - a)$

22. $g(c) = 0$
$2c^2 - 5c - 12 = 0$
$(2c + 3)(c - 4) = 0$
$2c + 3 = 0 \quad c - 4 = 0$
$2c = -3 \quad\quad c = 4$
$c = -\dfrac{3}{2}$

The zeros of g are $-\dfrac{3}{2}$ and 4.

23. $6x^2 = x + 1$
$6x^2 - x - 1 = 0$
$(2x - 1)(3x + 1) = 0$
$2x - 1 = 0 \quad 3x + 1 = 0$
$2x = 1 \quad\quad 3x = -1$
$x = \dfrac{1}{2} \quad\quad x = -\dfrac{1}{3}$

The solutions are $\dfrac{1}{2}$ and $-\dfrac{1}{3}$.

162 Chapter 5: Polynomials And Exponents

24.
$$6x^3 + x^2 - 6x - 1 = 0$$
$$x^2(6x+1) - 1(6x+1) = 0$$
$$(x^2 - 1)(6x+1) = 0$$
$$(x+1)(x-1)(6x+1) = 0$$
$$x+1=0 \quad x-1=0 \quad 6x+1=0$$
$$x=-1 \quad x=1 \quad 6x=-1$$
$$x=-\frac{1}{6}$$

The solutions are -1, 1, and $-\frac{1}{6}$.

25. Strategy
This is a geometry problem.
To find the area of the rectangle, replace the variables L and W in the equation $A = LW$ by the given values and solve for A.

Solution
$A = LW$
$A = (5x+1)(2x-1)$
$A = 10x^2 - 3x - 1$
The area of the rectangle is $(10x^2 - 3x - 1)$ ft^2.

26. Strategy
To find the time:
Use the equation $d = rt$, where d is the distance from Earth to the moon and r is the average velocity.

Solution
$d = rt$
$2.4 \times 10^5 = (2 \times 10^4)t$
$\frac{2.4 \times 10^5}{2 \times 10^4} = t$
$1.2 \times 10 = t$
$12 = t$
It takes 12 h for the space vehicle to reach the moon.

CUMULATIVE REVIEW EXERCISES

1.
$$8 - 2[-3-(-1)]^2 + 4 = 8 - 2[-3+1]^2 + 4$$
$$= 8 - 2[-2]^2 + 4$$
$$= 8 - 2(4) + 4$$
$$= 8 - 8 + 4$$
$$= 0 + 4$$
$$= 4$$

2.
$$\frac{2(4)-(-2)}{-2-6} = \frac{8+2}{-8} = \frac{10}{-8} = -\frac{5}{4}$$

3. The Inverse Property of Addition

4.
$$2x - 4[x - 2(3-2x) + 4] = 2x - 4[x - 6 + 4x + 4]$$
$$= 2x - 4(5x - 2)$$
$$= 2x - 20x + 8$$
$$= -18x + 8$$

5.
$$\frac{2}{3} - y = \frac{5}{6}$$
$$\frac{2}{3} - \frac{2}{3} - y = \frac{5}{6} - \frac{2}{3}$$
$$-y = \frac{1}{6}$$
$$(-1)(-y) = (-1)\frac{1}{6}$$
$$y = -\frac{1}{6}$$

6.
$$8x - 3 - x = -6 + 3x - 8$$
$$7x - 3 = 3x - 14$$
$$7x - 3 - 3x = 3x - 14 - 3x$$
$$4x - 3 = -14$$
$$4x - 3 + 3 = -14 + 3$$
$$4x = -11$$
$$\frac{4x}{4} = -\frac{11}{4}$$
$$x = -\frac{11}{4}$$

7.

```
3 | 1   0   0   -3
  |     3   9   27
  -------------------
    1   3   9   24
```

$$\frac{x^3 - 3}{x-3} = x^2 + 3x + 9 + \frac{24}{x-3}$$

8.
$$3 - |2 - 3x| = -2$$
$$-|2 - 3x| = -5$$
$$|2 - 3x| = 5$$
$$2 - 3x = 5 \quad 2 - 3x = -5$$
$$-3x = 3 \quad -3x = -7$$
$$x = -1 \quad x = \frac{7}{3}$$

The solutions are -1 and $\frac{7}{3}$.

9.
$$P(-2) = 3(-2)^2 - 2(-2) + 2$$
$$P(-2) = 12 + 4 + 2$$
$$P(-2) = 18$$

10. $x = -2$

11.
$$f(x) = 3x^2 - 4$$
$$f(-2) = 3(-2)^2 - 4 = 8$$
$$f(-1) = 3(-1)^2 - 4 = -1$$
$$f(0) = 3(0)^2 - 4 = -4$$
$$f(1) = 3(1)^2 - 4 = -1$$
$$f(2) = 3(2)^2 - 4 = 8$$
Range = $\{-4, -1, 8\}$

12.
$$m = \frac{y_2 - y_1}{x_2 - x_1} = \frac{2 - 3}{4 - (-2)} = -\frac{1}{6}$$

13. Use the point-slope formula.
$$y - y_1 = m(x - x_1)$$
$$y - 2 = -\frac{3}{2}[x - (-1)]$$
$$y - 2 = -\frac{3}{2}(x + 1)$$
$$y - 2 = -\frac{3}{2}x - \frac{3}{2}$$
$$y = -\frac{3}{2}x + \frac{1}{2}$$

14.
$$3x + 2y = 4$$
$$2y = -3x + 4$$
$$y = -\frac{3}{2}x + 2$$
$$m_1 = -\frac{3}{2}$$
$$m_1 \cdot m_2 = -1$$
$$-\frac{3}{2} \cdot m_2 = -1$$
$$m_2 = \frac{2}{3}$$
Now use the point-slope formula to find the equation of the line.
$$y - y_1 = m(x - x_1)$$
$$y - 4 = \frac{2}{3}[x - (-2)]$$
$$y - 4 = \frac{2}{3}(x + 2)$$
$$y - 4 = \frac{2}{3}x + \frac{4}{3}$$
$$y = \frac{2}{3}x + \frac{16}{3}$$
The equation of the perpendicular line is
$$y = \frac{2}{3}x + \frac{16}{3}.$$

15.
$$2x - 3y = 2$$
$$x + y = -3$$
$$D = \begin{vmatrix} 2 & -3 \\ 1 & 1 \end{vmatrix} = 2 - (-3) = 5$$
$$D_x = \begin{vmatrix} 2 & -3 \\ -3 & 1 \end{vmatrix} = 2 - 9 = -7$$
$$D_y = \begin{vmatrix} 2 & 2 \\ 1 & -3 \end{vmatrix} = -6 - 2 = -8$$
$$x = \frac{D_x}{D} = -\frac{7}{5}$$
$$y = \frac{D_y}{D} = -\frac{8}{5}$$
The solution is $\left(-\frac{7}{5}, -\frac{8}{5}\right)$.

16.
(1) $x - y + z = 0$
(2) $2x + y - 3z = -7$
(3) $-x + 2y + 2z = 5$

Add equations (1) and (3) to eliminate x.
$$x - y + z = 0$$
$$-x + 2y + 2z = 5$$
$$\overline{y + 3z = 5}$$

Add -2 times equation (1) and equation (2) to eliminate x.
$$-2x + 2y - 2z = 0$$
$$2x + y - 3z = -7$$
$$\overline{3y - 5z = -7}$$

Now solve the system in two variables.
$$y + 3z = 5$$
$$3y - 5z = -7$$

Add -3 times the first of these equations to the second.
$$-3y - 9z = -15$$
$$3y - 5z = -7$$
$$\overline{-14z = -22}$$
$$z = \frac{11}{7}$$

Next find y.
$$y + 3z = 5$$
$$y + 3\left(\frac{11}{7}\right) = 5$$
$$y + \frac{33}{7} = 5$$
$$y = \frac{2}{7}$$

Replace y and z in equation (1) and solve for x.
$$x - \frac{2}{7} + \frac{11}{7} = 0$$
$$x = -\frac{9}{7}$$

The solution is $\left(-\frac{9}{7}, \frac{2}{7}, \frac{11}{7}\right)$.

164 Chapter 5: Polynomials And Exponents

17. x-intercept: y-intercept:
$$3x - 4y = 12 \qquad 3x - 4y = 12$$
$$3x - 4(0) = 12 \qquad 3(0) - 4y = 12$$
$$3x = 12 \qquad\qquad -4y = 12$$
$$x = 4 \qquad\qquad\quad y = -3$$

The x-intercept is $(4, 0)$.
The y-intercept is $(0, -3)$.

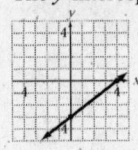

18. $-3x + 2y < 6$
$$2y < 3x + 6$$
$$y < \frac{3}{2}x + 3$$

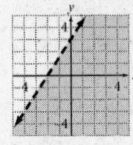

19.

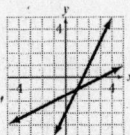

The lines intersect at $(1, -1)$.

20. Solve each inequality for y.
$$2x + y < 3 \qquad\qquad -6x + 3y \geq 4$$
$$y < 3 - 2x \qquad\qquad 3y \geq 4 + 6x$$
$$y \geq \frac{4}{3} + 2x$$

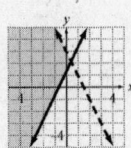

21.
$$(4a^{-2}b^3)(2a^3b^{-1})^{-2} = (4a^{-2}b^3)(2^{-2}a^{-6}b^2)$$
$$= 4 \cdot 2^{-2} a^{-2-6} b^{3+2}$$
$$= 4 \cdot \frac{1}{4} a^{-8} b^5$$
$$= \frac{b^5}{a^8}$$

22.
$$\frac{(5x^3 y^{-3} z)^{-2}}{y^4 z^{-2}} = \frac{5^{-2} x^{-6} y^6 z^{-2}}{y^4 z^{-2}}$$
$$= 5^{-2} x^{-6} y^{6-4} z^{-2-(-2)}$$
$$= \frac{1}{25} x^{-6} y^2$$
$$= \frac{y^2}{25x^6}$$

23.
$$3 - (3 - 3^{-1})^{-1} = 3 - \left(3 - \frac{1}{3}\right)^{-1}$$
$$= 3 - \left(\frac{8}{3}\right)^{-1}$$
$$= 3 - \frac{3}{8}$$
$$= \frac{21}{8}$$

24.
$$\begin{array}{r} 2x^2 - 3x + 1 \\ \times \qquad 2x + 3 \\ \hline 6x^2 - 9x + 3 \\ 4x^3 - 6x^2 + 2x \qquad\quad \\ \hline 4x^3 \qquad\quad - 7x + 3 \end{array}$$

25. $-4x^3 + 14x^2 - 12x = -2x(2x^2 - 7x + 6)$
$$= -2x(2x - 3)(x - 2)$$

26. $a(x - y) - b(y - x) = a(x - y) + b(x - y)$
$$= (x - y)(a + b)$$

27. $x^4 - 16 = (x^2 + 4)(x^2 - 4)$
$$= (x^2 + 4)(x + 2)(x - 2)$$

28. $2x^3 - 16 = 2(x^3 - 8)$
$$= 2(x - 2)(x^2 + 2x + 4)$$

29. Strategy
Smaller integer: x
Larger integer: $24 - x$
The difference between four times the smaller and nine is 3 less than twice the larger.
$4x - 9 = 2(24 - x) - 3$

Solution
$4x - 9 = 2(24 - x) - 3$
$4x - 9 = 48 - 2x - 3$
$4x - 9 = 45 - 2x$
$6x - 9 = 45$
$6x = 54$
$x = 9$
$24 - x = 15$
The integers are 9 and 15.

30. Strategy
The number of ounces of pure gold: x

	Amount	Cost	Value
Pure gold	x	360	$360x$
Alloy	80	120	80(120)
Mixture	$x + 80$	200	$200(x + 80)$

The sum of the values before mixing equals the value after mixing.
$360x + 80(120) = 200(x + 80)$

Solution
$360x + 80(120) = 200(x + 80)$
$360x + 9600 = 200x + 16{,}000$
$160x + 9600 = 16{,}000$
$160x = 6400$
$x = 40$
40 oz of pure gold must be mixed with the alloy.

31. Strategy
Faster cyclist: x
Slower cyclist: $\dfrac{2}{3}x$

	Rate	Time	Distance
Faster cyclist	x	2	$2x$
Slower cyclist	$\frac{2}{3}x$	2	$2\left(\frac{2}{3}x\right)$

The sum of the distances is 25 mi.

Solution
$2x + 2\left(\dfrac{2}{3}x\right) = 25$
$2x + \dfrac{4}{3}x = 25$
$\dfrac{10}{3}x = 25$
$x = 7.5$
$\dfrac{2}{3}x = 5$

The slower cyclist travels at 5 mph, the faster cyclist at 7.5 mph.

32. Strategy
Amount invested at 10%: x

	Principal	Rate	Interest
Amount at 7.5%	3000	0.075	0.075(3000)
Amount at 10%	x	0.10	$0.10x$
Amount at 9%	$3000 + x$	0.09	$0.09(3000 + x)$

The amount of interest earned at 9% equals the total amount of the interest earned at 7.5% and 10%.
$0.075(3000) + 0.10x = 0.09(3000 + x)$

Solution
$0.075(3000) + 0.10x = 0.09(3000 + x)$
$225 + 0.10x = 270 + 0.09x$
$0.01x + 225 = 270$
$0.01x = 45$
$x = 4500$
The additional investment is $4500.

33. $m = \dfrac{y_2 - y_1}{x_2 - x_1} = \dfrac{300 - 100}{6 - 2} = \dfrac{200}{4}$
$m = 50$
The slope represents the average speed of travel in miles per hour. The average speed was 50 mph.

CHAPTER 6: RATIONAL EXPRESSIONS

CHAPTER 6 PREP TEST

1. 50 [1.2.2]

2. $-\dfrac{3}{8} \cdot \dfrac{4}{9} = -\dfrac{1}{6}$ [1.2.2]

3. $-\dfrac{4}{5} \div \dfrac{8}{15} = -\dfrac{4}{5} \cdot \dfrac{15}{8} = -\dfrac{3}{2}$ [1.2.2]

4. $-\dfrac{5}{6} + \dfrac{7}{8} = -\dfrac{20}{24} + \dfrac{21}{24} = \dfrac{1}{24}$ [1.2.2]

5. $-\dfrac{3}{8} - \left(-\dfrac{7}{12}\right) = -\dfrac{9}{24} + \dfrac{14}{24} = \dfrac{5}{24}$ [1.2.2]

6. $\dfrac{\dfrac{2}{3} - \dfrac{1}{4}}{\dfrac{1}{8} - 2} = \dfrac{\dfrac{8}{12} - \dfrac{3}{12}}{\dfrac{1}{8} - \dfrac{16}{8}}$ [1.2.4]

   $= \dfrac{\dfrac{5}{12}}{-\dfrac{15}{8}}$

   $= \dfrac{5}{12} \cdot \left(-\dfrac{8}{15}\right)$

   $= -\dfrac{2}{9}$

7. $\dfrac{2(2) - 3}{(2)^2 - 2 + 1} = \dfrac{4 - 3}{4 - 1} = \dfrac{1}{3}$ [1.3.2]

8. $4(2x + 1) = 3(x - 2)$ [2.1.2]
   $8x + 4 = 3x - 6$
   $5x = -10$
   $x = -2$

9. $10\left(\dfrac{t}{2} + \dfrac{t}{5}\right) = 10(1)$ [2.1.2]
   $5t + 2t = 10$
   $7t = 10$
   $t = \dfrac{10}{7}$

10. Strategy
    Rate of the second plane: r
    Rate of the first plane: $r - 20$

	Rate	Time	Distance
2nd plane	r	2	$2r$
1st plane	$r-20$	2	$2(r-20)$

    The total distance traveled by the two planes is 480 mi. $2r + 2(r - 20) = 480$

    Solution
    $2r + 2(r - 20) = 480$
    $2r + 2r - 40 = 480$
    $4r = 520$
    $r = 130$
    $r - 20 = 130 - 20 = 110$
    The rate of the first plane is 110 mph.
    The rate of the second plane is 130 mph.

Section 6.1

Concept Review 6.1

1. Never true
   A rational expression is the quotient of polynomials. $x^{1/2} - 2x + 4$ is not a polynomial.

3. Always true

5. Never true
   The quotient $\dfrac{a+4}{a+4} = 1$. The correct solution is $\dfrac{1}{a+2}$.

Objective 6.1.1 Exercises

1. A rational function is a function that is written in terms of an expression in which the numerator and denominator are polynomials. An example is
   $f(x) = \dfrac{x^2 - 2x + 3}{7x - 4}$.

3. To find $f(-3)$, substitute -3 for x:
   $f(-3) = \dfrac{x+2}{1-2x} = \dfrac{-3+2}{1-2(-3)} = \dfrac{-1}{1+6} = -\dfrac{1}{7}$.

5. $f(x) = \dfrac{2}{x-3}$
   $f(4) = \dfrac{2}{4-3} = \dfrac{2}{1}$
   $f(4) = 2$

7. $f(x) = \dfrac{x-2}{x+4}$
   $f(-2) = \dfrac{-2-2}{-2+4} = \dfrac{-4}{2}$
   $f(-2) = -2$

9. $f(x) = \dfrac{1}{x^2 - 2x + 1}$
   $f(-2) = \dfrac{1}{(-2)^2 - 2(-2) + 1} = \dfrac{1}{9}$
   $f(-2) = \dfrac{1}{9}$

11. $f(x) = \dfrac{x-2}{2x^2+3x+8}$

$f(3) = \dfrac{3-2}{2(3)^2+3(3)+8} = \dfrac{1}{35}$

$f(3) = \dfrac{1}{35}$

13. $f(x) = \dfrac{x^2-2x}{x^3-x+4}$

$f(-1) = \dfrac{(-1)^2-2(-1)}{(-1)^3-(-1)+4} = \dfrac{3}{4}$

$f(-1) = \dfrac{3}{4}$

15. $x - 3 = 0$

$x = 3$

The domain of $H(x)$ is $\{x \mid x \neq 3\}$.

17. $x + 4 = 0$

$x = -4$

The domain $f(x)$ is $\{x \mid x \neq -4\}$.

19. $3x + 9 = 0$

$3x = -9$

$x = -3$

The domain of $R(x)$ is $\{x \mid x \neq -3\}$.

21. $(x-4)(x+2) = 0$

$x - 4 = 0 \qquad x + 2 = 0$

$x = 4 \qquad x = -2$

The domain of $q(x)$ is $\{x \mid x \neq -2, x \neq 4\}$.

23. $(2x+5)(3x-6) = 0$

$2x + 5 = 0 \qquad 3x - 6 = 0$

$2x = -5 \qquad 3x = 6$

$x = -\dfrac{5}{2} \qquad x = 2$

The domain of $V(x)$ is $\left\{x \mid x \neq -\dfrac{5}{2}, x \neq 2\right\}$.

25. $x = 0$

The domain of $f(x)$ is $\{x \mid x \neq 0\}$.

27. The domain must exclude values of x for which $x^2 + 1 = 0$. This is not possible, because $x^2 \geq 0$, and a positive number added to a number equal to or greater than zero cannot equal zero. Therefore, there are no real numbers that must be excluded from the domain of k.

The domain of $k(x)$ is $\{x \mid x \in \text{real numbers}\}$.

29. $x^2 + x - 6 = 0$

$(x+3)(x-2) = 0$

$x + 3 = 0 \qquad x - 2 = 0$

$x = -3 \qquad x = 2$

The domain of $f(x)$ is $\{x \mid x \neq -3, x \neq 2\}$.

31. $x^2 + 2x - 24 = 0$

$(x+6)(x-4) = 0$

$x + 6 = 0 \qquad x - 4 = 0$

$x = -6 \qquad x = 4$

The domain of $A(x)$ is $\{x \mid x \neq -6, x \neq 4\}$.

33. The domain must exclude values of x for which $3x^2 + 12 = 0$. This is not possible, because $3x^2 \geq 0$, and a positive number added to a number equal to or greater than zero cannot equal zero. Therefore, there are no real numbers that must be excluded from the domain of f.

The domain of $f(x)$ is $\{x \mid x \in \text{real numbers}\}$.

35. $6x^2 - 13x + 6 = 0$

$(2x-3)(3x-2) = 0$

$2x - 3 = 0 \qquad 3x - 2 = 0$

$x = \dfrac{3}{2} \qquad x = \dfrac{2}{3}$

The domain of $G(x)$ is $\left\{x \mid x \neq \dfrac{3}{2}, x \neq \dfrac{2}{3}\right\}$.

37. $2x^3 + 9x^2 - 5x = 0$

$x(2x^2 + 9x - 5) = 0$

$x(2x-1)(x+5) = 0$

$x = 0 \qquad 2x - 1 = 0 \qquad x + 5 = 0$

$\qquad\qquad x = \dfrac{1}{2} \qquad x = -5$

The domain of $f(x)$ is $\left\{x \mid x \neq 0, x \neq \dfrac{1}{2}, x \neq -5\right\}$.

39. Try evaluating the function $f(x) = -\dfrac{1}{x+1}$ for a few values of $x > -1$:

$f(0) = -\dfrac{1}{0+1} = -1 \quad f(1) = -\dfrac{1}{1+1} = -\dfrac{1}{2}$

$f(10) = -\dfrac{1}{10+1} = -\dfrac{1}{11}$

Because all of these function values are negative, the portion of the graph that is to the right of the vertical line $x = -1$ is below the x-axis.

Objective 6.1.2 Exercises

41. A rational function is in simplest form when the numerator and denominator have no common factors other than 1.

43. To simplify a rational expression, first write the numerator and denominator in <u>factored</u> form. Then simplify the rational expression by dividing its numerator and denominator by any <u>common</u> factors.

45. $\dfrac{4-8x}{4} = \dfrac{4(1-2x)}{4} = 1-2x$

47. $\dfrac{6x^2-2x}{2x} = \dfrac{2x(3x-1)}{2x} = 3x-1$

49. $\dfrac{8x^2(x-3)}{4x(x-3)} = \dfrac{8x^2}{4x} = 2x$

51. $\dfrac{2x-6}{3x-x^2} = \dfrac{2(x-3)}{x(3-x)} = \dfrac{2}{-x} = -\dfrac{2}{x}$

53. $\dfrac{6x^3-15x^2}{12x^2-30x} = \dfrac{3x^2(2x-5)}{6x(2x-5)} = \dfrac{3x^2}{6x} = \dfrac{x}{2}$

55. $\dfrac{a^2+4a}{4a-16} = \dfrac{a(a+4)}{4(a-4)}$
The expression is in simplest form.

57. $\dfrac{16x^3-8x^2+12x}{4x} = \dfrac{4x(4x^2-2x+3)}{4x}$
$= 4x^2-2x+3$

59. $\dfrac{-10a^4-20a^3+30a^2}{-10a^2} = \dfrac{-10a^2(a^2+2a-3)}{-10a^2}$
$= a^2+2a-3$

61. $\dfrac{3x^{3n}-9x^{2n}}{12x^{2n}} = \dfrac{3x^{2n}(x^n-3)}{12x^{2n}} = \dfrac{x^n-3}{4}$

63. $\dfrac{x^2-7x+12}{x^2-9x+20} = \dfrac{(x-3)(x-4)}{(x-4)(x-5)} = \dfrac{x-3}{x-5}$

65. $\dfrac{x^2-xy-2y^2}{x^2-3xy+2y^2} = \dfrac{(x+y)(x-2y)}{(x-y)(x-2y)} = \dfrac{x+y}{x-y}$

67. $\dfrac{6-x-x^2}{3x^2-10x+8} = \dfrac{(3+x)(2-x)}{(3x-4)(x-2)}$
$= \dfrac{-1(x+3)}{1(3x-4)}$
$= -\dfrac{x+3}{3x-4}$

69. $\dfrac{14-19x-3x^2}{3x^2-23x+14} = \dfrac{(7+x)(2-3x)}{(3x-2)(x-7)}$
$= \dfrac{-1(7+x)}{1(x-7)}$
$= -\dfrac{x+7}{x-7}$

71. $\dfrac{a^2-7a+10}{a^2+9a+14} = \dfrac{(a-5)(a-2)}{(a+7)(a+2)}$
The expression is in simplest form.

73. $\dfrac{a^2-b^2}{a^3+b^3} = \dfrac{(a+b)(a-b)}{(a+b)(a^2-ab+b^2)} = \dfrac{a-b}{a^2-ab+b^2}$

75. $\dfrac{x^3+y^3}{3x^3-3x^2y+3xy^2} = \dfrac{(x+y)(x^2-xy+y^2)}{3x(x^2-xy+y^2)}$
$= \dfrac{x+y}{3x}$

77. $\dfrac{x^3-4xy^2}{3x^3-2x^2y-8xy^2} = \dfrac{x(x^2-4y^2)}{x(3x^2-2xy-8y^2)}$
$= \dfrac{x(x+2y)(x-2y)}{x(3+4y)(x-2y)}$
$= \dfrac{x+2y}{3x+4y}$

79. $\dfrac{4x^3-14x^2+12x}{24x+4x^2-8x^3} = \dfrac{2x(2x^2-7x+6)}{4x(6+x-2x^2)}$
$= \dfrac{2x(2x-3)(x-2)}{4x(3+2x)(2-x)}$
$= \dfrac{2x(2x-3)(-1)}{4x(3+2x)(1)}$
$= -\dfrac{2x-3}{2(2x+3)}$

81. $\dfrac{x^2-4}{a(x+2)-b(x+2)} = \dfrac{(x+2)(x-2)}{(x+2)(a-b)}$
$= \dfrac{(1)(x-2)}{(1)(a-b)}$
$= \dfrac{x-2}{a-b}$

83. $\dfrac{x^4+3x^2+2}{x^4-1} = \dfrac{(x^2+1)(x^2+2)}{(x^2+1)(x^2-1)}$
$= \dfrac{(x^2+1)(x^2+2)}{(x^2+1)(x+1)(x-1)}$
$= \dfrac{(1)(x^2+2)}{(1)(x+1)(x-1)}$
$= \dfrac{x^2+2}{(x+1)(x-1)}$

85. $\dfrac{x^2y^2+4xy-21}{x^2y^2-10xy+21} = \dfrac{(xy+7)(xy-3)}{(xy-3)(xy-7)}$

$= \dfrac{(xy+7)(1)}{(1)(xy-7)}$

$= \dfrac{xy+7}{xy-7}$

87. $\dfrac{a^{2n}-a^n-2}{a^{2n}+3a^n+2} = \dfrac{(a^n+1)(a^n-2)}{(a^n+1)(a^n+2)}$

$= \dfrac{(1)(a^n-2)}{(1)(a^n+2)}$

$= \dfrac{a^n-2}{a^n+2}$

89. False. $\dfrac{4x^2-8x}{16-8x} = \dfrac{4x(x-2)}{-8(x-2)} = -\dfrac{x}{2}$

Applying Concepts 6.1

91. $h(x) = \dfrac{x+2}{x-3}$

$h(2.9) = -49$
$h(2.99) = -499$
$h(2.999) = -4999$
$h(2.9999) = -49,999$

As x becomes closer to 3, the values of $h(x)$ decrease.

93. a.

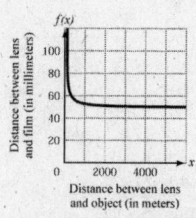

b. The ordered pair (2000, 51) means that when the distance between the object and the lens is 2000 m, the distance between the lens and the film is 51

c. For $x = 50$, the expression $\dfrac{50x}{x-50}$ is undefined. For $0 \le x < 50$, $f(x)$ is negative, and distance cannot be negative. Therefore, the domain is $x > 50$.

d. For $x > 1000$, $f(x)$ changes very little for large changes in x.

95. Let $x - a$ be a common factor of the numerator and denominator of a rational expression. Then

$\dfrac{f(x)(x-a)}{g(x)(x-a)} = \dfrac{f(x)}{g(x)} \cdot \dfrac{x-a}{x-a}$

$= \dfrac{f(x)}{g(x)} \cdot 1$

$= \dfrac{f(x)}{g(x)}$

These operations are valid as long as $x \ne a$. If $x = a$, then $x - a = 0$ and $\dfrac{x-a}{x-a}$ is undefined.

Section 6.2

Concept Review 6.2

1. Always true

3. Always true

Objective 6.2.1 Exercises

1. To multiply two rational expressions, find the product of the numerators and the product of the denominators. Then simplify by dividing the numerator and denominator by their common factors.

3. $\dfrac{27a^2b^5}{16xy^2} \cdot \dfrac{20x^2y^3}{9a^2b} = \dfrac{27a^2b^5 \cdot 20x^2y^3}{16xy^2 \cdot 9a^2b}$

$= \dfrac{15b^4xy}{4}$

5. $\dfrac{3x-15}{4x^2-2x} \cdot \dfrac{20x^2-10x}{15x-75} = \dfrac{3(x-5)}{2x(2x-1)} \cdot \dfrac{10x(2x-1)}{15(x-5)}$

$= \dfrac{3(x-5) \cdot 10x(2x-1)}{2x(2x-1) \cdot 15(x-5)}$

$= \dfrac{3(1) \cdot 10x(1)}{2x(1) \cdot 15(1)}$

$= 1$

7. $\dfrac{x^2y^3}{x^2-4x-5} \cdot \dfrac{2x^2-13x+15}{x^4y^3}$

$= \dfrac{x^2y^3}{(x+1)(x-5)} \cdot \dfrac{(2x-3)(x-5)}{x^4y^3}$

$= \dfrac{x^2y^3 \cdot (2x-3)(x-5)}{(x+1)(x-5) \cdot x^4y^3}$

$= \dfrac{x^2y^3 \cdot (2x-3)(1)}{(x+1)(1) \cdot x^4y^3}$

$= \dfrac{2x-3}{x^2(x+1)}$

9. $\dfrac{x^2-3x+2}{x^2-8x+15} \cdot \dfrac{x^2+x-12}{8-2x-x^2}$

$= \dfrac{(x-1)(x-2)}{(x-3)(x-5)} \cdot \dfrac{(x+4)(x-3)}{(4+x)(2-x)}$

$= \dfrac{(x-1)(x-2)(x+4)(x-3)}{(x-3)(x-5)(4+x)(2-x)}$

$= \dfrac{(x-1)(-1)(1)(1)}{(1)(x-5)(1)(1)}$

$= -\dfrac{x-1}{x-5}$

Chapter 6: Rational Expressions 169

11. $\dfrac{x^{n+1}+2x^n}{4x^2-6x} \cdot \dfrac{8x^2-12x}{x^{n+1}-x^n} = \dfrac{x^n(x+2)}{2x(2x-3)} \cdot \dfrac{4x(2x-3)}{x^n(x-1)}$

$= \dfrac{x^n(x+2) \cdot 4x(2x-3)}{2x(2x-3) \cdot x^n(x-1)}$

$= \dfrac{x^n(x+2) \cdot 4x(1)}{2x(1) \cdot x^n(x-1)}$

$= \dfrac{2(x+2)}{x-1}$

13. $\dfrac{12+x-6x^2}{6x^2+29x+28} \cdot \dfrac{2x^2+x-21}{4x^2-9}$

$= \dfrac{(4+3x)(3-2x)}{(2x+7)(3x+4)} \cdot \dfrac{(2x+7)(x-3)}{(2x+3)(2x-3)}$

$= \dfrac{(4+3x)(3-2x)(2x+7)(x-3)}{(2x+7)(3x+4)(2x+3)(2x-3)}$

$= \dfrac{(1)(-1)(1)(x-3)}{(1)(1)(2x+3)(1)}$

$= -\dfrac{x-3}{2x+3}$

15. $\dfrac{x^{2n}-x^n-6}{x^{2n}+x^n-2} \cdot \dfrac{x^{2n}-5x^n-6}{x^{2n}-2x^n-3}$

$= \dfrac{(x^n+2)(x^n-3)}{(x^n+2)(x^n-1)} \cdot \dfrac{(x^n+1)(x^n-6)}{(x^n+1)(x^n-3)}$

$= \dfrac{(x^n+2)(x^n-3)(x^n+1)(x^n-6)}{(x^n+2)(x^n-1)(x^n+1)(x^n-3)}$

$= \dfrac{(1)(1)(1)(x^n-6)}{(1)(x^n-1)(1)(1)}$

$= \dfrac{x^n-6}{x^n-1}$

17. $\dfrac{x^3-y^3}{2x^2+xy-3y^2} \cdot \dfrac{2x^2+5xy+3y^2}{x^2+xy+y^2}$

$= \dfrac{(x-y)(x^2+xy+y^2)}{(2x+3y)(x-y)} \cdot \dfrac{(2x+3y)(x+y)}{x^2+xy+y^2}$

$= \dfrac{(x-y)(x^2+xy+y^2)(2x+3y)(x+y)}{(2x+3y)(x-y)(x^2+xy+y^2)}$

$= \dfrac{(1)(1)(1)(x+y)}{(1)(1)(1)}$

$= x+y$

19. $\dfrac{6x^2y^4}{35a^2b^5} \div \dfrac{12x^3y^3}{7a^4b^5} = \dfrac{6x^2y^4}{35a^2b^5} \cdot \dfrac{7a^4b^5}{12x^3y^3}$

$= \dfrac{6x^2y^4 \cdot 7a^4b^5}{35a^2b^5 \cdot 12x^3y^3}$

$= \dfrac{a^2y}{10x}$

21. $\dfrac{2x-6}{6x^2-15x} \div \dfrac{4x^2-12x}{18x^3-45x^2}$

$= \dfrac{2x-6}{6x^2-15x} \cdot \dfrac{18x^3-45x^2}{4x^2-12x}$

$= \dfrac{2(x-3)}{3x(2x-5)} \cdot \dfrac{9x^2(2x-5)}{4x(x-3)}$

$= \dfrac{2(x-3) \cdot 9x^2(2x-5)}{3x(2x-5) \cdot 4x(x-3)}$

$= \dfrac{2(1) \cdot 9x^2(1)}{3x(1) \cdot 4x(1)}$

$= \dfrac{3}{2}$

23. $\dfrac{2x^2-2y^2}{14x^2y^4} \div \dfrac{x^2+2xy+y^2}{35xy^3}$

$= \dfrac{2x^2-2y^2}{14x^2y^4} \cdot \dfrac{35xy^3}{x^2+2xy+y^2}$

$= \dfrac{2(x^2-y^2)}{14x^2y^4} \cdot \dfrac{35xy^3}{(x+y)(x+y)}$

$= \dfrac{2(x+y)(x-y)}{14x^2y^4} \cdot \dfrac{35xy^3}{(x+y)(x+y)}$

$= \dfrac{2(x+y)(x-y) \cdot 35xy^3}{14x^2y^4(x+y)(x+y)}$

$= \dfrac{2(1)(x-y) \cdot 35xy^3}{14x^2y^4(1)(x+y)}$

$= \dfrac{5(x-y)}{xy(x+y)}$

25. $\dfrac{2x^2-5x-3}{2x^2+7x+3} \div \dfrac{2x^2-3x-20}{2x^2-x-15}$

$= \dfrac{2x^2-5x-3}{2x^2+7x+3} \cdot \dfrac{2x^2-x-15}{2x^2-3x-20}$

$= \dfrac{(2x+1)(x-3)}{(2x+1)(x+3)} \cdot \dfrac{(2x+5)(x-3)}{(2x+5)(x-4)}$

$= \dfrac{(1)(x-3)(1)(x-3)}{(1)(x+3)(1)(x-4)}$

$= \dfrac{(x-3)^2}{(x+3)(x-4)}$

27. $\dfrac{x^2-8x+15}{x^2+2x-35} \div \dfrac{15-2x-x^2}{x^2+9x+14}$

$= \dfrac{x^2-8x+15}{x^2+2x-35} \cdot \dfrac{x^2+9x+14}{15-2x-x^2}$

$= \dfrac{(x-3)(x-5)}{(x+7)(x-5)} \cdot \dfrac{(x+2)(x+7)}{(5+x)(3-x)}$

$= \dfrac{(x-3)(x-5)(x+2)(x+7)}{(x+7)(x-5)(5+x)(3-x)}$

$= \dfrac{(-1)(1)(x+2)(1)}{(1)(1)(5+x)(1)}$

$= -\dfrac{x+2}{x+5}$

29. $\dfrac{x^{2n}+x^n}{2x-2} \div \dfrac{4x^n+4}{x^{n+1}-x^n} = \dfrac{x^{2n}+x^n}{2x-2} \cdot \dfrac{x^{n+1}-x^n}{4x^n+4}$

$= \dfrac{x^n(x^n+1)}{2(x-1)} \cdot \dfrac{x^n(x-1)}{4(x^n+1)}$

$= \dfrac{x^n(x^n+1) \cdot x^n(x-1)}{2(x-1) \cdot 4(x^n+1)}$

$= \dfrac{x^n(1) \cdot x^n(1)}{2(1) \cdot 4(1)}$

$= \dfrac{x^{2n}}{8}$

31. $\dfrac{2x^2-13x+21}{2x^2+11x+15} \div \dfrac{2x^2+x-28}{3x^2+4x-15}$

$= \dfrac{2x^2-13x+21}{2x^2+11x+15} \cdot \dfrac{3x^2+4x-15}{2x^2+x-28}$

$= \dfrac{(2x-7)(x-3)}{(2x+5)(x+3)} \cdot \dfrac{(3x-5)(x+3)}{(2x-7)(x+4)}$

$= \dfrac{(2x-7)(x-3)(3x-5)(x+3)}{(2x+5)(x+3)(2x-7)(x+4)}$

$= \dfrac{(1)(x-3)(3x-5)(1)}{(2x+5)(1)(1)(x+4)}$

$= \dfrac{(x-3)(3x-5)}{(2x+5)(x+4)}$

33. $\dfrac{14+17x-6x^2}{3x^2+14x+8} \div \dfrac{4x^2-49}{2x^2+15x+28}$

$= \dfrac{14+17x-6x^2}{3x^2+14x+8} \cdot \dfrac{2x^2+15x+28}{4x^2-49}$

$= \dfrac{(2+3x)(7-2x)}{(3x+2)(x+4)} \cdot \dfrac{(2x+7)(x+4)}{(2x+7)(2x-7)}$

$= \dfrac{(2+3x)(7-2x)(2x+7)(x+4)}{(3x+2)(x+4)(2x+7)(2x-7)}$

$= \dfrac{(1)(-1)(1)(1)}{(1)(1)(1)(1)}$

$= -1$

35. $\dfrac{2x^{2n}-x^n-6}{x^{2n}-x^n-2} \div \dfrac{2x^{2n}+x^n-3}{x^{2n}-1}$

$= \dfrac{2x^{2n}-x^n-6}{x^{2n}-x^n-2} \cdot \dfrac{x^{2n}-1}{2x^{2n}+x^n-3}$

$= \dfrac{(2x^n+3)(x^n-2)}{(x^n+1)(x^n-2)} \cdot \dfrac{(x^n+1)(x^n-1)}{(2x^n+3)(x^n-1)}$

$= \dfrac{(2x^n+3)(x^n-2)(x^n+1)(x^n-1)}{(x^n+1)(x^n-2)(2x^n+3)(x^n-1)}$

$= \dfrac{(1)(1)(1)(1)}{(1)(1)(1)(1)}$

$= 1$

37. $\dfrac{6x^2+6x}{3x+6x^2+3x^3} \div \dfrac{x^2-1}{1-x^3}$

$= \dfrac{6x^2+6x}{3x+6x^2+3x^3} \cdot \dfrac{1-x^3}{x^2-1}$

$= \dfrac{6x(x+1)}{3x(1+2x+x^2)} \cdot \dfrac{(1-x)(1+x+x^2)}{(x+1)(x-1)}$

$= \dfrac{6x(x+1)}{3x(1+x)(1+x)} \cdot \dfrac{(1-x)(1+x+x^2)}{(x+1)(x-1)}$

$= \dfrac{6x(x+1)(1-x)(1+x+x^2)}{3x(1+x)(1+x)(x+1)(x-1)}$

$= \dfrac{6x(1)(-1)(1+x+x^2)}{3x(1)(1+x)(x+1)(1)}$

$= -\dfrac{2(1+x+x^2)}{(x+1)^2}$

39.
a. $\dfrac{x-1}{x+2} \div \dfrac{x-3}{x+5} = \dfrac{x-1}{x+2} \cdot \dfrac{x+5}{x-3} = \dfrac{x^2+4x-5}{x^2-x-6}$

b. $\dfrac{x-1}{x-3} \cdot \dfrac{x+5}{x+2} = \dfrac{x^2+4x-5}{x^2-x-6}$

c. $\dfrac{x-3}{x-1} \div \dfrac{x+5}{x+2} = \dfrac{x-3}{x-1} \cdot \dfrac{x+2}{x+5} = \dfrac{x^2-x-6}{x^2+4x-5}$

d. $\dfrac{x+5}{2x-6} \cdot \dfrac{2x-2}{x+2} = \dfrac{x+5}{2(x-3)} \cdot \dfrac{2(x-1)}{x+2} = \dfrac{x^2+4x-5}{x^2-x-6}$

Expression (c) is not equivalent to $\dfrac{x^2+4x-5}{x^2-x-6}$.

Chapter 6: Rational Expressions

Objective 6.2.2 Exercises

41. $x^2 - 6x + 9 = (x-3)^2$

    $x^2 - 9 = (x+3)(x-3)$

    The LCM will use the factor $x-3$ <u>two</u> times and the factor $x+3$ <u>one</u> time. The LCM is $(x-3)^2(x+3)$.

43. a. $x^2 - x - 12 = (x-4)(x+3)$ and $x^2 - 8x + 16 = (x-4)^2$. Two factors of $x-4$ in the LCM.

    b. $x^2 + x - 12 = (x+4)(x-3)$ and $x^2 + 8x + 16 = (x+4)^2$. Zero factors of $x-4$ in the LCM.

    c. $x^2 - 16 = (x+4)(x-4)$ and $x^2 - x - 12 = (x-4)(x+3)$. One factor of $x-4$ in the LCM.

45. The LCM is $12x^2y^4$.

    $\dfrac{3}{4x^2y} = \dfrac{3}{4x^2y} \cdot \dfrac{3y^3}{3y^3} = \dfrac{9y^3}{12x^2y^4}$

    $\dfrac{17}{12xy^4} = \dfrac{17}{12xy^4} \cdot \dfrac{x}{x} = \dfrac{17x}{12x^2y^4}$

47. The LCM is $(2x-3)(2x+3)$.

    $\dfrac{3x}{2x-3} = \dfrac{3x}{2x-3} \cdot \dfrac{2x+3}{2x+3} = \dfrac{6x^2+9x}{(2x-3)(2x+3)}$

    $\dfrac{5x}{2x+3} = \dfrac{5x}{2x+3} \cdot \dfrac{(2x-3)}{(2x-3)} = \dfrac{10x^2-15x}{(2x-3)(2x+3)}$

49. The LCM is $(x+3)(x-3)$.

    $\dfrac{2x}{x^2-9} = \dfrac{2x}{(x-3)(x+3)}$

    $\dfrac{x+1}{x-3} = \dfrac{x+1}{x-3} \cdot \dfrac{x+3}{x+3} = \dfrac{x^2+4x+3}{(x-3)(x+3)}$

51. The LCM is $(x+1)(x-1)^2$.

    $\dfrac{3x}{x^2-1} = \dfrac{3x}{(x+1)(x-1)} \cdot \dfrac{x-1}{x-1} = \dfrac{3x^2-3x}{(x+1)(x-1)^2}$

    $\dfrac{5x}{x^2-2x+1} = \dfrac{5x}{(x-1)^2} \cdot \dfrac{x+1}{x+1} = \dfrac{5x^2+5x}{(x+1)(x-1)^2}$

53. The LCM of the denominators is $(x-4)(x+3)$.

    $\dfrac{5}{x-4} - \dfrac{2}{x+3}$

    $= \dfrac{5}{x-4} \cdot \dfrac{x+3}{x+3} - \dfrac{2}{x+3} \cdot \dfrac{x-4}{x-4}$

    $= \dfrac{5x+15}{(x-4)(x+3)} - \dfrac{2x-8}{(x-4)(x+3)}$

    $= \dfrac{(5x+15)-(2x-8)}{(x-4)(x+3)} = \dfrac{5x+15-2x+8}{(x-4)(x+3)} = \dfrac{3x+23}{(x-4)(x+3)}$

55. The LCM is $2xy$.

    $\dfrac{3}{2xy} - \dfrac{7}{2xy} - \dfrac{9}{2xy} = \dfrac{3-7-9}{2xy} = -\dfrac{13}{2xy}$

57. The LCM is $x^2 - 3x + 2$.

    $\dfrac{x}{x^2-3x+2} - \dfrac{2}{x^2-3x+2} = \dfrac{x-2}{x^2-3x+2}$

    $= \dfrac{(1)}{(1)(x-1)}$

    $= \dfrac{1}{x-1}$

59. The LCM is $10x^2y$.

    $\dfrac{3}{2x^2y} - \dfrac{8}{5x} - \dfrac{9}{10xy}$

    $= \dfrac{3}{2x^2y} \cdot \dfrac{5}{5} - \dfrac{8}{5x} \cdot \dfrac{2xy}{2xy} - \dfrac{9}{10xy} \cdot \dfrac{x}{x}$

    $= \dfrac{15-16xy-9x}{10x^2y}$

61. The LCM is $30xy$.

    $\dfrac{2}{3x} - \dfrac{3}{2xy} + \dfrac{4}{5xy} - \dfrac{5}{6x}$

    $= \dfrac{2}{3x} \cdot \dfrac{10y}{10y} - \dfrac{3}{2xy} \cdot \dfrac{15}{15} + \dfrac{4}{5xy} \cdot \dfrac{6}{6} - \dfrac{5}{6x} \cdot \dfrac{5y}{5y}$

    $= \dfrac{20y-45+24-25y}{30xy}$

    $= \dfrac{-5y-21}{30xy}$

    $= -\dfrac{5y+21}{30xy}$

63. The LCM is $36x$.

$$\frac{2x-1}{12x} - \frac{3x+4}{9x} = \frac{2x-1}{12x} \cdot \frac{3}{3} - \frac{3x+4}{9x} \cdot \frac{4}{4}$$

$$= \frac{(6x-3)-(12x+16)}{36x}$$

$$= \frac{6x-3-12x-16}{36x}$$

$$= \frac{-6x-19}{36x}$$

$$= -\frac{6x+19}{36x}$$

65. The LCM is $12x^2y^2$.

$$\frac{3x+2}{4x^2y} - \frac{y-5}{6xy^2} = \frac{3x+2}{4x^2y} \cdot \frac{3y}{3y} - \frac{y-5}{6xy^2} \cdot \frac{2x}{2x}$$

$$= \frac{(9xy+6y)-(2xy-10x)}{12x^2y^2}$$

$$= \frac{9xy+6y-2xy+10x}{12x^2y^2}$$

$$= \frac{10x+7xy+6y}{12x^2y^2}$$

67. The LCM is $(x-3)(x-5)$.

$$\frac{2x}{x-3} - \frac{3x}{x-5} = \frac{2x}{x-3} \cdot \frac{x-5}{x-5} - \frac{3x}{x-5} \cdot \frac{x-3}{x-3}$$

$$= \frac{(2x^2-10x)-(3x^2-9x)}{(x-3)(x-5)}$$

$$= \frac{2x^2-10x-3x^2+9x}{(x-3)(x-5)}$$

$$= \frac{-x^2-x}{(x-3)(x-5)}$$

$$= -\frac{x^2+x}{(x-3)(x-5)}$$

69. $3-2a = -(2a-3)$

The LCM is $2a-3$.

$$\frac{3}{2a-3} + \frac{2a}{3-2a} = \frac{3}{2a-3} + \frac{2a}{3-2a} \cdot \frac{-1}{-1}$$

$$= \frac{3-2a}{2a-3}$$

$$= \frac{(-1)}{(1)}$$

$$= -1$$

71. $x^2 - 25 = (x+5)(x-5)$

The LCM is $(x+5)(x-5)$.

$$\frac{3}{x+5} + \frac{2x+7}{x^2-25} = \frac{3}{x+5} \cdot \frac{x-5}{x-5} + \frac{2x+7}{(x-5)(x+5)}$$

$$= \frac{3x-15}{(x+5)(x-5)} + \frac{2x+7}{(x-5)(x+5)}$$

$$= \frac{(3x-15)+(2x+7)}{(x-5)(x+5)}$$

$$= \frac{3x-15+2x+7}{(x-5)(x+5)}$$

$$= \frac{5x-8}{(x-5)(x+5)}$$

73. The LCM is $x(x-4)$.

$$\frac{2}{x} - 3 - \frac{10}{x-4} = \frac{2}{x} \cdot \frac{x-4}{x-4} - 3 \cdot \frac{x(x-4)}{x(x-4)} - \frac{10}{x-4} \cdot \frac{x}{x}$$

$$= \frac{(2x-8)-3x(x-4)-10x}{x(x-4)}$$

$$= \frac{2x-8-3x^2+12x-10x}{x(x-4)}$$

$$= \frac{-3x^2+4x-8}{x(x-4)}$$

$$= -\frac{3x^2-4x+8}{x(x-4)}$$

75. The LCM is $2x(2x-3)$.

$$\frac{1}{2x-3} - \frac{5}{2x} + 1$$

$$= \frac{1}{2x-3} \cdot \frac{2x}{2x} - \frac{5}{2x} \cdot \frac{2x-3}{2x-3} + \frac{2x(2x-3)}{2x(2x-3)}$$

$$= \frac{2x-5(2x-3)+4x^2-6x}{2x(2x-3)}$$

$$= \frac{2x-10x+15+4x^2-6x}{2x(2x-3)}$$

$$= \frac{4x^2-14x+15}{2x(2x-3)}$$

77. $x^2 - 1 = (x+1)(x-1)$

$x^2 + 2x + 1 = (x+1)^2$

The LCM is $(x-1)(x+1)^2$.

$$\frac{3}{x^2-1} + \frac{2x}{x^2+2x+1}$$

$$= \frac{3}{(x+1)(x-1)} \cdot \frac{x+1}{x+1} + \frac{2x}{(x+1)^2} \cdot \frac{x-1}{x-1}$$

$$= \frac{3x+3+2x^2-2x}{(x+1)^2(x-1)}$$

$$= \frac{2x^2+x+3}{(x-1)(x+1)^2}$$

79. $x^2 - 9 = (x+3)(x-3)$
The LCM is $(x+3)(x-3)$.
$\dfrac{x}{x+3} - \dfrac{3-x}{x^2-9} = \dfrac{x}{x+3} \cdot \dfrac{x-3}{x-3} - \dfrac{3-x}{(x+3)(x-3)}$
$= \dfrac{x^2 - 3x - (3-x)}{(x+3)(x-3)}$
$= \dfrac{x^2 - 3x - 3 + x}{(x+3)(x-3)}$
$= \dfrac{x^2 - 2x - 3}{(x+3)(x-3)}$
$= \dfrac{(x-3)(x+1)}{(x+3)(x-3)}$
$= \dfrac{(1)(x+1)}{(x+3)(1)}$
$= \dfrac{x+1}{x+3}$

81. $x^2 + 8x + 15 = (x+5)(x+3)$
The LCM is $(x+5)(x+3)$.
$\dfrac{2x-3}{x+5} - \dfrac{x^2 - 4x - 19}{x^2 + 8x + 15}$
$= \dfrac{2x-3}{x+5} \cdot \dfrac{x+3}{x+3} - \dfrac{x^2 - 4x - 19}{(x+5)(x+3)}$
$= \dfrac{2x^2 + 3x - 9 - (x^2 - 4x - 19)}{(x+3)(x+5)}$
$= \dfrac{2x^2 + 3x - 9 - x^2 + 4x + 19}{(x+3)(x+5)}$
$= \dfrac{x^2 + 7x + 10}{(x+3)(x+5)}$
$= \dfrac{(x+2)(x+5)}{(x+3)(x+5)}$
$= \dfrac{(x+2)(1)}{(x+3)(1)}$
$= \dfrac{x+2}{x+3}$

83. $x^{2n} - 1 = (x^n + 1)(x^n - 1)$
The LCM is $(x^n + 1)(x^n - 1)$.
$\dfrac{x^n}{x^{2n} - 1} - \dfrac{2}{x^n + 1}$
$= \dfrac{x^n}{(x^n + 1)(x^n - 1)} - \dfrac{2}{x^n + 1} \cdot \dfrac{x^n - 1}{x^n - 1}$
$= \dfrac{x^n - 2(x^n - 1)}{(x^n + 1)(x^n - 1)}$
$= \dfrac{x^n - 2x^n + 2}{(x^n + 1)(x^n - 1)}$
$= \dfrac{-x^n + 2}{(x^n + 1)(x^n - 1)}$
$= \dfrac{-(x^n - 2)}{(x^n + 1)(x^n - 1)}$
$= -\dfrac{x^n - 2}{(x^n + 1)(x^n - 1)}$

85. $4x^2 - 9 = (2x+3)(2x-3)$
$3 - 2x = -(2x - 3)$
The LCM is $(2x+3)(2x-3)$.
$\dfrac{2x-2}{4x^2 - 9} - \dfrac{5}{3 - 2x}$
$= \dfrac{2x-2}{(2x+3)(2x-3)} - \dfrac{(-5)}{2x-3} \cdot \dfrac{2x+3}{2x+3}$
$= \dfrac{2x - 2 + 5(2x+3)}{(2x-3)(2x+3)}$
$= \dfrac{2x - 2 + 10x + 15}{(2x+3)(2x-3)}$
$= \dfrac{12x + 13}{(2x+3)(2x-3)}$

87. $2x^2 - x - 3 = (2x-3)(x+1)$
The LCM is $(2x-3)(x+1)$.
$\dfrac{x-2}{x+1} - \dfrac{3 - 12x}{2x^2 - x - 3}$
$= \dfrac{x-2}{x+1} \cdot \dfrac{2x-3}{2x-3} - \dfrac{3-12x}{(2x-3)(x+1)}$
$= \dfrac{(x-2)(2x-3) - (3-12x)}{(2x-3)(x+1)}$
$= \dfrac{2x^2 - 7x + 6 - 3 + 12x}{(2x-3)(x+1)}$
$= \dfrac{2x^2 + 5x + 3}{(2x-3)(x+1)}$
$= \dfrac{(x+1)(2x+3)}{(2x-3)(x+1)}$
$= \dfrac{2x+3}{2x-3}$

89. $x^2 + x - 6 = (x+3)(x-2)$
$x^2 + 4x + 3 = (x+3)(x+1)$
The LCM is $(x+3)(x-2)(x+1)$.

$$\frac{x+1}{x^2+x-6} - \frac{x+2}{x^2+4x+3} = \frac{x+1}{(x+3)(x-2)} \cdot \frac{x+1}{x+1} - \frac{x+2}{(x+3)(x+1)} \cdot \frac{x-2}{x-2}$$
$$= \frac{x^2 + 2x + 1 - (x^2 - 4)}{(x+3)(x-2)(x+1)}$$
$$= \frac{x^2 + 2x + 1 - x^2 + 4}{(x+3)(x-2)(x+1)}$$
$$= \frac{2x+5}{(x+3)(x-2)(x+1)}$$

91. $2x^2 + 11x + 12 = (2x+3)(x+4)$
$2x^2 - 3x - 9 = (2x+3)(x-3)$
The LCM is $(x+4)(2x+3)(x-3)$.

$$\frac{x-1}{2x^2+11x+12} + \frac{2x}{2x^2-3x-9} = \frac{x-1}{(2x+3)(x+4)} \cdot \frac{x-3}{x-3} + \frac{2x}{(2x+3)(x-3)} \cdot \frac{x+4}{x+4}$$
$$= \frac{x^2 - 4x + 3 + 2x^2 + 8x}{(x+4)(2x+3)(x-3)}$$
$$= \frac{3x^2 + 4x + 3}{(x+4)(2x+3)(x-3)}$$

93. $x^2 + x - 12 = (x+4)(x-3)$
The LCM is $(x+4)(x-3)$.

$$\frac{x}{x-3} - \frac{2}{x+4} - \frac{14}{x^2+x-12}$$
$$= \frac{x}{x-3} \cdot \frac{x+4}{x+4} - \frac{2}{x+4} \cdot \frac{x-3}{x-3} - \frac{14}{(x+4)(x-3)}$$
$$= \frac{x^2 + 4x - 2x + 6 - 14}{(x+4)(x-3)}$$
$$= \frac{x^2 + 2x - 8}{(x+4)(x-3)}$$
$$= \frac{(x+4)(x-2)}{(x+4)(x-3)}$$
$$= \frac{x-2}{x-3}$$

95. $x^2 + 3x - 18 = (x+6)(x-3)$
$3 - x = -(x-3)$
The LCM is $(x+6)(x-3)$.

$$\frac{x^2+6x}{x^2+3x-18} - \frac{2x-1}{x+6} + \frac{x-2}{3-x}$$
$$= \frac{x^2+6x}{(x+6)(x-3)} - \frac{2x-1}{x+6} \cdot \frac{x-3}{x-3} + \frac{-(x-2)}{x-3} \cdot \frac{x+6}{x+6}$$
$$= \frac{x^2 + 6x - (2x-1)(x-3) - (x-2)(x+6)}{(x+6)(x-3)}$$
$$= \frac{x^2 + 6x - 2x^2 + 7x - 3 - x^2 - 4x + 12}{(x+6)(x-3)}$$
$$= \frac{-2x^2 + 9x + 9}{(x+6)(x-3)}$$
$$= \frac{-(2x^2 - 9x - 9)}{(x+6)(x-3)}$$
$$= -\frac{2x^2 - 9x - 9}{(x+6)(x-3)}$$

97. $6x^2 + 11x - 10 = (3x-2)(2x+5)$
$2 - 3x = -(3x-2)$
The LCM is $(3x-2)(2x+5)$.

$$\frac{4-20x}{6x^2+11x-10} - \frac{4}{2-3x} + \frac{x}{2x+5} = \frac{4-20x}{(3x-2)(2x+5)} - \frac{(-4)}{3x-2} \cdot \frac{2x+5}{2x+5} + \frac{x}{2x+5} \cdot \frac{3x-2}{3x-2}$$

$$= \frac{4-20x+8x+20+3x^2-2x}{(3x-2)(2x+5)}$$

$$= \frac{3x^2-14x+24}{(3x-2)(2x+5)}$$

99. $x^4 - 1 = (x+1)(x-1)(x^2+1)$
$x^2 - 1 = (x+1)(x-1)$
The LCM is $(x+1)(x-1)(x^2+1)$.

$$\frac{2x^2}{x^4-1} - \frac{1}{x^2-1} + \frac{1}{x^2+1} = \frac{2x^2}{(x+1)(x-1)(x^2+1)} - \frac{1}{(x+1)(x-1)} \cdot \frac{x^2+1}{x^2+1} + \frac{1}{x^2+1} \cdot \frac{(x+1)(x-1)}{(x+1)(x-1)}$$

$$= \frac{2x^2 - (x^2+1) + x^2 - 1}{(x+1)(x-1)(x^2+1)}$$

$$= \frac{2x^2 - x^2 - 1 + x^2 - 1}{(x+1)(x-1)(x^2+1)}$$

$$= \frac{2x^2 - 2}{(x+1)(x-1)(x^2+1)}$$

$$= \frac{2(x^2-1)}{(x+1)(x-1)(x^2+1)}$$

$$= \frac{2(x+1)(x-1)}{(x+1)(x-1)(x^2+1)}$$

$$= \frac{2}{x^2+1}$$

Applying Concepts 6.2

101. $\dfrac{(x+1)^2}{1-2x} \cdot \dfrac{2x-1}{x+1} = \dfrac{(x+1)(x+1)(2x-1)}{(1-2x)(x+1)}$
$= -(x+1)$
$= -x - 1$

103. $\left(\dfrac{y-2}{x^2}\right)^3 \cdot \left(\dfrac{x}{2-y}\right)^2 = \dfrac{(y-2)^3}{x^6} \cdot \dfrac{x^2}{(2-y)^2}$
$= \dfrac{(y-2)(y-2)(y-2)x^2}{x^6(2-y)(2-y)}$
$= \dfrac{y-2}{x^4}$

105. $\left(\dfrac{y+1}{y-1}\right)^2 - 1 = \left(\dfrac{y+1}{y-1}+1\right)\left(\dfrac{y+1}{y-1}-1\right)$
$= \left(\dfrac{y+1}{y-1}+\dfrac{y-1}{y-1}\right)\left(\dfrac{y+1}{y-1}-\dfrac{y-1}{y-1}\right)$
$= \left(\dfrac{y+1+y-1}{y-1}\right)\left(\dfrac{y+1-y+1}{y-1}\right)$
$= \left(\dfrac{2y}{y-1}\right)\left(\dfrac{2}{y-1}\right)$
$= \dfrac{4y}{(y-1)^2}$

107. $\dfrac{3x^2+6x}{4x^2-16} \cdot \dfrac{2x+8}{x^2+2x} \div \dfrac{3x-9}{5x-20} = \dfrac{3x^2+6x}{4x^2-16} \cdot \dfrac{2x+8}{x^2+2x} \cdot \dfrac{5x-20}{3x-9}$
$= \dfrac{3x(x+2) \cdot 2(x+4) \cdot 5(x-4)}{4(x+2)(x-2) \cdot x(x+2) \cdot 3(x-3)}$
$= \dfrac{5(x+4)(x-4)}{2(x+2)(x-2)(x-3)}$

109. $\dfrac{a^2+a-6}{4+11a-3a^2} \cdot \dfrac{15a^2-a-2}{4a^2+7a-2} \div \dfrac{6a^2-7a-3}{4-17a+4a^2} = \dfrac{a^2+a-6}{4+11a-3a^2} \cdot \dfrac{15a^2-a-2}{4a^2+7a-2} \cdot \dfrac{4-17a+4a^2}{6a^2-7a-3}$

$= \dfrac{(a+3)(a-2)(5a-2)(3a+1)(4-a)(1-4a)}{(4-a)(1+3a)(4a-1)(a+2)(3a+1)(2a-3)}$

$= -\dfrac{(a+3)(a-2)(5a-2)}{(a+2)(3a+1)(2a-3)}$

111. $\left(\dfrac{x+1}{2x-1} - \dfrac{x-1}{2x+1}\right) \cdot \left(\dfrac{2x-1}{x} - \dfrac{2x-1}{x^2}\right)$

$= \left[\dfrac{(x+1)(2x+1)}{(2x-1)(2x+1)} - \dfrac{(x-1)(2x-1)}{(2x+1)(2x-1)}\right] \cdot \left[\dfrac{x(2x-1)}{x^2} - \dfrac{2x-1}{x^2}\right]$

$= \dfrac{2x^2+3x+1-(2x^2-3x+1)}{(2x-1)(2x+1)} \cdot \dfrac{2x^2-x-(2x-1)}{x^2}$

$= \dfrac{6x}{(2x-1)(2x+1)} \cdot \dfrac{2x^2-3x+1}{x^2}$

$= \dfrac{6x(2x-1)(x-1)}{x^2(2x-1)(2x+1)}$

$= \dfrac{6(x-1)}{x(2x+1)}$

113. $\dfrac{1}{3} + \dfrac{1}{5} \neq \dfrac{1}{3+5}$

$\dfrac{5}{15} + \dfrac{3}{15} \neq \dfrac{1}{8}$

$\dfrac{8}{15} \neq \dfrac{1}{8}$

115. $\dfrac{3x+6y}{xy} = \dfrac{3x}{xy} + \dfrac{6y}{xy} = \dfrac{3}{y} + \dfrac{6}{x}$

117. $\dfrac{4a^2+3ab}{a^2b^2} = \dfrac{4a^2}{a^2b^2} + \dfrac{3ab}{a^2b^2} = \dfrac{4}{b^2} + \dfrac{3}{ab}$

119. a. $2x^2 + \dfrac{528}{x} = \dfrac{2x^3+528}{x}$

b.

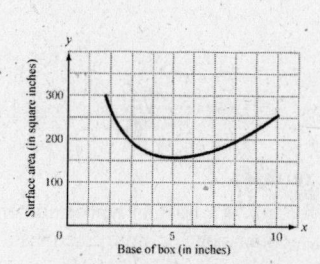

[Graph showing surface area (in square inches) on y-axis from 0 to 300, and base of box (in inches) on x-axis from 0 to 10, with a U-shaped curve with minimum near x=5]

c. The point whose coordinates are (4, 164) means that when the base of the box is 4 in., 164 in^2 of cardboard will be needed.

d. The base of the box that uses the minimum amount of cardboard is 5.1 in.

e. $f(x) = 2x^2 + \dfrac{528}{x}$

$f(5.1) = 2(5.1)^2 + \dfrac{528}{5.1} \approx 155.5$ in^2

121. When fractions are incorrectly added as in the inaccurate calculation $\frac{1}{5}+\frac{2}{3}=\frac{1+2}{5+3}=\frac{3}{8}$, the incorrect sum is always between the two fractions being added. Here is proof based on a, b, c, and d being positive real numbers. Assume $\frac{a}{b}<\frac{c}{d}$. Then $ad < bc$. Now consider $\frac{a+c}{b+d}$.

$$\frac{a+c}{b+d} = \frac{a+c}{b+d} \cdot \frac{d}{d} = \frac{ad+cd}{bd+d^2}$$
$$< \frac{bc+cd}{bd+d^2} = \frac{c(b+d)}{d(b+d)} = \frac{c}{d}$$

Therefore, $\frac{a+c}{b+c}<\frac{c}{d}$.

$$\frac{a+c}{b+d} = \frac{a+c}{b+d} \cdot \frac{a}{a} = \frac{a^2+ac}{ab+ad}$$
$$> \frac{a^2+ac}{ab+bc} = \frac{a(a+c)}{b(a+c)} = \frac{a}{b}$$

Therefore, $\frac{a+c}{b+d}>\frac{a}{b}$.

Section 6.3

Concept Review 6.3

1. Always true

3. Never true

$$\frac{c^{-1}}{a^{-1}+b^{-1}} = \frac{\frac{1}{c}}{\frac{1}{a}+\frac{1}{b}} = \frac{\frac{1}{c}}{\frac{b+a}{ab}} = \frac{ab}{c(b+a)}$$

Objective 6.3.1 Exercises

1. A complex fraction is a fraction whose numerator or denominator contains one or more fractions.

3. To simplify the complex fraction, multiply the numerator and denominator of the complex fraction by the LCM of the denominators of the fractions $\frac{2}{a}$ and $\frac{1}{a^2}$. The LCM is a^2.

5. When you multiply the denominator of the complex fraction by a^2, the denominator of the complex fraction simplifies to a^2+1.

7. The LCM is 3.
$$\frac{2-\frac{1}{3}}{4+\frac{11}{3}} = \frac{2-\frac{1}{3}}{4+\frac{11}{3}} \cdot \frac{3}{3} = \frac{2 \cdot 3 - \frac{1}{3} \cdot 3}{4 \cdot 3 + \frac{11}{3} \cdot 3} = \frac{6-1}{12+11} = \frac{5}{23}$$

9. The LCM is 6.
$$\frac{3-\frac{2}{3}}{5+\frac{5}{6}} = \frac{3-\frac{2}{3}}{5+\frac{5}{6}} \cdot \frac{6}{6} = \frac{3 \cdot 6 - \frac{2}{3} \cdot 6}{5 \cdot 6 + \frac{5}{6} \cdot 6} = \frac{18-4}{30+5} = \frac{14}{35} = \frac{2}{5}$$

11. The LCM of y^2 and y is y^2.
$$\frac{\frac{1}{y^2}-1}{1+\frac{1}{y}} = \frac{\frac{1}{y^2}-1}{1+\frac{1}{y}} \cdot \frac{y^2}{y^2}$$
$$= \frac{\frac{1}{y^2} \cdot y^2 - 1 \cdot y^2}{1 \cdot y^2 + \frac{1}{y} \cdot y^2}$$
$$= \frac{1-y^2}{y^2+y}$$
$$= \frac{(1+y)(1-y)}{y(y+1)}$$
$$= \frac{1-y}{y}$$

13. The LCM is a.
$$\frac{\frac{25}{a}-a}{5+a} = \frac{\frac{25}{a}-a}{5+a} \cdot \frac{a}{a}$$
$$= \frac{\frac{25}{a} \cdot a - a \cdot a}{5 \cdot a + a \cdot a}$$
$$= \frac{25-a^2}{5a+a^2}$$
$$= \frac{(5+a)(5-a)}{a(5+a)}$$
$$= \frac{5-a}{a}$$

15. The LCM of b, 2, and b^2 is $2b^2$.
$$\frac{\frac{1}{b}+\frac{1}{2}}{\frac{4}{b^2}-1} = \frac{\frac{1}{b}+\frac{1}{2}}{\frac{4}{b^2}-1} \cdot \frac{2b^2}{2b^2}$$
$$= \frac{\frac{1}{b} \cdot 2b^2 + \frac{1}{2} \cdot 2b^2}{\frac{4}{b^2} \cdot 2b^2 - 1 \cdot 2b^2}$$
$$= \frac{2b+b^2}{8-2b^2}$$
$$= \frac{b(2+b)}{2(4-b^2)}$$
$$= \frac{b(2+b)}{2(2+b)(2-b)}$$
$$= \frac{b}{2(2-b)}$$

17. The LCM is $2x-3$.
$$\frac{4+\frac{12}{2x-3}}{5+\frac{15}{2x-3}} = \frac{4+\frac{12}{2x-3}}{5+\frac{15}{2x-3}} \cdot \frac{2x-3}{2x-3}$$
$$= \frac{4(2x-3)+\frac{12}{2x-3}(2x-3)}{5(2x-3)+\frac{15}{2x-3}(2x-3)}$$
$$= \frac{8x-12+12}{10x-15+15}$$
$$= \frac{8x}{10x}$$
$$= \frac{4}{5}$$

19. The LCM is $b - 5$.

$$\frac{\frac{-5}{b-5} - 3}{\frac{10}{b-5} + 6} = \frac{\frac{-5}{b-5} - 3}{\frac{10}{b-5} + 6} \cdot \frac{b-5}{b-5}$$

$$= \frac{\frac{-5}{b-5} \cdot (b-5) - 3(b-5)}{\frac{10}{b-5} \cdot (b-5) + 6(b-5)}$$

$$= \frac{-5 - 3b + 15}{10 + 6b - 30}$$

$$= \frac{-3b + 10}{6b - 20}$$

$$= \frac{-3b + 10}{2(3b - 10)}$$

$$= -\frac{1}{2}$$

21. The LCM of $a - 1$ and a is $a(a - 1)$.

$$\frac{\frac{2a}{a-1} - \frac{3}{a}}{\frac{1}{a-1} + \frac{2}{a}} = \frac{\frac{2a}{a-1} - \frac{3}{a}}{\frac{1}{a-1} + \frac{2}{a}} \cdot \frac{a(a-1)}{a(a-1)}$$

$$= \frac{\frac{2a}{a-1} \cdot a(a-1) - \frac{3}{a} \cdot a(a-1)}{\frac{1}{a-1} \cdot a(a-1) + \frac{2}{a} \cdot a(a-1)}$$

$$= \frac{2a^2 - 3(a-1)}{a + 2(a-1)}$$

$$= \frac{2a^2 - 3a + 3}{a + 2a - 2}$$

$$= \frac{2a^2 - 3a + 3}{3a - 2}$$

23. The LCM of $3x - 2$ and $9x^2 - 4$ is $(3x+2)(3x-2)$.

$$\frac{\frac{x}{3x-2}}{\frac{x}{9x^2-4}} = \frac{\frac{x}{3x-2}}{\frac{x}{(3x+2)(3x-2)}}$$

$$= \frac{\frac{x}{3x-2}}{\frac{x}{(3x+2)(3x-2)}} \cdot \frac{(3x+2)(3x-2)}{(3x+2)(3x-2)}$$

$$= \frac{x(3x+2)}{x}$$

$$= 3x + 2$$

25. The LCM of x and x^2 is x^2.

$$\frac{1 - \frac{1}{x} - \frac{6}{x^2}}{1 - \frac{4}{x} + \frac{3}{x^2}} = \frac{1 - \frac{1}{x} - \frac{6}{x^2}}{1 - \frac{4}{x} + \frac{3}{x^2}} \cdot \frac{x^2}{x^2}$$

$$= \frac{1 \cdot x^2 - \frac{1}{x} \cdot x^2 - \frac{6}{x^2} \cdot x^2}{1 \cdot x^2 - \frac{4}{x} \cdot x^2 + \frac{3}{x^2} \cdot x^2}$$

$$= \frac{x^2 - x - 6}{x^2 - 4x + 3}$$

$$= \frac{(x+2)(x-3)}{(x-1)(x-3)}$$

$$= \frac{x+2}{x-1}$$

27. The LCM of x^2 and x is x^2.

$$\frac{\frac{15}{x^2} - \frac{2}{x} - 1}{\frac{4}{x^2} - \frac{5}{x} + 4} = \frac{\frac{15}{x^2} - \frac{2}{x} - 1}{\frac{4}{x^2} - \frac{5}{x} + 4} \cdot \frac{x^2}{x^2}$$

$$= \frac{\frac{15}{x^2} \cdot x^2 - \frac{2}{x} \cdot x^2 - 1 \cdot x^2}{\frac{4}{x^2} \cdot x^2 - \frac{5}{x} \cdot x^2 + 4 \cdot x^2}$$

$$= \frac{15 - 2x - x^2}{4 - 5x + 4x^2}$$

$$= -\frac{(x-3)(x+5)}{4x^2 - 5x + 4}$$

29. The LCM is $3x + 10$.

$$\frac{1 - \frac{12}{3x+10}}{x - \frac{8}{3x+10}} = \frac{1 - \frac{12}{3x+10}}{x - \frac{8}{3x+10}} \cdot \frac{3x+10}{3x+10}$$

$$= \frac{1 \cdot (3x+10) - \frac{12}{3x+10} \cdot (3x+10)}{x(3x+10) - \frac{8}{3x+10} \cdot (3x+10)}$$

$$= \frac{3x + 10 - 12}{3x^2 + 10x - 8}$$

$$= \frac{3x - 2}{(x+4)(3x-2)}$$

$$= \frac{1}{x+4}$$

31. The LCM is $x + 2$.

$$\frac{x - 5 - \frac{18}{x+2}}{x + 7 + \frac{6}{x+2}} = \frac{x - 5 - \frac{18}{x+2}}{x + 7 + \frac{6}{x+2}} \cdot \frac{x+2}{x+2}$$

$$= \frac{(x-5)(x+2) - \frac{18}{x+2} \cdot (x+2)}{(x+7)(x+2) + \frac{6}{x+2} \cdot (x+2)}$$

$$= \frac{x^2 - 3x - 10 - 18}{x^2 + 9x + 14 + 6}$$

$$= \frac{x^2 - 3x - 28}{x^2 + 9x + 20}$$

$$= \frac{(x+4)(x-7)}{(x+4)(x+5)}$$

$$= \frac{x-7}{x+5}$$

33. The LCM is $a(a - 2)$.

$$\frac{\frac{1}{a} - \frac{3}{a-2}}{\frac{2}{a} + \frac{5}{a-2}} = \frac{\frac{1}{a} - \frac{3}{a-2}}{\frac{2}{a} + \frac{5}{a-2}} \cdot \frac{a(a-2)}{a(a-2)}$$

$$= \frac{\frac{1}{a} \cdot a(a-2) - \frac{3}{a-2} \cdot a(a-2)}{\frac{2}{a} \cdot a(a-2) + \frac{5}{a-2} \cdot a(a-2)}$$

$$= \frac{a - 2 - 3a}{2a - 4 + 5a}$$

$$= \frac{-2a - 2}{7a - 4}$$

$$= \frac{-(2a+2)}{7a-4}$$

$$= -\frac{2a+2}{7a-4}$$

$$= -\frac{2(a+1)}{7a-4}$$

Chapter 6: Rational Expressions

35. The LCM of y^2, xy, and x^2 is x^2y^2.

$$\frac{\frac{1}{y^2} - \frac{1}{xy} - \frac{2}{x^2}}{\frac{1}{y^2} - \frac{3}{xy} + \frac{2}{x^2}} = \frac{\frac{1}{y^2} - \frac{1}{xy} - \frac{2}{x^2}}{\frac{1}{y^2} - \frac{3}{xy} + \frac{2}{x^2}} \cdot \frac{x^2y^2}{x^2y^2}$$

$$= \frac{\frac{1}{y^2} \cdot x^2y^2 - \frac{1}{xy} \cdot x^2y^2 - \frac{2}{x^2} \cdot x^2y^2}{\frac{1}{y^2} \cdot x^2y^2 - \frac{3}{xy} \cdot x^2y^2 + \frac{2}{x^2} \cdot x^2y^2}$$

$$= \frac{x^2 - xy - 2y^2}{x^2 - 3xy + 2y^2}$$

$$= \frac{(x+y)(x-2y)}{(x-y)(x-2y)}$$

$$= \frac{x+y}{x-y}$$

37. The LCM is $(x+1)(x-1)$.

$$\frac{\frac{x-1}{x+1} - \frac{x+1}{x-1}}{\frac{x-1}{x+1} + \frac{x+1}{x-1}}$$

$$= \frac{\frac{x-1}{x+1} - \frac{x+1}{x-1}}{\frac{x-1}{x+1} + \frac{x+1}{x-1}} \cdot \frac{(x+1)(x-1)}{(x+1)(x-1)}$$

$$= \frac{\frac{x-1}{x+1} \cdot (x+1)(x-1) - \frac{x+1}{x-1} \cdot (x+1)(x-1)}{\frac{x-1}{x+1} \cdot (x+1)(x-1) + \frac{x+1}{x-1} \cdot (x+1)(x-1)}$$

$$= \frac{(x-1)(x-1) - (x+1)(x+1)}{(x-1)(x-1) + (x+1)(x+1)}$$

$$= \frac{x^2 - 2x + 1 - (x^2 + 2x + 1)}{x^2 - 2x + 1 + x^2 + 2x + 1}$$

$$= \frac{x^2 - 2x + 1 - x^2 - 2x - 1}{2x^2 + 2}$$

$$= \frac{-4x}{2(x^2+1)}$$

$$= -\frac{2x}{x^2+1}$$

39. The LCM is x.

$$4 - \frac{2}{2 - \frac{3}{x}} = 4 - \frac{2}{2 - \frac{3}{x}} \cdot \frac{x}{x}$$

$$= 4 - \frac{2 \cdot x}{2 \cdot x - \frac{3}{x} \cdot x}$$

$$= 4 - \frac{2x}{2x - 3}$$

The LCM is $2x - 3$.

$$4 - \frac{2x}{2x-3} = 4 \cdot \frac{2x-3}{2x-3} - \frac{2x}{2x-3}$$

$$= \frac{4(2x-3) - 2x}{2x-3}$$

$$= \frac{8x - 12 - 2x}{2x-3}$$

$$= \frac{6x - 12}{2x-3}$$

$$= \frac{6(x-2)}{2x-3}$$

41. The expression $\dfrac{\frac{1}{x}}{\frac{x}{1}}$ is a complex fraction whose denominator is the reciprocal of its numerator. The simplified form is $\dfrac{\frac{1}{x}}{\frac{x}{1}} = \dfrac{1}{x} \div \dfrac{x}{1} = \dfrac{1}{x} \cdot \dfrac{1}{x} = \left(\dfrac{1}{x}\right)^2$, which is the square of the numerator of the complex fraction, as described in choice **c**.

Applying Concepts 6.3

43.
$$\frac{x^{-1}}{y^{-1}} + \frac{y}{x} = \frac{\frac{1}{x}}{\frac{1}{y}} + \frac{y}{x}$$

$$= \frac{\frac{1}{x}}{\frac{1}{y}} \cdot \frac{xy}{xy} + \frac{y}{x}$$

$$= \frac{y}{x} + \frac{y}{x}$$

$$= \frac{y+y}{x}$$

$$= \frac{2y}{x}$$

45.
$$\frac{x - \frac{1}{x}}{1 + \frac{1}{x}} = \frac{x - \frac{1}{x}}{1 + \frac{1}{x}} \cdot \frac{x}{x}$$

$$= \frac{x \cdot x - \frac{1}{x} \cdot x}{1 \cdot x + \frac{1}{x} \cdot x}$$

$$= \frac{x^2 - 1}{x + 1}$$

$$= \frac{(x+1)(x-1)}{x+1}$$

$$= x - 1$$

47.
$$2 - \frac{2}{2 - \frac{2}{c-1}} = 2 - \frac{2}{2 - \frac{2}{c-1}} \cdot \frac{c-1}{c-1}$$

$$= 2 - \frac{2(c-1)}{2(c-1) - \frac{2}{c-1}(c-1)}$$

$$= 2 - \frac{2(c-1)}{2c - 2 - 2}$$

$$= 2 - \frac{2c - 2}{2c - 4}$$

$$= 2 - \frac{c-1}{c-2}$$

$$= \frac{2c-4}{c-2} - \frac{c-1}{c-2}$$

$$= \frac{2c - 4 - (c-1)}{c-2}$$

$$= \frac{c-3}{c-2}$$

49.
$$a + \frac{a}{2+\frac{1}{1-\frac{2}{a}}} = a + \frac{a}{2+\frac{1}{1-\frac{2}{a}} \cdot \frac{a}{a}}$$
$$= a + \frac{a}{2+\frac{1 \cdot a}{1 \cdot a - \frac{2}{a} \cdot a}}$$
$$= a + \frac{a}{2+\frac{a}{a-2}}$$

The LCM is $a - 2$.

$$a + \frac{a}{\frac{2(a-2)}{a-2}+\frac{a}{a-2}} = a + \frac{a}{\frac{2a-4+a}{a-2}}$$
$$= a + \frac{a}{\frac{3a-4}{a-2}}$$
$$= a + \frac{a}{\frac{3a-4}{a-2}} \cdot \frac{a-2}{a-2}$$
$$= a + \frac{a(a-2)}{3a-4}$$

The LCM is $3a - 4$.

$$a \cdot \frac{(3a-4)}{3a-4} + \frac{a(a-2)}{3a-4} = \frac{3a^2 - 4a + a^2 - 2a}{3a-4}$$
$$= \frac{4a^2 - 6a}{3a-4}$$
$$= \frac{2a(2a-3)}{3a-4}$$

51.
$$\frac{\frac{1}{x+h}-\frac{1}{x}}{h} = \frac{\frac{1}{x+h}-\frac{1}{x}}{h} \cdot \frac{x(x+h)}{x(x+h)}$$
$$= \frac{\frac{1}{x+h} \cdot x(x+h) - \frac{1}{x} \cdot x(x+h)}{hx(x+h)}$$
$$= \frac{x-(x+h)}{hx(x+h)}$$
$$= \frac{-h}{hx(x+h)}$$
$$= -\frac{1}{x(x+h)}$$

53. Strategy
The first even integer: n
The second consecutive even integer: $n + 2$
The third consecutive even integer: $n + 4$
Add the reciprocals of the three integers.

Solution
$$\frac{1}{n}+\frac{1}{n+2}+\frac{1}{n+4}$$
$$= \frac{1}{n} \cdot \frac{(n+2)(n+4)}{(n+2)(n+4)} + \frac{1}{n+2} \cdot \frac{n(n+4)}{n(n+4)} + \frac{1}{n+4} \cdot \frac{n(n+2)}{n(n+2)}$$
$$= \frac{(n+2)(n+4)+n(n+4)+n(n+2)}{n(n+2)(n+4)}$$
$$= \frac{n^2+6n+8+n^2+4n+n^2+2n}{n^3+6n^2+8n}$$
$$= \frac{3n^2+12n+8}{n(n+2)(n+4)}$$

55. a.
$$\frac{Cx}{\left(1-\frac{1}{(x+1)^{60}}\right)} = \frac{Cx}{\left(\frac{(x+1)^{60}-1}{(x+1)^{60}}\right)} = \frac{Cx(x+1)^{60}}{(x+1)^{60}-1}$$

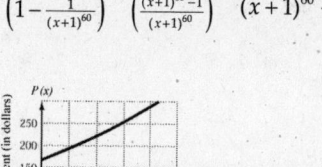

b.

c. $12(0) = 0$
$12(0.025) = 0.3$
The interval of annual interest rates is 0% to 30%.

d. The ordered pair (0.004, 386.66) means that when the monthly interest rate on a car loan is 0.5%, the monthly payment on the loan is $386.66.

e. The monthly payment with a loan amount of $20,000 and an annual interest rate of 8% is $406.

Section 6.4
Concept Review 6.4

1. Sometimes true
The variable expression cannot be zero.

3. Sometimes true
If the variable in the denominator is equivalent to zero, clearing the equation of the variable will not result in an equivalent equation.

Objective 6.4.1 Exercises

1. The first step in solving the equation is to clear denominators by multiplying each side of the equation by the LCM of the denominators x and $x+3$. The LCM is $x(x+3)$.

3. Write the equation that results from clearing denominators: $8(x+3) = 5x$. The solution of this equation is -8.

5.
$$\frac{x}{2}+\frac{5}{6}=\frac{x}{3}$$
$$12\left(\frac{x}{2}+\frac{5}{6}\right)=12\left(\frac{x}{3}\right)$$
$$12 \cdot \frac{x}{2}+12 \cdot \frac{5}{6}=4x$$
$$6x+10=4x$$
$$10=-2x$$
$$-5=x$$

The solution is -5.

182 Chapter 6: Rational Expressions

7.
$$1 - \frac{3}{y} = 4$$
$$y\left(1 - \frac{3}{y}\right) = y \cdot 4$$
$$y \cdot 1 - y \cdot \frac{3}{y} = 4y$$
$$y - 3 = 4y$$
$$-3 = 3y$$
$$-1 = y$$
The solution is -1.

9. $\frac{1}{3}$
The solution is $\frac{t}{3}$

11.
$$\frac{4}{x-4} = \frac{2}{x-2}$$
$$\frac{4}{x-4} \cdot (x-4)(x-2) = \frac{2}{x-2} \cdot (x-4)(x-2)$$
$$4(x-2) = 2(x-4)$$
$$4x - 8 = 2x - 8$$
$$2x - 8 = -8$$
$$2x = 0$$
$$x = 0$$
The solution is 0.

13. $\frac{7.5}{t}$
The solutions are -3 and 3.

15.
$$\frac{7.5}{20} + \frac{7.5}{t} = 1$$
$$20t\left(\frac{7.5}{20} + \frac{7.5}{t}\right) = 20t(1)$$
$$7.5t + 150 = 20t$$
$$150 = 12.5t$$
$$12 = t$$
The solution is 8.

17. $\frac{t}{30}$
The solution is 5.

19.
$$\frac{8}{x-5} = \frac{3}{x}$$
$$\frac{8}{x-5} \cdot x(x-5) = \frac{3}{x} \cdot x(x-5)$$
$$8x = 3(x-5)$$
$$8x = 3x - 15$$
$$5x = -15$$
$$x = -3$$
The solution is -3.

21.
$$5 + \frac{8}{a-2} = \frac{4a}{a-2}$$
$$(a-2)\left(5 + \frac{8}{a-2}\right) = (a-2) \cdot \frac{4a}{a-2}$$
$$(a-2)5 + (a-2) \cdot \frac{8}{a-2} = 4a$$
$$5a - 10 + 8 = 4a$$
$$5a - 2 = 4a$$
$$-2 = -a$$
$$2 = a$$
2 does not check as a solution. The equation has no solution.

23.
$$\frac{x}{2} + \frac{20}{x} = 7$$
$$2x\left(\frac{x}{2} + \frac{20}{x}\right) = 7(2x)$$
$$2x\left(\frac{x}{2}\right) + 2x\left(\frac{20}{x}\right) = 14x$$
$$x^2 + 40 = 14x$$
$$x^2 - 14x + 40 = 0$$
$$(x-10)(x-4) = 0$$
$$\frac{t}{36} + \frac{t}{45} = 1 \quad x - 4 = 0$$
$$x = 4$$
The solutions are 10 and 4.

25.
$$\frac{t}{5} + \frac{t}{7.5} = 1$$
$$15\left(\frac{t}{5} + \frac{t}{7.5}\right) = 15(1)$$
$$3t + 2t = 15$$
$$5t = 15$$
$$t = 3$$
The solution is -1.

27. $\frac{t}{3}$
$$\frac{1}{4.5} \quad x + 3 = 0$$
$$x = -3$$
The solutions are 4 and -3.

29.
$$-\frac{5}{x+7} + 1 = \frac{4}{x+7}$$
$$\frac{-5}{x+7} + 1 = \frac{4}{x+7}$$
$$(x+7)\left(\frac{-5}{x+7} + 1\right) = (x+7) \cdot \frac{4}{x+7}$$
$$(x+7) \cdot \frac{-5}{x+7} + (x+7)(1) = 4$$
$$-5 + x + 7 = 4$$
$$x + 2 = 4$$
$$x = 2$$
The solution is 2.

31.
$$\frac{2}{4y^2-9}+\frac{1}{2y-3}=\frac{3}{2y+3}$$
$$\frac{2}{(2y+3)(2y-3)}+\frac{1}{2y-3}=\frac{3}{2y+3}$$
$$(2y+3)(2y-3)\left(\frac{2}{(2y+3)(2y-3)}+\frac{1}{2y-3}\right)=(2y+3)(2y-3)\frac{3}{2y+3}$$
$$(2y+3)(2y-3)\frac{2}{(2y+3)(2y-3)}+(2y+3)(2y-3)\frac{1}{2y-3}=(2y-3)3$$
$$2+2y+3=6y-9$$
$$2y+5=6y-9$$
$$-4y+5=-9$$
$$-4y=-14$$
$$y=\frac{7}{2}$$

The solution is $\frac{7}{2}$.

33.
$$\frac{5}{x^2-7x+12}=\frac{2}{x-3}+\frac{5}{x-4}$$
$$\frac{5}{(x-3)(x-4)}=\frac{2}{x-3}+\frac{5}{x-4}$$
$$(x-3)(x-4)\frac{5}{(x-3)(x-4)}=(x-3)(x-4)\left(\frac{2}{x-3}+\frac{5}{x-4}\right)$$
$$5=(x-3)(x-4)\frac{2}{x-3}+(x-3)(x-4)\frac{5}{x-4}$$
$$5=(x-4)2+(x-3)5$$
$$5=2x-8+5x-15$$
$$5=7x-23$$
$$28=7x$$
$$4=x$$

4 does not check as a solution. The equation has no solution.

35. $x^2+x-12=0$
$(x+4)(x-3)=0$
$x+4=0 \quad x-3=0$
$x=-4 \quad x=3$

The values of −4 and 3 would result in division by zero when substituted into the original equation.

Objective 6.4.2 Exercises

37. It takes less than 2 h, because the first person can complete the task in that time when working alone.

39.

	Rate	•	Time	=	Part
Hot water hose	$\frac{1}{6}$	•	t	=	$\frac{t}{6}$
Cold water hose	$\frac{1}{4}$	•	t	=	$\frac{t}{4}$

184 Chapter 6: Rational Expressions

41. Strategy
Unknown time to process the data working together: t

	Rate	Time	Part
First computer	$\frac{1}{2}$	t	$\frac{t}{2}$
Second computer	$\frac{1}{3}$	t	$\frac{t}{3}$

The sum of the part of the task completed by the first computer and the part of the task completed by the second computer is 1.

$$\frac{t}{2} + \frac{t}{3} = 1$$

Solution
$$\frac{t}{2} + \frac{t}{3} = 1$$
$$6\left(\frac{t}{2} + \frac{t}{3}\right) = 6(1)$$
$$3t + 2t = 6$$
$$5t = 6$$
$$t = 1.2$$

With both computers, it would take 1.2 h to process the data.

43. Strategy
Unknown time to heat the water working together: t

	Rate	Time	Part
First panel	$\frac{1}{30}$	t	$\frac{t}{30}$
Second panel	$\frac{1}{45}$	t	$\frac{t}{45}$

The sum of the part of the task completed by the first panel and the part of the task completed by the second panel is 1.

$$\frac{t}{30} + \frac{t}{45} = 1$$

Solution
$$\frac{t}{30} + \frac{t}{45} = 1$$
$$90\left(\frac{t}{30} + \frac{t}{45}\right) = 90(1)$$
$$3t + 2t = 90$$
$$5t = 90$$
$$t = 18$$

With both panels working, it would take 18 min to raise the temperature 1°.

45. Strategy
Unknown time to wire the telephone lines working together: t

	Rate	Time	Part
First member	$\frac{1}{5}$	t	$\frac{t}{5}$
Second member	$\frac{1}{7.5}$	t	$\frac{t}{7.5}$

The sum of the part of the task completed by the first member and the part of the task completed by the second member is 1.

$$\frac{t}{5} + \frac{t}{7.5} = 1$$

Solution
$$\frac{t}{5} + \frac{t}{7.5} = 1$$
$$15\left(\frac{t}{5} + \frac{t}{7.5}\right) = 15(1)$$
$$3t + 2t = 15$$
$$5t = 15$$
$$t = 3$$

With both members working together, it would take 3 h to wire the telephone lines.

47. Strategy
Time for the new machine to package the transistors: t
Time for the old machine to package the transistors: $4t$

	Rate	Time	Part
New machine	$\frac{1}{t}$	8	$\frac{8}{t}$
Old machine	$\frac{1}{4t}$	8	$\frac{8}{4t}$

The sum of the parts of the task completed by each machine must equal 1.

$$\frac{8}{t} + \frac{8}{4t} = 1$$

Solution
$$\frac{8}{t} + \frac{8}{4t} = 1$$
$$4t\left(\frac{8}{t} + \frac{8}{4t}\right) = 4t(1)$$
$$32 + 8 = 4t$$
$$40 = 4t$$
$$10 = t$$

Working alone, the new machine would take 10 h to package the transistors.

49. **Strategy**
Time for the smaller printer to print the payroll: t

	Rate	Time	Part
Large printer	$\frac{1}{40}$	10	$\frac{10}{40}$
Smaller printer	$\frac{1}{t}$	60	$\frac{60}{t}$

The sum of the parts of the task completed by each printer must equal 1.
$$\frac{10}{40}+\frac{60}{t}=1$$

Solution
$$\frac{10}{40}+\frac{60}{t}=1$$
$$40t\left(\frac{10}{40}+\frac{60}{t}\right)=40t(1)$$
$$10t+2400=40t$$
$$2400=30t$$
$$80=t$$

Working alone, the smaller printer would take 80 min to print the payroll.

51. **Strategy**
Time for the apprentice to shingle the roof: t

	Rate	Time	Part
Roofer	$\frac{1}{12}$	3	$\frac{3}{12}$
Apprentice	$\frac{1}{t}$	15	$\frac{15}{t}$

The sum of the part of the task completed by the roofer and the part of the task completed by the apprentice is 1.
$$\frac{3}{12}+\frac{15}{t}=1$$

Solution
$$\frac{3}{12}+\frac{15}{t}=1$$
$$12t\left(\frac{3}{12}+\frac{15}{t}\right)=12t(1)$$
$$3t+180=12t$$
$$180=9t$$
$$20=t$$

It would take the apprentice 20 h to shingle the roof.

53. **Strategy**
Time, in seconds, to address envelopes working together: t

	Rate	Time	Part
First clerk	$\frac{1}{30}$	t	$\frac{t}{30}$
Second clerk	$\frac{1}{40}$	t	$\frac{t}{40}$

The sum of the parts of the job completed by each clerk is 1.
$$\frac{t}{30}+\frac{t}{40}=1$$

Solution
$$\frac{t}{30}+\frac{t}{40}=1$$
$$120\left(\frac{t}{30}+\frac{t}{40}\right)=120(1)$$
$$40t+30t=120$$
$$70t=120$$
$$t=\frac{12}{7}$$

Addressing 140 envelopes,
$$140 \cdot \frac{12}{7}=240$$

Converting from seconds to minutes,
$$\frac{240}{60}=40$$

It would take 40 min to address 140 envelopes if both clerks worked at the same time.

55. **Strategy**
Unknown time to fill the bathtub working together: t

	Rate	Time	Part
Faucets	$\frac{1}{10}$	t	$\frac{t}{10}$
Drain	$\frac{1}{15}$	t	$\frac{t}{15}$

The part of the task completed by the faucets minus the part of the task completed by the drain is 1.
$$\frac{t}{10}-\frac{t}{15}=1$$

Solution
$$\frac{t}{10}-\frac{t}{15}=1$$
$$30\left(\frac{t}{10}-\frac{t}{15}\right)=30(1)$$
$$3t-2t=30$$
$$t=30$$

In 30 min, the bathtub will start to overflow.

57. Strategy

Unknown time to fill the tank working together: t

	Rate	Time	Part
First inlet pipe	$\frac{1}{12}$	t	$\frac{t}{12}$
Second inlet pipe	$\frac{1}{20}$	t	$\frac{t}{20}$
Outlet pipe	$\frac{1}{10}$	t	$\frac{t}{10}$

The sum of the parts of the task completed by the inlet pipes minus the part of the task completed by the outlet pipe is 1.

$$\frac{t}{12}+\frac{t}{20}-\frac{t}{10}=1$$

Solution

$$\frac{t}{12}+\frac{t}{20}-\frac{t}{10}=1$$

$$60\left(\frac{t}{12}+\frac{t}{20}-\frac{t}{10}\right)=60(1)$$

$$5t+3t-6t=60$$

$$2t=60$$

$$t=30$$

When all three pipes are open, it will take 30 h to fill the tank.

59. The amount of time t that it takes them to weed a row of the garden together is less than both of their individual times, n and m. Therefore, $t < n$.

Objective 6.4.3 Exercises

61. a.

	Distance	÷	Rate	=	Time
Against the wind	1410	÷	$470-w$	=	$\frac{1410}{470-w}$
With the wind	1500	÷	$470+w$	=	$\frac{1500}{470+w}$

b. An equation that can be solved to find the rate of the wind: $\dfrac{1410}{470-w}=\dfrac{1500}{470+w}$

63. Strategy

Rate of the first skater: r
Rate of the second skater: $r-3$

	Distance	Rate	Time
First skater	15	r	$\frac{15}{r}$
Second skater	12	$r-3$	$\frac{12}{r-3}$

The time of the first skater is equal to the time of the second skater.

$$\frac{15}{r}=\frac{12}{r-3}$$

Solution

$$\frac{15}{r}=\frac{12}{r-3}$$

$$r(r-3)\frac{15}{r}=r(r-3)\frac{12}{r-3}$$

$$(r-3)(15)=(r)(12)$$

$$15r-45=12r$$

$$3r=45$$

$$r=15$$

$r-3=12$

The first skater's rate is 15 mph, and the second skater's rate is 12 mph.

65. Strategy

Rate of the freight train: r
Rate of the passenger train: $r+14$

	Distance	Rate	Time
Freight train	225	r	$\frac{225}{r}$
Passenger train	295	$r+14$	$\frac{295}{r+14}$

The time the freight train travels equals the time the passenger train travels.

$$\frac{225}{r}=\frac{295}{r+14}$$

Solution

$$\frac{225}{r}=\frac{295}{r+14}$$

$$r(r+14)\frac{225}{r}=r(r+14)\frac{295}{r+14}$$

$$(r+14)(225)=295r$$

$$225r+3150=295r$$

$$3150=70r$$

$$45=r$$

$r+14=45+14=59$

The rate of the freight train is 45 mph. The rate of the passenger train is 59 mph.

67. **Strategy**
Rate by foot: r
Rate by bicycle: $4r$

	Distance	Rate	Time
By foot	5	r	$\frac{5}{r}$
By bicycle	40	$4r$	$\frac{40}{4r}$

The total time spent walking and cycling was 5 h.
$$\frac{5}{r}+\frac{40}{4r}=5$$

Solution
$$\frac{5}{r}+\frac{40}{4r}=5$$
$$4r\left(\frac{5}{r}+\frac{40}{4r}\right)=4r(5)$$
$$20+40=20r$$
$$60=20r$$
$$3=r$$
$$4r=4(3)=12$$
The cyclist was riding 12 mph.

69. **Strategy**
Rate by foot: r
Rate by car: $12r$

	Distance	Rate	Time
By foot	4	r	$\frac{4}{r}$
By car	72	$12r$	$\frac{72}{12r}$

The total time spent riding and walking was 2.5 h.
$$\frac{4}{r}+\frac{72}{12r}=2.5$$

Solution
$$\frac{4}{r}+\frac{72}{12r}=2.5$$
$$12r\left(\frac{4}{r}+\frac{72}{12r}\right)=12r(2.5)$$
$$48+72=30r$$
$$120=30r$$
$$4=r$$
The motorist walks at the rate of 4 mph.

71. **Strategy**
Rate the executive can walk: r
Rate the executive walks using the moving sidewalk: $r+2$

	Distance	Rate	Time
Walking alone	360	r	$\frac{360}{r}$
Walking using moving sidewalk	480	$r+2$	$\frac{480}{r+2}$

It takes the executive the same time to walk 480 ft using the moving sidewalk as it takes to walk 360 ft without the moving sidewalk.
$$\frac{360}{r}=\frac{480}{r+2}$$

Solution
$$\frac{360}{r}=\frac{480}{r+2}$$
$$r(r+2)\frac{360}{r}=r(r+2)\frac{480}{r+2}$$
$$(r+2)(360)=(r)(480)$$
$$360r+720=480r$$
$$720=120r$$
$$6=r$$
The executive walks at a rate of 6 ft/s.

73. **Strategy**
Rate of the single-engine plane: r
Rate of the jet: $4r$

	Distance	Rate	Time
Single-engine plane	960	r	$\frac{960}{r}$
Jet	960	$4r$	$\frac{960}{4r}$

The time for the single-engine plane is 4 h more than the time for the jet.
$$\frac{960}{4r}+4=\frac{960}{r}$$

Solution
$$\frac{960}{4r}+4=\frac{960}{r}$$
$$4r\left(\frac{960}{4r}+4\right)=4r\left(\frac{960}{r}\right)$$
$$960+16r=3840$$
$$16r=2880$$
$$r=180$$
$$4r=4(180)=720$$
The rate of the single-engine plane is 180 mph. The rate of the jet is 720 mph.

Chapter 6: Rational Expressions

75. Strategy
Rate of the Gulf Stream: r
Rate sailing with the Gulf Stream: $28 + r$
Rate sailing against the Gulf Stream: $28 - r$

	Distance	Rate	Time
With Gulf Stream	170	$28 + r$	$\frac{170}{28+r}$
Against Gulf Stream	110	$28 - r$	$\frac{110}{28-r}$

It takes the same time to sail 170 mi with the Gulf Stream as it takes to sail 110 mi against the Gulf Stream.
$$\frac{170}{28+r} = \frac{110}{28-r}$$

Solution
$$\frac{170}{28+r} = \frac{110}{28-r}$$
$$(28-r)(28+r)\frac{170}{28+r} = (28-r)(28+r)\frac{110}{28-r}$$
$$(28-r)(170) = (28+r)(110)$$
$$4760 - 170r = 3080 + 110r$$
$$1680 = 280r$$
$$6 = r$$

The rate of the Gulf Stream is 6 mph.

77. Strategy
Rate of the current: r

	Distance	Rate	Time
With current	20	$7 + r$	$\frac{20}{7+r}$
Against current	8	$7 - r$	$\frac{8}{7-r}$

The time traveling with the current equals the time traveling against the current.
$$\frac{20}{7+r} = \frac{8}{7-r}$$

Solution
$$\frac{20}{7+r} = \frac{8}{7-r}$$
$$(7+r)(7-r)\frac{20}{7+r} = (7+r)(7-r)\frac{8}{7-r}$$
$$(7-r)20 = (7+r)8$$
$$140 - 20r = 56 + 8r$$
$$140 = 56 + 28r$$
$$84 = 28r$$
$$3 = r$$

The rate of the current is 3 mph.

Applying Concepts 6.4

79. Strategy
Numerator: n
Denominator: $n + 4$
Numerator increased by 3: $n + 3$
Denominator increased by 3: $(n + 4) + 3 = n + 7$
If the numerator and denominator of the fraction are increased by 3, the new fraction is $\frac{5}{6}$.

Solution
$$\frac{n+3}{n+7} = \frac{5}{6}$$
$$6(n+7)\left(\frac{n+3}{n+7}\right) = 6(n+7)\left(\frac{5}{6}\right)$$
$$6(n+3) = (n+7)(5)$$
$$6n + 18 = 5n + 35$$
$$n + 18 = 35$$
$$n = 17$$
$$n + 4 = 17 + 4 = 21$$

The original fraction is $\frac{17}{21}$.

81. Strategy
Time to complete the job: t

	Rate	Time	Part
First printer	$\frac{1}{24}$	t	$\frac{t}{24}$
Second printer	$\frac{1}{16}$	t	$\frac{t}{16}$
Third printer	$\frac{1}{12}$	t	$\frac{t}{12}$

The sum of the parts of the task completed by each printer is 1.

Solution
$$\frac{t}{24} + \frac{t}{16} + \frac{t}{12} = 1$$
$$48\left(\frac{t}{24} + \frac{t}{16} + \frac{t}{12}\right) = 48(1)$$
$$2t + 3t + 4t = 48$$
$$9t = 48$$
$$t = \frac{48}{9} = \frac{16}{3}$$

It would take $5\frac{1}{3}$ min to print the checks.

83. **Strategy**
Usual speed: r
Reduced speed: $r - 5$

	Distance	Rate	Time
Usual rate	165	r	$\frac{165}{r}$
Reduced speed	165	$r-5$	$\frac{165}{r-5}$

At the decreased speed, the drive takes 15 min more than usual. $\left(Note: 15 \text{ min} = \frac{1}{4} \text{ h}\right)$

Solution
$$\frac{165}{r} = \frac{165}{r-5} - \frac{1}{4}$$
$$4r(r-5)\left(\frac{165}{r}\right) = 4r(r-5)\left(\frac{165}{r-5} - \frac{1}{4}\right)$$
$$4(r-5)(165) = 4r(165) - r(r-5)$$
$$660(r-5) = 660r - r^2 + 5r$$
$$660r - 3300 = 665r - r^2$$
$$r^2 - 5r - 3300 = 0$$
$$(r-60)(r+55) = 0$$
$$r - 60 = 0 \quad r + 55 = 0$$
$$r = 60 \quad r = -55$$

Because the rate cannot be negative, −55 is not a solution. The bus usually travels at 60 mph.

85. **Strategy**
The rate running from front of parade to back of parade is the sum of the runner's rate and the rate of the parade (5 mph + 3 mph).
The rate running from back of parade to front of parade is the difference between the runner's rate and the rate of the parade (5 mph − 3 mph).

	Distance	Rate	Time
Back to front	1	2	$\frac{1}{2}$
Front to back	1	8	$\frac{1}{8}$

The sum of the time running from front to back and the time running from back to front is the total time t that the runner runs.
$$\frac{1}{2} + \frac{1}{8} = t$$

Solution
$$\frac{1}{2} + \frac{1}{8} = t$$
$$\frac{5}{8} = t$$

The total time is $\frac{5}{8}$ h.

87. A unit fraction is a fraction in which the numerator is 1. In Egyptian mathematics, all fractions $\left(\text{except } \frac{2}{3}\right)$ had to be decomposed into unit fractions. The Rhind papyrus gave a table that showed the decomposition of fractions with numerator 2 and an odd-number denominator.

Section 6.5

Concept Review 6.5

1. Never true
If x varies inversely as y, then when x is doubled, y is halved.

3. Never true
If a varies jointly as b and c, then $a = kbc$.

5. Never true
If the area of a triangle is held constant, then the length varies inversely as the width.

Objective 6.5.1 Exercises

1. A ratio is the quotient of two quantities that have the same units. A rate is the quotient of two quantities that have different units.

3. The scale on a map shows that a distance of 2 cm on the map represents an actual distance of 5 mi. This rate can be expressed as the quotient $\frac{5 \text{ mi}}{2 \text{ cm}}$, or as the quotient $\frac{2 \text{ cm}}{5 \text{ mi}}$.

5.
$$\frac{x+1}{10} = \frac{2}{5}$$
$$50 \cdot \frac{x+1}{10} = 50 \cdot \frac{2}{5}$$
$$5(x+1) = 10(2)$$
$$5x + 5 = 20$$
$$5x = 15$$
$$x = 3$$

7.
$$\frac{x}{4} = \frac{x-2}{8}$$
$$8 \cdot \frac{x}{4} = 8 \cdot \frac{x-2}{8}$$
$$2x = x - 2$$
$$x = -2$$

9.
$$\frac{8}{x-2} = \frac{4}{x+1}$$
$$(x+1)(x-2)\frac{8}{x-2} = (x+1)(x-2)\frac{4}{x+1}$$
$$(x+1)8 = (x-2)4$$
$$8x + 8 = 4x - 8$$
$$4x = -16$$
$$x = -4$$

11.
$$\frac{8}{3x-2} = \frac{2}{2x+1}$$
$$(3x-2)(2x+1)\frac{8}{3x-2} = (3x-2)(2x+1)\frac{2}{2x+1}$$
$$(2x+1)8 = (3x-2)2$$
$$16x+8 = 6x-4$$
$$10x = -12$$
$$x = -\frac{6}{5}$$

13. True

15. Strategy
To find the number of ducks in the preserve, write and solve a proportion using x to represent the number of ducks in the preserve.

Solution
$$\frac{60}{x} = \frac{3}{200}$$
$$200x \cdot \frac{60}{x} = \frac{3}{200} \cdot 200x$$
$$12{,}000 = 3x$$
$$4000 = x$$
There are 4000 ducks in the preserve.

17. Strategy
To find the amount of additional fruit punch, write and solve a proportion using x to represent the additional amount of fruit punch. Then $x + 2$ is the total amount of fruit punch.

Solution
$$\frac{2}{30} = \frac{x+2}{75}$$
$$\frac{2}{30} \cdot 150 = \frac{x+2}{75} \cdot 150$$
$$2 \cdot 5 = (x+2)2$$
$$10 = 2x+4$$
$$6 = 2x$$
$$3 = x$$
To serve 75 people, 3 additional gallons of fruit punch are necessary.

19. Strategy
To find the dimensions of the room, write and solve two proportions using L to represent the length of the room and W to represent the width of the room.

Solution
$$\frac{\frac{1}{4}}{1} = \frac{4\frac{1}{4}}{W} \qquad \frac{\frac{1}{4}}{1} = \frac{5\frac{1}{2}}{L}$$
$$\frac{\frac{1}{4}}{1} \cdot W = \frac{4\frac{1}{4}}{W} \cdot W \qquad \frac{\frac{1}{4}}{1} \cdot L = \frac{5\frac{1}{2}}{L} \cdot L$$
$$\frac{1}{4}W = 4\frac{1}{4} \qquad \frac{1}{4}L = 5\frac{1}{2}$$
$$W = 17 \qquad L = 22$$
The dimensions of the room are 17 ft by 22 ft.

21. Strategy
To find the number of miles to walk, write and solve a proportion using x to represent the number of miles.

Solution
$$\frac{4}{650} = \frac{x}{3500}$$
$$(3500)(650)\frac{4}{650} = (3500)(650)\frac{x}{3500}$$
$$4(3500) = 650x$$
$$14{,}000 = 650x$$
$$21.54 \approx x$$
A person would have to walk about 21.54 mi to lose one pound.

23. Strategy
To find the additional amount of medicine, write and solve a proportion using x to represent the additional amount of medicine. Then $x + 0.75$ is the total amount of medicine.

Solution
$$\frac{0.75}{120} = \frac{x+0.75}{200}$$
$$\frac{0.75}{120} \cdot 600 = \frac{x+0.75}{200} \cdot 600$$
$$0.75(5) = (x+0.75)3$$
$$3.75 = 3x + 2.25$$
$$1.5 = 3x$$
$$0.5 = x$$
An additional 0.5 oz of medicine is required.

25. Strategy
To find the additional amount of insecticide, write and solve a proportion using x to represent the additional amount of insecticide. Then $x + 6$ is the total amount of insecticide.

Solution
$$\frac{6}{15} = \frac{x+6}{100}$$
$$\frac{6}{15} \cdot 300 = \frac{x+6}{100} \cdot 300$$
$$6 \cdot 20 = (x+6)3$$
$$120 = 3x + 18$$
$$102 = 3x$$
$$34 = x$$
An additional 34 oz of insecticide are required.

27. Strategy
Let x represent the number of seconds needed to download a 5 megabyte file.

Solution
$$\frac{60}{4} = \frac{x}{5}$$
$$4(5)\frac{60}{4} = 4(5)\frac{x}{5}$$
$$5(60) = 4x$$
$$300 = 4x$$
$$75 = x$$

It will take 75 s to download a 5 megabyte file.

29. Strategy
To find the actual length of the whale, write and solve a proportion using x to represent the actual length of the whale. The length of the whale in the picture is 2 in.

Solution
$$\frac{1}{48} = \frac{2}{x}$$
$$\frac{1}{48}(48)x = \frac{2}{x}(48)x$$
$$x = 96$$

The whale is 96 ft long.

Objective 6.5.2 Exercises

31. A direct variation is a function that can be expressed as the equation $y = kx$, where k is a constant.

33. $250 \div 30 = 8.\overline{3}$ and $800 \div 96 = 8.\overline{3}$
The relationship between the circumference of a ball made out of this material and its weight is a direct variation.

35. Strategy
To find the pressure:
Write the basic direct variation equation, replace the variables by the given values, and solve for k.
Write the direct variation equation, replacing k by its value. Substitute 30 for d and solve for p.

Solution
$$p = kd$$
$$3.6 = k(8)$$
$$0.45 = k$$
$$p = 0.45d = 0.45(30) = 13.5$$

The pressure is 13.5 pounds per square inch.

37. Strategy
To find the profit:
Write the basic direct variation equation, replace the variables by the given values, and solve for k.
Write the direct variation equation, replacing k by its value. Substitute 300 for s and solve for P.

Solution
$$P = ks$$
$$2500 = k(20)$$
$$125 = k$$
$$P = 125s = 125(300) = 37{,}500$$

When the company sells 300 products, the profit is $37,500.

39. Strategy
To find the length of the person's face:
Write the basic direct variation equation, replace the variables by the given values, and solve for k.
Write the direct variation equation, replacing k by its value. Substitute 1.7 for x and solve for y.

Solution
$$y = kx \qquad y = 6x$$
$$9 = k(1.5) \qquad y = 6(1.7)$$
$$6 = k \qquad y = 10.2$$

The person's face is 10.2 in. long.

41. Strategy
To find the distance:
Write the basic direct variation equation, replace the variables by the given values, and solve for k.
Write the direct variation equation, replacing k by its value. Substitute 800 for H and solve for d.

Solution
$$d = k\sqrt{H}$$
$$19 = k\sqrt{500}$$
$$19 = k(22.36)$$
$$0.85 = k$$
$$d = 0.85\sqrt{H} = 0.85\sqrt{800} = 0.85(28.28) = 24.04$$

The horizon is 24.04 mi away from a point that is 800 ft high.

192 Chapter 6: Rational Expressions

43. Strategy
To find the distance:
Write the basic direct variation equation, replace the variables by the given values, and solve for k.
Write the direct variation equation, replacing k by its value. Substitute 4 for t and solve for s.

Solution
$s = kt^2$
$5 = k(1)^2$
$5 = k$
$s = 5t^2 = 5(4)^2 = 5(16) = 80$
In 4 s, the ball will roll 80 ft.

45. Strategy
To find the pressure:
Write the basic inverse variation equation, replace the variables by the given values, and solve for k.
Write the inverse variation equation, replacing k by its value. Substitute 150 for V and solve for P.

Solution
$P = \dfrac{k}{V}$
$25 = \dfrac{k}{400}$
$10{,}000 = k$
$P = \dfrac{10{,}000}{V} = \dfrac{10{,}000}{150} = 66\dfrac{2}{3}$
When the volume is 150 ft^3, the pressure is $66\dfrac{2}{3}$ pounds per square inch.

47. Strategy
To find the pressure:
Write the basic combined variation equation, replace the variables by the given values, and solve for k.
Write the combined variation equation, replacing k by its value. Substitute 60 for d and 1.2 for D, and solve for p.

Solution
$p = kdD$
$37.5 = k(100)(1.2)$
$37.5 = 120k$
$0.3125 = k$
$p = 0.3125dD = 0.3125(60)(1.2) = 22.5$
The pressure is 22.5 pounds per square inch.

49. Strategy
To find the repulsive force:
Write the basic inverse variation equation, replace the variables by the given values, and solve for k.
Write the inverse variation equation, replacing k by its value. Substitute 1.2 for d and solve for f.

Solution
$f = \dfrac{k}{d^2}$
$18 = \dfrac{k}{3^2}$
$18 = \dfrac{k}{9}$
$162 = k$
$f = \dfrac{162}{d^2} = \dfrac{162}{1.2^2} = \dfrac{162}{1.44} = 112.5$
The repulsive force is 112.5 lb when the distance is 1.2 in.

51. Strategy
To find the resistance:
Write the basic combined variation equation, replace the variables by the given values, and solve for k.
Write the combined variation equation, replacing k by its value. Substitute 50 for L and 0.02 for d, and solve for R.

Solution
$R = \dfrac{kL}{d^2}$
$9 = \dfrac{k(50)}{(0.05)^2}$
$9 = \dfrac{50k}{0.0025}$
$0.0225 = 50k$
$0.00045 = k$
$R = \dfrac{0.00045L}{d^2}$
$= \dfrac{0.00045(50)}{(0.02)^2}$
$= \dfrac{0.00045(50)}{0.0004}$
$= 56.25$
The resistance is 56.25 ohms.

53. Strategy

To find the wind force:
Write the basic combined variation equation, replace the variables by the given values, and solve for k. Write the combined variation equation, replacing k by its value. Substitute 10 for A and 60 for v, and solve for w.

Solution
$$w = kAv^2$$
$$45 = k(10)(30)^2$$
$$45 = k(10)(900)$$
$$45 = 9000k$$
$$0.005 = k$$
$$w = 0.005Av^2$$
$$= 0.005(10)(60)^2$$
$$= 0.005(10)(3600)$$
$$= 180$$

The wind force is 180 lb.

Applying Concepts 6.5

55. a.

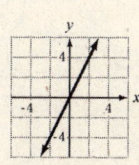

b. The graph represents a linear function.

57. a.

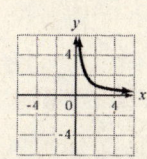

b. The graph is the graph of a function.

59. If y doubles, x is halved.

61. If a varies directly as b and inversely as c, then c varies <u>directly</u> as b and <u>inversely</u> as a.

63. If the width of a rectangle is held constant, the area of the rectangle varies <u>directly</u> as the length.

65. a. The average of a and b is $\dfrac{1}{h} = \dfrac{\frac{1}{a}+\frac{1}{b}}{2}$. Solving for h,

$$\frac{1}{h} = \frac{\frac{1}{a}+\frac{1}{b}}{2}$$

$$2h\frac{1}{h} = 2h\frac{\frac{1}{a}+\frac{1}{b}}{2}$$

$$2 = h\left(\frac{1}{a}+\frac{1}{b}\right)$$

$$2ab = abh\left(\frac{1}{a}+\frac{1}{b}\right)$$

$$2ab = bh + ah$$

$$2ab = h(b+a)$$

$$\frac{2ab}{a+b} = h$$

b. Using the expression, $\dfrac{2ab}{a+b}$

$$\frac{2(10)(15)}{10+15} = \frac{300}{25} = 12$$

Section 6.6

Concept Review 6.6

1. Always true

3. Never true

$$R_2 = \frac{R_1 R}{R_1 - R}$$

Objective 6.6.1 Exercises

1. Multiply each side of the equation by <u>n</u>.
$$P = \frac{R-C}{n}$$
$$(n)P = (n)\frac{R-C}{n}$$
$$nP = R - C \qquad \text{Subtract } \underline{R} \text{ from each side.}$$
$$nP - R = -C$$
$$R - nP = C$$

3.
$$P = 2L + 2W$$
$$P - 2L = 2W$$
$$\frac{P-2L}{2} = W$$

5.
$$S = C - rC$$
$$S = C(1-r)$$
$$\frac{S}{1-r} = C$$

7. $PV = nRT$
$\dfrac{PV}{nT} = R$

9. $F = \dfrac{Gm_1m_2}{r^2}$
$Fr^2 = Gm_1m_2$
$\dfrac{Fr^2}{Gm_1} = m_2$

11. $I = \dfrac{E}{R+r}$
$I(R+r) = E$
$IR + Ir = E$
$IR = E - Ir$
$R = \dfrac{E - Ir}{I}$

13. $A = \dfrac{1}{2}h(b_1 + b_2)$
$2A = h(b_1 + b_2)$
$2A = hb_1 + hb_2$
$2A - hb_1 = hb_2$
$\dfrac{2A - hb_1}{h} = b_2$

15. $\dfrac{1}{R} = \dfrac{1}{R_1} + \dfrac{1}{R_2}$
$RR_1R_2\left(\dfrac{1}{R}\right) = RR_1R_2\left(\dfrac{1}{R_1} + \dfrac{1}{R_2}\right)$
$R_1R_2 = RR_1R_2\left(\dfrac{1}{R_1}\right) + RR_1R_2\left(\dfrac{1}{R_2}\right)$
$R_1R_2 = RR_2 + RR_1$
$R_1R_2 - RR_2 = RR_1$
$R_2(R_1 - R) = RR_1$
$R_2 = \dfrac{RR_1}{R_1 - R}$

17. $a_n = a_1 + (n-1)d$
$a_n - a_1 = (n-1)d$
$\dfrac{a_n - a_1}{n-1} = d$

19. $S = 2WH + 2WL + 2LH$
$S - 2WL = 2WH + 2LH$
$S - 2WL = H(2W + 2L)$
$\dfrac{S - 2WL}{2W + 2L} = H$

21. $ax + by + c = 0$
$ax + by = -c$
$ax = -by - c$
$x = \dfrac{-by - c}{a}$

23. $ax + b = cx + d$
$ax + b - cx = d$
$ax - cx = d - b$
$x(a - c) = d - b$
$x = \dfrac{d - b}{a - c}$

25. $\dfrac{a}{x} = \dfrac{b}{c}$
$xc \cdot \dfrac{a}{x} = xc \cdot \dfrac{b}{c}$
$ac = xb$
$\dfrac{ac}{b} = x$

27. $\dfrac{1}{a} + \dfrac{1}{b} = \dfrac{1}{x}$
$abx\left(\dfrac{1}{a} + \dfrac{1}{b}\right) = abx\left(\dfrac{1}{x}\right)$
$bx + ax = ab$
$x(a + b) = ab$
$x = \dfrac{ab}{a + b}$

29. Yes, they are equivalent:
$-\dfrac{b}{a-c} = \dfrac{b}{-1(a-c)} = \dfrac{b}{-a+c} = \dfrac{b}{c-a}$

Applying Concepts 6.6

31. $\dfrac{x - y}{y} = \dfrac{x + 5}{2y}$
$2y\left(\dfrac{x-y}{y}\right) = 2y\left(\dfrac{x+5}{2y}\right)$
$2(x - y) = x + 5$
$2x - 2y = x + 5$
$x - 2y = 5$
$x = 2y + 5$
The solution is $2y + 5$.

33.

$$\frac{x}{x+y} = \frac{2x}{4y}$$

$$(4y)(x+y)\left(\frac{x}{x+y}\right) = (4y)(x+y)\left(\frac{2x}{4y}\right)$$

$$(4y)(x) = (x+y)(2x)$$

$$4xy = 2x^2 + 2xy$$

$$2xy = 2x^2$$

$$0 = 2x^2 - 2xy$$

$$0 = 2x(x-y)$$

$2x = 0 \quad\quad x - y = 0$
$x = 0 \quad\quad x = y$

The solutions are 0 and y.

35.

$$\frac{x-y}{2x} = \frac{x-3y}{5y}$$

$$(2x)(5y)\left(\frac{x-y}{2x}\right) = (2x)(5y)\left(\frac{x-3y}{5y}\right)$$

$$(5y)(x-y) = (2x)(x-3y)$$

$$5xy - 5y^2 = 2x^2 - 6xy$$

$$-5y^2 = 2x^2 - 11xy$$

$$0 = 2x^2 - 11xy + 5y^2$$

$$0 = (2x-y)(x-5y)$$

$2x - y = 0 \quad\quad x - 5y = 0$
$2x = y \quad\quad x = 5y$

$$x = \frac{y}{2}$$

The solutions are $\frac{y}{2}$ and $5y$.

37.

$$\frac{w_1}{w_2} = \frac{f_2 - f}{f - f_1}$$

$$w_2(f - f_1)\left(\frac{w_1}{w_2}\right) = \left(\frac{f_2 - f}{f - f_1}\right)w_2(f - f_1)$$

$$(f - f_1)w_1 = (f_2 - f)w_2$$

$$fw_1 - f_1 w_1 = f_2 w_2 - fw_2$$

$$fw_1 + fw_2 = f_2 w_2 + f_1 w_1$$

$$f(w_1 + w_2) = f_2 w_2 + f_1 w_1$$

$$f = \frac{f_2 w_2 + f_1 w_1}{w_1 + w_2}$$

CHAPTER 6 REVIEW EXERCISES

1. $P(x) = \dfrac{x}{x-3}$

$P(4) = \dfrac{4}{4-3} = \dfrac{4}{1}$

$P(4) = 4$

2. $P(x) = \dfrac{x^2 - 2}{3x^2 - 2x + 5}$

$P(-2) = \dfrac{(-2)^2 - 2}{3(-2)^2 - 2(-2) + 5} = \dfrac{4-2}{12+4+5} = \dfrac{2}{21}$

$P(-2) = \dfrac{2}{21}$

3. $g(x) = \dfrac{2x}{x-3}$

$x - 3 = 0$

$x = 3$

The domain is $\{x | x \ne 3\}$.

4. $f(x) = \dfrac{2x-7}{3x^2 + 3x - 18}$

$3x^2 + 3x - 18 = 0$

$3(x^2 + x - 6) = 0$

$3(x+3)(x-2) = 0$

$x + 3 = 0 \quad\quad x - 2 = 0$

$x = -3 \quad\quad x = 2$

The domain is $\{x | x \ne -3, x \ne 2\}$.

5. The domain must exclude values of x for which $3x^2 + 4 = 0$. This is not possible, because $3x^2 \ge 0$, and a positive number added to a number equal to or greater than zero cannot equal zero. Therefore, there are no real numbers that must be excluded from the domain of F.

The domain of $F(x)$ is $\{x | x \in \text{real numbers}\}$.

6. $\dfrac{6a^{5n} + 4a^{4n} - 2a^{3n}}{2a^{3n}} = 3a^{2n} + 2a^n - 1$

7. $\dfrac{16 - x^2}{x^3 - 2x^2 - 8x} = \dfrac{(4+x)(4-x)}{x(x^2 - 2x - 8)}$

$= \dfrac{(4+x)(4-x)}{x(x-4)(x+2)}$

$= \dfrac{(4+x)(-1)}{x(1)(x+2)}$

$= -\dfrac{x+4}{x(x+2)}$

8. $\dfrac{x^3 - 27}{x^2 - 9} = \dfrac{(x-3)(x^2 + 3x + 9)}{(x+3)(x-3)} = \dfrac{x^2 + 3x + 9}{x+3}$

9. $\dfrac{a^6 b^4 + a^4 b^6}{a^5 b^4 - a^4 b^4} \cdot \dfrac{a^2 - b^2}{a^4 - b^4}$

$= \dfrac{a^4 b^4 (a^2 + b^2)}{a^4 b^4 (a-1)} \cdot \dfrac{(a+b)(a-b)}{(a^2 + b^2)(a+b)(a-b)}$

$= \dfrac{a^4 b^4 (a^2 + b^2)(a+b)(a-b)}{a^4 b^4 (a-1)(a^2 + b^2)(a+b)(a-b)}$

$= \dfrac{1}{a-1}$

Chapter 6: Rational Expressions

10. $\dfrac{x^3-8}{x^3+2x^2+4x} \cdot \dfrac{x^3+2x^2}{x^2-4}$

$= \dfrac{(x-2)(x^2+2x+4)}{x(x^2+2x+4)} \cdot \dfrac{x^2(x+2)}{(x+2)(x-2)}$

$= \dfrac{(x-2)(x^2+2x+4) \cdot x^2(x+2)}{x(x^2+2x+4)(x+2)(x-2)}$

$= x$

11. $\dfrac{16-x^2}{6x-6} \cdot \dfrac{x^2+5x+6}{x^2-8x+16}$

$= \dfrac{(4+x)(4-x)}{6(x-1)} \cdot \dfrac{(x+3)(x+2)}{(x-4)(x-4)}$

$= \dfrac{(4+x)(4-x)(x+3)(x+2)}{6(x-1)(x-4)(x-4)}$

$= -\dfrac{(x+4)(x+3)(x+2)}{6(x-1)(x-4)}$

12. $\dfrac{x^{2n}-5x^n+4}{x^{2n}-2x^n-8} \div \dfrac{x^{2n}-4x^n+3}{x^{2n}+8x^n+12}$

$= \dfrac{x^{2n}-5x^n+4}{x^{2n}-2x^n-8} \cdot \dfrac{x^{2n}+8x^n+12}{x^{2n}-4x^n+3}$

$= \dfrac{(x^n-4)(x^n-1)}{(x^n-4)(x^n+2)} \cdot \dfrac{(x^n+6)(x^n+2)}{(x^n-3)(x^n-1)}$

$= \dfrac{(x^n-4)(x^n-1)(x^n+6)(x^n+2)}{(x^n-4)(x^n+2)(x^n-3)(x^n-1)}$

$= \dfrac{x^n+6}{x^n-3}$

13. $\dfrac{27x^3-8}{9x^3+6x^2+4x} \div \dfrac{9x^2-12x+4}{9x^2-4}$

$= \dfrac{27x^3-8}{9x^3+6x^2+4x} \cdot \dfrac{9x^2-4}{9x^2-12x+4}$

$= \dfrac{(3x-2)(9x^2+6x+4)}{x(9x^2+6x+4)} \cdot \dfrac{(3x+2)(3x-2)}{(3x-2)(3x-2)}$

$= \dfrac{(3x-2)(9x^2+6x+4)(3x+2)(3x-2)}{x(9x^2+6x+4)(3x-2)(3x-2)}$

$= \dfrac{3x+2}{x}$

14. $\dfrac{3-x}{x^2+3x+9} \div \dfrac{x^2-9}{x^3-27}$

$= \dfrac{3-x}{x^2+3x+9} \cdot \dfrac{x^3-27}{x^2-9}$

$= \dfrac{3-x}{x^2+3x+9} \cdot \dfrac{(x-3)(x^2+3x+9)}{(x+3)(x-3)}$

$= \dfrac{(3-x)(x-3)(x^2+3x+9)}{(x^2+3x+9)(x+3)(x-3)}$

$= \dfrac{3-x}{x+3}$

$= \dfrac{(-1)(x-3)}{x+3}$

$= -\dfrac{x-3}{x+3}$

15. The LCM is $24a^2b^4$.

$\dfrac{5}{3a^2b^3} + \dfrac{7}{8ab^4} = \dfrac{5}{3a^2b^3} \cdot \dfrac{8b}{8b} + \dfrac{7}{8ab^4} \cdot \dfrac{3a}{3a}$

$= \dfrac{40b+21a}{24a^2b^4}$

16. $\dfrac{3x^2+2}{x^2-4} - \dfrac{9x-x^2}{x^2-4} = \dfrac{3x^2+2-(9x-x^2)}{x^2-4}$

$= \dfrac{3x^2+2-9x+x^2}{x^2-4}$

$= \dfrac{4x^2-9x+2}{x^2-4}$

$= \dfrac{(4x-1)(x-2)}{(x+2)(x-2)}$

$= \dfrac{4x-1}{x+2}$

17. The LCM is $(3x+2)(3x-2)$.

$\dfrac{8}{9x^2-4} + \dfrac{5}{3x-2} - \dfrac{4}{3x+2} = \dfrac{8}{(3x+2)(3x-2)} + \dfrac{5}{3x-2} - \dfrac{4}{3x+2}$

$= \dfrac{8}{(3x+2)(3x-2)} + \dfrac{5}{3x-2} \cdot \dfrac{3x+2}{3x+2} - \dfrac{4}{3x+2} \cdot \dfrac{3x-2}{3x-2}$

$= \dfrac{8+5(3x+2)-4(3x-2)}{(3x+2)(3x-2)}$

$= \dfrac{8+15x+10-12x+8}{(3x+2)(3x-2)}$

$= \dfrac{3x+26}{(3x+2)(3x-2)}$

18. $3x^2 - 7x + 2 = (3x-1)(x-2)$
The LCM is $(3x-1)(x-2)$.

$$\frac{6x}{3x^2-7x+2} - \frac{2}{3x-1} + \frac{3x}{x-2} = \frac{6x}{(3x-1)(x-2)} - \frac{2}{3x-1} \cdot \frac{x-2}{x-2} + \frac{3x}{x-2} \cdot \frac{3x-1}{3x-1}$$
$$= \frac{6x - 2(x-2) + 3x(3x-1)}{(3x-1)(x-2)}$$
$$= \frac{6x - 2x + 4 + 9x^2 - 3x}{(3x-1)(x-2)}$$
$$= \frac{9x^2 + x + 4}{(3x-1)(x-2)}$$

19. The LCM is $(x-3)(x+2)$.

$$\frac{x}{x-3} - 4 - \frac{2x-5}{x+2} = \frac{x}{x-3} \cdot \frac{x+2}{x+2} - 4 \cdot \frac{(x-3)(x+2)}{(x-3)(x+2)} - \frac{2x-5}{x+2} \cdot \frac{x-3}{x-3}$$
$$= \frac{x^2 + 2x - 4(x^2 - x - 6) - (2x^2 - 11x + 15)}{(x-3)(x+2)}$$
$$= \frac{x^2 + 2x - 4x^2 + 4x + 24 - 2x^2 + 11x - 15}{(x-3)(x+2)}$$
$$= \frac{-5x^2 + 17x + 9}{(x-3)(x+2)}$$
$$= \frac{-(5x^2 - 17x - 9)}{(x-3)(x+2)}$$
$$= -\frac{5x^2 - 17x - 9}{(x-3)(x+2)}$$

20. The LCM is $x - 1$.

$$\frac{x - 6 + \frac{6}{x-1}}{x + 3 - \frac{12}{x-1}} = \frac{x - 6 + \frac{6}{x-1}}{x + 3 - \frac{12}{x-1}} \cdot \frac{x-1}{x-1}$$
$$= \frac{(x-6)(x-1) + 6}{(x+3)(x-1) - 12}$$
$$= \frac{x^2 - 7x + 6 + 6}{x^2 + 2x - 3 - 12}$$
$$= \frac{x^2 - 7x + 12}{x^2 + 2x - 15}$$
$$= \frac{(x-3)(x-4)}{(x+5)(x-3)}$$
$$= \frac{x-4}{x+5}$$

21. The LCM is $x - 4$.

$$\frac{x + \frac{3}{x-4}}{3 + \frac{x}{x-4}} = \frac{x + \frac{3}{x-4}}{3 + \frac{x}{x-4}} \cdot \frac{x-4}{x-4}$$
$$= \frac{x(x-4) + 3}{3(x-4) + x}$$
$$= \frac{x^2 - 4x + 3}{3x - 12 + x}$$
$$= \frac{(x-3)(x-1)}{4x - 12}$$
$$= \frac{(x-3)(x-1)}{4(x-3)}$$
$$= \frac{x-1}{4}$$

22.
$$\frac{5x}{2x-3} + 4 = \frac{3}{2x-3}$$
$$(2x-3)\left(\frac{5x}{2x-3} + 4\right) = \frac{3}{2x-3}(2x-3)$$
$$5x + 4(2x-3) = 3$$
$$5x + 8x - 12 = 3$$
$$13x - 12 = 3$$
$$13x = 15$$
$$x = \frac{15}{13}$$

The solution is $\frac{15}{13}$.

23.
$$\frac{x}{x-3} = \frac{2x+5}{x+1}$$
$$(x-3)(x+1)\left(\frac{x}{x-3}\right) = \left(\frac{2x+5}{x+1}\right)(x-3)(x+1)$$
$$(x+1)x = (2x+5)(x-3)$$
$$x^2 + x = 2x^2 - x - 15$$
$$0 = x^2 - 2x - 15$$
$$0 = (x-5)(x+3)$$
$$x - 5 = 0 \quad x + 3 = 0$$
$$x = 5 \quad x = -3$$

The solutions are 5 and −3.

24.
$$\frac{6}{x-3} - \frac{1}{x+3} = \frac{51}{x^2-9}$$
$$\frac{6}{x-3} - \frac{1}{x+3} = \frac{51}{(x+3)(x-3)}$$
$$(x+3)(x-3)\left(\frac{6}{x-3} - \frac{1}{x+3}\right) = \frac{51}{(x+3)(x-3)} \cdot (x+3)(x-3)$$
$$6(x+3) - 1(x-3) = 51$$
$$6x + 18 - x + 3 = 51$$
$$5x + 21 = 51$$
$$5x = 30$$
$$x = 6$$

The solution is 6.

25.
$$\frac{30}{x^2+5x+4} + \frac{10}{x+4} = \frac{4}{x+1}$$
$$\frac{30}{(x+4)(x+1)} + \frac{10}{x+4} = \frac{4}{x+1}$$
$$(x+4)(x+1)\left(\frac{30}{(x+4)(x+1)} + \frac{10}{x+4}\right) = \frac{4}{x+1} \cdot (x+4)(x+1)$$
$$30 + 10(x+1) = 4(x+4)$$
$$30 + 10x + 10 = 4x + 16$$
$$10x + 40 = 4x + 16$$
$$6x + 40 = 16$$
$$6x = -24$$
$$x = -4$$

−4 does not check as a solution. The equation has no solution.

26.
$$I = \frac{1}{R}V$$
$$R \cdot I = R \cdot \frac{1}{R}V$$
$$RI = V$$
$$\frac{RI}{I} = \frac{V}{I}$$
$$R = \frac{V}{I}$$

27.
$$Q = \frac{N-S}{N}$$
$$Q \cdot N = \frac{N-S}{N} \cdot N$$
$$QN = N - S$$
$$QN - N = -S$$
$$N(Q-1) = -S$$
$$N = \frac{-S}{Q-1}$$
$$N = \frac{S}{1-Q}$$

28.
$$S = \frac{a}{1-r}$$
$$S(1-r) = \frac{a}{1-r}(1-r)$$
$$S - Sr = a$$
$$-Sr = a - S$$
$$r = \frac{a-S}{-S}$$
$$r = \frac{S-a}{S}$$

29. Strategy
To find the number of tanks of fuel, write and solve a proportion using x to represent the number of tanks of fuel.

Solution
$$\frac{4}{1800} = \frac{x}{3000}$$
$$9000 \cdot \frac{4}{1800} = \frac{x}{3000} \cdot 9000$$
$$5 \cdot 4 = 3x$$
$$20 = 3x$$
$$x = \frac{20}{3} = 6\frac{2}{3}$$

The number of tanks of fuel is $6\frac{2}{3}$.

30. Strategy
To find the number of miles represented, write and solve a proportion using x to represent the number of miles.

Solution
$$\frac{2.5}{10} = \frac{12}{x}$$
$$\frac{2.5}{10} \cdot 10x = \frac{12}{x} \cdot 10x$$
$$2.5x = 120$$
$$x = 48$$

The number of miles is 48.

31. Strategy
Unknown time for apprentice, working alone, to install fan: t

	Rate	Time	Part
Electrician	$\frac{1}{65}$	40	$\frac{40}{65}$
Apprentice	$\frac{1}{t}$	40	$\frac{40}{t}$

The sum of the part of the task completed by the electrician and the part of the task completed by the apprentice is 1.
$$\frac{40}{65} + \frac{40}{t} = 1$$

Solution
$$\frac{40}{65} + \frac{40}{t} = 1$$
$$65t\left(\frac{40}{65} + \frac{40}{t}\right) = 1 \cdot 65t$$
$$40t + 2600 = 65t$$
$$2600 = 25t$$
$$104 = t$$

The apprentice would take 104 min to complete the job alone.

32. Strategy
Unknown time to empty a full tub when both pipes are open: t

	Rate	Time	Part
Inlet pipe	$\frac{1}{24}$	t	$\frac{t}{24}$
Drain pipe	$\frac{1}{15}$	t	$\frac{t}{15}$

The difference between the part of the job done by the drain and the part of the job done by the inlet pipe is 1.

Solution
$$\frac{t}{15} - \frac{t}{24} = 1$$
$$120\left(\frac{t}{15} - \frac{t}{24}\right) = 1(120)$$
$$8t - 5t = 120$$
$$3t = 120$$
$$t = 40$$

It takes 40 min to empty the tub.

33. Strategy
Unknown time for 3 students to paint dormitory room working together: t

	Rate	Time	Part
1st painter	$\frac{1}{8}$	t	$\frac{t}{8}$
2nd painter	$\frac{1}{16}$	t	$\frac{t}{16}$
3rd painter	$\frac{1}{16}$	t	$\frac{t}{16}$

The sum of the parts of the task completed by 3 painters is 1.

Solution
$$\frac{t}{8} + \frac{t}{16} + \frac{t}{16} = 1$$
$$16\left(\frac{t}{8} + \frac{t}{16} + \frac{t}{16}\right) = 1(16)$$
$$2t + t + t = 16$$
$$4t = 16$$
$$t = 4$$

It takes 4 h for the 3 painters to paint the dormitory room

Chapter 6: Rational Expressions

34. Strategy
Rate of the current: c

	Distance	Rate	Time
With the current	60	$10+c$	$\frac{60}{10+c}$
Against the current	40	$10-c$	$\frac{40}{10-c}$

The time traveling with the current is equal to the time traveling against the current.
$$\frac{60}{10+c} = \frac{40}{10-c}$$

Solution
$$\frac{60}{10+c} = \frac{40}{10-c}$$
$$(10+c)(10-c)\frac{60}{10+c} = (10+c)(10-c)\frac{40}{10-c}$$
$$(10-c)60 = 40(10+c)$$
$$600-60c = 400+40c$$
$$200 = 100c$$
$$2 = c$$
The rate of the current is 2 mph.

35. Strategy
Rate of cyclist: r
Rate of bus: $3r$

	Distance	Rate	Time
Bus	90	$3r$	$\frac{90}{3r}$
Cyclist	90	r	$\frac{90}{r}$

The cyclist arrives 4 h after the bus.
$$\frac{90}{3r} + 4 = \frac{90}{r}$$

Solution
$$\frac{90}{3r} + 4 = \frac{90}{r}$$
$$3r\left(\frac{90}{3r} + 4\right) = \frac{90}{r} \cdot 3r$$
$$90 + 12r = 270$$
$$12r = 180$$
$$r = 15$$
$3r = 3(15) = 45$
The rate of the bus is 45 mph.

36. Strategy
Rate of the car: r
Rate of the tractor: $r-15$

	Distance	Rate	Time
Car	15	r	$\frac{15}{r}$
Tractor	10	$r-15$	$\frac{10}{r-15}$

The time that the car travels equals the time that the tractor travels.
$$\frac{15}{r} = \frac{10}{r-15}$$

Solution
$$\frac{15}{r} = \frac{10}{r-15}$$
$$r(r-15)\left(\frac{15}{r}\right) = \left(\frac{10}{r-15}\right)r(r-15)$$
$$15(r-15) = 10r$$
$$15r - 225 = 10r$$
$$-225 = -5r$$
$$45 = r$$
$r - 15 = 45 - 15 = 30$
The rate of the tractor is 30 mph.

37. Strategy
To find the pressure:
Write the basic joint variation equation, replace the variables with the given values, and solve for k. Write the joint variation equation, replacing k with its value. Substitute 22 ft² for A and 20 mph for v.

Solution
$$P = kAv^2$$
$$10 = k \cdot 22 \cdot 10^2$$
$$\frac{1}{220} = k$$
$$P = \frac{1}{220}Av^2$$
$$P = \frac{1}{220} \cdot 22(20^2) = 40$$
The pressure is 40 lb.

38. Strategy
To find the illumination:
Write the basic inverse variation equation, replace the variables with the given values, and solve for k. Write the inverse variation equation, replacing k with its value. Substitute d for 2 and solve for T.

Solution
$$I = \frac{k}{d^2}$$
$$12 = \frac{k}{(10)^2}$$
$$1200 = k$$
$$I = \frac{1200}{d^2} = \frac{1200}{(2)^2} = 300$$

The illumination is 300 lumens.

39. Strategy
To find the resistance:
Write the basic combined variation equation, replace the variables with the given values, and solve for k. Write the combined variation equation, replacing k with its value. Substitute 8000 for l and $\frac{1}{2}$ for d, and solve for r.

Solution
$$r = \frac{kl}{d^2}$$
$$3.2 = \frac{k \cdot 16{,}000}{\left(\frac{1}{4}\right)^2}$$
$$0.0000125 = k$$
$$r = \frac{0.0000125 l}{d^2} = \frac{0.0000125(8000)}{\left(\frac{1}{2}\right)^2} = 0.4$$

The resistance is 0.4 ohm.

CHAPTER 6 TEST

1.
$$\frac{v^3 - 4v}{2v^2 - 5v + 2} = \frac{v(v^2 - 4)}{(2v-1)(v-2)}$$
$$= \frac{v(v+2)(v-2)}{(2v-1)(v-2)}$$
$$= \frac{v(v+2)}{2v-1}$$

2.
$$\frac{2a^2 - 8a + 8}{4 + 4a - 3a^2} = \frac{2(a^2 - 4a + 4)}{(2-a)(2+3a)}$$
$$= \frac{2(a-2)(a-2)}{(2-a)(2+3a)}$$
$$= -\frac{2(a-2)}{3a+2}$$

3.
$$\frac{3x^2 - 12}{5x - 15} \cdot \frac{2x^2 - 18}{x^2 + 5x + 6}$$
$$= \frac{3(x+2)(x-2)}{5(x-3)} \cdot \frac{2(x+3)(x-3)}{(x+3)(x+2)}$$
$$= \frac{3(x+2)(x-2)2(x+3)(x-3)}{5(x-3)(x+3)(x+2)}$$
$$= \frac{6(x-2)}{5}$$

4. $P(x) = \dfrac{3 - x^2}{x^3 - 2x^2 + 4}$

$P(-1) = \dfrac{3 - (-1)^2}{(-1)^3 - 2(-1)^2 + 4} = \dfrac{3-1}{-1-2+4} = \dfrac{2}{1}$

$P(-1) = 2$

5.
$$\frac{2x^2 - x - 3}{2x^2 - 5x + 3} \div \frac{3x^2 - x - 4}{x^2 - 1}$$
$$= \frac{2x^2 - x - 3}{2x^2 - 5x + 3} \cdot \frac{x^2 - 1}{3x^2 - x - 4}$$
$$= \frac{(2x-3)(x+1)}{(2x-3)(x-1)} \cdot \frac{(x+1)(x-1)}{(3x-4)(x+1)}$$
$$= \frac{(2x-3)(x+1)(x+1)(x-1)}{(2x-3)(x-1)(3x-4)(x+1)}$$
$$= \frac{x+1}{3x-4}$$

6.
$$\frac{x^{2n} - x^n - 2}{x^{2n} + x^n} \cdot \frac{x^{2n} - x^n}{x^{2n} - 4}$$
$$= \frac{(x^n - 2)(x^n + 1)}{x^n(x^n + 1)} \cdot \frac{x^n(x^n - 1)}{(x^n + 2)(x^n - 2)}$$
$$= \frac{(x^n - 2)(x^n + 1)x^n(x^n - 1)}{x^n(x^n + 1)(x^n + 2)(x^n - 2)}$$
$$= \frac{x^n - 1}{x^n + 2}$$

7. The LCM is $2x^2 y^2$.

$$\frac{2}{x^2} + \frac{3}{y^2} - \frac{5}{2xy}$$
$$= \frac{2}{x^2} \cdot \frac{2y^2}{2y^2} + \frac{3}{y^2} \cdot \frac{2x^2}{2x^2} - \frac{5}{2xy} \cdot \frac{xy}{xy}$$
$$= \frac{4y^2}{2x^2 y^2} + \frac{6x^2}{2x^2 y^2} - \frac{5xy}{x^2 y^2}$$
$$= \frac{4y^2 + 6x^2 - 5xy}{2x^2 y^2}$$

Chapter 6: Rational Expressions

8. The LCM is $(x-2)(x+2)$.

$$\frac{3x}{x-2} - 3 + \frac{4}{x+2} = \frac{3x}{x-2} \cdot \frac{x+2}{x+2} - 3\frac{(x-2)(x+2)}{(x-2)(x+2)} + \frac{4}{x+2} \cdot \frac{x-2}{x-2}$$

$$= \frac{3x(x+2) - 3(x-2)(x+2) + 4(x-2)}{(x+2)(x-2)}$$

$$= \frac{3x^2 + 6x - 3(x^2 - 4) + 4x - 8}{(x+2)(x-2)}$$

$$= \frac{3x^2 + 6x - 3x^2 + 12 + 4x - 8}{(x+2)(x-2)}$$

$$= \frac{10x + 4}{(x+2)(x-2)}$$

$$= \frac{2(5x+2)}{(x+2)(x-2)}$$

9. $f(x) = \dfrac{3x^2 - x + 1}{x^2 - 9}$

$x^2 - 9 = 0$

$(x+3)(x-3) = 0$

$x + 3 = 0 \quad x - 3 = 0$

$x = -3 \quad\quad x = 3$

The domain is $\{x | x \neq -3, 3\}$.

10. $x^2 + 3x - 4 = (x+4)(x-1)$

$x^2 - 1 = (x+1)(x-1)$

The LCM is $(x+4)(x-1)(x+1)$.

$$\frac{x+2}{x^2+3x-4} - \frac{2x}{x^2-1} = \frac{x+2}{(x+4)(x-1)} \cdot \frac{x+1}{x+1} - \frac{2x}{(x+1)(x-1)} \cdot \frac{x+4}{x+4}$$

$$= \frac{(x+2)(x+1) - 2x(x+4)}{(x+4)(x-1)(x+1)}$$

$$= \frac{x^2 + 3x + 2 - 2x^2 - 8x}{(x+4)(x-1)(x+1)}$$

$$= \frac{-x^2 - 5x + 2}{(x+1)(x+4)(x-1)}$$

$$= -\frac{x^2 + 5x - 2}{(x+1)(x+4)(x-1)}$$

11.

$$\frac{1 - \frac{1}{x} - \frac{12}{x^2}}{1 + \frac{6}{x} + \frac{9}{x^2}} = \frac{1 - \frac{1}{x} - \frac{12}{x^2}}{1 + \frac{6}{x} + \frac{9}{x^2}} \cdot \frac{x^2}{x^2}$$

$$= \frac{x^2 - x - 12}{x^2 + 6x + 9}$$

$$= \frac{(x-4)(x+3)}{(x+3)(x+3)}$$

$$= \frac{x-4}{x+3}$$

12.

$$\frac{1 - \frac{1}{x+2}}{1 - \frac{3}{x+4}} = \frac{1 - \frac{1}{x+2}}{1 - \frac{3}{x+4}} \cdot \frac{(x+2)(x+4)}{(x+2)(x+4)}$$

$$= \frac{(x+2)(x+4) - (x+4)}{(x+2)(x+4) - 3(x+2)}$$

$$= \frac{x^2 + 6x + 8 - x - 4}{x^2 + 6x + 8 - 3x - 6}$$

$$= \frac{x^2 + 5x + 4}{x^2 + 3x + 2}$$

$$= \frac{(x+4)(x+1)}{(x+2)(x+1)}$$

$$= \frac{x+4}{x+2}$$

13. $$\frac{3}{x+1} = \frac{2}{x}$$
$$\frac{3}{x+1} \cdot x(x+1) = \frac{2}{x} \cdot x(x+1)$$
$$3x = 2(x+1)$$
$$3x = 2x + 2$$
$$x = 2$$
The solution is 2.

14. $$\frac{4x}{2x-1} = 2 - \frac{1}{2x-1}$$
$$(2x-1)\frac{4x}{2x-1} = \left(2 - \frac{1}{2x-1}\right)(2x-1)$$
$$4x = 2(2x-1) - 1$$
$$4x = 4x - 2 - 1$$
$$4x = 4x - 3$$
$$0 = -3$$
There is no solution.

15. $$ax = bx + c$$
$$ax - bx = c$$
$$x(a-b) = c$$
$$x = \frac{c}{a-b}$$

16. **Strategy**
Unknown time to empty the tank with both pipes open: t

	Rate	Time	Part
Inlet pipe	$\frac{1}{48}$	t	$\frac{t}{48}$
Outlet pipe	$\frac{1}{30}$	t	$\frac{t}{30}$

The difference between the part of the task completed by the outlet pipe and the part of the task completed by the inlet pipe is 1.
$$\frac{t}{30} - \frac{t}{48} = 1$$

Solution
$$\frac{t}{30} - \frac{t}{48} = 1$$
$$240\left(\frac{t}{30} - \frac{t}{48}\right) = 240(1)$$
$$8t - 5t = 240$$
$$3t = 240$$
$$t = 80$$
It will take 80 min to empty the full tank with both pipes open.

17. **Strategy**
To find the number of rolls of wallpaper, write and solve a proportion using x to represent the number of rolls.

Solution
$$\frac{2}{45} = \frac{x}{315}$$
$$\left(\frac{2}{45}\right)315 = \left(\frac{x}{315}\right)315$$
$$14 = x$$
The office requires 14 rolls of wallpaper.

18. **Strategy**
Unknown time for both landscapers working together: t

	Rate	Time	Part
First landscaper	$\frac{1}{30}$	t	$\frac{t}{30}$
Second landscaper	$\frac{1}{15}$	t	$\frac{t}{15}$

The sum of the part of the task completed by the first landscaper and the part of the task completed by the second landscaper is 1.
$$\frac{t}{30} + \frac{t}{15} = 1$$

Solution
$$\frac{t}{30} + \frac{t}{15} = 1$$
$$30\left(\frac{t}{30} + \frac{t}{15}\right) = 1(30)$$
$$t + 2t = 30$$
$$3t = 30$$
$$t = 10$$
Working together, the landscapers can complete the task in 10 min.

19. Strategy
Rate of hiker: r
Rate of cyclist: $r + 7$

	Distance	Rate	Time
Hiker	6	r	$\frac{6}{r}$
Cyclist	20	$r+7$	$\frac{20}{r+7}$

The time the hiker hikes equals the time the cyclist cycles.
$$\frac{6}{r} = \frac{20}{r+7}$$

Solution
$$\frac{6}{r} = \frac{20}{r+7}$$
$$\frac{6}{r}[r(r+7)] = \frac{20}{r+7}[r(r+7)]$$
$$6(r+7) = 20r$$
$$6r + 42 = 20r$$
$$42 = 14r$$
$$3 = r$$
$r + 7 = 3 + 7 = 10$
The rate of the cyclist is 10 mph.

20. Strategy
To find the stopping distance:
Write the general direct variation equation, replace the variables by the given values, and solve for k. Write the direct variation equation, replacing k by its value. Substitute 30 for v and solve for s.

Solution
$$s = kv^2$$
$$170 = k(50)^2$$
$$170 = k(2500)$$
$$0.068 = k$$

$s = kv^2$
$= 0.068v^2$
$= 0.068(30)^2$
$= 0.068(900)$
$= 61.2$

The stopping distance for a car traveling at 30 mph is 61.2 ft.

CUMULATIVE REVIEW EXERCISES

1. $8 - 4[-3 - (-2)]^2 \div 5$
$= 8 - 4[-3 + 2]^2 \div 5$
$= 8 - 4[-1]^2 \div 5$
$= 8 - 4(1) \div 5$
$= 8 - 4 \div 5$
$= 8 - \frac{4}{5}$
$= \frac{36}{5}$

2. $\frac{2x-3}{6} - \frac{x}{9} = \frac{x-4}{3}$
$18\left(\frac{2x-3}{6} - \frac{x}{9}\right) = \left(\frac{x-4}{3}\right)18$
$3(2x-3) - 2x = 6(x-4)$
$6x - 9 - 2x = 6x - 24$
$4x - 9 = 6x - 24$
$-2x = -15$
$x = \frac{15}{2}$

3. $5 - |x-4| = 2$
$-|x-4| = -3$
$|x-4| = 3$
$x - 4 = 3 \quad x - 4 = -3$
$x = 7 \quad x = 1$
The solutions are 7 and 1.

4. $\frac{x}{x-3}$
$x - 3 = 0$
$x = 3$
The domain is $\{x | x \neq 3\}$.

5. $P(x) = \frac{x-1}{2x-3}$
$P(-2) = \frac{-2-1}{2(-2)-3} = \frac{-3}{-4-3} = \frac{-3}{-7}$
$P(-2) = \frac{3}{7}$

6. $0.000000035 = 3.5 \times 10^{-8}$

7. $\frac{x}{x+1} = 1$
$(x+1)\frac{x}{x+1} = 1(x+1)$
$x = x + 1$
$0 \neq 1$
There is no solution.

8. $(9x-1)(x-4) = 0$
   $9x-1 = 0 \quad x-4 = 0$
   $9x = 1 \quad\quad x = 4$
   $x = \dfrac{1}{9}$

   The solutions are $\dfrac{1}{9}$ and 4.

9. $\dfrac{(2a^{-2}b^3)}{(4a)^{-1}} = 2a^{-2}b^3 \cdot 4a$

   $= \dfrac{2b^3 \cdot 4a}{a^2}$

   $= \dfrac{8b^3 a}{a^2}$

   $= \dfrac{8b^3}{a}$

10. $x - 3(1-2x) \geq 1 - 4(2-2x)$
   $x - 3 + 6x \geq 1 - 8 + 8x$
   $7x - 3 \geq 8x - 7$
   $-x - 3 \geq -7$
   $-x \geq -4$
   $(-1)(-x) \leq (-1)(-4)$
   $x \leq 4$
   $\{x \mid x \leq 4\}$

11. $(2a^2 - 3a + 1)(-2a^2) = -4a^4 + 6a^3 - 2a^2$

12. Let $x^n = u$.
   $2x^{2n} + 3x^n - 2 = 2u^2 + 3u - 2$
   $= (2u-1)(u+2)$
   $= (2x^n - 1)(x^n + 2)$

13. $x^3 y^3 - 27 = (xy)^3 - (3)^3$
   $= (xy - 3)(x^2 y^2 + 3xy + 9)$

14. $\dfrac{x^4 + x^3 y - 6x^2 y^2}{x^3 - 2x^2 y} = \dfrac{x^2(x^2 + xy - 6y^2)}{x^2(x - 2y)}$

   $= \dfrac{x^2(x+3y)(x-2y)}{x^2(x-2y)}$

   $= x + 3y$

15. $3x - 2y = 6$
   $-2y = -3x + 6$
   $y = \dfrac{3}{2}x - 3$
   $m = \dfrac{3}{2}$
   $y - y_1 = m(x - x_1)$
   $y - (-1) = \dfrac{3}{2}[x - (-2)]$
   $y + 1 = \dfrac{3}{2}(x+2)$
   $y + 1 = \dfrac{3}{2}x + 3$
   $y = \dfrac{3}{2}x + 2$

   The equation of the line is $y = \dfrac{3}{2}x + 2$.

16. $(x - x^{-1})^{-1} = \dfrac{1}{x - x^{-1}}$

   $= \dfrac{1}{x - \frac{1}{x}}$

   $= \dfrac{1}{x - \frac{1}{x}} \cdot \dfrac{x}{x}$

   $= \dfrac{x}{x^2 - 1}$

17. $-3x + 5y = -15$
   x-intercept: $(5, 0)$
   y-intercept: $(0, -3)$

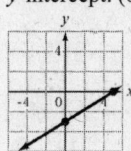

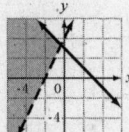

18. $x + y \leq 3 \quad\quad -2x + y > 4$
   $y \leq 3 - x \quad\quad y > 4 + 2x$

19. $\dfrac{4x^3 + 2x^2 - 10x + 1}{x - 2}$

$$\begin{array}{r|rrrr} 2 & 4 & 2 & -10 & 1 \\ & & 8 & 20 & 20 \\ \hline & 4 & 10 & 10 & 21 \end{array}$$

The simplified form is $4x^2 + 10x + 10 + \dfrac{21}{x-2}$.

20.
$$\frac{16x^2 - 9y^2}{16x^2y - 12xy^2} \div \frac{4x^2 - xy - 3y^2}{12x^2y^2}$$
$$= \frac{16x^2 - 9y^2}{16x^2y - 12xy^2} \cdot \frac{12x^2y^2}{4x^2 - xy - 3y^2}$$
$$= \frac{(4x - 3y)(4x + 3y)}{4xy(4x - 3y)} \cdot \frac{12x^2y^2}{(4x + 3y)(x - y)}$$
$$= \frac{(4x - 3y)(4x + 3y) \cdot 12x^2y^2}{4xy(4x - 3y)(4x + 3y)(x - y)}$$
$$= \frac{3xy}{x - y}$$

21. The domain must exclude values of x for which $3x^2 + 5 = 0$. This is not possible, because $3x^2 \geq 0$, and a positive number added to a number equal to or greater than zero cannot equal zero. Therefore, there are no real numbers that must be excluded from the domain of f.

The domain of $f(x)$ is $\{x | x \in \text{real numbers}\}$.

22. $3x^2 - x - 2 = (3x + 2)(x - 1)$
$x^2 - 1 = (x + 1)(x - 1)$
The LCM is $(3x + 2)(x + 1)(x - 1)$.
$$\frac{5x}{3x^2 - x - 2} - \frac{2x}{x^2 - 1}$$
$$= \frac{5x}{(3x + 2)(x - 1)} \cdot \frac{x + 1}{x + 1} - \frac{2x}{(x + 1)(x - 1)} \cdot \frac{3x + 2}{3x + 2}$$
$$= \frac{5x(x + 1) - 2x(3x + 2)}{(3x + 2)(x - 1)(x + 1)}$$
$$= \frac{5x^2 + 5x - 6x^2 - 4x}{(3x + 2)(x - 1)(x + 1)}$$
$$= \frac{-x^2 + x}{(3x + 2)(x - 1)(x + 1)}$$
$$= \frac{-x(x - 1)}{(3x + 2)(x - 1)(x + 1)}$$
$$= -\frac{x}{(3x + 2)(x + 1)}$$

23. $\begin{vmatrix} 6 & 5 \\ 2 & -3 \end{vmatrix} = 6(-3) - 5 \cdot 2 = -18 - 10 = -28$

24.
$$\frac{x - 4 + \frac{5}{x+2}}{x + 2 - \frac{1}{x+2}} = \frac{x - 4 + \frac{5}{x+2}}{x + 2 - \frac{1}{x+2}} \cdot \frac{x + 2}{x + 2}$$
$$= \frac{(x - 4)(x + 2) + 5}{(x + 2)^2 - 1}$$
$$= \frac{x^2 - 2x - 8 + 5}{x^2 + 4x + 4 - 1}$$
$$= \frac{x^2 - 2x - 3}{x^2 + 4x + 3}$$
$$= \frac{(x - 3)(x + 1)}{(x + 3)(x + 1)}$$
$$= \frac{x - 3}{x + 3}$$

25. $x + y + z = 3$
$-2x + y + 3z = 2$
$2x - 4y + z = -1$

$D = \begin{vmatrix} 1 & 1 & 1 \\ -2 & 1 & 3 \\ 2 & -4 & 1 \end{vmatrix} = 27$

$D_x = \begin{vmatrix} 3 & 1 & 1 \\ 2 & 1 & 3 \\ -1 & -4 & 1 \end{vmatrix} = 27$

$D_y = \begin{vmatrix} 1 & 3 & 1 \\ -2 & 2 & 3 \\ 2 & -1 & 1 \end{vmatrix} = 27$

$D_z = \begin{vmatrix} 1 & 1 & 3 \\ -2 & 1 & 2 \\ 2 & -4 & -1 \end{vmatrix} = 27$

$x = \dfrac{D_x}{D} = \dfrac{27}{27} = 1$

$y = \dfrac{D_y}{D} = \dfrac{27}{27} = 1$

$z = \dfrac{D_z}{D} = \dfrac{27}{27} = 1$

The solution is $(1, 1, 1)$.

26. $f(x) = x^2 - 3x + 3$
$f(c) = c^2 - 3c + 3$
$1 = c^2 - 3c + 3$
$c^2 - 3c + 2 = 0$
$(c - 2)(c - 1) = 0$
$c - 2 = 0 \quad c - 1 = 0$
$c = 2 \quad c = 1$
The solutions are 1 and 2.

27.
$$\frac{2}{x-3} = \frac{5}{2x-3}$$
$$\left(\frac{2}{x-3}\right)(x-3)(2x-3) = \left(\frac{5}{2x-3}\right)(x-3)(2x-3)$$
$$2(2x-3) = 5(x-3)$$
$$4x - 6 = 5x - 15$$
$$-x - 6 = -15$$
$$-x = -9$$
$$x = 9$$
The solution is 9.

28.
$$\frac{3}{x^2 - 36} = \frac{2}{x-6} - \frac{5}{x+6}$$
$$\frac{3}{(x+6)(x-6)} = \frac{2}{x-6} - \frac{5}{x+6}$$
$$(x+6)(x-6)\left(\frac{3}{(x+6)(x-6)}\right) = \left(\frac{2}{x-6} - \frac{5}{x+6}\right)(x+6)(x-6)$$
$$3 = 2(x+6) - 5(x-6)$$
$$3 = 2x + 12 - 5x + 30$$
$$3 = -3x + 42$$
$$-39 = -3x$$
$$13 = x$$
The solution is 13.

29. $(a+5)(a^3 - 3a + 4) = a(a^3 - 3a + 4) + 5(a^3 - 3a + 4)$
$= a^4 - 3a^2 + 4a + 5a^3 - 15a + 20$
$= a^4 + 5a^3 - 3a^2 - 11a + 20$

30.
$$I = \frac{E}{R+r}$$
$$I(R+r) = \frac{E}{R+r}(R+r)$$
$$IR + Ir = E$$
$$Ir = E - IR$$
$$r = \frac{E - IR}{I}$$

31. $4x + 3y = 12$ $3x + 4y = 16$
     $3y = -4x + 12$ $4y = -3x + 16$
     $y = -\frac{4}{3}x + 4$ $y = -\frac{3}{4}x + 4$

The lines are not perpendicular, since the slopes are not negative reciprocals of each other.

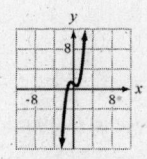

32.

33. Strategy
Smaller integer: x
Larger integer: $15 - x$

Five times the smaller is five more than twice the larger.
$5x = 5 + 2(15 - x)$

Solution
$5x = 5 + 2(15 - x)$
$5x = 5 + 30 - 2x$
$7x = 35$
$x = 5$
$15 - x = 10$
The smaller integer is 5 and larger integer is 10.

34. Strategy
The unknown number of pounds of almonds: x

	Amount	Cost	Total
Almonds	x	5.40	$5.40x$
Peanuts	50	2.60	$50(2.60)$
Mixture	$x + 50$	4.00	$4(x + 50)$

The sum of the values before mixing equals the value after mixing.
$5.40x + 50(2.60) = 4(x + 50)$

Solution
$$5.40x + 50(2.60) = 4(x + 50)$$
$$5.4x + 130 = 4x + 200$$
$$1.4x + 130 = 200$$
$$1.4x = 70$$
$$x = 50$$
The number of pounds of almonds is 50.

35. Strategy
To find the number of people expected to vote, write and solve a proportion using x to represent the number of people expected to vote.

Solution
$$\frac{3}{5} = \frac{x}{125,000}$$
$$\frac{3}{5} \cdot 125,000 = \frac{x}{125,000} \cdot 125,000$$
$$75,000 = x$$
The number of people expected to vote is 75,000.

36. Strategy
Time it takes older computer: $6r$
Time it takes new computer: r

	Rate	Time	Part
Older computer	$\frac{1}{6r}$	12	$\frac{12}{6r}$
New computer	$\frac{1}{r}$	12	$\frac{12}{r}$

The sum of the parts of the task completed by the older computer and the part of the task completed by the new computer is 1.
$$\frac{12}{6r} + \frac{12}{r} = 1$$

Solution
$$\frac{12}{6r} + \frac{12}{r} = 1$$
$$6r\left(\frac{12}{6r} + \frac{12}{r}\right) = (1)6r$$
$$12 + 72 = 6r$$
$$84 = 6r$$
$$14 = r$$
It takes the new computer 14 minutes to do the job working alone.

37. Strategy
Unknown rate of the wind: r

	Distance	Rate	Time
With the wind	900	$300 + r$	$\frac{900}{300+r}$
Against the wind	600	$300 - r$	$\frac{600}{300-r}$

The time traveled with the wind equals the time traveled against the wind.
$$\frac{900}{300 + r} = \frac{600}{300 - r}$$

Solution
$$\frac{900}{300 + r} = \frac{600}{300 - r}$$
$$(300 + r)(300 - r)\left(\frac{900}{300+r}\right) = \left(\frac{600}{300-r}\right)(300+r)(300-r)$$
$$(300 - r)(900) = 600(300 + r)$$
$$270,000 - 900r = 180,000 + 600r$$
$$-1500r = -90,000$$
$$r = 60$$
The rate of the wind is 60 mph.

38. Strategy
To find the frequency:
Write the basic inverse variation equation, replace the variables by the given values, and solve for k. Write the inverse variation equation, replacing k by its value. Substitute 1.5 for L and solve for f.

Solution
$$f = \frac{k}{L}$$
$$60 = \frac{k}{2}$$
$$120 = k$$
$$f = \frac{120}{L} = \frac{120}{1.5} = 80$$
The frequency is 80 vibrations per minute.

CHAPTER 7: RATIONAL EXPONENTS AND RADICALS

CHAPTER 7 PREP TEST

1. $48 \div 3 = 16$ [1.2.1]
   $48 = 16 \cdot 3$

2. $2^5 = 2 \cdot 2 \cdot 2 \cdot 2 \cdot 2 = 32$ [1.3.2]

3. $6\left(\dfrac{3}{2}\right) = 9$ [1.2.2]

4. $\dfrac{1}{2} - \dfrac{2}{3} + \dfrac{1}{4} = \dfrac{6}{12} - \dfrac{8}{12} + \dfrac{3}{12} = \dfrac{1}{12}$ [1.2.2]

5. $(3 - 7x) - (4 - 2x)$ [1.3.3]
   $= 3 - 7x - 4 + 2x$
   $= -5x - 1$

6. $\dfrac{3x^5 y^6}{12 x^4 y} = \dfrac{xy^5}{4}$ [5.1.2]

7. $(3x - 2)^2$ [5.3.3]
   $= (3x - 2)(3x - 2)$
   $= 9x^2 - 12x + 4$

8. $(2 + 4x)(5 - 3x)$ [5.3.2]
   $= 10 + 14x - 12x^2$
   $= -12x^2 + 14x + 10$

9. $(6x - 1)(6x + 1) = 36x^2 - 1$ [5.3.3]

10. $x^2 - 14x - 5 = 10$ [5.7.1
    $x^2 - 14x - 15 = 0$
    $(x - 15)(x + 1) = 0$
    $x - 15 = 0 \quad x + 1 = 0$
    $\quad x = 15 \quad\quad x = -1$
    The solutions are -1, 15.

Section 7.1

Concept Review 7.1

1. Sometimes true
   $\sqrt{x^2} = x$ is true for positive numbers, false for negative numbers.

3. Always true

5. Always true

Objective 7.1.1 Exercises

1. $125^{\frac{1}{3}}$ is the number whose <u>third</u> power is <u>125</u>, so $125^{\frac{1}{3}} = \underline{5}$.

3. $64^{-\frac{3}{2}} = (8^2)^{-\frac{3}{2}}$
   $= 8^{-3}$
   $= \dfrac{1}{8^3}$
   $= \dfrac{1}{512}$

5. d (the square root of a negative number is undefined)

7. $8^{1/3} = (2^3)^{1/3} = 2$

9. $9^{3/2} = (3^2)^{3/2} = 3^3 = 27$

11. $27^{-2/3} = (3^3)^{-2/3} = 3^{-2} = \dfrac{1}{3^2} = \dfrac{1}{9}$

13. $32^{2/5} = (2^5)^{2/5} = 2^2 = 4$

15. $(-25)^{5/2}$
    The base of the exponential expression is a negative number, while the denominator of the exponent is a positive even number.
    Therefore, $(-25)^{5/2}$ is not a real number.

17. $\left(\dfrac{25}{49}\right)^{-3/2} = \left(\dfrac{5^2}{7^2}\right)^{-3/2} = \left[\left(\dfrac{5}{7}\right)^2\right]^{-3/2}$
    $= \left(\dfrac{5}{7}\right)^{-3} = \dfrac{5^{-3}}{7^{-3}} = \dfrac{7^3}{5^3} = \dfrac{343}{125}$

19. $x^{1/2} x^{1/2} = x$

21. $y^{-1/4} y^{3/4} = y^{1/2}$

23. $x^{-2/3} \cdot x^{3/4} = x^{1/12}$

25. $a^{1/3} \cdot a^{3/4} \cdot a^{-1/2} = a^{7/12}$

27. $\dfrac{a^{1/2}}{a^{3/2}} = a^{-1} = \dfrac{1}{a}$

29. $\dfrac{y^{-3/4}}{y^{1/4}} = y^{-1} = \dfrac{1}{y}$

31. $\dfrac{y^{2/3}}{y^{-5/6}} = y^{9/6} = y^{3/2}$

33. $(x^2)^{-1/2} = x^{-1} = \dfrac{1}{x}$

35. $(x^{-2/3})^6 = x^{-4} = \dfrac{1}{x^4}$

37. $(a^{-1/2})^{-2} = a$

39. $(x^{-3/8})^{-4/5} = x^{3/10}$

41. $(a^{1/2} \cdot a)^2 = (a^{3/2})^2 = a^3$

43. $(x^{-1/2} x^{3/4})^{-2} = (x^{1/4})^{-2}$
$= x^{-1/2}$
$= \dfrac{1}{x^{1/2}}$

45. $(y^{-1/2} y^{3/2})^{2/3} = y^{2/3}$

47. $(x^8 y^2)^{1/2} = x^4 y$

49. $(x^4 y^2 z^6)^{3/2} = x^6 y^3 z^9$

51. $(x^{-3} y^6)^{-1/3} = xy^{-2} = \dfrac{x}{y^2}$

53. $(x^{-2} y^{1/3})^{-3/4} = x^{3/2} y^{-1/4} = \dfrac{x^{3/2}}{y^{1/4}}$

55. $\left(\dfrac{x^{1/2}}{y^{-2}}\right)^4 = \dfrac{x^2}{y^{-8}} = x^2 y^8$

57. $\dfrac{x^{1/4} \cdot x^{-1/2}}{x^{2/3}} = \dfrac{x^{-1/4}}{x^{2/3}} = x^{-11/12} = \dfrac{1}{x^{11/12}}$

59. $\left(\dfrac{y^{2/3} \cdot y^{-5/6}}{y^{1/9}}\right)^9 = \left(\dfrac{y^{-1/6}}{y^{1/9}}\right)^9$
$= (y^{-5/18})^9$
$= y^{-5/2}$
$= \dfrac{1}{y^{5/2}}$

61. $\left(\dfrac{b^2 \cdot b^{-3/4}}{b^{-1/2}}\right)^{-1/2} = \left(\dfrac{b^{5/4}}{b^{-1/2}}\right)^{-1/2}$
$= (b^{7/4})^{-1/2}$
$= b^{-7/8}$
$= \dfrac{1}{b^{7/8}}$

63. $(a^{2/3} b^2)^6 (a^3 b^3)^{1/3} = (a^4 b^{12})(ab) = a^5 b^{13}$

65. $(16 m^{-2} n^4)^{-1/2} (mn^{1/2}) = (2^4)^{-1/2} mn^{-2} \cdot mn^{1/2}$
$= 2^{-2} m^2 n^{-3/2}$
$= \dfrac{m^2}{2^2 n^{3/2}}$
$= \dfrac{m^2}{4n^{3/2}}$

67. $\left(\dfrac{x^{1/2} y^{-3/4}}{y^{2/3}}\right)^{-6} = (x^{1/2} y^{-17/12})^{-6}$
$= x^{-3} y^{17/2}$
$= \dfrac{y^{17/2}}{x^3}$

69. $\left(\dfrac{2^{-6} b^{-3}}{a^{-1/2}}\right)^{-2/3} = \dfrac{2^4 b^2}{a^{1/3}} = \dfrac{16 b^2}{a^{1/3}}$

71. $\dfrac{(x^{-2} y^4)^{1/2}}{(x^{1/2})^4} = \dfrac{x^{-1} y^2}{x^2} = \dfrac{y^2}{x^3}$

73. $a^{-1/4}(a^{5/4} - a^{9/4}) = a^1 - a^2 = a - a^2$

75. $y^{2/3}(y^{1/3} + y^{-2/3}) = y^1 + y^0 = y + 1$

77. $a^{1/6}(a^{5/6} - a^{-7/6}) = a^1 - a^{-1} = a - \dfrac{1}{a}$

79. $(a^{2/n})^{-5n} = a^{-10} = \dfrac{1}{a^{10}}$

81. $a^{n/2} \cdot a^{-n/3} = a^{n/6}$

83. $\dfrac{b^{m/3}}{b^m} = b^{-2m/3} = \dfrac{1}{b^{2m/3}}$

85. $(x^{5n})^{2n} = x^{10n^2}$

87. $(x^{n/2} y^{n/3})^6 = x^{3n} y^{2n}$

Objective 7.1.2 Exercises

89. To rewrite $\sqrt[5]{a^4}$ as an exponential expression, find the exponent for the base a. The rational exponent will be $\dfrac{\text{power on the radicand}}{\text{index}} = \dfrac{4}{5}$. $\sqrt[5]{a^4} = a^{\frac{4}{5}}$.

91. True

93. $5^{1/2} = \sqrt{5}$

95. $b^{4/3} = (b^4)^{1/3} = \sqrt[3]{b^4}$

97. $(3x)^{2/3} = \sqrt[3]{(3x)^2} = \sqrt[3]{9x^2}$

99. $-3a^{2/5} = -3(a^2)^{1/5} = -3\sqrt[5]{a^2}$

101. $(x^2 y^3)^{3/4} = \sqrt[4]{(x^2 y^3)^3} = \sqrt[4]{x^6 y^9}$

103. $(a^3 b^7)^{3/2} = \sqrt{(a^3 b^7)^3}$
$= \sqrt{a^9 b^{21}}$

105. $(3x - 2)^{1/3} = \sqrt[3]{3x - 2}$

107. $\sqrt{14} = 14^{1/2}$

109. $\sqrt[3]{x} = x^{1/3}$

111. $\sqrt[3]{x^4} = x^{4/3}$

113. $\sqrt[5]{b^3} = b^{3/5}$

115. $\sqrt[3]{2x^2} = (2x^2)^{1/3}$

117. $-\sqrt{3x^5} = -(3x^5)^{1/2}$

119. $3x\sqrt[3]{y^2} = 3xy^{2/3}$

121. $\sqrt{a^2+2} = (a^2+2)^{1/2}$

Objective 7.1.3 Exercises

123. $32a^{10}b^{15}$ is a perfect fifth power because $32 = (\underline{2})^5$, $32 = (\underline{a^2})^5$, and $b^{15} = (\underline{b^3})^5$.

125. Negative

127. Positive

129. $\sqrt{y^{14}} = y^7$

131. $-\sqrt{a^6} = -a^3$

133. $\sqrt{a^{14}b^6} = a^7b^3$

135. $\sqrt{121y^{12}} = 11y^6$

137. $\sqrt[3]{a^6b^{12}} = a^2b^4$

139. $-\sqrt[3]{a^9b^9} = -a^3b^3$

141. $\sqrt[3]{125b^{15}} = \sqrt[3]{5^3 b^{15}} = 5b^5$

143. $\sqrt[3]{-a^6b^9} = -a^2b^3$

145. $\sqrt{25x^8y^2} = \sqrt{5^2 x^8 y^2} = 5x^4y$

147. $\sqrt{-9a^6b^8}$
The square root of a negative number is not a real number, since the square of a real number must be positive. Therefore, $\sqrt{-9a^6b^8}$ is not a real number.

149. $\sqrt[3]{8a^{21}b^6} = \sqrt[3]{2^3 a^{21} b^6} = 2a^7b^2$

151. $\sqrt[3]{-27a^3b^{15}} = \sqrt[3]{(-3)^3 a^3 b^{15}} = -3ab^5$

153. $\sqrt[4]{y^{12}} = y^3$

155. $\sqrt[4]{81a^{20}} = \sqrt[4]{3^4 a^{20}} = 3a^5$

157. $-\sqrt[4]{a^{16}b^4} = -a^4b$

159. $\sqrt[5]{a^5b^{25}} = ab^5$

161. $\sqrt[4]{16a^8b^{20}} = \sqrt[4]{2^4 a^8 b^{20}} = 2a^2b^5$

163. $\sqrt[5]{-32x^{15}y^{20}} = \sqrt[5]{(-2)^5 x^{15} y^{20}} = -2x^3y^4$

165. $\sqrt{\dfrac{16x^2}{y^{14}}} = \sqrt{\dfrac{2^4 x^2}{y^{14}}} = \dfrac{2^2 x}{y^7} = \dfrac{4x}{y^7}$

167. $\sqrt[3]{\dfrac{27b^3}{a^9}} = \sqrt[3]{\dfrac{3^3 b^3}{a^9}} = \dfrac{3b}{a^3}$

169. $\sqrt{(2x+3)^2} = 2x+3$

171. $\sqrt{x^2+2x+1} = \sqrt{(x+1)^2} = x+1$

Applying Concepts 7.1

173. $\sqrt[3]{\sqrt{x^6}} = \sqrt[3]{x^3} = x$

175. $\sqrt[5]{\sqrt[3]{b^{15}}} = \sqrt[5]{b^5} = b$

177. $\sqrt[5]{\sqrt{a^{10}b^{20}}} = \sqrt[5]{a^5b^{10}} = ab^2$

179. $y^p y^{2/5} = y$

$y^p = \dfrac{y^1}{y^{2/5}}$

$y^p = y^{3/5}$

$p = \dfrac{3}{5}$

When the value of p is $\dfrac{3}{5}$, the equation is true.

181. $x^p x^{-1/2} = x^{1/4}$

$x^{p-(1/2)} = x^{1/4}$

$p - \dfrac{1}{2} = \dfrac{1}{4}$

$p = \dfrac{3}{4}$

When the value of p is $\dfrac{3}{4}$, the equation is true.

Section 7.2

Concept Review 7.2

1. Sometimes true
If a is a positive number, $\sqrt{a}$ is a real number. If a is negative, $\sqrt{a}$ is not a real number.

3. Never true
The index of each radical must be the same in order to multiply radical expressions.

5. Always true

Objective 7.2.1 Exercises

1. $\sqrt{50} = \sqrt{25 \cdot 2}$
$= \sqrt{25}\sqrt{2}$
$= 5\sqrt{2}$

3. a. $\sqrt{19}$ is irrational.
    b. $-\sqrt{81}$ is rational
    c. $\sqrt[3]{25}$ is irrational
    d. $\sqrt[3]{-27}$ is rational

5. The radical expression $\sqrt{8}$ is not in simplest form because the radicand contains a perfect square factor of 4.

7. $\sqrt{18} = \sqrt{3^2 \cdot 2}$
    $= \sqrt{3^2}\sqrt{2}$
    $= 3\sqrt{2}$

9. $\sqrt{98} = \sqrt{7^2 \cdot 2}$
    $= \sqrt{7^2}\sqrt{2}$
    $= 7\sqrt{2}$

11. $\sqrt[3]{72} = \sqrt[3]{2^3 \cdot 3^2}$
    $= \sqrt[3]{2^3}\sqrt[3]{3^2}$
    $= 2\sqrt[3]{9}$

13. $\sqrt[3]{16} = \sqrt[3]{2^3 \cdot 2}$
    $= \sqrt[3]{2^3}\sqrt[3]{2}$
    $= 2\sqrt[3]{2}$

15. $\sqrt{x^4 y^3 z^5} = \sqrt{x^4 y^2 z^4 (yz)}$
    $= \sqrt{x^4 y^2 z^4}\sqrt{yz}$
    $= x^2 y z^2 \sqrt{yz}$

17. $\sqrt{8a^3 b^8} = \sqrt{2^3 a^3 b^8}$
    $= \sqrt{2^2 a^2 b^8 (2a)}$
    $= \sqrt{2^2 a^2 b^8}\sqrt{2a}$
    $= 2ab^4 \sqrt{2a}$

19. $\sqrt{45 x^2 y^3 z^5} = \sqrt{3^2 \cdot 5 x^2 y^3 z^5}$
    $= \sqrt{3^2 x^2 y^2 z^4 (5yz)}$
    $= \sqrt{3^2 x^2 y^2 z^4}\sqrt{5yz}$
    $= 3xyz^2 \sqrt{5yz}$

21. $\sqrt[3]{-125 x^2 y^4} = \sqrt[3]{(-5)^3 x^2 y^4}$
    $= \sqrt[3]{(-5)^3 y^3 (x^2 y)}$
    $= \sqrt[3]{(-5)^3 y^3}\sqrt[3]{x^2 y}$
    $= -5y\sqrt[3]{x^2 y}$

23. $\sqrt[3]{-216 x^5 y^9} = \sqrt[3]{(-6)^3 x^5 y^9}$
    $= \sqrt[3]{(-6)^3 x^3 y^9 (x^2)}$
    $= \sqrt[3]{(-6)^3 x^3 y^9}\sqrt[3]{x^2}$
    $= -6xy^3 \sqrt[3]{x^2}$

25. $\sqrt[3]{a^5 b^8} = \sqrt[3]{a^3 b^6 (a^2 b^2)}$
    $= \sqrt[3]{a^3 b^6}\sqrt[3]{a^2 b^2}$
    $= ab^2 \sqrt[3]{a^2 b^2}$

Objective 7.2.2 Exercises

27. The Distributive Property can be used to add or subtract radicals that have the same <u>radicand</u> and the same <u>index</u>.

29. a. No, the expressions have different indices.
    b. No, the expressions have unlike radicands.
    c. Yes, the expressions have the same radicands.
    d. Yes, after simplifying the radicals, the expressions have the same radicands.

31. $\sqrt{2} + \sqrt{2} = 2\sqrt{2}$

33. $4\sqrt[3]{7} - \sqrt[3]{7} = 3\sqrt[3]{7}$

35. $2\sqrt{x} - 8\sqrt{x} = -6\sqrt{x}$

37. $\sqrt{8} - \sqrt{32} = \sqrt{2^3} - \sqrt{2^5}$
    $= \sqrt{2^2}\sqrt{2} - \sqrt{2^4}\sqrt{2}$
    $= 2\sqrt{2} - 2^2 \sqrt{2}$
    $= 2\sqrt{2} - 4\sqrt{2}$
    $= -2\sqrt{2}$

39. $\sqrt{128x} - \sqrt{98x} = \sqrt{2^7 x} - \sqrt{2 \cdot 7^2 x}$
    $= \sqrt{2^6}\sqrt{2x} - \sqrt{7^2}\sqrt{2x}$
    $= 2^3 \sqrt{2x} - 7\sqrt{2x}$
    $= 8\sqrt{2x} - 7\sqrt{2x}$
    $= \sqrt{2x}$

41. $\sqrt{27a} - \sqrt{8a} = \sqrt{3^3 a} - \sqrt{2^3 a}$
    $= \sqrt{3^2}\sqrt{3a} - \sqrt{2^2}\sqrt{2a}$
    $= 3\sqrt{3a} - 2\sqrt{2a}$

43. $2\sqrt{2x^3} + 4x\sqrt{8x} = 2\sqrt{2x^3} + 4x\sqrt{2^3 x}$
    $= 2\sqrt{x^2}\sqrt{2x} + 4x\sqrt{2^2}\sqrt{2x}$
    $= 2x\sqrt{2x} + 2 \cdot 4x\sqrt{2x}$
    $= 2x\sqrt{2x} + 8x\sqrt{2x}$
    $= 10x\sqrt{2x}$

45. $x\sqrt{75xy} - \sqrt{27x^3y} = x\sqrt{3 \cdot 5^2 xy} - \sqrt{3^3 x^3 y}$
$\phantom{x\sqrt{75xy} - \sqrt{27x^3y}} = x\sqrt{5^2}\sqrt{3xy} - \sqrt{3^2 x^2}\sqrt{3xy}$
$\phantom{x\sqrt{75xy} - \sqrt{27x^3y}} = 5x\sqrt{3xy} - 3x\sqrt{3xy}$
$\phantom{x\sqrt{75xy} - \sqrt{27x^3y}} = 2x\sqrt{3xy}$

47. $2\sqrt{32x^2y^3} - xy\sqrt{98y}$
$= 2\sqrt{2^5 x^2 y^3} - xy\sqrt{2 \cdot 7^2 y}$
$= 2\sqrt{2^4 x^2 y^2}\sqrt{2y} - xy\sqrt{7^2}\sqrt{2y}$
$= 2 \cdot 2^2 xy\sqrt{2y} - 7xy\sqrt{2y}$
$= 8xy\sqrt{2y} - 7xy\sqrt{2y}$
$= xy\sqrt{2y}$

49. $7b\sqrt{a^5b^3} - 2ab\sqrt{a^3b^3}$
$= 7b\sqrt{a^4 b^2}\sqrt{ab} - 2ab\sqrt{a^2 b^2}\sqrt{ab}$
$= 7b \cdot a^2 b\sqrt{ab} - 2ab \cdot ab\sqrt{ab}$
$= 7a^2 b^2 \sqrt{ab} - 2a^2 b^2 \sqrt{ab}$
$= 5a^2 b^2 \sqrt{ab}$

51. $\sqrt[3]{128} + \sqrt[3]{250} = \sqrt[3]{2^7} + \sqrt[3]{2 \cdot 5^3}$
$\phantom{\sqrt[3]{128} + \sqrt[3]{250}} = \sqrt[3]{2^6}\sqrt[3]{2} + \sqrt[3]{5^3}\sqrt[3]{2}$
$\phantom{\sqrt[3]{128} + \sqrt[3]{250}} = 2^2 \sqrt[3]{2} + 5\sqrt[3]{2}$
$\phantom{\sqrt[3]{128} + \sqrt[3]{250}} = 4\sqrt[3]{2} + 5\sqrt[3]{2}$
$\phantom{\sqrt[3]{128} + \sqrt[3]{250}} = 9\sqrt[3]{2}$

53. $2\sqrt[3]{3a^4} - 3a\sqrt[3]{81a} = 2\sqrt[3]{3a^4} - 3a\sqrt[3]{3^4 a}$
$\phantom{2\sqrt[3]{3a^4} - 3a\sqrt[3]{81a}} = 2\sqrt[3]{a^3}\sqrt[3]{3a} - 3a\sqrt[3]{3^3}\sqrt[3]{3a}$
$\phantom{2\sqrt[3]{3a^4} - 3a\sqrt[3]{81a}} = 2a\sqrt[3]{3a} - 3a \cdot 3\sqrt[3]{3a}$
$\phantom{2\sqrt[3]{3a^4} - 3a\sqrt[3]{81a}} = 2a\sqrt[3]{3a} - 9a\sqrt[3]{3a}$
$\phantom{2\sqrt[3]{3a^4} - 3a\sqrt[3]{81a}} = -7a\sqrt[3]{3a}$

55. $3\sqrt[3]{x^5 y^7} - 8xy\sqrt[3]{x^2 y^4}$
$= 3\sqrt[3]{x^3 y^6}\sqrt[3]{x^2 y} - 8xy\sqrt[3]{y^3}\sqrt[3]{x^2 y}$
$= 3xy^2 \sqrt[3]{x^2 y} - 8xy \cdot y\sqrt[3]{x^2 y}$
$= 3xy^2 \sqrt[3]{x^2 y} - 8xy^2 \sqrt[3]{x^2 y}$
$= -5xy^2 \sqrt[3]{x^2 y}$

57. $2a\sqrt[4]{16ab^5} + 3b\sqrt[4]{256a^5 b}$
$= 2a\sqrt[4]{2^4 ab^5} + 3b\sqrt[4]{2^8 a^5 b}$
$= 2a\sqrt[4]{2^4 b^4}\sqrt[4]{ab} + 3b\sqrt[4]{2^8 a^4}\sqrt[4]{ab}$
$= 2a \cdot 2b\sqrt[4]{ab} + 3b \cdot 2^2 a\sqrt[4]{ab}$
$= 4ab\sqrt[4]{ab} + 12ab\sqrt[4]{ab}$
$= 16ab\sqrt[4]{ab}$

59. $3\sqrt{108} - 2\sqrt{18} - 3\sqrt{48}$
$= 3\sqrt{2^2 \cdot 3^3} - 2\sqrt{2 \cdot 3^2} - 3\sqrt{2^4 \cdot 3}$
$= 3\sqrt{2^2 \cdot 3^2}\sqrt{3} - 2\sqrt{3^2}\sqrt{2} - 3\sqrt{2^4}\sqrt{3}$
$= 3 \cdot 2 \cdot 3\sqrt{3} - 2 \cdot 3\sqrt{2} - 3 \cdot 2^2 \sqrt{3}$
$= 18\sqrt{3} - 6\sqrt{2} - 12\sqrt{3}$
$= 6\sqrt{3} - 6\sqrt{2}$

61. $\sqrt{4x^7 y^5} + 9x^2 \sqrt{x^3 y^5} - 5xy\sqrt{x^5 y^3} = \sqrt{2^2 x^7 y^5} + 9x^2 \sqrt{x^3 y^5} - 5xy\sqrt{x^5 y^3}$
$\phantom{\sqrt{4x^7 y^5} + 9x^2 \sqrt{x^3 y^5} - 5xy\sqrt{x^5 y^3}} = \sqrt{2^2 x^6 y^4}\sqrt{xy} + 9x^2 \sqrt{x^2 y^4}\sqrt{xy} - 5xy\sqrt{x^4 y^2}\sqrt{xy}$
$\phantom{\sqrt{4x^7 y^5} + 9x^2 \sqrt{x^3 y^5} - 5xy\sqrt{x^5 y^3}} = 2x^3 y^2 \sqrt{xy} + 9x^2 \cdot xy^2 \sqrt{xy} - 5xy \cdot x^2 y\sqrt{xy}$
$\phantom{\sqrt{4x^7 y^5} + 9x^2 \sqrt{x^3 y^5} - 5xy\sqrt{x^5 y^3}} = 2x^3 y^2 \sqrt{xy} + 9x^3 y^2 \sqrt{xy} - 5x^3 y^2 \sqrt{xy}$
$\phantom{\sqrt{4x^7 y^5} + 9x^2 \sqrt{x^3 y^5} - 5xy\sqrt{x^5 y^3}} = 6x^3 y^2 \sqrt{xy}$

63. $5a\sqrt{3a^3 b} + 2a^2 \sqrt{27ab} - 4\sqrt{75a^5 b} = 5a\sqrt{3a^3 b} + 2a^2 \sqrt{3^3 ab} - 4\sqrt{3 \cdot 5^2 a^5 b}$
$\phantom{5a\sqrt{3a^3 b} + 2a^2 \sqrt{27ab} - 4\sqrt{75a^5 b}} = 5a\sqrt{a^2}\sqrt{3ab} + 2a^2 \sqrt{3^2}\sqrt{3ab} - 4\sqrt{5^2 a^4}\sqrt{3ab}$
$\phantom{5a\sqrt{3a^3 b} + 2a^2 \sqrt{27ab} - 4\sqrt{75a^5 b}} = 5a \cdot a\sqrt{3ab} + 2a^2 \cdot 3\sqrt{3ab} - 4 \cdot 5a^2 \sqrt{3ab}$
$\phantom{5a\sqrt{3a^3 b} + 2a^2 \sqrt{27ab} - 4\sqrt{75a^5 b}} = 5a^2 \sqrt{3ab} + 6a^2 \sqrt{3ab} - 20a^2 \sqrt{3ab}$
$\phantom{5a\sqrt{3a^3 b} + 2a^2 \sqrt{27ab} - 4\sqrt{75a^5 b}} = -9a^2 \sqrt{3ab}$

Copyright © Houghton Mifflin Company. All rights reserved.

Objective 7.2.3 Exercises

65. $\sqrt{3}(\sqrt{6}-6\sqrt{x}) = \sqrt{3}\cdot\sqrt{6}-\sqrt{3}\cdot 6\sqrt{x}$
$= \sqrt{18}-6\sqrt{3x}$
$= 3\sqrt{2}-6\sqrt{3x}$

67. $\sqrt{8}\sqrt{32} = \sqrt{256} = \sqrt{2^8} = 2^4 = 16$

69. $\sqrt[3]{4}\sqrt[3]{8} = \sqrt[3]{32} = \sqrt[3]{2^5} = \sqrt[3]{2^3}\sqrt[3]{2^2} = 2\sqrt[3]{4}$

71. $\sqrt{x^2y^5}\sqrt{xy} = \sqrt{x^3y^6} = \sqrt{x^2y^6}\sqrt{x} = xy^3\sqrt{x}$

73. $\sqrt{2x^2y}\sqrt{32xy} = \sqrt{64x^3y^2} = \sqrt{2^6x^3y^2} = \sqrt{2^6x^2y^2}\sqrt{x} = 2^3xy\sqrt{x} = 8xy\sqrt{x}$

75. $\sqrt[3]{x^2y}\sqrt[3]{16x^4y^2} = \sqrt[3]{16x^6y^3} = \sqrt[3]{2^4x^6y^3} = \sqrt[3]{2^3x^6y^3}\sqrt[3]{2} = 2x^2y\sqrt[3]{2}$

77. $\sqrt[4]{12ab^3}\sqrt[4]{4a^5b^2} = \sqrt[4]{48a^6b^5} = \sqrt[4]{2^4\cdot 3a^6b^5} = \sqrt[4]{2^4a^4b^4}\sqrt[4]{3a^2b} = 2ab\sqrt[4]{3a^2b}$

79. $\sqrt{3}(\sqrt{27}-\sqrt{3}) = \sqrt{81}-\sqrt{9}$
$= \sqrt{3^4}-\sqrt{3^2}$
$= 3^2-3$
$= 9-3$
$= 6$

81. $\sqrt{x}(\sqrt{x}-\sqrt{2}) = \sqrt{x^2}-\sqrt{2x} = x-\sqrt{2x}$

83. $\sqrt{2x}(\sqrt{8x}-\sqrt{32}) = \sqrt{16x^2}-\sqrt{64x}$
$= \sqrt{2^4x^2}-\sqrt{2^6x}$
$= 2^2x-2^3\sqrt{x}$
$= 4x-8\sqrt{x}$

85. $(\sqrt{x}-3)^2 = (\sqrt{x})^2-3\sqrt{x}-3\sqrt{x}+9$
$= x-6\sqrt{x}+9$

87. $(4\sqrt{5}+2)^2 = (4\sqrt{5})^2+8\sqrt{5}+8\sqrt{5}+4$
$= 16\cdot 5+16\sqrt{5}+4$
$= 80+16\sqrt{5}+4$
$= 84+16\sqrt{5}$

89. $2\sqrt{14xy}\cdot 4\sqrt{7x^2y}\cdot 3\sqrt{8xy^2} = 24\sqrt{784x^4y^4}$
$= 24\sqrt{2^4\cdot 7^2x^4y^4}$
$= 24\cdot 2^2\cdot 7x^2y^2$
$= 672x^2y^2$

91. $\sqrt[3]{2a^2b}\sqrt[3]{4a^3b^2}\sqrt[3]{8a^5b^6} = \sqrt[3]{64a^{10}b^9}$
$= \sqrt[3]{2^6a^{10}b^9}$
$= \sqrt[3]{2^6a^9b^9}\sqrt[3]{a}$
$= 2^2a^3b^3\sqrt[3]{a}$
$= 4a^3b^3\sqrt[3]{a}$

93. $(\sqrt{5}-5)(2\sqrt{5}+2) = 2\sqrt{5^2}+2\sqrt{5}-10\sqrt{5}-10$
$= 2\cdot 5-8\sqrt{5}-10$
$= 10-8\sqrt{5}-10$
$= -8\sqrt{5}$

95. $(\sqrt{x}-y)(\sqrt{x}+y) = \sqrt{x^2}-y^2 = x-y^2$

97. $(2\sqrt{3x}-\sqrt{y})(2\sqrt{3x}+\sqrt{y}) = 4\sqrt{3^2x^2}-\sqrt{y^2}$
$= 4\cdot 3x-y$
$= 12x-y$

99. $(\sqrt{x}+4)(\sqrt{x}-7) = \sqrt{x^2}-7\sqrt{x}+4\sqrt{x}-28$
$= x-3\sqrt{x}-28$

101. $(\sqrt[3]{x}-4)(\sqrt[3]{x}+5) = \sqrt[3]{x^2}+5\sqrt[3]{x}-4\sqrt[3]{x}-20$
$= \sqrt[3]{x^2}+\sqrt[3]{x}-20$

103. False. $(\sqrt{a}-1)(\sqrt{a}+1) = a-1$, which is less than, not greater than, a.

Objective 7.2.4 Exercises

105. Use the Quotient Property of Radicals.
$\dfrac{\sqrt[4]{80x^{10}}}{\sqrt[4]{5x}} = \sqrt[4]{\dfrac{80x^{10}}{5x}}$
$= \sqrt[4]{16x^9}$
$= \sqrt[4]{16x^8}\sqrt[4]{x}$
$= 2x^2\sqrt[4]{x}$

107. a. Yes, the expression is in simplest form.

b. No, the expression is not in simplest form because there is a perfect square factor in the radicand.

c. No, the expression is not in simplest form because there is a radical in the denominator.

d. No, the expression is not in simplest form because there is a radical in the denominator.

109. $\dfrac{\sqrt{32x^2}}{\sqrt{2x}} = \sqrt{\dfrac{32x^2}{2x}}$
$= \sqrt{16x}$
$= \sqrt{2^4 x}$
$= \sqrt{2^4}\sqrt{x}$
$= 2^2\sqrt{x}$
$= 4\sqrt{x}$

111. $\dfrac{\sqrt{42a^3b^5}}{\sqrt{14a^2b}} = \sqrt{\dfrac{42a^3b^5}{14a^2b}}$
$= \sqrt{3ab^4}$
$= \sqrt{b^4}\sqrt{3a}$
$= b^2\sqrt{3a}$

113. $\dfrac{1}{\sqrt{5}} = \dfrac{1}{\sqrt{5}}\cdot\dfrac{\sqrt{5}}{\sqrt{5}} = \dfrac{\sqrt{5}}{\sqrt{5^2}} = \dfrac{\sqrt{5}}{5}$

115. $\dfrac{1}{\sqrt{2x}} = \dfrac{1}{\sqrt{2x}}\cdot\dfrac{\sqrt{2x}}{\sqrt{2x}} = \dfrac{\sqrt{2x}}{\sqrt{2^2 x^2}} = \dfrac{\sqrt{2x}}{2x}$

117. $\dfrac{5}{\sqrt{5x}} = \dfrac{5}{\sqrt{5x}}\cdot\dfrac{\sqrt{5x}}{\sqrt{5x}} = \dfrac{5\sqrt{5x}}{\sqrt{5^2 x^2}} = \dfrac{5\sqrt{5x}}{5x} = \dfrac{\sqrt{5x}}{x}$

119. $\sqrt{\dfrac{x}{5}} = \dfrac{\sqrt{x}}{\sqrt{5}} = \dfrac{\sqrt{x}}{\sqrt{5}}\cdot\dfrac{\sqrt{5}}{\sqrt{5}} = \dfrac{\sqrt{5x}}{\sqrt{5^2}} = \dfrac{\sqrt{5x}}{5}$

121. $\dfrac{3}{\sqrt[3]{2}} = \dfrac{3}{\sqrt[3]{2}}\cdot\dfrac{\sqrt[3]{2^2}}{\sqrt[3]{2^2}} = \dfrac{3\sqrt[3]{2^2}}{\sqrt[3]{2^3}} = \dfrac{3\sqrt[3]{4}}{2}$

123. $\dfrac{3}{\sqrt[3]{4x^2}} = \dfrac{3}{\sqrt[3]{2^2 x^2}}\cdot\dfrac{\sqrt[3]{2x}}{\sqrt[3]{2x}} = \dfrac{3\sqrt[3]{2x}}{\sqrt[3]{2^3 x^3}} = \dfrac{3\sqrt[3]{2x}}{2x}$

125. $\dfrac{\sqrt{40x^3 y^2}}{\sqrt{80x^2 y^3}} = \sqrt{\dfrac{40x^3 y^2}{80x^2 y^3}}$
$= \sqrt{\dfrac{x}{2y}}$
$= \dfrac{\sqrt{x}}{\sqrt{2y}}\cdot\dfrac{\sqrt{2y}}{\sqrt{2y}}$
$= \dfrac{\sqrt{2xy}}{\sqrt{2^2 y^2}}$
$= \dfrac{\sqrt{2xy}}{2y}$

127. $\dfrac{\sqrt{24a^2 b}}{\sqrt{18ab^4}} = \sqrt{\dfrac{24a^2 b}{18ab^4}}$
$= \sqrt{\dfrac{4a}{3b^3}}$
$= \dfrac{\sqrt{4a}}{\sqrt{3b^3}}$
$= \dfrac{\sqrt{2^2}\sqrt{a}}{\sqrt{b^2}\sqrt{3b}}$
$= \dfrac{2\sqrt{a}}{b\sqrt{3b}}\cdot\dfrac{\sqrt{3b}}{\sqrt{3b}}$
$= \dfrac{2\sqrt{3ab}}{b\sqrt{3^2 b^2}}$
$= \dfrac{2\sqrt{3ab}}{b\cdot 3b}$
$= \dfrac{2\sqrt{3ab}}{3b^2}$

129. $\dfrac{2}{\sqrt{5}+2} = \dfrac{2}{\sqrt{5}+2}\cdot\dfrac{\sqrt{5}-2}{\sqrt{5}-2}$
$= \dfrac{2\sqrt{5}-4}{(\sqrt{5})^2 - 2^2}$
$= \dfrac{2\sqrt{5}-4}{5-4}$
$= \dfrac{2\sqrt{5}-4}{1}$
$= 2\sqrt{5}-4$

131. $\dfrac{3}{\sqrt{y}-2} = \dfrac{3}{\sqrt{y}-2}\cdot\dfrac{\sqrt{y}+2}{\sqrt{y}+2}$
$= \dfrac{3\sqrt{y}+6}{(\sqrt{y})^2 - 2^2}$
$= \dfrac{3\sqrt{y}+6}{y-4}$

133. $\dfrac{\sqrt{2}-\sqrt{3}}{\sqrt{2}+\sqrt{3}} = \dfrac{\sqrt{2}-\sqrt{3}}{\sqrt{2}+\sqrt{3}}\cdot\dfrac{\sqrt{2}-\sqrt{3}}{\sqrt{2}-\sqrt{3}}$
$= \dfrac{(\sqrt{2})^2 - \sqrt{6} - \sqrt{6} + (\sqrt{3})^2}{(\sqrt{2})^2 - (\sqrt{3})^2}$
$= \dfrac{2 - 2\sqrt{6} + 3}{2-3}$
$= \dfrac{5 - 2\sqrt{6}}{-1}$
$= -5 + 2\sqrt{6}$

135. $\dfrac{4-\sqrt{2}}{2-\sqrt{3}} = \dfrac{4-\sqrt{2}}{2-\sqrt{3}} \cdot \dfrac{2+\sqrt{3}}{2+\sqrt{3}}$

$= \dfrac{8+4\sqrt{3}-2\sqrt{2}-\sqrt{6}}{(2)^2-(\sqrt{3})^2}$

$= \dfrac{8+4\sqrt{3}-2\sqrt{2}-\sqrt{6}}{4-3}$

$= \dfrac{8+4\sqrt{3}-2\sqrt{2}-\sqrt{6}}{1}$

$= 8+4\sqrt{3}-2\sqrt{2}-\sqrt{6}$

137. $\dfrac{\sqrt{3}-\sqrt{5}}{\sqrt{2}+\sqrt{5}} = \dfrac{\sqrt{3}-\sqrt{5}}{\sqrt{2}+\sqrt{5}} \cdot \dfrac{\sqrt{2}-\sqrt{5}}{\sqrt{2}-\sqrt{5}}$

$= \dfrac{\sqrt{6}-\sqrt{15}-\sqrt{10}+\sqrt{5^2}}{(\sqrt{2})^2-(\sqrt{5})^2}$

$= \dfrac{\sqrt{6}-\sqrt{15}-\sqrt{10}+5}{2-5}$

$= \dfrac{\sqrt{6}-\sqrt{15}-\sqrt{10}+5}{-3}$

$= \dfrac{\sqrt{15}+\sqrt{10}-\sqrt{6}-5}{3}$

139. $\dfrac{3}{\sqrt[4]{8x^3}} = \dfrac{3}{\sqrt[4]{2^3 x^3}} \cdot \dfrac{\sqrt[4]{2x}}{\sqrt[4]{2x}} = \dfrac{3\sqrt[4]{2x}}{\sqrt[4]{2^4 x^4}} = \dfrac{3\sqrt[4]{2x}}{2x}$

141. $\dfrac{4}{\sqrt[5]{16a^2}} = \dfrac{4}{\sqrt[5]{2^4 a^2}} \cdot \dfrac{\sqrt[5]{2a^3}}{\sqrt[5]{2a^3}}$

$= \dfrac{4\sqrt[5]{2a^3}}{\sqrt[5]{2^5 a^5}}$

$= \dfrac{4\sqrt[5]{2a^3}}{2a}$

$= \dfrac{2\sqrt[5]{2a^3}}{a}$

143. $\dfrac{2x}{\sqrt[5]{64x^3}} = \dfrac{2x}{\sqrt[5]{2^6 x^3}} \cdot \dfrac{\sqrt[5]{2^4 x^2}}{\sqrt[5]{2^4 x^2}}$

$= \dfrac{2x\sqrt[5]{2^4 x^2}}{\sqrt[5]{2^{10} x^5}}$

$= \dfrac{2x\sqrt[5]{16x^2}}{2^2 x}$

$= \dfrac{\sqrt[5]{16x^2}}{2}$

145. $\dfrac{\sqrt{a}+a\sqrt{b}}{\sqrt{a}-a\sqrt{b}} = \dfrac{\sqrt{a}+a\sqrt{b}}{\sqrt{a}-a\sqrt{b}} \cdot \dfrac{\sqrt{a}+a\sqrt{b}}{\sqrt{a}+a\sqrt{b}}$

$= \dfrac{\sqrt{a^2}+a\sqrt{ab}+a\sqrt{ab}+a^2\sqrt{b^2}}{(\sqrt{a})^2-(a\sqrt{b})^2}$

$= \dfrac{a+2a\sqrt{ab}+a^2 b}{a-a^2 b}$

$= \dfrac{a(1+2\sqrt{ab}+ab)}{a(1-ab)}$

$= \dfrac{1+2\sqrt{ab}+ab}{1-ab}$

147. $\dfrac{3\sqrt{xy}+2\sqrt{xy}}{\sqrt{x}-\sqrt{y}} = \dfrac{5\sqrt{xy}}{\sqrt{x}-\sqrt{y}}$

$= \dfrac{5\sqrt{xy}}{\sqrt{x}-\sqrt{y}} \cdot \dfrac{\sqrt{x}+\sqrt{y}}{\sqrt{x}+\sqrt{y}}$

$= \dfrac{5\sqrt{x^2 y}+5\sqrt{xy^2}}{(\sqrt{x})^2-(\sqrt{y})^2}$

$= \dfrac{5\sqrt{x^2}\sqrt{y}+5\sqrt{y^2}\sqrt{x}}{x-y}$

$= \dfrac{5x\sqrt{y}+5y\sqrt{x}}{x-y}$

Applying Concepts 7.2

149. $(\sqrt{8}-\sqrt{2})^3 = (\sqrt{2^3}-\sqrt{2})^3 = (2\sqrt{2}-\sqrt{2})^3$

$= (\sqrt{2})^3 = \sqrt{2}\cdot\sqrt{2}\cdot\sqrt{2}$

$= 2\sqrt{2}$

151. $(\sqrt{2}-3)^3 = (\sqrt{2}-3)(\sqrt{2}-3)(\sqrt{2}-3)$

$= (2-6\sqrt{2}+9)(\sqrt{2}-3)$

$= (-6\sqrt{2}+11)(\sqrt{2}-3)$

$= -12+18\sqrt{2}+11\sqrt{2}-33$

$= 29\sqrt{2}-45$

153. $\dfrac{3}{\sqrt{y+1}+1} = \dfrac{3}{\sqrt{y+1}+1} \cdot \dfrac{\sqrt{y+1}-1}{\sqrt{y+1}-1}$

$= \dfrac{3\sqrt{y+1}-3}{(\sqrt{y+1})^2-1^2}$

$= \dfrac{3\sqrt{y+1}-3}{y+1-1}$

$= \dfrac{3\sqrt{y+1}-3}{y}$

155. $\dfrac{\sqrt[3]{(x+y)^2}}{\sqrt{x+y}} = \dfrac{(x+y)^{2/3}}{(x+y)^{1/2}}$

$= (x+y)^{1/6}$

$= \sqrt[6]{x+y}$

157. $\sqrt[4]{2y}\sqrt{x+3} = (2y)^{1/4}(x+3)^{1/2}$

$= (2y)^{1/4}(x+3)^{2/4}$

$= \sqrt[4]{2y}\sqrt[4]{(x+3)^2}$

$= \sqrt[4]{2y(x+3)^2}$

159. $\sqrt{a}\sqrt[3]{a+3} = a^{1/2}(a+3)^{1/3}$

$= a^{3/6}(a+3)^{2/6}$

$= \sqrt[6]{a^3}\sqrt[6]{(a+3)^2}$

$= \sqrt[6]{a^3(a+3)^2}$

161. $\sqrt{16^{1/2}} = (16^{1/2})^{1/2} = 16^{1/4} = 2$

163. $\sqrt[4]{32^{-4/5}} = (32^{-4/5})^{1/4}$

$= 32^{-1/5}$

$= (2^5)^{-1/5}$

$= 2^{-1}$

$= \dfrac{1}{2}$

165. a. $D(x) = 3.06x^{0.3}$

$D(x) = 3.06x^{3/10}$

$D(x) = 3.06\sqrt[10]{x^3}$

b. For the year 1995, $x = 5$.

$D(5) = 3.06\sqrt[10]{5^3}$

$D(5) \approx 4.96$

The model predicted a debt of \$4.96 trillion by 1995.

c. For the year 2003, $x = 13$.

$D(13) = 3.06\sqrt[10]{13^3}$

$D(13) \approx 6.61$

The model predicts a debt of \$6.61 trillion by the year 2003.

Section 7.3
Concept Review 7.3
1. Always true
3. Never true
The domain is $\{x | x \in \text{real numbers.}\}$.

Objective 7.3.1 Exercises

1. A radical function is one that contains a variable underneath a radical sign or contains a fractional exponent.

3. a. When n is an even number, the domain of $f(x) = x^{\frac{1}{n}}$ is $\{x | x \ge 0\}$.

   b. When n is an odd number, the domain of $f(x) = x^{\frac{1}{n}}$ is $\{x | x \in \text{real numbers}\}$.

5. a. $(-\infty, 0]$

   b. empty set

   c. $[0, \infty)$

   d. all real numbers

7. The domain of $f(x) = 2x^{1/3}$ is $\{x | x \text{ is a real number}\}$.

9. $x + 1 \ge 0$

$x \ge -1$

The domain of $g(x) = -2\sqrt{x+1}$ is $\{x | x \ge -1\}$.

11. $x \ge 0$

The domain of $f(x) = 2x\sqrt{x} - 3$ is $\{x | x \ge 0\}$.

13. $x^{3/4} = \sqrt[4]{x^3}$

$x^3 \ge 0$

$x \ge 0$

The domain of $C(x) = -3x^{3/4} + 1$ is $\{x | x \ge 0\}$.

15. $(3x - 6)^{1/2} = \sqrt{3x-6}$

$3x - 6 \ge 0$

$3x \ge 6$

$x \ge 2$

The domain of $F(x) = 4(3x-6)^{1/2}$ is $\{x | x \ge 2\}$.

17. The domain of $g(x) = 2(2x - 10)^{2/3}$ is $(-\infty, \infty)$.

19. $12 - 4x \ge 0$

$-4x \ge -12$

$x \le 3$

The domain of $V(x) = x - \sqrt{12 - 4x}$ is $(-\infty, 3]$.

21. $(x-2)^2 \ge 0$

$x - 2 \ge 0$

$x \ge 2$

The domain of $h(x) = 3\sqrt[4]{(x-2)^3}$ is $[2, \infty)$.

218 Chapter 7: Rational Exponents And Radicals

23. $(4-6x)^{1/2} = \sqrt{4-6x}$
$4 - 6x \geq 0$
$-6x \geq -4$
$x \leq \dfrac{2}{3}$

The domain of $f(x) = x - (4-6x)^{1/2}$ is $\left(-\infty, \dfrac{2}{3}\right]$

Objective 7.3.2 Exercises

25. The domain of f is $\{x \mid x \geq 5\}$.

27. $f(14) = -\sqrt{14-5} = -\sqrt{9} = -3$

29.

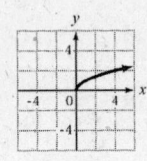

31.

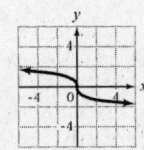

33.

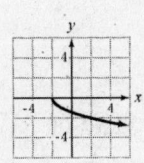

35.

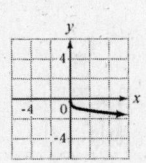

37.

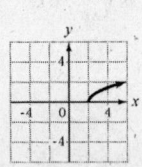

39.

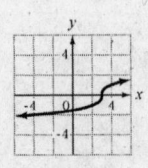

41.

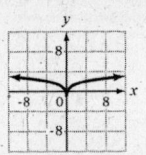

43.

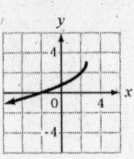

45.

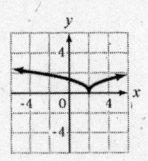

47.

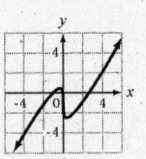

49.

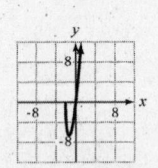

51. a. Both
    b. Below
    c. Above
    d. Above

Applying Concepts 7.3

53. a. In the context of the problem, the domain of the function is the numbers 1, 5, 10, 20, 50, 100. 0 is not in the domain because there is no $0 bill. Negative numbers are not in the domain because there are no bills whose denominators are negative.

    b. The model predicts a life span of approximately 1.8 years for a $2 bill. This estimate is not reasonable. The $2 bill is not in circulation as much as the other denominations listed in the table. Therefore, its life span would not be accurately depicted by the mode.

Section 7.4

Concept Review 7.4

1. Sometimes true
   $(-2)^2 = 2^2$, but -2 does not equal 2.

3. Sometimes true
   The Pythagorean Theorem is valid only for right triangles.

Objective 7.4.1 Exercises

1. $\sqrt{2x+1} = 7$

3. The first step is to subtract 3 from both sides of the equation. This achieves the goal of getting the radical expression alone on one side of the equation.

5. $\sqrt{x} = 5$
   $(\sqrt{x})^2 = 5^2$
   $x = 25$
   Check:
   $$\begin{array}{c|c} \sqrt{x} = 5 \\ \hline \sqrt{25} & 5 \\ 5 = 5 \end{array}$$
   The solution is 25.

7. $\sqrt[3]{a} = 3$
   $(\sqrt[3]{a})^3 = 3$
   $a = 27$
   Check:
   $$\begin{array}{c|c} \sqrt[3]{a} = 3 \\ \hline \sqrt[3]{27} & 3 \\ 3 = 3 \end{array}$$
   The solution is 27.

9. $\sqrt{3x} = 12$
   $(\sqrt{3x})^2 = 12^2$
   $3x = 144$
   $x = 48$
   Check:
   $$\begin{array}{c|c} \sqrt{3x} = 12 \\ \hline \sqrt{3(48)} & 12 \\ \sqrt{144} & 12 \\ 12 = 12 \end{array}$$
   The solution is 48.

11. $\sqrt[3]{4x} = -2$
    $(\sqrt[3]{4x})^3 = (-2)^3$
    $4x = -8$
    $x = -2$
    Check:
    $$\begin{array}{c|c} \sqrt[3]{4x} = -2 \\ \hline \sqrt[3]{4(-2)} & -2 \\ \sqrt[3]{-8} & -2 \\ -2 = -2 \end{array}$$
    The solution is -2.

13. $\sqrt{2x} = -4$
    $(\sqrt{2x})^2 = (-4)^2$
    $2x = 16$
    $x = 8$
    Check:
    $$\begin{array}{c|c} \sqrt{2x} = -4 \\ \hline \sqrt{2(8)} & -4 \\ \sqrt{16} & -4 \\ 4 \neq -4 \end{array}$$
    8 does not check as a solution.
    The equation has no solution.

15. $\sqrt{3x-2} = 5$
    $(\sqrt{3x-2})^2 = 5^2$
    $3x - 2 = 25$
    $3x = 27$
    $x = 9$
    Check:
    $$\begin{array}{c|c} \sqrt{3x-2} = 5 \\ \hline \sqrt{3(9)-2} & 5 \\ \sqrt{27-2} & 5 \\ \sqrt{25} & 5 \\ 5 = 5 \end{array}$$
    The solution is 9.

17. $\sqrt{3-2x} = 7$
    $(\sqrt{3-2x})^2 = 7^2$
    $3 - 2x = 49$
    $-2x = 46$
    $x = -23$
    Check:
    $$\begin{array}{c|c} \sqrt{3-2x} = 7 \\ \hline \sqrt{3-2(-23)} & 7 \\ \sqrt{3+46} & 7 \\ \sqrt{49} & 7 \\ 7 = 7 \end{array}$$
    The solution is -23.

19.
$7 = \sqrt{1-3x}$
$(7)^2 = (\sqrt{1-3x})^2$
$49 = 1 - 3x$
$48 = -3x$
$-16 = x$
Check:

$7 = \sqrt{1-3x}$	
7	$\sqrt{1-3(-16)}$
7	$\sqrt{1+48}$
7	$\sqrt{49}$
$7 = 7$	

The solution is −16.

21.
$\sqrt[3]{4x-1} = 2$
$(\sqrt[3]{4x-1})^3 = 2^3$
$4x - 1 = 8$
$4x = 9$
$4x = \dfrac{9}{4}$
Check:

$\sqrt[3]{4x-1} = 2$	
$\sqrt[3]{4\left(\dfrac{9}{4}\right)-1}$	2
$\sqrt[3]{9-1}$	2
$\sqrt[3]{8}$	2
	$2 = 2$

The solution is $\dfrac{9}{4}$.

23.
$\sqrt[3]{1-2x} = -3$
$(\sqrt[3]{1-2x})^3 = (-3)^3$
$1 - 2x = -27$
$-2x = -28$
$x = 14$
Check:

$\sqrt[3]{1-2x} = -3$	
$\sqrt[3]{1-2(14)}$	-3
$\sqrt[3]{1-28}$	-3
$\sqrt[3]{-27}$	-3
	$-3 = -3$

The solution is 14.

25.
$\sqrt[3]{9x+1} = 4$
$(\sqrt[3]{9x+1})^3 = 4^3$
$9x + 1 = 64$
$9x = 63$
$x = 7$
Check:

$\sqrt[3]{9x+1} = 4$	
$\sqrt[3]{9(7)+1}$	4
$\sqrt[3]{63+1}$	4
$\sqrt[3]{64}$	4
	$4 = 4$

The solution is 7.

27.
$\sqrt{4x-3} - 5 = 0$
$\sqrt{4x-3} = 5$
$(\sqrt{4x-3})^2 = 5^2$
$4x - 3 = 25$
$4x = 28$
$x = 7$
Check:

$\sqrt{4x-3} - 5 = 0$	
$\sqrt{4(7)-3} - 5$	0
$\sqrt{28-3} - 5$	0
$\sqrt{25} - 5$	0
$5 - 5$	0
	$0 = 0$

The solution is 7.

29.
$\sqrt[3]{x-3} + 5 = 0$
$\sqrt[3]{x-3} = -5$
$(\sqrt[3]{x-3})^3 = (-5)^5$
$x - 3 = -125$
$x = -122$
Check:

$\sqrt[3]{x-3} + 5 = 0$	
$\sqrt[3]{-122-3} + 5$	0
$\sqrt[3]{-125} + 5$	0
$-5 + 5$	0
	$0 = 0$

The solution is −122.

31. $\sqrt[3]{2x-6} = 4$
$(\sqrt[3]{2x-6})^3 = 4^3$
$2x - 6 = 64$
$2x = 70$
$x = 35$
Check:

$\sqrt[3]{2x-6} = 4$	
$\sqrt[3]{2(35)-6}$	4
$\sqrt[3]{70-6}$	4
$\sqrt[3]{64}$	4
$4 = 4$	

The solution is 35.

33. $\sqrt[4]{2x-9} = 3$
$(\sqrt[4]{2x-9})^4 = 3^4$
$2x - 9 = 81$
$2x = 90$
$x = 45$
Check:

$\sqrt[4]{2x-9} = 3$	
$\sqrt[4]{2(45)-9}$	3
$\sqrt[4]{90-9}$	3
$\sqrt[4]{81}$	3
$3 = 3$	

The solution is 45.

35. $\sqrt{3x-5} - 5 = 3$
$\sqrt{3x-5} = 8$
$(\sqrt{3x-5})^2 = 8^2$
$3x - 5 = 64$
$3x = 69$
$x = 23$
Check:

$\sqrt{3x-5} - 5 = 3$	
$\sqrt{3(23)-5} - 5$	3
$\sqrt{69-5} - 5$	3
$\sqrt{64} - 5$	3
$8 - 5$	3
$3 = 3$	

The solution is 23.

37. $\sqrt[3]{x-4} + 7 = 5$
$\sqrt[3]{x-4} = -2$
$(\sqrt[3]{x-4})^3 = (-2)^3$
$x - 4 = -8$
$x = -4$
Check:

$\sqrt[3]{x-4} + 7 = 5$	
$\sqrt[3]{-4-4} + 7$	5
$\sqrt[3]{-8} + 7$	5
$-2 + 7$	5
$5 = 5$	

The solution is –4.

39. $\sqrt{3x-5} - 2 = 3$
$(\sqrt{3x-5})^2 = (5)^2$
$3x - 5 = 25$
$3x = 30$
$x = 10$
Check:

$\sqrt{3x-5} - 2 = 3$	
$\sqrt{3(10)-5} - 2$	3
$\sqrt{30-5} - 2$	3
$\sqrt{25} - 2$	3
$5 - 2$	3
$3 = 3$	

The solution is 10.

41. $\sqrt{7x+2} - 10 = -7$
$(\sqrt{7x+2})^2 = (3)^2$
$7x + 2 = 9$
$7x = 7$
$x = 1$
Check:

$\sqrt{7x+2} - 10 = -7$	
$\sqrt{7(1)+2} - 10$	-7
$\sqrt{7+2} - 10$	-7
$\sqrt{9} - 10$	-7
$3 - 10$	-7
$-7 = -7$	

The solution is 1.

43. $\sqrt[3]{1-3x}+5=3$
$(\sqrt[3]{1-3x})^3=(-2)^3$
$1-3x=-8$
$-3x=-9$
$x=3$

Check:

$\sqrt[3]{1-3x}+5=3$	
$\sqrt[3]{1-3(3)}+5$	3
$\sqrt[3]{1-9}+5$	3
$\sqrt[3]{-8}+5$	3
$-2+5$	3
$3=3$	

The solution is 3.

45. $7-\sqrt{3x+1}=-1$
$(-\sqrt{3x+1})^2=(-8)^2$
$3x+1=64$
$3x=63$
$x=21$

Check:

$7-\sqrt{3x+1}=-1$	
$7-\sqrt{3(21)+1}$	-1
$7-\sqrt{63+1}$	-1
$7-\sqrt{64}$	-1
$7-8$	-1
$-1=-1$	

The solution is 21.

47. $\sqrt{2x+4}=3-\sqrt{2x}$
$(\sqrt{2x+4})^2=(3-\sqrt{2x})^2$
$2x+4=9-6\sqrt{2x}+2x$
$6\sqrt{2x}=5$
$(6\sqrt{2x})^2=(5)^2$
$36(2x)=25$
$72x=25$
$x=\dfrac{25}{72}$

Check:

$\sqrt{2x+4}=3-\sqrt{2x}$	
$\sqrt{2\left(\dfrac{25}{72}\right)+4}$	$3-\sqrt{2\left(\dfrac{25}{72}\right)}$
$\sqrt{\dfrac{25}{36}+4}$	$3-\sqrt{\dfrac{25}{36}}$
$\sqrt{\dfrac{169}{36}}$	$3-\dfrac{5}{6}$
$\dfrac{13}{6}=\dfrac{13}{6}$	

The solution is $\dfrac{25}{72}$.

49. $\sqrt{x^2-4x-1}+3=x$
$\sqrt{x^2-4x-1}=x-3$
$(\sqrt{x^2-4x-1})^2=(x-3)^2$
$x^2-4x-1=x^2-6x+9$
$2x=10$
$x=5$

Check:

$\sqrt{x^2-4x-1}+3=x$	
$\sqrt{5^2-4(5)-1}+3$	5
$\sqrt{25-20-1}+3$	5
$\sqrt{4}+3$	5
$2+3=5$	

The solution is 5.

51. $\sqrt{x^2-2x+1}=3$
$(\sqrt{x^2-2x+1})^2=3^2$
$x^2-2x+1=9$
$x^2-2x-8=0$
$(x-4)(x+2)=0$
$x-4=0 \quad x+2=0$
$x=4 \quad\quad x=-2$

Check:

$\sqrt{x^2-2x+1}=3$		$\sqrt{x^2-2x+1}=3$	
$\sqrt{4^2-2(4)+1}$	3	$\sqrt{(-2)^2-2(-2)+1}$	3
$\sqrt{16-8+1}$	3	$\sqrt{4+4+1}$	3
$\sqrt{9}$	3	$\sqrt{9}$	3
$3=3$		$3=3$	

The solutions are 4 and -2.

53. $\sqrt{4x+1} - \sqrt{2x+4} = 1$
$$(\sqrt{4x+1})^2 = (1+\sqrt{2x+4})^2$$
$$4x+1 = 1 + 2\sqrt{2x+4} + 2x + 4$$
$$4x+1 = 5 + 2x + 2\sqrt{2x+4}$$
$$(2x-4)^2 = (2\sqrt{2x+4})^2$$
$$4x^2 - 16x + 16 = 4(2x+4)$$
$$4x^2 - 16x + 16 = 8x + 16$$
$$4x^2 - 24x = 0$$
$$4x(x-6) = 0$$
$$x = 0 \quad x - 6 = 0$$
$$x = 6$$

Check:

$\sqrt{4x+1} - \sqrt{2x+4} = 1$		$\sqrt{4x+1} - \sqrt{2x+4} = 1$	
$\sqrt{4(0)+1} - \sqrt{2(0)+4}$	1	$\sqrt{4(6)+1} - \sqrt{2(6)+4}$	1
$\sqrt{1} - \sqrt{4}$	1	$\sqrt{24+1} - \sqrt{12+4}$	1
$1 - 2$	1	$\sqrt{25} - \sqrt{16}$	1
$-1 \neq 1$		$5 - 4$	1
		$1 = 1$	

The solution is 6.

55. $\sqrt{5x+4} - \sqrt{3x+1} = 1$
$$(\sqrt{5x+4})^2 = (1+\sqrt{3x+1})^2$$
$$5x+4 = 1 + 2\sqrt{3x+1} + 3x + 1$$
$$5x+4 = 2 + 3x + 2\sqrt{3x+1}$$
$$(2x+2)^2 = (2\sqrt{3x+1})^2$$
$$4x^2 + 8x + 4 = 4(3x+1)$$
$$4x^2 + 8x + 4 = 12x + 4$$
$$4x^2 - 4x = 0$$
$$4x(x-1) = 0$$
$$x = 0 \quad x - 1 = 0$$
$$x = 1$$

Check:

$\sqrt{5x+4} - \sqrt{3x+1} = 1$		$\sqrt{5x+4} - \sqrt{3x+1} = 1$	
$\sqrt{5(0)+4} - \sqrt{3(0)+1}$	1	$\sqrt{5(1)+4} - \sqrt{3(1)+1}$	1
$\sqrt{4} - \sqrt{1}$	1	$\sqrt{5+4} - \sqrt{3+1}$	1
$2 - 1$	1	$\sqrt{9} - \sqrt{4}$	1
$1 = 1$		$3 - 2$	1
		$1 = 1$	

The solutions are 0 and 1.

57. $\sqrt[4]{x^2+x-1}-1=0$

$\sqrt[4]{x^2+x-1}=1$

$(\sqrt[4]{x^2+x-1})^4=1^4$

$x^2+x-1=1$

$x^2+x-2=0$

$(x+2)(x-1)=0$

$x+2=0 \quad x-1=0$

$x=-2 \quad x=1$

Check:

$\sqrt[4]{x^2+x-1}-1=0$		$\sqrt[4]{x^2+x-1}-1=0$	
$\sqrt[4]{(-2)^2+(-2)-1}-1$	0	$\sqrt[4]{1^2+1-1}-1$	0
$\sqrt[4]{4-2-1}-1$	0	$\sqrt[4]{1+1-1}-1$	0
$\sqrt[4]{1}-1$	0	$\sqrt[4]{1}-1$	0
$1-1$	0	$1-1$	0
$0=0$		$0=0$	

The solutions are -2 and 1.

59. $3\sqrt{x-2}+2=x$

$3\sqrt{x-2}=x-2$

$(3\sqrt{x-2})^2=(x-2)^2$

$9(x-2)=x^2-4x+4$

$9x-18=x^2-4x+4$

$0=x^2-13x+22$

$0=(x-2)(x-11)$

$x-2=0 \quad x-11=0$

$x=2 \quad x=11$

Check:

$3\sqrt{x-2}+2=x$		$3\sqrt{x-2}+2=x$	
$3\sqrt{2-2}+2$	2	$3\sqrt{11-2}+2$	11
$3\sqrt{0}+2$	2	$3\sqrt{9}+2$	11
$3(0)+2$	2	$3(3)+2$	11
$0+2$	2	$9+2$	11
$2=2$		$11=11$	

The solutions are 2 and 11.

61.
$$x + 2\sqrt{x+1} = 7$$
$$2\sqrt{x+1} = 7 - x$$
$$(2\sqrt{x+1})^2 = (7-x)^2$$
$$4(x+1) = 49 - 14x + x^2$$
$$4x + 4 = 49 - 14x + x^2$$
$$0 = x^2 - 18x + 45$$
$$0 = (x-3)(x-15)$$
$$x - 3 = 0 \quad x - 15 = 0$$
$$x = 3 \qquad x = 15$$

Check:

$x + 2\sqrt{x+1} = 7$		$x + 2\sqrt{x+1} = 7$	
$3 + 2\sqrt{3+1}$	7	$15 + 2\sqrt{15+1}$	7
$3 + 2\sqrt{4}$	7	$15 + 2\sqrt{16}$	7
$3 + 2(2)$	7	$15 + 2(4)$	7
$3 + 4$	7	$15 + 8$	7
$7 = 7$		$23 \neq 7$	

The solution is 3.

63. a. One time

b. Two times

Objective 7.4.2 Exercises

65. **d** is not a possible value for h because a 15-foot ladder cannot reach a height on the building greater than 15 feet.

67. Strategy
To find the air pressure, replace v in the equation by 64 and solve for p.

Solution
$$v = 6.3\sqrt{1013-p}$$
$$64 = 6.3\sqrt{1013-p}$$
$$\frac{64}{6.3} = \sqrt{1013-p}$$
$$\left(\frac{64}{5.3}\right)^2 = 1013 - p$$
$$p = 1013 - \left(\frac{64}{6.3}\right)^2$$
$$p = 909.8$$

The air pressure is 909.8 mb.
As air pressure decreases, the wind speed increases.

69. Strategy
To find the mean distance of Tethys from Saturn, replace T in the equation by 1.89 and solve for d.

Solution
$$T = 0.373\sqrt{d^3}$$
$$1.89 = 0.373\sqrt{d^3}$$
$$\frac{1.89}{0.373} = d^{3/2}$$
$$\left(\frac{1.89}{0.373}\right)^{2/3} = d$$
$$2.95 = d$$

Tethys is 295,000 km from Saturn.

71. Strategy
To find the weight, replace M in the equation by 60,000 and solve for W.

Solution
$$M = 126.4\sqrt[4]{W^3}$$
$$60,000 = 126.4\sqrt[4]{W^3}$$
$$\frac{60,000}{126.4} = W^{3/4}$$
$$\left(\frac{60,000}{126.4}\right)^{4/3} = W$$
$$3700 = W$$

The elephant weighs 3700 lb.

Chapter 7: Rational Exponents And Radicals

73. Strategy
To find the distance, use the Pythagorean Theorem. The hypotenuse is the length of the ladder. The distance along the ground from the building to the ladder is the unknown leg.

Solution
$$c^2 = a^2 + b^2$$
$$26^2 = 24^2 + b^2$$
$$676 = 576 + b^2$$
$$100 = b^2$$
$$\sqrt{100} = \sqrt{b^2}$$
$$10 = b$$
The distance is 10 ft.

75. Strategy
To find the distance, use the Pythagorean Theorem. The distance from the starting point to the jogger traveling east (3 m/s for 180 sec = 540 m) is the first leg. The distance from the starting point to the jogger traveling south (3.5 m/s for 180 sec = 630 ft) is the other leg. The hypotenuse, the distance between the joggers, is unknown.

Solution
$$c^2 = a^2 + b^2$$
$$c^2 = 540^2 + 630^2$$
$$c^2 = 291{,}600 + 396{,}900$$
$$c^2 = 688{,}500$$
$$\sqrt{c^2} = \sqrt{688{,}500}$$
$$c \approx 829.76$$
The distance is 829.76 m.

77. Strategy
To find the distance, use the Pythagorean Theorem. The distance from vertex to the corner is the first leg. The distance from the vertex to the opposite corner is the other leg. The hypotenuse, the length of the diagonal, is unknown.

Solution
$$c^2 = a^2 + b^2$$
$$c^2 = 2.5^2 + 8.5^2$$
$$c^2 = 6.25 + 72.25$$
$$c^2 = 78.5$$
$$\sqrt{c^2} = \sqrt{78.5}$$
$$c \approx 8.9$$
The length is 8.9 cm.

79. Strategy
To find the distance above the water, replace d in the equation by the given value and solve for h.

Solution
$$d = \sqrt{1.5h}$$
$$3.5 = \sqrt{1.5h}$$
$$(3.5)^2 = (\sqrt{1.5h})^2$$
$$(3.5)^2 = 1.5h$$
$$\frac{(3.5)^2}{1.5} = h$$
$$8.17 = h$$
The periscope must be 8.17 ft above the water.

81. Strategy
To find the distance, replace the variable v in the equation by 120 and solve for d.

Solution
$$v = 8\sqrt{d}$$
$$120 = 8\sqrt{d}$$
$$\frac{120}{8} = \sqrt{d}$$
$$\left(\frac{120}{8}\right)^2 = (\sqrt{d})^2$$
$$225 = d$$
The distance is 225 ft.

83. Strategy
To find the distance, replace the variables v and a in the equation by their given values and solve for s.

Solution
$$v = \sqrt{2as}$$
$$48 = \sqrt{24s}$$
$$(48)^2 = (\sqrt{24s})^2$$
$$2304 = 24s$$
$$96 = s$$
The distance is 96 ft.

85. Strategy

To find the height of the satellite above Earth's surface, replace v in the equation by the given value and solve for h.

Solution

$$v = \sqrt{\frac{4 \times 10^{14}}{h + 6.4 \times 10^6}}$$

$$7500 = \sqrt{\frac{4 \times 10^{14}}{h + 6.4 \times 10^6}}$$

$$(7500)^2 = \left(\sqrt{\frac{4 \times 10^{14}}{h + 6.4 \times 10^6}}\right)^2$$

$$5.625 \times 10^7 = \frac{4 \times 10^{14}}{h + 6.4 \times 10^6}$$

$$(5.625 \times 10^7)(h + 6.4 \times 10^6) = (4 \times 10^{14})$$

$$h + 6.4 \times 10^6 = \frac{4 \times 10^{14}}{5.625 \times 10^7}$$

$$h = \frac{4 \times 10^{14}}{5.625 \times 10^7} - (6.4 \times 10^6)$$

$$h = 7.\overline{1} \times 10^5$$

$$h \approx 711{,}000$$

The height of the satellite above Earth's surface is 711,000 m.

Applying Concepts 7.4

87. $x^{2/3} = 9$

$(x^{2/3})^{3/2} = 9^{3/2}$

$x = (\sqrt{9})^3$

$x = 3^3$

$x = 27$

89. $v = \sqrt{64d}$

$v^2 = (\sqrt{64d})^2$

$v^2 = 64d$

$\dfrac{v^2}{64} = d$

91. $V = \pi r^2 h$

$\dfrac{V}{\pi h} = r^2$

$\sqrt{\dfrac{V}{\pi h}} = \sqrt{r^2}$

$\sqrt{\dfrac{V}{\pi h}} = r$

$r = \dfrac{\sqrt{V}}{\sqrt{\pi h}} \cdot \dfrac{\sqrt{\pi h}}{\sqrt{\pi h}} = \dfrac{\sqrt{V \pi h}}{\pi h}$

93.
$\sqrt{3x-2} = \sqrt{2x-3} + \sqrt{x-1}$

$(\sqrt{3x-2})^2 = (\sqrt{2x-3} + \sqrt{x-1})^2$

$3x - 2 = 2x - 3 + 2\sqrt{2x-3}\sqrt{x-1} + x - 1$

$2 = 2\sqrt{2x-3}\sqrt{x-1}$

$1 = \sqrt{2x-3}\sqrt{x-1}$

$1^2 = (\sqrt{2x-3}\sqrt{x-1})^2$

$1 = (2x-3)(x-1)$

$1 = 2x^2 - 5x + 3$

$0 = 2x^2 - 5x + 2$

$0 = (2x-1)(x-2)$

$2x - 1 = 0 \quad x - 2 = 0$

$2x = 1 \quad\quad x = 2$

$x = \dfrac{1}{2}$

Check:

$\sqrt{3x-2} = \sqrt{2x-3} + \sqrt{x-1}$	
$\sqrt{3\left(\frac{1}{2}\right)-2}$	$\sqrt{2\left(\frac{1}{2}\right)-3} + \sqrt{\frac{1}{2}-1}$
$\sqrt{-\frac{1}{2}} \neq \sqrt{-2} + \sqrt{-\frac{1}{2}}$	

$\sqrt{3x-2} = \sqrt{2x-3} + \sqrt{x-1}$	
$\sqrt{3(2)-2}$	$\sqrt{2(2)-3} + \sqrt{2-1}$
$\sqrt{4}$	$\sqrt{1} + \sqrt{1}$
2	$1 + 1$
$2 = 2$	

The solution is 2.

95. Impossible; at least one of the integers must be even.

97. The equation $\sqrt{a^2 + b^2} = a + b$, where both a and b are nonnegative, is true when both a and b are zero and when either a or b is zero. In all other cases, the equation is false.

Section 7.5

Concept Review 7.5

1. Always true

3. Always true

Objective 7.5.1 Exercises

1. An imaginary number is a number whose square is a negative number. Imaginary numbers are defined in terms of i, the number whose square is -1.
A complex number is of the form $a + bi$, where a and b are real numbers and $i = \sqrt{-1}$.

Chapter 7: Rational Exponents And Radicals

3. $i^2 = \left(\sqrt{-1}\right)^2 = -1$

5. $\sqrt{-54} = i\sqrt{54} = i\sqrt{(9)(6)} = 3i\sqrt{6}$

7. $10 - 3i = 10 - 3\sqrt{i^2} = 10 - 3\sqrt{-1} = 10 - \sqrt{-1(3)^2} = 10 - \sqrt{-9}$

9. $\sqrt{-4} = i\sqrt{4} = i\sqrt{2^2} = 2i$

11. $\sqrt{-98} = i\sqrt{98} = i\sqrt{2 \cdot 7^2} = 7i\sqrt{2}$

13. $\sqrt{-27} = i\sqrt{27} = i\sqrt{3^2 \cdot 3} = 3i\sqrt{3}$

15. $\sqrt{16} + \sqrt{-4} = \sqrt{16} + i\sqrt{4} = \sqrt{2^4} + i\sqrt{2^2} = 4 + 2i$

17. $\sqrt{12} - \sqrt{-18} = \sqrt{12} - i\sqrt{18}$
$= \sqrt{2^2 \cdot 3} - i\sqrt{3^2 \cdot 2}$
$= 2\sqrt{3} - 3i\sqrt{2}$

19. $\sqrt{160} - \sqrt{-147} = \sqrt{160} - i\sqrt{147}$
$= \sqrt{2^4 \cdot 2 \cdot 5} - i\sqrt{7^2 \cdot 3}$
$= 4\sqrt{10} - 7i\sqrt{3}$

21. $\sqrt{-4a^2} = i\sqrt{4a^2} = i\sqrt{2^2 a^2} = 2ai$

23. $\sqrt{-49x^{12}} = i\sqrt{49x^{12}} = i\sqrt{7^2 x^{12}} = 7x^6 i$

25. $\sqrt{-144a^3 b^5} = i\sqrt{2^4 \cdot 3^2 a^3 b^5}$
$= 2^2 \cdot 3 iab^2 \sqrt{ab}$
$= 12ab^2 i\sqrt{ab}$

27. $\sqrt{4a} + \sqrt{-12a^2} = \sqrt{2^2 a} + i\sqrt{2^2 \cdot 3a^2}$
$= 2\sqrt{a} + 2ai\sqrt{3}$

29. $\sqrt{18b^5} - \sqrt{-27b^3} = \sqrt{2 \cdot 3^2 b^5} - i\sqrt{3 \cdot 3^2 b^3}$
$= 3b^2 \sqrt{2b} - 3bi\sqrt{3b}$

31. $\sqrt{-50x^3 y^3} + x\sqrt{25x^4 y^3}$
$= i\sqrt{5^2 \cdot 2x^3 y^3} + x\sqrt{5^2 x^4 y^3}$
$= 5ixy\sqrt{2xy} + 5x^3 y\sqrt{y}$
$= 5x^3 y\sqrt{y} + 5xyi\sqrt{2xy}$

33. $\sqrt{-49a^5 b^2} - ab\sqrt{-25a^3}$
$= i\sqrt{7^2 a^5 b^2} - iab\sqrt{5^2 a^3}$
$= 7ia^2 b\sqrt{a} - 5ia^2 b\sqrt{a}$
$= 2a^2 bi\sqrt{a}$

35. $\sqrt{12a^3} + \sqrt{-27b^3}$
$= \sqrt{2^2 \cdot 3a^3} + i\sqrt{3^2 \cdot 3b^3}$
$= 2a\sqrt{3a} + 3bi\sqrt{3b}$

Objective 7.5.2 Exercises

37. $12 - (6 - \sqrt{-64}) = 12 - (6 - 8i)$
$= (12 - 6) - (-8i)$
$= 6 + 8i$

39. $(6 - 9i) + (4 + 2i) = 10 - 7i$

41. $(3 - 5i) + (8 - 2i) = 11 - 7i$

43. $(5 - \sqrt{-25}) - (11 - \sqrt{-36})$
$= (5 - i\sqrt{25}) - (11 - i\sqrt{36})$
$= (5 - i\sqrt{5^2}) - (11 - i\sqrt{2^2 \cdot 3^2})$
$= (5 - 5i) - (11 - 6i)$
$= -6 + i$

45. $(5 - \sqrt{-12}) - (9 + \sqrt{-108})$
$= (5 - i\sqrt{12}) - (9 + i\sqrt{108})$
$= (5 - i\sqrt{2^2 \cdot 3}) - (9 + i\sqrt{2^2 \cdot 3^2 \cdot 3})$
$= (5 - 2i\sqrt{3}) - (9 + 6i\sqrt{3})$
$= -4 - 8i\sqrt{3}$

47. $(\sqrt{40} - \sqrt{-98}) - (\sqrt{90} + \sqrt{-32})$
$= (\sqrt{40} - i\sqrt{98}) - (\sqrt{90} + i\sqrt{32})$
$= (\sqrt{2^2 \cdot 2 \cdot 5} - i\sqrt{7^2 \cdot 2}) - (\sqrt{3^2 \cdot 2 \cdot 5} + i\sqrt{2^4 \cdot 2})$
$= (2\sqrt{10} - 7i\sqrt{2}) - (3\sqrt{10} + 4i\sqrt{2})$
$= -\sqrt{10} - 11i\sqrt{2}$

49. $(6 - 8i) + 4i = 6 - 4i$

51. $(8 - 3i) + (-8 + 3i) = 0$

53. $(4 + 6i) + 7 = (4 + 6i) + (7 + 0i) = 11 + 6i$

55. True

57. $m - n = (a + bi) - (c - di) = (a - c) + (b + d)i$. If $a < c$, then the real part of $m - n$ is negative.

Objective 7.5.3 Exercises

59. $(-6i)(-4i) = 24i^2 = 24(-1) = -24$

61. $\sqrt{-5}\sqrt{-45} = i\sqrt{5} \cdot i\sqrt{45}$
$= i^2 \sqrt{225}$
$= -\sqrt{3^2 \cdot 5^2}$
$= -15$

63. $\sqrt{-5}\sqrt{-10} = i\sqrt{5} \cdot i\sqrt{10} = i^2\sqrt{50}$
$= -\sqrt{5^2 \cdot 2} = -5\sqrt{2}$

65. $-3i(4-5i) = -12i + 15i^2$
$= -12i + 15(-1)$
$= -15 - 12i$

67. $(1+9i)(1-9i) = (1)^2 + (9)^2$
$= 1 + 81 = 82$

69. $\sqrt{-3}(\sqrt{12} - \sqrt{-6}) = i\sqrt{3}(\sqrt{12} - i\sqrt{6})$
$= i\sqrt{36} - i^2\sqrt{18}$
$= i\sqrt{2^2 \cdot 3^2} + \sqrt{3^2 \cdot 2}$
$= 6i + 3\sqrt{2}$
$= 3\sqrt{2} + 6i$

71. $(2-4i)(2-i) = 4 - 2i - 8i + 4i^2$
$= 4 - 10i + 4i^2$
$= 4 - 10i + 4(-1)$
$= -10i$

73. $(4-7i)(2+3i) = 8 + 12i - 14i - 21i^2$
$= 8 - 2i - 21i^2$
$= 8 - 2i - 21(-1)$
$= 29 - 2i$

75. $\left(\frac{4}{5} - \frac{2}{5}i\right)\left(1 + \frac{1}{2}i\right) = \frac{4}{5} + \frac{2}{5}i - \frac{2}{5}i - \frac{1}{5}i^2$
$= \frac{4}{5} - \frac{1}{5}i^2$
$= \frac{4}{5} - \frac{1}{5}(-1)$
$= \frac{4}{5} + \frac{1}{5}$
$= 1$

77. $(2-i)\left(\frac{2}{5} + \frac{1}{5}i\right) = \frac{4}{5} + \frac{2}{5}i - \frac{2}{5}i - \frac{1}{5}i^2$
$= \frac{4}{5} - \frac{1}{5}i^2$
$= \frac{4}{5} - \frac{1}{5}(-1)$
$= \frac{4}{5} + \frac{1}{5}$
$= 1$

79. $(8-5i)(8+5i) = 8^2 + 5^2 = 64 + 25 = 89$

81. $(7-i)(7+i) = 7^2 + 1^2 = 49 + 1 = 50$

83. $(4-3i)^2 = (4-3i)(4-3i)$
$= 16 - 24i + 9i^2$
$= 16 - 24i - 9$
$= 7 - 24i$

85. $(-1+i)^2 = (-1+i)(-1+i)$
$= 1 - 2i + i^2$
$= 1 - 2i - 1$
$= -2i$

87. False. For example, $-3i$ and $2i$ are two imaginary numbers whose product is a positive real number: $(-3i)(2i) = -6(i^2) = -6(-1) = 6$.

Objective 7.5.4 Exercises

89. To simplify $\frac{2i}{2-3i}$, multiply the numerator and denominator by $2+3i$.

91. $\frac{4}{5i} = \frac{4}{5i} \cdot \frac{i}{i} = \frac{4i}{5i^2} = \frac{4i}{5(-1)} = \frac{4i}{-5} = -\frac{4}{5}i$

93. $\frac{16+5i}{-3i} = \frac{16+5i}{-3i} \cdot \frac{i}{i}$
$= \frac{16i + 5i^2}{-3i^2}$
$= \frac{16i + 5(-1)}{-3(-1)}$
$= \frac{-5 + 16i}{3}$
$= -\frac{5}{3} + \frac{16}{3}i$

95. $\frac{6}{5+2i} = \frac{6}{5+2i} \cdot \frac{5-2i}{5-2i}$
$= \frac{30 - 12i}{25 + 4}$
$= \frac{30 - 12i}{29}$
$= \frac{30}{29} - \frac{12}{29}i$

97. $\frac{5}{4-i} = \frac{5}{4-i} \cdot \frac{4+i}{4+i}$
$= \frac{20 + 5i}{16 + 1}$
$= \frac{20 + 5i}{17}$
$= \frac{20}{17} + \frac{5}{17}i$

99.
$$\frac{2+12i}{5+i} = \frac{2+12i}{5+i} \cdot \frac{5-i}{5-i}$$
$$= \frac{10 - 2i + 60i - 12i^2}{25 + 1}$$
$$= \frac{10 + 58i - 12(-1)}{26}$$
$$= \frac{22 + 58i}{26}$$
$$= \frac{11 + 29i}{13}$$
$$= \frac{11}{13} + \frac{29}{13}i$$

101.
$$\frac{\sqrt{-2}}{\sqrt{12} - \sqrt{-8}} = \frac{i\sqrt{2}}{\sqrt{12} - i\sqrt{8}}$$
$$= \frac{i\sqrt{2}}{\sqrt{2^2 \cdot 3} - i\sqrt{2^2 \cdot 2}}$$
$$= \frac{i\sqrt{2}}{2\sqrt{3} - 2i\sqrt{2}} \cdot \frac{2\sqrt{3} + 2i\sqrt{2}}{2\sqrt{3} + 2i\sqrt{2}}$$
$$= \frac{2i^2\sqrt{6} + 2i^2\sqrt{2^2}}{(2\sqrt{3})^2 + (2\sqrt{2})^2}$$
$$= \frac{2i\sqrt{6} - 2 \cdot 2}{12 + 8}$$
$$= \frac{-4 + 2i\sqrt{6}}{20}$$
$$= \frac{-2 + i\sqrt{6}}{10}$$
$$= -\frac{1}{5} + \frac{\sqrt{6}}{10}i$$

103.
$$\frac{3+5i}{1-i} = \frac{3+5i}{1-i} \cdot \frac{1+i}{1+i}$$
$$= \frac{3 + 3i + 5i + 5i^2}{1+1}$$
$$= \frac{3 + 8i + 5i^2}{2}$$
$$= \frac{3 + 8i + 5(-1)}{2}$$
$$= \frac{-2 + 8i}{2}$$
$$= -1 + 4i$$

105. True

Applying Concepts 7.5

107. $i^6 = i^4 \cdot i^2 = 1(-1) = -1$

109. $i^{57} = i^{56} \cdot i = 1 \cdot i = i$

111. $i^{-6} = \frac{1}{i^6} = \frac{1}{i^4 \cdot i^2} = \frac{1}{1 \cdot -1} = \frac{1}{-1} = -1$

113. $i^{-58} = \frac{1}{i^{58}} = \frac{1}{i^{56} \cdot i^2} = \frac{1}{1(-1)} = \frac{1}{-1} = -1$

115. a. $2x^2 + 18 = 0$
$2(3i)^2 + 18 \quad 0$
$2(9i^2) + 18 \quad 0$
$18i^2 + 18 \quad 0$
$-18 + 18 \quad 0$
$0 = 0$
Yes, $3i$ is a solution of $2x^2 + 18 = 0$.

b. $2x^2 + 18 = 0$
$2(-3i)^2 + 18 \quad 0$
$2(9i^2) + 18 \quad 0$
$18i^2 + 18 \quad 0$
$-18 + 18 \quad 0$
$0 = 0$
Yes, $-3i$ is a solution of $2x^2 + 18 = 0$.

c. $x^2 - 2x - 10 = 0$
$(1-3i)^2 - 2(1-3i) - 10 \quad 0$
$1 - 6i + 9i^2 - 2 + 6i - 10 \quad 0$
$9i^2 - 11 \quad 0$
$-9 - 11 \quad 0$
$-20 \neq 0$
No, $1 - 3i$ is not a solution of $x^2 - 2x - 10 = 0$.

117. a. $x^2 + 3 = 7$
$x^2 = 4$
$x = \pm 2$
x is an integer, a rational number, and a real number.

b. $x^2 + 1 = 0$
$x^2 = -1$
$x = \pm i$
x is an imaginary number.

c. $\frac{5}{8}x = \frac{2}{3}$
$\frac{8}{5} \cdot \frac{5}{8}x = \frac{8}{5} \cdot \frac{2}{3}$
$x = \frac{16}{15}$
x is a rational number and a real number.

d. $x^2 + 1 = 9$
$x^2 = 8$
$x = \pm 2\sqrt{2}$
x is a irrational number and a real number.

e. $x^{3/4} = 8$
$(x^{3/4})^{4/3} = (8)^{4/3}$
$x = (\sqrt[3]{8})^4 = 2^4 = 16$

x is an integer, a rational number, and a real number.

f. $\sqrt[3]{x} = -27$
$(\sqrt[3]{x})^3 = (-27)^3$
$x = -19{,}683$

x is an integer, a rational number, and a real number.

CHAPTER 7 REVIEW EXERCISES

1. $81^{-1/4} = (3^4)^{-1/4} = 3^{-1} = \dfrac{1}{3}$

2. $\dfrac{x^{-3/2}}{x^{7/2}} = x^{-10/2} = x^{-5} = \dfrac{1}{x^5}$

3. $(a^{16})^{-5/8} = a^{-10} = \dfrac{1}{a^{10}}$

4. $(16x^{-4}y^{12})(100x^6y^{-2})^{1/2}$
$= 16x^{-4}y^{12} \cdot (10^2)^{1/2} x^3 y^{-1}$
$= 160x^{-1}y^{11}$
$= \dfrac{160y^{11}}{x}$

5. $3x^{3/4} = 3\sqrt[4]{x^3}$

6. $7y\sqrt[3]{x^2} = 7x^{2/3}y$

7. $\sqrt[4]{81a^8b^{12}} = \sqrt[4]{3^4 a^8 b^{12}} = 3a^2b^3$

8. $-\sqrt{49x^6y^{16}} = -\sqrt{7^2 x^6 y^{16}} = -7x^3y^8$

9. $\sqrt[3]{-8a^6b^{12}} = \sqrt[3]{(-2)^3 a^6 b^{12}} = -2a^2b^4$

10. $\sqrt{18a^3b^6} = \sqrt{3^2 a^2 b^6 (2a)} = 3ab^3\sqrt{2a}$

11. $\sqrt[5]{-64a^8b^{12}} = \sqrt[5]{(-2)^5 a^5 b^{10} (2a^3b^2)}$
$= -2ab^2\sqrt[5]{2a^3b^2}$

12. $\sqrt[4]{x^6y^8z^{10}} = \sqrt[4]{x^4 y^8 z^8 \cdot x^2 z^2}$
$= xy^2z^2\sqrt[4]{x^2z^2}$

13. $\sqrt{54} + \sqrt{24} = \sqrt{3^2 \cdot 6} + \sqrt{2^2 \cdot 6}$
$= 3\sqrt{6} + 2\sqrt{6}$
$= 5\sqrt{6}$

14. $\sqrt{48x^5y} - x\sqrt{80x^3y}$
$= \sqrt{4^2 x^4 (3xy)} - x\sqrt{4^2 x^2 (5xy)}$
$= 4x^2\sqrt{3xy} - 4x^2\sqrt{5xy}$

15. $\sqrt{50a^4b^3} - ab\sqrt{18a^2b}$
$= \sqrt{5^2 a^4 b^2 (2b)} - ab\sqrt{3^2 a^2 (2b)}$
$= 5a^2b\sqrt{2b} - 3a^2b\sqrt{2b}$
$= 2a^2b\sqrt{2b}$

16. $4x\sqrt{12x^2y} + \sqrt{3x^4y} - x^2\sqrt{27y}$
$= 4x\sqrt{2^2 x^2 (3y)} + \sqrt{x^4(3y)} - x^2\sqrt{3^2(3y)}$
$= 8x^2\sqrt{3y} + x^2\sqrt{3y} - 3x^2\sqrt{3y}$
$= 6x^2\sqrt{3y}$

17. $\sqrt{32}\sqrt{50} = \sqrt{1600} = \sqrt{40^2} = 40$

18. $\sqrt[3]{16x^4y}\sqrt[3]{4xy^5} = \sqrt[3]{64x^5y^6}$
$= \sqrt[3]{4^3 x^3 y^6 (x^2)}$
$= 4xy^2\sqrt[3]{x^2}$

19. $\sqrt{3x}(3 + \sqrt{3x}) = 3\sqrt{3x} + (\sqrt{3x})^2$
$= 3\sqrt{3x} + 3x$
$= 3x + 3\sqrt{3x}$

20. $(5 - \sqrt{6})^2 = 25 - 10\sqrt{6} + \sqrt{6}^2$
$= 25 - 10\sqrt{6} + 6$
$= 31 - 10\sqrt{6}$

21. $(\sqrt{3} + 8)(\sqrt{3} - 2) = \sqrt{3}^2 + 6\sqrt{3} - 16$
$= 3 + 6\sqrt{3} - 16$
$= -13 + 6\sqrt{3}$

22. $\dfrac{\sqrt{125x^6}}{\sqrt{5x^3}} = \sqrt{\dfrac{125x^6}{5x^3}}$
$= \sqrt{25x^3}$
$= \sqrt{5^2 x^2 (x)}$
$= 5x\sqrt{x}$

23. $\dfrac{8}{\sqrt{3y}} = \dfrac{8}{\sqrt{3y}} \cdot \dfrac{\sqrt{3y}}{\sqrt{3y}} = \dfrac{8\sqrt{3y}}{\sqrt{3^2 y^2}} = \dfrac{8\sqrt{3y}}{3y}$

Chapter 7: Rational Exponents And Radicals

24. $\dfrac{x+2}{\sqrt{x}+\sqrt{2}} = \dfrac{x+2}{\sqrt{x}+\sqrt{2}} \cdot \dfrac{\sqrt{x}-\sqrt{2}}{\sqrt{x}-\sqrt{2}}$

    $= \dfrac{x\sqrt{x} - x\sqrt{2} + 2\sqrt{x} - 2\sqrt{2}}{\sqrt{x^2} - \sqrt{2^2}}$

    $= \dfrac{x\sqrt{x} - x\sqrt{2} + 2\sqrt{x} - 2\sqrt{2}}{x-2}$

25. $\dfrac{\sqrt{x}+\sqrt{y}}{\sqrt{x}-\sqrt{y}} = \dfrac{\sqrt{x}+\sqrt{y}}{\sqrt{x}-\sqrt{y}} \cdot \dfrac{\sqrt{x}+\sqrt{y}}{\sqrt{x}+\sqrt{y}}$

    $= \dfrac{\sqrt{x^2} + \sqrt{xy} + \sqrt{xy} + \sqrt{y^2}}{\sqrt{x^2} - \sqrt{y^2}}$

    $= \dfrac{x + 2\sqrt{xy} + y}{x - y}$

26. $\sqrt{-36} = \sqrt{-1}\sqrt{36} = i\sqrt{6^2} = 6i$

27. $\sqrt{-50} = i\sqrt{50} = i\sqrt{5^2 \cdot 2} = 5i\sqrt{2}$

28. $\sqrt{49} - \sqrt{-16} = \sqrt{7^2} - \sqrt{-1}\sqrt{16}$

    $= 7 - i\sqrt{4^2}$

    $= 7 - 4i$

29. $\sqrt{200} + \sqrt{-12} = \sqrt{10^2 \cdot 2} + \sqrt{-1}\sqrt{2^2}\sqrt{3}$

    $= 10\sqrt{2} + 2i\sqrt{3}$

30. $(5+2i)+(4-3i) = (5+4)+(2+(-3))i$

    $= 9 - i$

31. $(-8+3i)-(4-7i)$

    $= (-8+3i)+(-4+7i)$

    $= (-8+(-4))+(3+7)i$

    $= -12 + 10i$

32. $(9-\sqrt{-16})+(5+\sqrt{-36})$

    $= (9-4i)+(5+6i)$

    $= (9+5)+(-4+6)i$

    $= 14 + 2i$

33. $(\sqrt{50}+\sqrt{-72})-(\sqrt{162}-\sqrt{-8})$

    $= (\sqrt{5^2 \cdot 2} + i\sqrt{6^2 \cdot 2}) - (\sqrt{9^2 \cdot 2} - i\sqrt{2^2 \cdot 2})$

    $= (5\sqrt{2} + 6i\sqrt{2}) - (9\sqrt{2} - 2i\sqrt{2})$

    $= -4\sqrt{2} + 8i\sqrt{2}$

34. $(3-9i)+7 = 10 - 9i$

35. $(8i)(2i) = 16i^2 = 16(-1) = -16$

36. $i(3-7i) = 3i - 7i^2 = 3i - 7(-1) = 7 + 3i$

37. $\sqrt{-12}\sqrt{-6} = i\sqrt{12} \cdot i\sqrt{6}$

    $= i^2\sqrt{72}$

    $= (-1)\sqrt{6^2 \cdot 2}$

    $= -6\sqrt{2}$

38. $(6-5i)(4+3i) = 24 + 18i - 20i - 15i^2$

    $= 24 - 2i - 15(-1)$

    $= 24 + 15 - 2i$

    $= 39 - 2i$

39. $\dfrac{-6}{i} = -\dfrac{6}{i} \cdot \dfrac{i}{i} = \dfrac{-6i}{i^2} = \dfrac{-6i}{-1} = 6i$

40. $\dfrac{5+2i}{3i} = \dfrac{5+2i}{3i} \cdot \dfrac{-3i}{-3i}$

    $= \dfrac{-15i - 6i^2}{-9i^2}$

    $= \dfrac{-15i - 6(-1)}{-9(-1)}$

    $= \dfrac{-15i + 6}{9}$

    $= \dfrac{6}{9} - \dfrac{15}{9}i$

    $= \dfrac{2}{3} - \dfrac{5}{3}i$

41. $\dfrac{7}{2-i} = \dfrac{7}{2-i} \cdot \dfrac{2+i}{2+i}$

    $= \dfrac{14 + 7i}{2^2 + 1}$

    $= \dfrac{14 + 7i}{4 + 1}$

    $= \dfrac{14 + 7i}{5}$

    $= \dfrac{14}{5} + \dfrac{7}{5}i$

42. $\dfrac{\sqrt{16}}{\sqrt{4}-\sqrt{-4}} = \dfrac{4}{2 - 2i}$

    $= \dfrac{4}{2-2i} \cdot \dfrac{2+2i}{2+2i}$

    $= \dfrac{8 + 8i}{4 - 4i^2}$

    $= \dfrac{8 + 8i}{8}$

    $= 1 + i$

43. $\dfrac{5+9i}{1-i} = \dfrac{5+9i}{1-i} \cdot \dfrac{1+i}{1+i}$

$= \dfrac{5+5i+9i+9i^2}{1+1}$

$= \dfrac{5+14i+9(-1)}{2}$

$= \dfrac{5-9+14i}{2}$

$= \dfrac{-4+14i}{2}$

$= -2+7i$

44. $\sqrt[3]{9x} = -6$

$(\sqrt[3]{9x})^3 = (-6)^3$

$9x = -216$

$x = -24$

Check:

$\sqrt[3]{9x} = -6$	
$\sqrt[3]{9(-24)}$	-6
$\sqrt[3]{-216}$	-6
$-6 = -6$	

The solution is -24.

45. The function f contains an even root. The radicand must be greater than or equal to zero.

$3x - 2 \geq 0$

$3x \geq 2$

$x \geq \dfrac{2}{3}$

The domain is $\left\{ x \middle| x \geq \dfrac{2}{3} \right\}$.

46. The function f contains an odd root. The radicand may be positive or negative.

The domain is $\{x | x \text{ is a real number}\}$.

47.

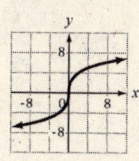

48.

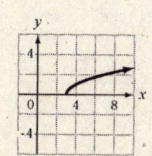

49. $\sqrt[3]{3x-5} = 2$

$(\sqrt[3]{3x-5})^3 = 2^3$

$3x - 5 = 8$

$3x = 13$

$x = \dfrac{13}{3}$

Check:

$\sqrt[3]{3x-5} = 2$	
$\sqrt[3]{3 \cdot \dfrac{13}{3} - 5}$	2
$\sqrt[3]{13-5}$	2
$\sqrt[3]{8}$	2
$2 = 2$	

The solution is $\dfrac{13}{3}$.

50. $\sqrt{4x+9} + 10 = 11$

$\sqrt{4x+9} = 1$

$(\sqrt{4x+9})^2 = 1^2$

$4x + 9 = 1$

$4x = -8$

$x = -2$

Check:

$\sqrt{4x+9} + 10 = 11$	
$\sqrt{4(-2)+9} + 10$	11
$\sqrt{1} + 10$	11
$1 + 10$	11
$11 = 11$	

The solution is -2.

51. **Strategy**
To find the width of the rectangle, use the Pythagorean Theorem. The unknown width is one leg, the length is the other leg, and the diagonal is the hypotenuse.

Solution

$a^2 + b^2 = c^2$

$a^2 + (12)^2 = (13)^2$

$a^2 + 144 = 169$

$a^2 = 25$

$a = 5$

The width is 5 in.

52. **Strategy**
To find the amount of power, replace v in the equation with the given value and solve for p.

Solution
$$v = 4.05\sqrt[3]{P}$$
$$20 = 4.05\sqrt[3]{P}$$
$$\frac{20}{4.05} \approx \sqrt[3]{P}$$
$$\left(\frac{20}{4.05}\right)^3 = \left(\sqrt[3]{P}\right)^3$$
$$120 \approx P$$
The amount of power is 120 watts.

53. **Strategy**
To find the distance required, replace v and a in the equation with the given values and solve for s.

Solution
$$v = \sqrt{2as}$$
$$88 = \sqrt{2 \cdot 16s}$$
$$7744 = 32s$$
$$242 = s$$
The distance required is 242 feet.

54. **Strategy**
To find the distance, use the Pythagorean Theorem. The hypotenuse is the length of the ladder (12 ft). One leg is the height on the building that the ladder reaches (10 ft). The distance from the bottom of the ladder to the building is the other leg.

Solution
$$c^2 = a^2 + b^2$$
$$12^2 = 10^2 + b^2$$
$$144 = 100 + b^2$$
$$44 = b^2$$
$$44^{1/2} = (b^2)^{1/2}$$
$$\sqrt{44} = b$$
$$6.63 = b$$
The distance is 6.63 feet.

CHAPTER 7 TEST

1. $\dfrac{r^{2/3}r^{-1}}{r^{-1/2}} = \dfrac{r^{-1/3}}{r^{-1/2}} = r^{1/6}$

2. $\dfrac{(2x^{1/3}y^{-2/3})^6}{(x^{-4}y^8)^{1/4}} = \dfrac{2^6 x^2 y^{-4}}{x^{-1}y^2} = 2^6 x^3 y^{-6} = \dfrac{64x^3}{y^6}$

3. $\left(\dfrac{4a^4}{b^2}\right)^{-3/2} = \dfrac{4^{-3/2}a^{-6}}{b^{-3}}$
$= (2^2)^{-3/2} a^{-6} b^3$
$= 2^{-3} a^{-6} b^3$
$= \dfrac{b^3}{8a^6}$

4. $3y^{2/5} = 3\sqrt[5]{y^2}$

5. $\dfrac{1}{2}\sqrt[4]{x^3} = \dfrac{1}{2}x^{3/4}$

6. The function f contains an even root. The radicand must be greater than or equal to zero.
$$4 - x \geq 0$$
$$-x \geq -4$$
$$x \leq 4$$
The domain is $\{x \mid x \leq 4\}$.

7. The function f contains an odd root. The radicand may be positive or negative.

The domain is $(-\infty, \infty)$.

8. $\sqrt[3]{27a^4 b^3 c^7} = \sqrt[3]{3^3 a^3 b^3 c^6 (ac)} = 3abc^2 \sqrt[3]{ac}$

9. $\sqrt{18a^3} + a\sqrt{50a} = \sqrt{3^2 a^2 (2a)} + a\sqrt{5^2 (2a)}$
$= 3a\sqrt{2a} + 5a\sqrt{2a}$
$= 8a\sqrt{2a}$

10. $\sqrt[3]{54x^7 y^3} - x\sqrt[3]{128x^4 y^3} - x^2 \sqrt[3]{2xy^3}$
$= \sqrt[3]{3^3 x^6 y^3 (2x)} - x\sqrt[3]{4^3 x^3 y^3 (2x)} - x^2 \sqrt[3]{y^3 (2x)}$
$= 3x^2 y\sqrt[3]{2x} - 4x^2 y\sqrt[3]{2x} - x^2 y\sqrt[3]{2x}$
$= -2x^2 y\sqrt[3]{2x}$

11. $\sqrt{3x}(\sqrt{x} - \sqrt{25x}) = \sqrt{3x^2} - \sqrt{75x^2}$
$= \sqrt{x^2 (3)} - \sqrt{5^2 x^2 (3)}$
$= x\sqrt{3} - 5x\sqrt{3}$
$= -4x\sqrt{3}$

12. $(2\sqrt{3} + 4)(3\sqrt{3} - 1) = 6\sqrt{3^2} - 2\sqrt{3} + 12\sqrt{3} - 4$
$= 18 + 10\sqrt{3} - 4$
$= 14 + 10\sqrt{3}$

13. $(\sqrt{a} - 3\sqrt{b})(2\sqrt{a} + 5\sqrt{b})$
$= 2\sqrt{a^2} + 5\sqrt{ab} - 6\sqrt{ab} - 15\sqrt{b^2}$
$= 2a - \sqrt{ab} - 15b$

14. $(2\sqrt{x} + \sqrt{y})^2 = 4\sqrt{x^2} + 4\sqrt{xy} + \sqrt{y^2}$
$= 4x + 4\sqrt{xy} + y$

15. $\dfrac{\sqrt{32x^5 y}}{\sqrt{2xy^3}} = \sqrt{\dfrac{32x^5 y}{2xy^3}} = \sqrt{\dfrac{16x^4}{y^2}} = \sqrt{\dfrac{4^2 x^4}{y^2}} = \dfrac{4x^2}{y}$

16. $\dfrac{4-2\sqrt{5}}{2-\sqrt{5}} = \dfrac{4-2\sqrt{5}}{2-\sqrt{5}} \cdot \dfrac{2+\sqrt{5}}{2+\sqrt{5}}$

$= \dfrac{8+4\sqrt{5}-4\sqrt{5}-2\sqrt{5}^2}{2^2 - \sqrt{5}^2}$

$= \dfrac{8-2\cdot 5}{4-5}$

$= \dfrac{8-10}{-1}$

$= \dfrac{-2}{-1}$

$= 2$

17. $\dfrac{\sqrt{x}}{\sqrt{x}-\sqrt{y}} = \dfrac{\sqrt{x}}{\sqrt{x}-\sqrt{y}} \cdot \dfrac{\sqrt{x}+\sqrt{y}}{\sqrt{x}+\sqrt{y}}$

$= \dfrac{\sqrt{x^2}+\sqrt{xy}}{\sqrt{x^2}-\sqrt{y^2}}$

$= \dfrac{x+\sqrt{xy}}{x-y}$

18. $(\sqrt{-8})(\sqrt{-2}) = i\sqrt{8}\cdot i\sqrt{2} = i^2\sqrt{16} = -1\cdot 4 = -4$

19. $(5-2i)-(8-4i) = -3+2i$

20. $(2+5i)(4-2i) = 8-4i+20i-10i^2$

$= 8+16i-10(-1)$

$= 8+16i+10$

$= 18+16i$

21. $\dfrac{2+3i}{1-2i} = \dfrac{2+3i}{1-2i} \cdot \dfrac{1+2i}{1+2i}$

$= \dfrac{2+4i+3i+6i^2}{1+4}$

$= \dfrac{2+7i+6(-1)}{5}$

$= \dfrac{2-6+7i}{5} = \dfrac{-4+7i}{5}$

$= -\dfrac{4}{5}+\dfrac{7}{5}i$

22. $(2+i)+(2-i) = 4$

23. $\sqrt{x+12}-\sqrt{x} = 2$

$\sqrt{x+12} = 2+\sqrt{x}$

$(\sqrt{x+12})^2 = (2+\sqrt{x})^2$

$x+12 = 4+4\sqrt{x}+x$

$12 = 4+4\sqrt{x}$

$8 = 4\sqrt{x}$

$2 = \sqrt{x}$

$2^2 = (\sqrt{x})^2$

$4 = x$

Check:

$\sqrt{x+12}-\sqrt{x} = 2$	
$\sqrt{4+12}-\sqrt{4}$	2
$\sqrt{16}-\sqrt{4}$	2
$4-2$	2
$2 = 2$	

The solution is 4.

24. $\sqrt[3]{2x-2}+4 = 2$

$\sqrt[3]{2x-2} = -2$

$(\sqrt[3]{2x-2})^3 = (-2)^3$

$2x-2 = -8$

$2x = -6$

$x = -3$

Check:

$\sqrt[3]{2x-2}+4 = 2$	
$\sqrt[3]{2(-3)-2}+4$	2
$\sqrt[3]{-8}+4$	2
$-2+4$	2
$2 = 2$	

The solution is −3.

25. **Strategy**
To find the distance, use the Pythagorean Theorem. The hypotenuse is the length of the guy wire. The distance along the ground from the pole to the wire (6 ft) is one leg.

Solution
$c^2 = a^2 + b^2$
$c^2 = 6^2 + 30^2$
$c^2 = 36 + 900$
$c^2 = 936$
$\sqrt{c^2} = \sqrt{936}$
$c \approx 30.6$
The length of the wire is 30.6 ft.

CUMULATIVE REVIEW EXERCISES

1. The Distributive Property

2. $2x - 3[x - 2(x - 4) + 2x]$
   $= 2x - 3[x - 2x + 8 + 2x]$
   $= 2x - 3[x + 8]$
   $= 2x - 3x - 24$
   $= -x - 24$

3. $A \cap B = \emptyset$

4. $\sqrt[3]{2x - 5} + 3 = 6$
   $\sqrt[3]{2x - 5} = 3$
   $(\sqrt[3]{2x - 5})^3 = 3^3$
   $2x - 5 = 27$
   $2x = 32$
   $x = 16$
   Check:
   $$\begin{array}{c|c} \sqrt[3]{2x - 5} + 3 = 6 \\ \hline \sqrt[3]{2(16) - 5} + 3 & 6 \\ \sqrt[3]{32 - 5} + 3 & 6 \\ \sqrt[3]{27} + 3 & 6 \\ 6 = 6 \end{array}$$
   The solution is 16.

5. $5 - \dfrac{2}{3}x = 4$
   $5 - \dfrac{2}{3}x - 5 = 4 - 5$
   $-\dfrac{2}{3}x = -1$
   $\left(-\dfrac{3}{2}\right)\left(-\dfrac{2}{3}\right)x = -1\left(-\dfrac{3}{2}\right)$
   $x = \dfrac{3}{2}$
   The solution is $\dfrac{3}{2}$.

6. $2[4 - 2(3 - 2x)] = 4(1 - x)$
   $2[4 - 6 + 4x] = 4 - 4x$
   $2[-2 + 4x] = 4 - 4x$
   $-4 + 8x = 4 - 4x$
   $-4 + 8x + 4x = 4 - 4x + 4x$
   $12x - 4 = 4$
   $12x - 4 + 4 = 4 + 4$
   $12x = 8$
   $\left(\dfrac{1}{12}\right)12x = \dfrac{1}{12}(8)$
   $x = \dfrac{2}{3}$
   The solution is $\dfrac{2}{3}$.

7. $3x - 4 \le 8x + 1$
   $-5x - 4 \le 1$
   $-5x \le 5$
   $x \ge -1$
   $\{x | x \ge -1\}$

8. $5 < 2x - 3 < 7$
   $5 + 3 < 2x - 3 + 3 < 7 + 3$
   $8 < 2x < 10$
   $\dfrac{1}{2} \cdot 8 < \dfrac{1}{2} \cdot 2x < \dfrac{1}{2} \cdot 10$
   $4 < x < 5$
   $\{x | 4 < x < 5\}$

9. $|7 - 3x| > 1$
   $7 - 3x < -1$ or $7 - 3x > 1$
   $-3x < -8$ $\quad$ $-3x > -6$
   $x > \dfrac{8}{3}$ $\quad$ $x < 2$
   $\left\{x | x > \dfrac{8}{3}\right\}$ $\quad$ $\{x | x < 2\}$
   $\left\{x | x > \dfrac{8}{3}\right\} \cup \{x | x < 2\} = \left\{x | x < 2 \text{ or } x > \dfrac{8}{3}\right\}$

10. $64a^2 - b^2 = (8a)^2 - b^2 = (8a + b)(8a - b)$

11. $x^5 + 2x^3 - 3x = x(x^4 + 2x^2 - 3)$
    $= x(x^2 + 3)(x^2 - 1)$
    $= x(x^2 + 3)(x + 1)(x - 1)$

12. $3x^2 + 13x - 10 = 0$
    $(3x - 2)(x + 5) = 0$
    $3x - 2 = 0 \quad x + 5 = 0$
    $3x = 2 \quad\quad x = -5$
    $x = \dfrac{2}{3}$
    The solutions are $\dfrac{2}{3}$ and -5.

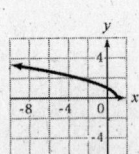

13.

14. $x - 2y = 4 \quad\quad 2x + y = 4$
    $-2y = -x + 4 \quad y = -2x + 4$
    $y = \dfrac{1}{2}x - 2$
    $m_1 m_2 = \dfrac{1}{2}(-2) = -1$
    Yes, the lines are perpendicular.

15. $(3^{-1}x^3y^{-5})(3^{-1}y^{-2})^{-2} = (3^{-1}x^3y^{-5})(3^2y^4)$
$= \dfrac{x^3}{3y^5}(3^2y^4)$
$= \dfrac{3x^3}{y}$

16. $\left(\dfrac{x^{-1/2}y^{3/4}}{y^{-5/4}}\right)^4 = \left(\dfrac{y^2}{x^{1/2}}\right)^4 = \dfrac{y^8}{x^2}$

17. $\sqrt{20x^3} - x\sqrt{45x} = \sqrt{2^2x^2 \cdot 5x} - x\sqrt{3^2 \cdot 5x}$
$= \sqrt{2^2x^2}\sqrt{5x} - x\sqrt{3^2}\sqrt{5x}$
$= 2x\sqrt{5x} - 3x\sqrt{5x}$
$= -x\sqrt{5x}$

18. $(\sqrt{5}-3)(\sqrt{5}-2) = 5 - 2\sqrt{5} - 3\sqrt{5} + 6$
$= 11 - 5\sqrt{5}$

19. $\dfrac{\sqrt[3]{4x^5y^4}}{\sqrt[3]{8x^2y^5}} = \sqrt[3]{\dfrac{4x^5y^4}{8x^2y^5}}$
$= \sqrt[3]{\dfrac{x^3}{2y}}$
$= \dfrac{\sqrt[3]{x^3}}{\sqrt[3]{2y}}$
$= \dfrac{x}{\sqrt[3]{2y}} \cdot \dfrac{\sqrt[3]{4y^2}}{\sqrt[3]{4y^2}}$
$= \dfrac{x\sqrt[3]{4y^2}}{\sqrt[3]{2^3y^3}}$
$= \dfrac{x\sqrt[3]{4y^2}}{2y}$

20. $\dfrac{3i}{2-i} = \dfrac{3i}{2-i} \cdot \dfrac{2+i}{2+i}$
$= \dfrac{6i + 3i^2}{2^2 - i^2}$
$= \dfrac{6i + 3(-1)}{4 - (-1)}$
$= \dfrac{-3 + 6i}{5}$
$= -\dfrac{3}{5} + \dfrac{6}{5}i$

21.

22. The function g contains an odd root. The radicand may be positive or negative.
The domain is $\{x \mid x \in \text{ real numbers.}\}$.

23. $f(x) = 3x^2 - 2x + 1$
$f(-3) = 3(-3)^2 - 2(-3) + 1$
$f(-3) = 27 + 6 + 1$
$f(-3) = 34$
The value of $f(-3)$ is 34.

24. First find the slope of the line.
$m = \dfrac{y_2 - y_1}{x_2 - x_1} = \dfrac{2-3}{-1-2} = \dfrac{-1}{-3} = \dfrac{1}{3}$
Use the point-slope form to find an equation of the line.
$y - y_1 = m(x - x_1)$
$y - 3 = \dfrac{1}{3}(x - 2)$
$y - 3 = \dfrac{1}{3}x - \dfrac{2}{3}$
$y = \dfrac{1}{3}x + \dfrac{7}{3}$

An equation of the line is $y = \dfrac{1}{3}x + \dfrac{7}{3}$.

25. $\begin{vmatrix} 1 & 2 & -3 \\ 0 & -1 & 2 \\ 3 & 1 & -2 \end{vmatrix} = 1 \cdot \begin{vmatrix} -1 & 2 \\ 1 & -2 \end{vmatrix} - 2\begin{vmatrix} 0 & 2 \\ 3 & -2 \end{vmatrix} - 3\begin{vmatrix} 0 & -1 \\ 3 & 1 \end{vmatrix}$
$= 1 \cdot 0 - 2(-6) - 3 \cdot 3$
$= 3$

26. $2x - y = 4$
$-2x + 3y = 5$
$D = \begin{vmatrix} 2 & -1 \\ -2 & 3 \end{vmatrix} = 4$
$D_x = \begin{vmatrix} 4 & -1 \\ 5 & 3 \end{vmatrix} = 17$
$D_y = \begin{vmatrix} 2 & 4 \\ -2 & 5 \end{vmatrix} = 18$
$x = \dfrac{D_x}{D} = \dfrac{17}{4}$
$y = \dfrac{D_y}{D} = \dfrac{18}{4} = \dfrac{9}{2}$
The solution is $\left(\dfrac{17}{4}, \dfrac{9}{2}\right)$.

27. Find the *y*-intercept at $x = 0$.
$3(0) - 2y = -6$
$y = 3$ The *y*-intercept is (0, 3).
To find the slope, find the *x*-intercept and use it to get the slope
$3x - 2(0) = -6$
$x = -2$ The *x*-intercept is (-2, 0).
$m = \dfrac{y_2 - y_1}{x_2 - x_1}$
$m = \dfrac{3 - 0}{0 - (-2)} = \dfrac{3}{2}$
The slope is $\dfrac{3}{2}$, and the *y*-intercept is (0, 3).

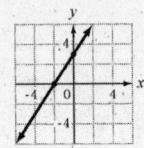

28. $3x + 2y \leq 4$
$2y \leq -3x + 4$
$y \leq -\dfrac{3}{2}x + 2$
Sketch the solid line $y = -\dfrac{3}{2}x + 2$. Shade below the solid line.

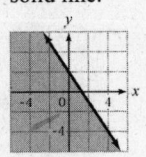

29. Strategy
Number of 18¢ stamps: *x*
Number of 13¢ stamps: 30 − *x*

Stamps	Number	Value	Total Value
18¢	*x*	18	18*x*
13¢	30 − *x*	13	13(30 − *x*)

The sum of the total values of each type of stamp equals the total value of the stamps (485¢).
$18x + 13(30 - x) = 485$

Solution
$18x + 13(30 - x) = 485$
$18x + 390 - 13x = 485$
$5x + 390 = 485$
$5x = 95$
$x = 19$
There are nineteen 18¢ stamps.

30. Strategy
Amount invested at 8.4%: *x*

	Principal	Rate	Interest
Amount invested at 7.2%	2500	0.072	0.072(2500)
Amount invested at 8.4%	*x*	0.084	0.084*x*

The total amount of interest earned is $516.
$0.072(2500) + 0.084x = 516$

Solution
$0.072(2500) + 0.084x = 516$
$180 + 0.084x = 516$
$0.084x = 336$
$x = 4000$
The additional investment must be $4000.

31. Strategy
Length of the rectangle: *x*
Width of the rectangle: *x* − 6
Use the equation for the area of a rectangle, $A = L \cdot W$.

Solution
$A = L \cdot W$
$72 = x(x - 6)$
$72 = x^2 - 6x$
$0 = x^2 - 6x - 72$
$0 = (x - 12)(x + 6)$
$x - 12 = 0 \quad x + 6 = 0$
$\quad x = 12 \quad\quad x = -6$
The length cannot be negative, so −6 is not a solution.
$x - 6 = 12 - 6 = 6$
The length is 12 ft and the width is 6 ft.

32. **Strategy**
Unknown rate of the car: x
Unknown rate of the plane: $5x$

	Distance	Rate	Time
Car	25	x	$\dfrac{25}{x}$
Plane	625	$5x$	$\dfrac{625}{5x}$

The total time of the trip was 3 h.
$$\frac{25}{x} + \frac{625}{5x} = 3$$

Solution
$$5x\left(\frac{25}{x} + \frac{625}{5x}\right) = 3(5x)$$
$$125 + 625 = 15x$$
$$750 = 15x$$
$$50 = x$$
$$250 = 5x$$
The rate of the plane is 250 mph.

33. **Strategy**
To find the time it takes light to travel from the earth to the moon, use the formula $RT = D$, substituting for R and D and solving for T.

Solution
$$RT = D$$
$$1.86 \times 10^5 \cdot T = 232,500$$
$$1.86 \times 10^5 \cdot T = 2.325 \times 10^5$$
$$T = 1.25 \times 10^0$$
$$T = 1.25$$
The time is 1.25 seconds.

34. **Strategy**
To find the height of the periscope, replace d in the equation by the given value and solve for h.

Solution
$$d = \sqrt{1.5h}$$
$$7 = \sqrt{1.5h}$$
$$7^2 = 1.5h$$
$$\frac{7^2}{1.5} = h$$
$$32.7 = h$$
The height of the periscope is 32.7 ft.

35. Slope $m = \dfrac{y_2 - y_1}{x_2 - x_1} = \dfrac{400 - 0}{5000 - 0} = \dfrac{400}{5000} = 0.08$

The slope represents the simple interest on the investment. The interest rate is 8%.

CHAPTER 8: QUADRATIC EQUATIONS AND INEQUALITIES

CHAPTER 8 PREP TEST

1. $\sqrt{18} = \sqrt{2 \cdot 9} = 3\sqrt{2}$ [7.2.1]

2. $\sqrt{-9} = 3i$ [7.2.1]

3. $\dfrac{3x-2}{x-1} - 1 = \dfrac{3x-2}{x-1} - \dfrac{x-1}{x-1}$ [6.2.2]
   $= \dfrac{3x-2-x+1}{x-1}$
   $= \dfrac{2x-1}{x-1}$

4. $b^2 - 4ac$
   $(-4)^2 - 4(2)(1) = 16 - 8 = 8$ [1.3.2]

5. $4x^2 + 28x + 49 = (2x+7)^2$ [5.6.1]
   Yes

6. $4x^2 - 4x + 1 = (2x-1)^2$ [5.6.1]

7. $9x^2 - 4 = (3x+2)(3x-2)$ [5.6.1]

8. $\{x \mid x < -1\} \cap \{x \mid x < 4\}$ [1.1.2]

9. $x(x-1) = x + 15$ [5.7.1]
   $x^2 - x = x + 15$
   $x^2 - 2x - 15 = 0$
   $(x-5)(x+3) = 0$
   $x - 5 = 0 \quad x + 3 = 0$
   $x = 5 \quad x = -3$
   The solutions are $-3, 5$.

10. $\dfrac{4}{x-3} = \dfrac{16}{x}$ [6.4.1]
    $x(x-3)\dfrac{4}{x-3} = x(x-3)\dfrac{16}{x}$
    $4x = 16x - 48$
    $-12x = -48$
    $x = 4$
    The solution is 4.

Section 8.1

Concept Review 8.1

1. Sometimes true
   A quadratic equation may have two real roots, one real root, or two imaginary roots.

3. Never true
   The Principle of Zero Products states that at least one of the factors must be zero if the product is zero. In this case the product is 8.

Objective 8.1.1 Exercises

1. If $a = 0$ in $ax^2 + bx + c = 0$, then there is no second-degree term in the equation and it is, therefore, not a quadratic equation.

3. $2x^2 - 4x - 5 = 0$
   $a = 2, b = -4, c = -5$

5. $4x^2 - 5x - 6 = 0$
   $a = 4, b = -5, c = -6$

7. • The equation is in standard form.
   $x^2 - 49 = 0$
   $(x-7)(x+7) = 0$
   • Use the Principle of Zero Products to set each factor equal to 0.
   $x + 7 = 0 \quad\quad x - 7 = 0$
   $x = -7 \quad\quad x = 7$

9. $x^2 - 4x = 0$
   $x(x-4) = 0$
   $x = 0 \quad\quad x - 4 = 0$
   $\quad\quad\quad\quad\quad x = 4$
   The solutions are 0 and 4.

11. $t^2 - 25 = 0$
    $(t-5)(t+5) = 0$
    $t - 5 = 0 \quad\quad t + 5 = 0$
    $t = 5 \quad\quad\quad t = -5$
    The solutions are 5 and -5.

13. $s^2 - s - 6 = 0$
    $(s-3)(s+2) = 0$
    $s - 3 = 0 \quad\quad s + 2 = 0$
    $s = 3 \quad\quad\quad s = -2$
    The solutions are 3 and -2.

15. $y^2 - 6y + 9 = 0$
    $(y-3)(y-3) = 0$
    $y - 3 = 0 \quad\quad y - 3 = 0$
    $y = 3 \quad\quad\quad y = 3$
    The solution is 3.

17. $9z^2 - 18z = 0$
    $9z(z-2) = 0$
    $9z = 0 \quad\quad z - 2 = 0$
    $z = 0 \quad\quad\quad z = 2$
    The solutions are 0 and 2.

19. $r^2 - 3r = 10$
$r^2 - 3r - 10 = 0$
$(r-5)(r+2) = 0$
$r - 5 = 0 \quad\quad r + 2 = 0$
$r = 5 \quad\quad r = -2$
The solutions are 5 and –2.

21. $v^2 + 10 = 7v$
$v^2 - 7v + 10 = 0$
$(v-2)(v-5) = 0$
$v - 2 = 0 \quad\quad v - 5 = 0$
$v = 2 \quad\quad v = 5$
The solutions are 2 and 5.

23. $2x^2 - 9x - 18 = 0$
$(x-6)(2x+3) = 0$
$x - 6 = 0 \quad\quad 2x + 3 = 0$
$x = 6 \quad\quad 2x = -3$
$\quad\quad\quad\quad\quad x = -\dfrac{3}{2}$
The solutions are 6 and $-\dfrac{3}{2}$.

25. $4z^2 - 9z + 2 = 0$
$(z-2)(4z-1) = 0$
$z - 2 = 0 \quad\quad 4z - 1 = 0$
$z = 2 \quad\quad 4z = 1$
$\quad\quad\quad\quad\quad z = \dfrac{1}{4}$
The solutions are 2 and $\dfrac{1}{4}$.

27. $3w^2 + 11w = 4$
$3w^2 + 11w - 4 = 0$
$(3w-1)(w+4) = 0$
$3w - 1 = 0 \quad\quad w + 4 = 0$
$3w = 1 \quad\quad w = -4$
$w = \dfrac{1}{3}$
The solutions are $\dfrac{1}{3}$ and –4.

29. $6x^2 = 23x + 18$
$6x^2 - 23x - 18 = 0$
$(2x-9)(3x+2) = 0$
$2x - 9 = 0 \quad\quad 3x + 2 = 0$
$2x = 9 \quad\quad 3x = -2$
$x = \dfrac{9}{2} \quad\quad x = -\dfrac{2}{3}$
The solutions are $\dfrac{9}{2}$ and $-\dfrac{2}{3}$.

31. $4 - 15u - 4u^2 = 0$
$(1-4u)(4+u) = 0$
$1 - 4u = 0 \quad\quad 4 + u = 0$
$-4u = -1 \quad\quad u = -4$
$u = \dfrac{1}{4}$
The solutions are $\dfrac{1}{4}$ and –4.

33. $x + 18 = x(x-6)$
$x + 18 = x^2 - 6x$
$0 = x^2 - 7x - 18$
$0 = (x-9)(x+2)$
$x - 9 = 0 \quad\quad x + 2 = 0$
$x = 9 \quad\quad x = -2$
The solutions are 9 and –2.

35. $4s(s+3) = s - 6$
$4s^2 + 12s = s - 6$
$4s^2 + 11s + 6 = 0$
$(s+2)(4s+3) = 0$
$s + 2 = 0 \quad\quad 4s + 3 = 0$
$s = -2 \quad\quad 4s = -3$
$\quad\quad\quad\quad\quad s = -\dfrac{3}{4}$
The solutions are –2 and $-\dfrac{3}{4}$.

37. $u^2 - 2u + 4 = (2u-3)(u+2)$
$u^2 - 2u + 4 = 2u^2 + u - 6$
$0 = u^2 + 3u - 10$
$0 = (u-2)(u+5)$
$u - 2 = 0 \quad\quad u + 5 = 0$
$u = 2 \quad\quad u = -5$
The solutions are 2 and –5.

39. $(3x-4)(x+4) = x^2 - 3x - 28$
$3x^2 + 8x - 16 = x^2 - 3x - 28$
$2x^2 + 11x + 12 = 0$
$(x+4)(2x+3) = 0$
$x + 4 = 0 \quad\quad 2x + 3 = 0$
$x = -4 \quad\quad 2x = -3$
$\quad\quad\quad\quad\quad x = -\dfrac{3}{2}$
The solutions are –4 and $-\dfrac{3}{2}$.

242 Chapter 8: Quadratic Equations And Inequalities

41. $x^2 - 9bx + 14b^2 = 0$
$(x - 2b)(x - 7b) = 0$
$x - 2b = 0 \qquad x - 7b = 0$
$x = 2b \qquad x = 7b$
The solutions are $2b$ and $7b$.

43. $x^2 - 6cx - 7c^2 = 0$
$(x - 7c)(x + c) = 0$
$x - 7c = 0 \qquad x + c = 0$
$x = 7c \qquad x = -c$
The solutions are $7c$ and $-c$.

45. $2x^2 + 3bx + b^2 = 0$
$(2x + b)(x + b) = 0$
$2x + b = 0 \qquad x + b = 0$
$2x = -b \qquad x = -b$
$x = -\dfrac{b}{2}$
The solutions are $-\dfrac{b}{2}$ and $-b$.

47. $3x^2 - 14ax + 8a^2 = 0$
$(x - 4a)(3x - 2a) = 0$
$x - 4a = 0 \qquad 3x - 2a = 0$
$x = 4a \qquad 3x = 2a$
$x = \dfrac{2a}{3}$
The solutions are $4a$ and $\dfrac{2a}{3}$.

49. a. One positive and one negative solution
  b. Two negative solutions
  c. Two positive solutions

Objective 8.1.2 Exercises

51. If r is a solution to $ax^2 + bx + c = 0$, then $x - r$ is a factor of $ax^2 + bx + c$.

53. $(x - r_1)(x - r_2) = 0$
$(x - 2)(x - 5) = 0$
$x^2 - 7x + 10 = 0$

55. $(x - r_1)(x - r_2) = 0$
$[x - (-2)][x - (-4)] = 0$
$(x + 2)(x + 4) = 0$
$x^2 + 6x + 8 = 0$

57. $(x - r_1)(x - r_2) = 0$
$(x - 6)[x - (-1)] = 0$
$(x - 6)(x + 1) = 0$
$x^2 - 5x - 6 = 0$

59. $(x - r_1)(x - r_2) = 0$
$(x - 3)[x - (-3)] = 0$
$(x - 3)(x + 3) = 0$
$x^2 - 9 = 0$

61. $(x - r_1)(x - r_2) = 0$
$(x - 4)(x - 4) = 0$
$x^2 - 8x + 16 = 0$

63. $(x - r_1)(x - r_2) = 0$
$(x - 0)(x - 5) = 0$
$x(x - 5) = 0$
$x^2 - 5x = 0$

65. $(x - r_1)(x - r_2) = 0$
$(x - 0)(x - 3) = 0$
$x(x - 3) = 0$
$x^2 - 3x = 0$

67. $(x - r_1)(x - r_2) = 0$
$(x - 3)\left(x - \dfrac{1}{2}\right) = 0$
$x^2 - \dfrac{7}{2}x + \dfrac{3}{2} = 0$
$2\left(x^2 - \dfrac{7}{2}x + \dfrac{3}{2}\right) = 2 \cdot 0$
$2x^2 - 7x + 3 = 0$

69. $(x - r_1)(x - r_2) = 0$
$\left[x - \left(-\dfrac{3}{4}\right)\right](x - 2) = 0$
$\left(x + \dfrac{3}{4}\right)(x - 2) = 0$
$x^2 - \dfrac{5}{4}x - \dfrac{3}{2} = 0$
$4\left(x^2 - \dfrac{5}{4}x - \dfrac{3}{2}\right) = 4 \cdot 0$
$4x^2 - 5x - 6 = 0$

71.
$(x-r_1)(x-r_2) = 0$
$\left[x-\left(-\frac{5}{3}\right)\right][x-(-2)] = 0$
$\left(x+\frac{5}{3}\right)(x+2) = 0$
$x^2 + \frac{11}{3}x + \frac{10}{3} = 0$
$3\left(x^2 + \frac{11}{3}x + \frac{10}{3}\right) = 3 \cdot 0$
$3x^2 + 11x + 10 = 0$

73.
$(x-r_1)(x-r_2) = 0$
$\left[x-\left(-\frac{2}{3}\right)\right]\left(x-\frac{2}{3}\right) = 0$
$\left(x+\frac{2}{3}\right)\left(x-\frac{2}{3}\right) = 0$
$x^2 - \frac{4}{9} = 0$
$9\left(x^2 - \frac{4}{9}\right) = 9 \cdot 0$
$9x^2 - 4 = 0$

75.
$(x-r_1)(x-r_2) = 0$
$\left(x-\frac{1}{2}\right)\left(x-\frac{1}{3}\right) = 0$
$x^2 - \frac{5}{6}x + \frac{1}{6} = 0$
$6\left(x^2 - \frac{5}{6}x + \frac{1}{6}\right) = 6 \cdot 0$
$6x^2 - 5x + 1 = 0$

77.
$(x-r_1)(x-r_2) = 0$
$\left(x-\frac{6}{5}\right)\left[x-\left(-\frac{1}{2}\right)\right] = 0$
$\left(x-\frac{6}{5}\right)\left(x+\frac{1}{2}\right) = 0$
$x^2 - \frac{7}{10}x - \frac{3}{5} = 0$
$10\left(x^2 - \frac{7}{10}x - \frac{3}{5}\right) = 10 \cdot 0$
$10x^2 - 7x - 6 = 0$

79.
$(x-r_1)(x-r_2) = 0$
$\left[x-\left(-\frac{1}{4}\right)\right]\left[x-\left(-\frac{1}{2}\right)\right] = 0$
$\left(x+\frac{1}{4}\right)\left(x+\frac{1}{2}\right) = 0$
$x^2 + \frac{3}{4}x + \frac{1}{8} = 0$
$8\left(x^2 + \frac{3}{4}x + \frac{1}{8}\right) = 8 \cdot 0$
$8x^2 + 6x + 1 = 0$

81.
$(x-r_1)(x-r_2) = 0$
$\left(x-\frac{3}{5}\right)\left[x-\left(-\frac{1}{10}\right)\right] = 0$
$\left(x-\frac{3}{5}\right)\left(x+\frac{1}{10}\right) = 0$
$x^2 - \frac{1}{2}x - \frac{3}{50} = 0$
$50\left(x^2 - \frac{1}{2}x - \frac{3}{50}\right) = 50 \cdot 0$
$50x^2 - 25x - 3 = 0$

83. If $r_1 < 0$ and $r_2 < 0$, then b is positive and c is positive.

85. If $r_1 > 0$, $r_2 < 0$, and $r_1 < |r_2|$, then b is positive and c is negative.

Objective 8.1.3 Exercises

87. The notation $x = \pm 6$ means $x = 6$ or $x = -6$.

89.
$y^2 = 49$
$\sqrt{y^2} = \sqrt{49}$
$|y| = 7$
$y = \pm 7$
The solutions are 7 and –7.

91.
$z^2 = -4$
$\sqrt{z^2} = \sqrt{-4}$
$|z| = 2i$
$z = \pm 2i$
The solutions are $2i$ and $-2i$.

93.
$s^2 - 4 = 0$
$s^2 = 4$
$\sqrt{s^2} = \sqrt{4}$
$|s| = 2$
$s = \pm 2$
The solutions are 2 and –2.

95.
$4x^2 - 81 = 0$
$4x^2 = 81$
$x^2 = \frac{81}{4}$
$\sqrt{x^2} = \sqrt{\frac{81}{4}}$
$|x| = \frac{9}{2}$
$x = \pm \frac{9}{2}$
The solutions are $\frac{9}{2}$ and $-\frac{9}{2}$.

97. $y^2 + 49 = 0$
$y^2 = -49$
$\sqrt{y^2} = \sqrt{-49}$
$|y| = 7i$
$y = \pm 7i$
The solutions are $7i$ and $-7i$.

99. $v^2 - 48 = 0$
$v^2 = 48$
$\sqrt{v^2} = \sqrt{48}$
$|v| = 4\sqrt{3}$
$v = \pm 4\sqrt{3}$
The solutions are $4\sqrt{3}$ and $-4\sqrt{3}$.

101. $r^2 - 75 = 0$
$r^2 = 75$
$\sqrt{r^2} = \sqrt{75}$
$|r| = 5\sqrt{3}$
$r = \pm 5\sqrt{3}$
The solutions are $5\sqrt{3}$ and $-5\sqrt{3}$.

103. $z^2 + 18 = 0$
$z^2 = -18$
$\sqrt{z^2} = \sqrt{-18}$
$|z| = 3i\sqrt{2}$
$z = \pm 3i\sqrt{2}$
The solutions are $3i\sqrt{2}$ and $-3i\sqrt{2}$.

105. $(x-1)^2 = 36$
$\sqrt{(x-1)^2} = \sqrt{36}$
$|x-1| = 6$
$x - 1 = \pm 6$
$x - 1 = 6 \qquad x - 1 = -6$
$x = 7 \qquad x = -5$
The solutions are 7 and −5.

107. $3(y+3)^2 = 27$
$(y+3)^2 = 9$
$\sqrt{(y+3)^2} = \sqrt{9}$
$|y+3| = 3$
$y + 3 = \pm 3$
$y + 3 = 3 \qquad y + 3 = -3$
$y = 0 \qquad y = -6$
The solutions are 0 and −6.

109. $5(z+2)^2 = 125$
$(z+2)^2 = 25$
$\sqrt{(z+2)^2} = \sqrt{25}$
$|z+2| = 5$
$z + 2 = \pm 5$
$z + 2 = 5 \qquad z + 2 = -5$
$z = 3 \qquad z = -7$
The solutions are 3 and −7.

111. $(x+5)^2 = -25$
$\sqrt{(x+5)^2} = \sqrt{-25}$
$|x+5| = 5i$
$x + 5 = \pm 5i$
$x + 5 = 5i \qquad x + 5 = -5i$
$x = -5 + 5i \qquad x = -5 - 5i$
The solutions are $-5 + 5i$ and $-5 - 5i$.

113. $3(x-4)^2 = -12$
$(x-4)^2 = -4$
$\sqrt{(x-4)^2} = \sqrt{-4}$
$|x-4| = 2i$
$x - 4 = \pm 2i$
$x - 4 = 2i \qquad x - 4 = -2i$
$x = 4 + 2i \qquad x = 4 - 2i$
The solutions are $4 + 2i$ and $4 - 2i$.

115. $3(x-9)^2 = -27$
$(x-9)^2 = -9$
$\sqrt{(x-9)^2} = \sqrt{-9}$
$|x-9| = 3i$
$x - 9 = \pm 3i$
$x - 9 = 3i \qquad x - 9 = -3i$
$x = 9 + 3i \qquad x = 9 - 3i$
The solutions are $9 + 3i$ and $9 - 3i$.

117. $\left(v - \dfrac{1}{2}\right)^2 = \dfrac{1}{4}$
$\sqrt{\left(v-\dfrac{1}{2}\right)^2} = \sqrt{\dfrac{1}{4}}$
$\left|v - \dfrac{1}{2}\right| = \dfrac{1}{2}$
$v - \dfrac{1}{2} = \pm \dfrac{1}{2}$
$v - \dfrac{1}{2} = \dfrac{1}{2} \qquad v - \dfrac{1}{2} = -\dfrac{1}{2}$
$v = 1 \qquad v = 0$
The solutions are 1 and 0.

119. $\left(x-\dfrac{2}{5}\right)^2 = \dfrac{9}{25}$

$\sqrt{\left(x-\dfrac{2}{5}\right)^2} = \sqrt{\dfrac{9}{25}}$

$\left|x-\dfrac{2}{5}\right| = \dfrac{3}{5}$

$x-\dfrac{2}{5} = \pm\dfrac{3}{5}$

$x-\dfrac{2}{5} = \dfrac{3}{5}$ $\qquad$ $x-\dfrac{2}{5} = -\dfrac{3}{5}$

$x = \dfrac{2}{5}+\dfrac{3}{5} = \dfrac{5}{5}$ $\qquad$ $x = \dfrac{2}{5}-\dfrac{3}{5}$

$x = 1$ $\qquad\qquad\qquad$ $x = -\dfrac{1}{5}$

The solutions are 1 and $-\dfrac{1}{5}$.

121. $\left(a+\dfrac{3}{4}\right)^2 = \dfrac{9}{16}$

$\sqrt{\left(a+\dfrac{3}{4}\right)^2} = \sqrt{\dfrac{9}{16}}$

$\left|a+\dfrac{3}{4}\right| = \dfrac{3}{4}$

$a+\dfrac{3}{4} = \pm\dfrac{3}{4}$

$a+\dfrac{3}{4} = \dfrac{3}{4}$ $\qquad$ $a+\dfrac{3}{4} = -\dfrac{3}{4}$

$a = \dfrac{3}{4}-\dfrac{3}{4}$ $\qquad$ $a = -\dfrac{3}{4}-\dfrac{3}{4} = -\dfrac{6}{4}$

$a = 0$ $\qquad\qquad\qquad$ $a = -\dfrac{3}{2}$

The solutions are 0 and $-\dfrac{3}{2}$.

123. $3\left(x-\dfrac{5}{3}\right)^2 = \dfrac{4}{3}$

$\left(x-\dfrac{5}{3}\right)^2 = \dfrac{4}{9}$

$\sqrt{\left(x-\dfrac{5}{3}\right)^2} = \sqrt{\dfrac{4}{9}}$

$\left|x-\dfrac{5}{3}\right| = \dfrac{2}{3}$

$x-\dfrac{5}{3} = \pm\dfrac{2}{3}$

$x-\dfrac{5}{3} = \dfrac{2}{3}$ $\qquad$ $x-\dfrac{5}{3} = -\dfrac{2}{3}$

$x = \dfrac{5}{3}+\dfrac{2}{3}$ $\qquad$ $x = \dfrac{5}{3}-\dfrac{2}{3} = \dfrac{3}{3}$

$x = \dfrac{7}{3}$ $\qquad\qquad\qquad$ $x = 1$

The solutions are $\dfrac{7}{3}$ and 1.

125. $(x+5)^2 - 6 = 0$

$(x+5)^2 = 6$

$\sqrt{(x+5)^2} = \sqrt{6}$

$|x+5| = \sqrt{6}$

$x+5 = \pm\sqrt{6}$

$x+5 = \sqrt{6}$ $\qquad$ $x+5 = -\sqrt{6}$

$x = -5+\sqrt{6}$ $\qquad$ $x = -5-\sqrt{6}$

The solutions are $-5+\sqrt{6}$ and $-5-\sqrt{6}$.

127. $(s-2)^2 - 24 = 0$

$(s-2)^2 = 24$

$\sqrt{(s-2)^2} = \sqrt{24}$

$|s-2| = 2\sqrt{6}$

$s-2 = \pm 2\sqrt{6}$

$s-2 = 2\sqrt{6}$ $\qquad$ $s-2 = -2\sqrt{6}$

$s = 2+2\sqrt{6}$ $\qquad$ $s = 2-2\sqrt{6}$

The solutions are $2+2\sqrt{6}$ and $2-2\sqrt{6}$.

129. $(z+1)^2 + 12 = 0$

$(z+1)^2 = -12$

$\sqrt{(z+1)^2} = \sqrt{-12}$

$|z+1| = 2i\sqrt{3}$

$z+1 = \pm 2i\sqrt{3}$

$z+1 = 2i\sqrt{3}$ $\qquad$ $z+1 = -2i\sqrt{3}$

$z = -1+2i\sqrt{3}$ $\qquad$ $z = -1-2i\sqrt{3}$

The solutions are $-1+2i\sqrt{3}$ and $-1-2i\sqrt{3}$.

131. $(v-3)^2 + 45 = 0$
$(v-3)^2 = -45$
$\sqrt{(v-3)^2} = \sqrt{-45}$
$|v-3| = 3i\sqrt{5}$
$v-3 = \pm 3i\sqrt{5}$

$v - 3 = 3i\sqrt{5}$ $v - 3 = -3i\sqrt{5}$
$v = 3 + 3i\sqrt{5}$ $v = 3 - 3i\sqrt{5}$

The solutions are $3 + 3i\sqrt{5}$ and $3 - 3i\sqrt{5}$.

133. $\left(u + \dfrac{2}{3}\right)^2 - 18 = 0$
$\left(u + \dfrac{2}{3}\right)^2 = 18$
$\sqrt{\left(u + \dfrac{2}{3}\right)^2} = \sqrt{18}$
$\left|u + \dfrac{2}{3}\right| = 3\sqrt{2}$
$u + \dfrac{2}{3} = \pm 3\sqrt{2}$

$u + \dfrac{2}{3} = 3\sqrt{2}$ $u + \dfrac{2}{3} = -3\sqrt{2}$
$u = -\dfrac{2}{3} + 3\sqrt{2}$ $u = -\dfrac{2}{3} - 3\sqrt{2}$
$u = \dfrac{-2 + 9\sqrt{2}}{3}$ $u = \dfrac{-2 - 9\sqrt{2}}{3}$

The solutions are $\dfrac{-2 + 9\sqrt{2}}{3}$ and $\dfrac{-2 - 9\sqrt{2}}{3}$.

135. $\left(x + \dfrac{1}{2}\right)^2 + 40 = 0$
$\left(x + \dfrac{1}{2}\right)^2 = -40$
$\sqrt{\left(x + \dfrac{1}{2}\right)^2} = \sqrt{-40}$
$\left|x + \dfrac{1}{2}\right| = 2i\sqrt{10}$
$x + \dfrac{1}{2} = \pm 2i\sqrt{10}$

$x + \dfrac{1}{2} = 2i\sqrt{10}$ $x + \dfrac{1}{2} = -2i\sqrt{10}$
$x = -\dfrac{1}{2} + 2i\sqrt{10}$ $x = -\dfrac{1}{2} - 2i\sqrt{10}$

The solutions are $-\dfrac{1}{2} + 2i\sqrt{10}$ and $-\dfrac{1}{2} - 2i\sqrt{10}$.

137. Two complex solutions

139. One real solution

Applying Concepts 8.1

141. $(x - r_1)(x - r_2) = 0$
$(x - \sqrt{2})[x - (-\sqrt{2})] = 0$
$(x - \sqrt{2})(x + \sqrt{2}) = 0$
$x^2 - 2 = 0$

143. $(x - r_1)(x - r_2) = 0$
$(x - i)[x - (-i)] = 0$
$(x - i)(x + i) = 0$
$x^2 - i^2 = 0$
$x^2 + 1 = 0$

145. $(x - r_1)(x - r_2) = 0$
$(x - 2\sqrt{2})[x - (-2\sqrt{2})] = 0$
$(x - 2\sqrt{2})(x + 2\sqrt{2}) = 0$
$x^2 - 4(2) = 0$
$x^2 - 8 = 0$

147. $(x - r_1)(x - r_2) = 0$
$(x - 2\sqrt{3})[x - (-2\sqrt{3})] = 0$
$(x - 2\sqrt{3})(x + 2\sqrt{3}) = 0$
$x^2 - 4(3) = 0$
$x^2 - 12 = 0$

149. $(x - r_1)(x - r_2) = 0$
$(x - 2i\sqrt{3})[x - (-2i\sqrt{3})] = 0$
$(x - 2i\sqrt{3})(x + 2i\sqrt{3}) = 0$
$x^2 - 4i^2(3) = 0$
$x^2 - 12i^2 = 0$
$x^2 + 12 = 0$

151. $5y^2x^2 = 125z^2$
$y^2x^2 = 25z^2$
$y^2x^2 - 25z^2 = 0$
$(xy + 5z)(xy - 5z) = 0$

$xy + 5z = 0$ $xy - 5z = 0$
$xy = -5z$ $xy = 5z$
$x = -\dfrac{5z}{y}$ $x = \dfrac{5z}{y}$

The solutions are $-\dfrac{5z}{y}$ and $\dfrac{5z}{y}$.

153.
$$2(x-y)^2 - 8 = 0$$
$$(x-y)^2 - 4 = 0$$
$$(x-y)^2 = 4$$
$$\sqrt{(x-y)^2} = \sqrt{4}$$
$$|x-y| = 2$$
$$x - y = \pm 2$$
$$x = y \pm 2$$
The solutions are $y + 2$ and $y - 2$.

155.
$$(x-4)^2 = (x+2)^2$$
$$x^2 - 8x + 16 = x^2 + 4x + 4$$
$$x^2 - x^2 - 8x + 16 = x^2 - x^2 + 4x + 4$$
$$-8x + 16 = 4x + 4$$
$$-12x + 16 = 4$$
$$-12x = -12$$
$$x = 1$$
The solution is 1.

157.
$$ax^2 + c = 0$$
$$ax^2 = -c$$
$$x^2 = -\frac{c}{a}$$
$$\sqrt{x^2} = \sqrt{-\frac{c}{a}}$$
$$|x| = i\sqrt{\frac{c}{a}}$$
$$x = \pm i\sqrt{\frac{c}{a} \cdot \frac{a}{a}}$$
$$x = \pm i\sqrt{\frac{ca}{a^2}}$$
$$x = \pm i\frac{\sqrt{ca}}{a}$$
$$x = \pm \frac{\sqrt{ca}}{a}i$$
The solutions are $\frac{\sqrt{ac}}{a}i$ and $-\frac{\sqrt{ac}}{a}i$.

159. There are a couple of methods to prove this result. Here is one that does not depend on properties of conjugates. Because $z = z_1 + z_2 i$ is a solution of $ax^2 + bx + c = 0$ $a \neq 0$ and all coefficients are real numbers, $a(z_1 + z_2 i)^2 + b(z_1 + z_2 i) + c = 0$. Multiplying this out and collecting the real and imaginary parts, we have
$$\left[a(z_1^2 - z_2^2) + bz_1 + c\right] + (2az_1 z_2 + bz_2)i = 0.$$ This implies $a(z_1^2 - z_2^2) + bz_1 + c = 0$ and $2az_1 z_2 + bz_2 = 0$. Now consider the complex conjugate.
$$a(z_1 - z_2 i)^2 + b(z_1 - z_2 i) + c$$
$$= \left[a(z_1^2 - z_2^2) + bz_1 + c\right] - (2az_1 z_2 + bz_2)i$$
$$= 0 - 0i$$
$$= 0$$
Thus the complex conjugate is also a solution.

Section 8.2

Concept Review 8.2

1. Always true

3. Never true
If $b^2 > 4ac$, then $b^2 - 4ac > 0$ and the quadratic equation has two real roots.

5. Never true
$\left[\left(\frac{1}{2}\right)5\right]^2 = \frac{25}{4}$, so the last term is not correct.

Objective 8.2.1 Exercises

1. a. To complete the square on $x^2 + 14x$, find the constant c that makes $x^2 + 14x + c$ a <u>perfect-square</u> trinomial. The constant term c will be the square of half the coefficient of x, so
$$c = \left[\frac{1}{2}(\underline{7})\right]^2 = \underline{49}.$$

b. Complete the square on $x^2 + 14x$ and write the result as the square of a binomial:
$$x^2 + 14x + \underline{49} = (\underline{x+7})^2$$

3. The next step is to add, to each side of the equation, the constant term that completes the square on
$x^2 + 6x$. $\left[\frac{1}{2}(6)\right]^2 = 9$

Chapter 8: Quadratic Equations And Inequalities

5. $x^2 - 4x - 5 = 0$
$x^2 - 4x = 5$
Complete the square.
$x^2 - 4x + 4 = 5 + 4$
$(x - 2)^2 = 9$
$\sqrt{(x-2)^2} = \sqrt{9}$
$|x - 2| = 3$
$x - 2 = \pm 3$

$x - 2 = 3 \qquad x - 2 = -3$
$x = 5 \qquad x = -1$
The solutions are 5 and –1.

7. $v^2 + 8v - 9 = 0$
$v^2 + 8v = 9$
Complete the square.
$v^2 + 8v + 16 = 9 + 16$
$(v + 4)^2 = 25$
$\sqrt{(v+4)^2} = \sqrt{25}$
$|v + 4| = 5$
$v + 4 = \pm 5$

$v + 4 = 5 \qquad v + 4 = -5$
$v = 1 \qquad v = -9$
The solutions are 1 and –9.

9. $z^2 - 6z + 9 = 0$
$z^2 - 6z = -9$
Complete the square.
$z^2 - 6z + 9 = -9 + 9$
$(z - 3)^2 = 0$
$\sqrt{(z-3)^2} = \sqrt{0}$
$|z - 3| = 0$
$z - 3 = 0$
$z = 3$
The solution is 3.

11. $r^2 + 4r - 7 = 0$
$r^2 + 4r = 7$
Complete the square.
$r^2 + 4r + 4 = 7 + 4$
$(r + 2)^2 = 11$
$\sqrt{(r+2)^2} = \sqrt{11}$
$|r + 2| = \sqrt{11}$
$r + 2 = \pm\sqrt{11}$

$r + 2 = \sqrt{11} \qquad r + 2 = -\sqrt{11}$
$r = -2 + \sqrt{11} \qquad r = -2 - \sqrt{11}$
The solutions are $-2 + \sqrt{11}$ and $-2 - \sqrt{11}$.

13. $x^2 - 6x + 7 = 0$
$x^2 - 6x = -7$
Complete the square.
$x^2 - 6x + 9 = -7 + 9$
$(x - 3)^2 = 2$
$\sqrt{(x-3)^2} = \sqrt{2}$
$|x - 3| = \sqrt{2}$
$x - 3 = \pm\sqrt{2}$

$x - 3 = \sqrt{2} \qquad x - 3 = -\sqrt{2}$
$x = 3 + \sqrt{2} \qquad x = 3 - \sqrt{2}$
The solutions are $3 + \sqrt{2}$ and $3 - \sqrt{2}$.

15. $z^2 - 2z + 2 = 0$
$z^2 - 2z = -2$
Complete the square.
$z^2 - 2z + 1 = -2 + 1$
$(z - 1)^2 = -1$
$\sqrt{(z-1)^2} = \sqrt{-1}$
$|z - 1| = i$
$z - 1 = \pm i$

$z - 1 = i \qquad z - 1 = -i$
$z = 1 + i \qquad z = 1 - i$
The solutions are $1 + i$ and $1 - i$.

17. $t^2 - t - 1 = 0$
$t^2 - t = 1$
Complete the square.
$t^2 - t + \dfrac{1}{4} = 1 + \dfrac{1}{4}$
$\left(t - \dfrac{1}{2}\right)^2 = \dfrac{5}{4}$
$\sqrt{\left(t - \dfrac{1}{2}\right)^2} = \sqrt{\dfrac{5}{4}}$
$\left|t - \dfrac{1}{2}\right| = \dfrac{\sqrt{5}}{2}$
$t - \dfrac{1}{2} = \pm\dfrac{\sqrt{5}}{2}$

$t - \dfrac{1}{2} = \dfrac{\sqrt{5}}{2} \qquad t - \dfrac{1}{2} = -\dfrac{\sqrt{5}}{2}$
$t = \dfrac{1}{2} + \dfrac{\sqrt{5}}{2} \qquad t = \dfrac{1}{2} - \dfrac{\sqrt{5}}{2}$
The solutions are $\dfrac{1 + \sqrt{5}}{2}$ and $\dfrac{1 - \sqrt{5}}{2}$.

Chapter 8: Quadratic Equations And Inequalities

19. $y^2 - 6y = 4$
Complete the square.
$y^2 - 6y + 9 = 4 + 9$
$(y-3)^2 = 13$
$\sqrt{(y-3)^2} = \sqrt{13}$
$|y-3| = \sqrt{13}$
$y - 3 = \pm\sqrt{13}$

$y - 3 = \sqrt{13}$ $\qquad y - 3 = -\sqrt{13}$
$y = 3 + \sqrt{13}$ $\qquad y = 3 - \sqrt{13}$
The solutions are $3 + \sqrt{13}$ and $3 - \sqrt{13}$.

21. $x^2 = 8x - 15$
$x^2 - 8x = -15$
Complete the square.
$x^2 - 8x + 16 = -15 + 16$
$(x-4)^2 = 1$
$\sqrt{(x-4)^2} = \sqrt{1}$
$|x-4| = 1$
$x - 4 = \pm 1$

$x - 4 = 1$ $\qquad x - 4 = -1$
$x = 5$ $\qquad x = 3$
The solutions are 5 and 3.

23. $v^2 = 4v - 13$
$v^2 - 4v = -13$
Complete the square.
$v^2 - 4v + 4 = -13 + 4$
$(v-2)^2 = -9$
$\sqrt{(v-2)^2} = \sqrt{-9}$
$|v-2| = 3i$
$v - 2 = \pm 3i$

$v - 2 = 3i$ $\qquad v - 2 = -3i$
$v = 2 + 3i$ $\qquad v = 2 - 3i$
The solutions are $2 + 3i$ and $2 - 3i$.

25. $p^2 + 6p = -13$
Complete the square.
$p^2 + 6p + 9 = -13 + 9$
$(p+3)^2 = -4$
$\sqrt{(p+3)^2} = \sqrt{-4}$
$|p+3| = 2i$
$p + 3 = \pm 2i$

$p + 3 = 2i$ $\qquad p + 3 = -2i$
$p = -3 + 2i$ $\qquad p = -3 - 2i$
The solutions are $-3 + 2i$ and $-3 - 2i$.

27. $y^2 - 2y = 17$
Complete the square.
$y^2 - 2y + 1 = 17 + 1$
$(y-1)^2 = 18$
$\sqrt{(y-1)^2} = \sqrt{18}$
$|y-1| = 3\sqrt{2}$
$y - 1 = \pm 3\sqrt{2}$

$y - 1 = 3\sqrt{2}$ $\qquad y - 1 = -3\sqrt{2}$
$y = 1 + 3\sqrt{2}$ $\qquad y = 1 - 3\sqrt{2}$
The solutions are $1 + 3\sqrt{2}$ and $1 - 3\sqrt{2}$.

29. $z^2 = z + 4$
$z^2 - z = 4$
Complete the square.
$z^2 - z + \dfrac{1}{4} = 4 + \dfrac{1}{4}$
$\left(z - \dfrac{1}{2}\right)^2 = \dfrac{17}{4}$
$\sqrt{\left(z - \dfrac{1}{2}\right)^2} = \sqrt{\dfrac{17}{4}}$
$\left|z - \dfrac{1}{2}\right| = \dfrac{\sqrt{17}}{2}$
$z - \dfrac{1}{2} = \pm \dfrac{\sqrt{17}}{2}$

$z - \dfrac{1}{2} = \dfrac{\sqrt{17}}{2}$ $\qquad z - \dfrac{1}{2} = -\dfrac{\sqrt{17}}{2}$
$z = \dfrac{1}{2} + \dfrac{\sqrt{17}}{2}$ $\qquad z = \dfrac{1}{2} - \dfrac{\sqrt{17}}{2}$
The solutions are $\dfrac{1+\sqrt{17}}{2}$ and $\dfrac{1-\sqrt{17}}{2}$.

31. $x^2 + 13 = 2x$
$x^2 - 2x = -13$
Complete the square.
$x^2 - 2x + 1 = -13 + 1$
$(x-1)^2 = -12$
$\sqrt{(x-1)^2} = \sqrt{-12}$
$|x-1| = 2i\sqrt{3}$
$x - 1 = \pm 2i\sqrt{3}$

$x - 1 = 2i\sqrt{3}$ $\qquad x - 1 = -2i\sqrt{3}$
$x = 1 + 2i\sqrt{3}$ $\qquad x = 1 - 2i\sqrt{3}$
The solutions are $1 + 2i\sqrt{3}$ and $1 - 2i\sqrt{3}$.

Chapter 8: Quadratic Equations And Inequalities

33. $2y^2 + 3y + 1 = 0$
$2y^2 + 3y = -1$
$\frac{1}{2}(2y^2 + 3y) = \frac{1}{2}(-1)$
$y^2 + \frac{3}{2}y = -\frac{1}{2}$
Complete the square.
$y^2 + \frac{3}{2}y + \frac{9}{16} = -\frac{1}{2} + \frac{9}{16}$
$\left(y + \frac{3}{4}\right)^2 = \frac{1}{16}$
$\sqrt{\left(y + \frac{3}{4}\right)^2} = \sqrt{\frac{1}{16}}$
$\left|y + \frac{3}{4}\right| = \frac{1}{4}$
$y + \frac{3}{4} = \pm\frac{1}{4}$

$y + \frac{3}{4} = \frac{1}{4} \qquad y + \frac{3}{4} = -\frac{1}{4}$
$y = -\frac{2}{4} = -\frac{1}{2} \qquad y = -\frac{4}{4} = -1$

The solutions are $-\frac{1}{2}$ and -1.

35. $4r^2 - 8r = -3$
$\frac{1}{4}(4r^2 - 8r) = \frac{1}{4}(-3)$
$r^2 - 2r = -\frac{3}{4}$
Complete the square.
$r^2 - 2r + 1 = -\frac{3}{4} + 1$
$(r-1)^2 = \frac{1}{4}$
$\sqrt{(r-1)^2} = \sqrt{\frac{1}{4}}$
$|r - 1| = \frac{1}{2}$
$r - 1 = \pm\frac{1}{2}$

$r - 1 = \frac{1}{2} \qquad r - 1 = -\frac{1}{2}$
$r = \frac{3}{2} \qquad r = \frac{1}{2}$

The solutions are $\frac{3}{2}$ and $\frac{1}{2}$.

37. $6y^2 - 5y = 4$
$\frac{1}{6}(6y^2 - 5y) = \frac{1}{6}(4)$
$y^2 - \frac{5}{6}y = \frac{2}{3}$
Complete the square.
$y^2 - \frac{5}{6}y + \frac{25}{144} = \frac{2}{3} + \frac{25}{144}$
$\left(y - \frac{5}{12}\right)^2 = \frac{121}{144}$
$\sqrt{\left(y - \frac{5}{12}\right)^2} = \sqrt{\frac{121}{144}}$
$\left|y - \frac{5}{12}\right| = \frac{11}{12}$
$y - \frac{5}{12} = \pm\frac{11}{12}$

$y - \frac{5}{12} = \frac{11}{12} \qquad y - \frac{5}{12} = -\frac{11}{12}$
$y = \frac{16}{12} = \frac{4}{3} \qquad y = -\frac{6}{12} = -\frac{1}{2}$

The solutions are $\frac{4}{3}$ and $-\frac{1}{2}$.

39. $4x^2 - 4x + 5 = 0$
$4x^2 - 4x = -5$
$\frac{1}{4}(4x^2 - 4x) = \frac{1}{4}(-5)$
$x^2 - x = -\frac{5}{4}$
Complete the square.
$x^2 - x + \frac{1}{4} = -\frac{5}{4} + \frac{1}{4}$
$\left(x - \frac{1}{2}\right)^2 = -1$
$\sqrt{\left(x - \frac{1}{2}\right)^2} = \sqrt{-1}$
$\left|x - \frac{1}{2}\right| = i$
$x - \frac{1}{2} = \pm i$

$x - \frac{1}{2} = i \qquad x - \frac{1}{2} = -i$
$x = \frac{1}{2} + i \qquad x = \frac{1}{2} - i$

The solutions are $\frac{1}{2} + i$ and $\frac{1}{2} - i$.

41. $9x^2 - 6x + 2 = 0$
$9x^2 - 6x = -2$
$\frac{1}{9}(9x^2 - 6x) = \frac{1}{9}(-2)$
$x^2 - \frac{2}{3}x = -\frac{2}{9}$
Complete the square.
$x^2 - \frac{2}{3}x + \frac{1}{9} = -\frac{2}{9} + \frac{1}{9}$
$\left(x - \frac{1}{3}\right)^2 = -\frac{1}{9}$
$\sqrt{\left(x - \frac{1}{3}\right)^2} = \sqrt{-\frac{1}{9}}$
$\left|x - \frac{1}{3}\right| = \frac{1}{3}i$
$x - \frac{1}{3} = \pm\frac{1}{3}i$

$x - \frac{1}{3} = \frac{1}{3}i \qquad x - \frac{1}{3} = -\frac{1}{3}i$
$x = \frac{1}{3} + \frac{1}{3}i \qquad x = \frac{1}{3} - \frac{1}{3}i$

The solutions are $\frac{1}{3} + \frac{1}{3}i$ and $\frac{1}{3} - \frac{1}{3}i$.

43. $2s^2 = 4s + 5$
$2s^2 - 4s = 5$
$\frac{1}{2}(2s^2 - 4s) = \frac{1}{2}(5)$
$s^2 - 2s = \frac{5}{2}$
Complete the square.
$s^2 - 2s + 1 = \frac{5}{2} + 1$
$(s - 1)^2 = \frac{7}{2}$
$\sqrt{(s - 1)^2} = \sqrt{\frac{7}{2}}$
$|s - 1| = \frac{\sqrt{14}}{2}$
$s - 1 = \pm\frac{\sqrt{14}}{2}$

$s - 1 = \frac{\sqrt{14}}{2} \qquad s - 1 = -\frac{\sqrt{14}}{2}$
$s = \frac{2}{2} + \frac{\sqrt{14}}{2} \qquad x = \frac{2}{2} - \frac{\sqrt{14}}{2}$

The solutions are $\frac{2 + \sqrt{14}}{2}$ and $\frac{2 - \sqrt{14}}{2}$.

45. $2r^2 = 3 - r$
$2r^2 + r = 3$
$\frac{1}{2}(2r^2 + r) = \frac{1}{2}(3)$
$r^2 + \frac{1}{2}r = \frac{3}{2}$
Complete the square.
$r^2 + \frac{1}{2}r + \frac{1}{16} = \frac{3}{2} + \frac{1}{16}$
$\left(r + \frac{1}{4}\right)^2 = \frac{25}{16}$
$\sqrt{\left(r + \frac{1}{4}\right)^2} = \sqrt{\frac{25}{16}}$
$\left|r + \frac{1}{4}\right| = \frac{5}{4}$
$r + \frac{1}{4} = \pm\frac{5}{4}$

$r + \frac{1}{4} = \frac{5}{4} \qquad r + \frac{1}{4} = -\frac{5}{4}$
$r = \frac{4}{4} = 1 \qquad r = -\frac{6}{4} = -\frac{3}{2}$

The solutions are 1 and $-\frac{3}{2}$.

47. $y - 2 = (y - 3)(y + 2)$
$y - 2 = y^2 - y - 6$
$y^2 - 2y = 4$
Complete the square.
$y^2 - 2y + 1 = 4 + 1$
$(y - 1)^2 = 5$
$\sqrt{(y - 1)^2} = \sqrt{5}$
$|y - 1| = \sqrt{5}$
$y - 1 = \pm\sqrt{5}$

$y - 1 = \sqrt{5} \qquad y - 1 = -\sqrt{5}$
$y = 1 + \sqrt{5} \qquad y = 1 - \sqrt{5}$

The solutions are $1 + \sqrt{5}$ and $1 - \sqrt{5}$.

49.
$$6t - 2 = (2t-3)(t-1)$$
$$6t - 2 = 2t^2 - 5t + 3$$
$$2t^2 - 11t = -5$$
$$\frac{1}{2}(2t^2 - 11t) = \frac{1}{2}(-5)$$
$$t^2 - \frac{11}{2}t = -\frac{5}{2}$$
Complete the square.
$$t^2 - \frac{11}{2}t + \frac{121}{16} = -\frac{5}{2} + \frac{121}{16}$$
$$\left(t - \frac{11}{4}\right)^2 = \frac{81}{16}$$
$$\sqrt{\left(t - \frac{11}{4}\right)^2} = \sqrt{\frac{81}{16}}$$
$$\left|t - \frac{11}{4}\right| = \frac{9}{4}$$
$$t - \frac{11}{4} = \pm\frac{9}{4}$$

$$t - \frac{11}{4} = \frac{9}{4} \qquad t - \frac{11}{4} = -\frac{9}{4}$$
$$t = \frac{20}{4} = 5 \qquad t = \frac{2}{4} = \frac{1}{2}$$

The solutions are 5 and $\frac{1}{2}$.

51.
$$(x-4)(x+1) = x - 3$$
$$x^2 - 3x - 4 = x - 3$$
$$x^2 - 4x = 1$$
Complete the square.
$$x^2 - 4x + 4 = 1 + 4$$
$$(x-2)^2 = 5$$
$$\sqrt{(x-2)^2} = \sqrt{5}$$
$$|x - 2| = \sqrt{5}$$
$$x - 2 = \pm\sqrt{5}$$
$$x - 2 = \sqrt{5} \qquad x - 2 = -\sqrt{5}$$
$$x = 2 + \sqrt{5} \qquad x = 2 - \sqrt{5}$$
The solutions are $2 + \sqrt{5}$ and $2 - \sqrt{5}$.

53.
$$z^2 + 2z = 4$$
Complete the square.
$$z^2 + 2z + 1 = 4 + 1$$
$$(z+1)^2 = 5$$
$$\sqrt{(z+1)^2} = \sqrt{5}$$
$$|z + 1| = \sqrt{5}$$
$$z + 1 = \pm 2.236$$
$$z + 1 \approx 2.236 \qquad z + 1 \approx -2.236$$
$$z \approx -1 + 2.236 \qquad z \approx -1 - 2.236$$
$$z \approx 1.236 \qquad z \approx -3.236$$
The solutions are approximately 1.236 and −3.236.

55.
$$2x^2 = 4x - 1$$
$$2x^2 - 4x = -1$$
$$\frac{1}{2}(2x^2 - 4x) = \frac{1}{2}(-1)$$
$$x^2 - 2x = -\frac{1}{2}$$
Complete the square.
$$x^2 - 2x + 1 = -\frac{1}{2} + 1$$
$$(x-1)^2 = \frac{1}{2}$$
$$\sqrt{(x-1)^2} = \sqrt{\frac{1}{2}}$$
$$|x-1| = \sqrt{\frac{1}{2}}$$
$$x - 1 \approx \pm 0.707$$
$$x - 1 \approx 0.707 \qquad x - 1 \approx -0.707$$
$$x \approx 1 + 0.707 \qquad x \approx 1 - 0.707$$
$$x \approx 1.707 \qquad x \approx 0.293$$
The solutions are approximately 1.707 and 0.293.

57.
$$4z^2 + 2z - 1 = 0$$
$$4z^2 + 2z = 1$$
$$\frac{1}{4}(4z^2 + 2z) = \frac{1}{4}(1)$$
$$z^2 + \frac{1}{2}z = \frac{1}{4}$$
Complete the square.
$$z^2 + \frac{1}{2}z + \frac{1}{16} = \frac{1}{4} + \frac{1}{16}$$
$$\left(z + \frac{1}{4}\right)^2 = \frac{5}{16}$$
$$\sqrt{\left(z + \frac{1}{4}\right)^2} = \sqrt{\frac{5}{16}}$$
$$\left|z + \frac{1}{4}\right| = \frac{\sqrt{5}}{4}$$
$$z + \frac{1}{4} \approx \pm 0.559$$
$$z + \frac{1}{4} \approx 0.559 \qquad z + \frac{1}{4} \approx -0.559$$
$$z \approx -\frac{1}{4} + 0.559 \qquad z \approx -\frac{1}{4} - 0.559$$
$$z \approx 0.309 \qquad z \approx -0.809$$
The solutions are approximately 0.309 and −0.809.

59. False. For example, if $c = 6$, the solutions are *not* real numbers.
$$x^2 + 4x + 6 = 0$$
$$x^2 + 4x = -6$$
$$x^2 + 4x + 4 = -6 + 4$$
$$(x+2)^2 = -2$$
$$\sqrt{(x+2)^2} = \sqrt{-2}$$
$$|x+2| = i\sqrt{2}$$
$$x + 2 = \pm i\sqrt{2}$$
$$x = -2 \pm i\sqrt{2}$$

Objective 8.2.2 Exercises

61. The quadratic formula is $x = \dfrac{-b \pm \sqrt{b^2 - 4ac}}{2a}$. In this formula, a is the coefficient of x^2, b is the coefficient of x, and c is the constant term in the quadratic equation.

63. No, it does not matter which way you write the equation. There will be differences in the values of a, b, and c in the quadratic equation, but the solutions will be the same.

65. To write the equation $x^2 = 6x - 10$ in standard form, subtract $\underline{6x}$ from and add $\underline{10}$ to each side of the equation. The resulting equation is $\underline{x^2 - 6x + 10} = 0$. Then $a = \underline{1}$, $b = \underline{-6}$, and $c = \underline{10}$.

67. Based on the value of the discriminant found in the previous problem, the equation $x^2 = 6x - 10$ must have two $\underline{\text{complex}}$ number solutions.

69.
$$x^2 - 3x - 10 = 0$$
$$a = 1, b = -3, c = -10$$
$$x = \frac{-b \pm \sqrt{b^2 - 4ac}}{2a}$$
$$= \frac{-(-3) \pm \sqrt{(-3)^2 - 4(1)(-10)}}{2(1)}$$
$$= \frac{3 \pm \sqrt{9 + 40}}{2}$$
$$= \frac{3 \pm \sqrt{49}}{2}$$
$$= \frac{3 \pm 7}{2}$$

$x = \dfrac{3+7}{2} \qquad x = \dfrac{3-7}{2}$
$= \dfrac{10}{2} \qquad\quad = \dfrac{-4}{2}$
$= 5 \qquad\qquad = -2$

The solutions are 5 and –2.

71.
$$y^2 + 5y - 36 = 0$$
$$a = 1, b = 5, c = -36$$
$$y = \frac{-b \pm \sqrt{b^2 - 4ac}}{2a}$$
$$= \frac{-5 \pm \sqrt{(5)^2 - 4(1)(-36)}}{2(1)}$$
$$= \frac{-5 \pm \sqrt{25 + 144}}{2}$$
$$= \frac{-5 \pm \sqrt{169}}{2}$$
$$= \frac{-5 \pm 13}{2}$$

$y = \dfrac{-5+13}{2} \qquad y = \dfrac{-5-13}{2}$
$= \dfrac{8}{2} \qquad\qquad = \dfrac{-18}{2}$
$= 4 \qquad\qquad\quad = -9$

The solutions are 4 and –9.

73.
$$w^2 = 8w + 72$$
$$w^2 - 8w - 72 = 0$$
$$a = 1, b = -8, c = -72$$
$$w = \frac{-b \pm \sqrt{b^2 - 4ac}}{2a}$$
$$= \frac{-(-8) \pm \sqrt{(-8)^2 - 4(1)(-72)}}{2(1)}$$
$$= \frac{8 \pm \sqrt{64 + 288}}{2}$$
$$= \frac{8 \pm \sqrt{352}}{2}$$
$$= \frac{8 \pm 4\sqrt{22}}{2}$$
$$= 4 \pm 2\sqrt{22}$$

The solutions are $4 + 2\sqrt{22}$ and $4 - 2\sqrt{22}$.

254 Chapter 8: Quadratic Equations And Inequalities

75. $v^2 = 24 - 5v$
$v^2 + 5v - 24 = 0$
$a = 1, b = 5, c = -24$
$v = \dfrac{-b \pm \sqrt{b^2 - 4ac}}{2a}$
$= \dfrac{-5 \pm \sqrt{(5)^2 - 4(1)(-24)}}{2(1)}$
$= \dfrac{-5 \pm \sqrt{25 + 96}}{2}$
$= \dfrac{-5 \pm \sqrt{121}}{2}$
$= \dfrac{-5 \pm 11}{2}$

$v = \dfrac{-5 + 11}{2}$ $\quad v = \dfrac{-5 - 11}{2}$
$= \dfrac{6}{2}$ $\qquad\quad = \dfrac{-16}{2}$
$= 3$ $\qquad\qquad = -8$

The solutions are 3 and −8.

77. $2y^2 + 5y - 3 = 0$
$a = 2, b = 5, c = -3$
$y = \dfrac{-b \pm \sqrt{b^2 - 4ac}}{2a}$
$= \dfrac{-5 \pm \sqrt{(5)^2 - 4(2)(-3)}}{2(2)}$
$= \dfrac{-5 \pm \sqrt{25 + 24}}{4}$
$= \dfrac{-5 \pm \sqrt{49}}{4}$
$= \dfrac{-5 \pm 7}{4}$

$y = \dfrac{-5 + 7}{4}$ $\quad y = \dfrac{-5 - 7}{4}$
$= \dfrac{2}{4}$ $\qquad\quad = \dfrac{-12}{4}$
$= \dfrac{1}{2}$ $\qquad\quad = -3$

The solutions are $\dfrac{1}{2}$ and −3.

79. $8s^2 = 10s + 3$
$8s^2 - 10s - 3 = 0$
$a = 8, b = -10, c = -3$
$s = \dfrac{-b \pm \sqrt{b^2 - 4ac}}{2a}$
$= \dfrac{-(-10) \pm \sqrt{(-10)^2 - 4(8)(-3)}}{2(8)}$
$= \dfrac{10 \pm \sqrt{100 + 96}}{16}$
$= \dfrac{10 \pm \sqrt{196}}{16}$
$= \dfrac{10 \pm 14}{16}$

$s = \dfrac{10 + 14}{16}$ $\quad s = \dfrac{10 - 14}{16}$
$= \dfrac{24}{16}$ $\qquad\quad = \dfrac{-4}{16}$
$= \dfrac{3}{2}$ $\qquad\quad = -\dfrac{1}{4}$

The solutions are $\dfrac{3}{2}$ and $-\dfrac{1}{4}$.

81. $v^2 - 2v - 7 = 0$
$a = 1, b = -2, c = -7$
$v = \dfrac{-b \pm \sqrt{b^2 - 4ac}}{2a}$
$= \dfrac{-(-2) \pm \sqrt{(-2)^2 - 4(1)(-7)}}{2(1)}$
$= \dfrac{2 \pm \sqrt{4 + 28}}{2}$
$= \dfrac{2 \pm \sqrt{32}}{2}$
$= \dfrac{2 \pm 4\sqrt{2}}{2}$
$= 1 \pm 2\sqrt{2}$

The solutions are $1 + 2\sqrt{2}$ and $1 - 2\sqrt{2}$.

Chapter 8: Quadratic Equations And Inequalities 255

83. $y^2 - 8y - 20 = 0$
$a = 1, b = -8, c = -20$
$y = \dfrac{-b \pm \sqrt{b^2 - 4ac}}{2a}$
$= \dfrac{-(-8) \pm \sqrt{(-8)^2 - 4(1)(-20)}}{2(1)}$
$= \dfrac{8 \pm \sqrt{64 + 80}}{2}$
$= \dfrac{8 \pm \sqrt{144}}{2}$
$= \dfrac{8 \pm 12}{2}$

$y = \dfrac{8 + 12}{2} \quad y = \dfrac{8 - 12}{2}$
$= \dfrac{20}{2} \quad\quad = \dfrac{-4}{2}$
$= 10 \quad\quad = -2$

The solutions are 10 and –2.

85. $v^2 = 12v - 24$
$v^2 - 12v + 24 = 0$
$a = 1, b = -12, c = 24$
$v = \dfrac{-b \pm \sqrt{b^2 - 4ac}}{2a}$
$= \dfrac{-(-12) \pm \sqrt{(-12)^2 - 4(1)(24)}}{2(1)}$
$= \dfrac{12 \pm \sqrt{144 - 96}}{2}$
$= \dfrac{12 \pm \sqrt{48}}{2}$
$= \dfrac{12 \pm 4\sqrt{3}}{2}$
$= 6 \pm 2\sqrt{3}$

The solutions are $6 + 2\sqrt{3}$ and $6 - 2\sqrt{3}$.

87. $4x^2 - 4x - 7 = 0$
$a = 4, b = -4, c = -7$
$x = \dfrac{-b \pm \sqrt{b^2 - 4ac}}{2a}$
$= \dfrac{-(-4) \pm \sqrt{(-4)^2 - 4(4)(-7)}}{2(4)}$
$= \dfrac{4 \pm \sqrt{16 + 112}}{8}$
$= \dfrac{4 \pm \sqrt{128}}{8}$
$= \dfrac{4 \pm 8\sqrt{2}}{8}$
$= \dfrac{1 \pm 2\sqrt{2}}{2}$

The solutions are $\dfrac{1 + 2\sqrt{2}}{2}$ and $\dfrac{1 - 2\sqrt{2}}{2}$.

89. $2s^2 - 3s + 1 = 0$
$a = 2, b = -3, c = 1$
$s = \dfrac{-b \pm \sqrt{b^2 - 4ac}}{2a}$
$= \dfrac{-(-3) \pm \sqrt{(-3)^2 - 4(2)(1)}}{2(2)}$
$= \dfrac{3 \pm \sqrt{9 - 8}}{4}$
$= \dfrac{3 \pm \sqrt{1}}{4}$
$= \dfrac{3 \pm 1}{4}$

$s = \dfrac{3 + 1}{4} \quad s = \dfrac{3 - 1}{4}$
$= \dfrac{4}{4} \quad\quad = \dfrac{2}{4}$
$= 1 \quad\quad = \dfrac{1}{2}$

The solutions are 1 and $\dfrac{1}{2}$.

91. $3x^2 + 10x + 6 = 0$
$a = 3, b = 10, c = 6$
$x = \dfrac{-b \pm \sqrt{b^2 - 4ac}}{2a}$
$= \dfrac{-10 \pm \sqrt{(10)^2 - 4(3)(6)}}{2(3)}$
$= \dfrac{-10 \pm \sqrt{100 - 72}}{6}$
$= \dfrac{-10 \pm \sqrt{28}}{6}$
$= \dfrac{-10 \pm 2\sqrt{7}}{6}$
$= \dfrac{-5 \pm \sqrt{7}}{3}$

The solutions are $\dfrac{-5 + \sqrt{7}}{3}$ and $\dfrac{-5 - \sqrt{7}}{3}$.

Copyright © Houghton Mifflin Company. All rights reserved.

Chapter 8: Quadratic Equations And Inequalities

93. $6w^2 = 19w - 10$
$6w^2 - 19w + 10 = 0$
$a = 6, b = -19, c = 10$
$w = \dfrac{-b \pm \sqrt{b^2 - 4ac}}{2a}$
$= \dfrac{-(-19) \pm \sqrt{(-19)^2 - 4(6)(10)}}{2(6)}$
$= \dfrac{19 \pm \sqrt{361 - 240}}{12}$
$= \dfrac{19 \pm \sqrt{121}}{12}$
$= \dfrac{19 \pm 11}{12}$

$w = \dfrac{19 + 11}{12} \qquad w = \dfrac{19 - 11}{12}$
$= \dfrac{30}{12} \qquad\qquad = \dfrac{8}{12}$
$= \dfrac{5}{2} \qquad\qquad = \dfrac{2}{3}$

The solutions are $\dfrac{5}{2}$ and $\dfrac{2}{3}$.

95. $p^2 - 4p + 5 = 0$
$a = 1, b = -4, c = 5$
$p = \dfrac{-b \pm \sqrt{b^2 - 4ac}}{2a}$
$= \dfrac{-(-4) \pm \sqrt{(-4)^2 - 4(1)(5)}}{2(1)}$
$= \dfrac{4 \pm \sqrt{16 - 20}}{2}$
$= \dfrac{4 \pm \sqrt{-4}}{2}$
$= \dfrac{4 \pm 2i}{2}$
$= 2 \pm i$

The solutions are $2 + i$ and $2 - i$.

97. $x^2 + 6x + 13 = 0$
$a = 1, b = 6, c = 13$
$x = \dfrac{-b \pm \sqrt{b^2 - 4ac}}{2a}$
$= \dfrac{-6 \pm \sqrt{(6)^2 - 4(1)(13)}}{2(1)}$
$= \dfrac{-6 \pm \sqrt{36 - 52}}{2}$
$= \dfrac{-6 \pm \sqrt{-16}}{2}$
$= \dfrac{-6 \pm 4i}{2}$
$= -3 \pm 2i$

The solutions are $-3 + 2i$ and $-3 - 2i$.

99. $t^2 - 6t + 10 = 0$
$a = 1, b = -6, c = 10$
$t = \dfrac{-b \pm \sqrt{b^2 - 4ac}}{2a}$
$= \dfrac{-(-6) \pm \sqrt{(-6)^2 - 4(1)(10)}}{2(1)}$
$= \dfrac{6 \pm \sqrt{36 - 40}}{2}$
$= \dfrac{6 \pm \sqrt{-4}}{2}$
$= \dfrac{6 \pm 2i}{2}$
$= 3 \pm i$

The solutions are $3 + i$ and $3 - i$.

101. $4v^2 + 8v + 3 = 0$
$a = 4, b = 8, c = 3$
$v = \dfrac{-b \pm \sqrt{b^2 - 4ac}}{2a}$
$= \dfrac{-8 \pm \sqrt{(8)^2 - 4(4)(3)}}{2(4)}$
$= \dfrac{-8 \pm \sqrt{64 - 48}}{8}$
$= \dfrac{-8 \pm \sqrt{16}}{8}$
$= \dfrac{-8 \pm 4}{8}$

$v = \dfrac{-8 + 4}{8} \qquad v = \dfrac{-8 - 4}{8}$
$= \dfrac{-4}{8} \qquad\qquad = \dfrac{-12}{8}$
$= -\dfrac{1}{2} \qquad\qquad = -\dfrac{3}{2}$

The solutions are $-\dfrac{1}{2}$ and $-\dfrac{3}{2}$.

103. $2y^2 + 2y + 13 = 0$
$a = 2, b = 2, c = 13$
$y = \dfrac{-b \pm \sqrt{b^2 - 4ac}}{2a}$
$= \dfrac{-2 \pm \sqrt{(2)^2 - 4(2)(13)}}{2(2)}$
$= \dfrac{-2 \pm \sqrt{4 - 104}}{4}$
$= \dfrac{-2 \pm \sqrt{-100}}{4}$
$= \dfrac{-2 \pm 10i}{4}$
$= \dfrac{-1 \pm 5i}{2}$

The solutions are $-\dfrac{1}{2} + \dfrac{5}{2}i$ and $-\dfrac{1}{2} - \dfrac{5}{2}i$.

105. $3v^2 + 6v + 1 = 0$
$a = 3, b = 6, c = 1$
$x = \dfrac{-b \pm \sqrt{b^2 - 4ac}}{2a}$
$= \dfrac{-6 \pm \sqrt{(6)^2 - 4(3)(1)}}{2(3)}$
$= \dfrac{-6 \pm \sqrt{36 - 12}}{6}$
$= \dfrac{-6 \pm \sqrt{24}}{6}$
$= \dfrac{-6 \pm 2\sqrt{6}}{6}$
$= \dfrac{-3 \pm \sqrt{6}}{3}$

The solutions are $\dfrac{-3 + \sqrt{6}}{3}$ and $\dfrac{-3 - \sqrt{6}}{3}$.

107. $3y^2 = 6y - 5$
$3y^2 - 6y + 5 = 0$
$a = 3, b = -6, c = 5$
$y = \dfrac{-b \pm \sqrt{b^2 - 4ac}}{2a}$
$= \dfrac{-(-6) \pm \sqrt{(-6)^2 - 4(3)(5)}}{2(3)}$
$= \dfrac{6 \pm \sqrt{36 - 60}}{6}$
$= \dfrac{6 \pm \sqrt{-24}}{6}$
$= \dfrac{6 \pm 2i\sqrt{6}}{6}$
$= \dfrac{3 \pm i\sqrt{6}}{3}$

The solutions are $1 + \dfrac{\sqrt{6}}{3}i$ and $1 - \dfrac{\sqrt{6}}{3}i$.

109. $10y(y + 4) = 15y - 15$
$10y^2 + 40y = 15y - 15$
$10y^2 + 25y + 15 = 0$
$5(2y^2 + 5y + 3) = 0$
$2y^2 + 5y + 3 = 0$
$a = 2, b = 5, c = 3$
$y = \dfrac{-b \pm \sqrt{b^2 - 4ac}}{2a}$
$= \dfrac{-5 \pm \sqrt{(5)^2 - 4(2)(3)}}{2(2)}$
$= \dfrac{-5 \pm \sqrt{25 - 24}}{4}$
$= \dfrac{-5 \pm \sqrt{1}}{4}$
$= \dfrac{-5 \pm 1}{4}$

$y = \dfrac{-5 + 1}{4}$ $\quad y = \dfrac{-5 - 1}{4}$
$= \dfrac{-4}{4}$ $\quad\quad = \dfrac{-6}{4}$
$= -1$ $\quad\quad\quad = -\dfrac{3}{2}$

The solutions are -1 and $-\dfrac{3}{2}$.

111. $(2t + 1)(t - 3) = 9$
$2t^2 - 5t - 3 = 9$
$2t^2 - 5t - 12 = 0$
$a = 2, b = -5, c = -12$
$t = \dfrac{-b \pm \sqrt{b^2 - 4ac}}{2a}$
$= \dfrac{-(-5) \pm \sqrt{(-5)^2 - 4(2)(-12)}}{2(2)}$
$= \dfrac{5 \pm \sqrt{25 + 96}}{4}$
$= \dfrac{5 \pm \sqrt{121}}{4}$
$= \dfrac{5 \pm 11}{4}$

$t = \dfrac{5 + 11}{4}$ $\quad t = \dfrac{5 - 11}{4}$
$= \dfrac{16}{4}$ $\quad\quad = \dfrac{-6}{4}$
$= 4$ $\quad\quad\quad = -\dfrac{3}{2}$

The solutions are 4 and $-\dfrac{3}{2}$.

113. $p^2 - 8p + 3 = 0$
$a = 1, b = -8, c = 3$
$p = \dfrac{-b \pm \sqrt{b^2 - 4ac}}{2a}$
$= \dfrac{-(-8) \pm \sqrt{(-8)^2 - 4(1)(3)}}{2(1)}$
$= \dfrac{8 \pm \sqrt{64 - 12}}{2}$
$= \dfrac{8 \pm \sqrt{52}}{2}$
$= \dfrac{8 \pm 2\sqrt{13}}{2}$
$= 4 \pm \sqrt{13}$
$\approx 4 \pm 3.606$
$p \approx 4 + 3.606 \qquad p \approx 4 - 3.606$
$\approx 7.606 \qquad\qquad \approx 0.394$
The solutions are approximately 7.606 and 0.394.

115. $w^2 + 4w - 1 = 0$
$a = 1, b = 4, c = -1$
$w = \dfrac{-b \pm \sqrt{b^2 - 4ac}}{2a}$
$= \dfrac{-4 \pm \sqrt{(4)^2 - 4(1)(-1)}}{2(1)}$
$= \dfrac{-4 \pm \sqrt{16 + 4}}{2}$
$= \dfrac{-4 \pm \sqrt{20}}{2}$
$= \dfrac{-4 \pm 2\sqrt{5}}{2}$
$= -2 \pm \sqrt{5}$
$\approx -2 \pm 2.236$
$w \approx -2 + 2.236 \qquad w \approx -2 - 2.236$
$\approx 0.236 \qquad\qquad \approx -4.236$
The solutions are approximately 0.236 and –4.236.

117. $2y^2 = y + 5$
$2y^2 - y - 5 = 0$
$a = 2, b = -1, c = -5$
$y = \dfrac{-b \pm \sqrt{b^2 - 4ac}}{2a}$
$= \dfrac{-(-1) \pm \sqrt{(-1)^2 - 4(2)(-5)}}{2(2)}$
$= \dfrac{1 \pm \sqrt{1 + 40}}{4}$
$= \dfrac{1 \pm \sqrt{41}}{4}$
$\approx \dfrac{1 \pm 6.403}{4}$
$y \approx \dfrac{1 + 6.403}{4} \qquad y \approx \dfrac{1 - 6.403}{4}$
$\approx \dfrac{7.403}{4} \qquad\qquad \approx \dfrac{-5.403}{4}$
$\approx 1.851 \qquad\qquad \approx -1.351$
The solutions are approximately 1.851 and –1.351.

119. $3y^2 + y + 1 = 0$
$a = 3, b = 1, c = 1$
$b^2 - 4ac$
$1^2 - 4(3)(1) = 1 - 12 = -11$
$-11 < 0$
Since the discriminant is less than zero, the equation has two complex number solutions.

121. $4x^2 + 20x + 25 = 0$
$a = 4, b = 20, c = 25$
$b^2 - 4ac$
$20^2 - 4(4)(25) = 400 - 400 = 0$
Since the discriminant is equal to zero, the equation has one real number solution, a double root.

123. $3w^2 + 3w - 2 = 0$
$a = 3, b = 3, c = -2$
$b^2 - 4ac$
$3^2 - 4(3)(-2) = 9 + 24 = 33$
$33 > 0$
Since the discriminant is greater than zero, the equation has two real number solutions that are not equal.

125. $2t^2 + 9t + 3 = 0$
$a = 2, b = 9, c = 3$
$b^2 - 4ac$
$9^2 - 4(2)(3) = 81 - 24 = 57$
$57 > 0$
Since the discriminant is greater than zero, the equation has two real number solutions that are not equal.

127. True

129. $2x^2 + 3x = 1$
quadratic

131. $4x - 2 = 5$
linear

133. $6x(x-2) = 7$
$6x^2 - 12x = 7$
quadratic

Applying Concepts 8.2

135. $\sqrt{2}y^2 + 3y - 2\sqrt{2} = 0$
$a = \sqrt{2}, b = 3, c = -2\sqrt{2}$
$y = \dfrac{-b \pm \sqrt{b^2 - 4ac}}{2a}$
$= \dfrac{-3 \pm \sqrt{(3)^2 - 4(\sqrt{2})(-2\sqrt{2})}}{2(\sqrt{2})}$
$= \dfrac{-3 \pm \sqrt{9 + 16}}{2\sqrt{2}}$
$= \dfrac{-3 \pm \sqrt{25}}{2\sqrt{2}} = \dfrac{-3 \pm 5}{2\sqrt{2}}$
$y = \dfrac{-3 + 5}{2\sqrt{2}}$
$= \dfrac{2}{2\sqrt{2}} \cdot \dfrac{\sqrt{2}}{\sqrt{2}}$
$= \dfrac{2\sqrt{2}}{4}$
$= \dfrac{\sqrt{2}}{2}$
$= \dfrac{-3 - 5}{2\sqrt{2}}$
$= \dfrac{-8}{2\sqrt{2}}$
$= -\dfrac{4}{\sqrt{2}}$
$= -\dfrac{4}{\sqrt{2}} \cdot \dfrac{\sqrt{2}}{\sqrt{2}}$
$= -\dfrac{4\sqrt{2}}{2}$
$= -2\sqrt{2}$

The solutions are $\dfrac{\sqrt{2}}{2}$ and $-2\sqrt{2}$.

137. $\sqrt{2}x^2 + 5x - 3\sqrt{2} = 0$
$a = \sqrt{2}, b = 5, c = -3\sqrt{2}$
$x = \dfrac{-b \pm \sqrt{b^2 - 4ac}}{2a}$
$= \dfrac{-5 \pm \sqrt{(5)^2 - 4(\sqrt{2})(-3\sqrt{2})}}{2(\sqrt{2})}$
$= \dfrac{-5 \pm \sqrt{25 + 24}}{2\sqrt{2}}$
$= \dfrac{-5 \pm \sqrt{49}}{2\sqrt{2}}$
$= \dfrac{-5 \pm 7}{2\sqrt{2}}$
$x = \dfrac{-5 + 7}{2\sqrt{2}}$
$= \dfrac{2}{2\sqrt{2}}$
$= \dfrac{1}{\sqrt{2}}$
$= \dfrac{1}{\sqrt{2}} \cdot \dfrac{\sqrt{2}}{\sqrt{2}}$
$= \dfrac{\sqrt{2}}{2}$
$= \dfrac{-5 - 7}{2\sqrt{2}}$
$= \dfrac{-12}{2\sqrt{2}}$
$= \dfrac{-6}{\sqrt{2}}$
$= -\dfrac{6}{\sqrt{2}} \cdot \dfrac{\sqrt{2}}{\sqrt{2}}$
$= -\dfrac{6\sqrt{2}}{2}$
$= -3\sqrt{2}$

The solutions are $\dfrac{\sqrt{2}}{2}$ and $-3\sqrt{2}$.

260 Chapter 8: Quadratic Equations And Inequalities

139. $t^2 - t\sqrt{3} + 1 = 0$
$a = 1, b = -\sqrt{3}, c = 1$
$t = \dfrac{-b \pm \sqrt{b^2 - 4ac}}{2a}$
$= \dfrac{-(-\sqrt{3}) \pm \sqrt{(-\sqrt{3})^2 - 4(1)(1)}}{2(1)}$
$= \dfrac{\sqrt{3} \pm \sqrt{3 - 4}}{2}$
$= \dfrac{\sqrt{3} \pm \sqrt{-1}}{2}$
$= \dfrac{\sqrt{3} \pm i}{2}$
$= \dfrac{\sqrt{3}}{2} \pm \dfrac{1}{2}i$
The solutions are $\dfrac{\sqrt{3}}{2} + \dfrac{1}{2}i$ and $\dfrac{\sqrt{3}}{2} - \dfrac{1}{2}i$.

141. $x^2 - ax - 2a^2 = 0$
$(x - 2a)(x + a) = 0$
$x - 2a = 0 \qquad x + a = 0$
$\quad x = 2a \qquad\quad x = -a$
The solutions are $2a$ and $-a$.

143. $2x^2 + 3ax - 2a^2 = 0$
$(2x - a)(x + 2a) = 0$
$2x - a = 0 \qquad x + 2a = 0$
$\quad 2x = a \qquad\quad x = -2a$
$\quad x = \dfrac{a}{2}$
The solutions are $\dfrac{a}{2}$ and $-2a$.

145. $x^2 - 2x - y = 0$
$a = 1, b = -2, c = -y$
$x = \dfrac{-b \pm \sqrt{b^2 - 4ac}}{2a}$
$= \dfrac{-(-2) \pm \sqrt{(-2)^2 - 4(1)(-y)}}{2(1)}$
$= \dfrac{2 \pm \sqrt{4 + 4y}}{2}$
$= \dfrac{2 \pm 2\sqrt{1 + y}}{2}$
$= 1 \pm \sqrt{1 + y}$
The solutions are $1 + \sqrt{y+1}$ and $1 - \sqrt{y+1}$.

147. $x^2 - 6x + p = 0$
$a = 1, b = -6, c = p$
$b^2 - 4ac > 0$
$(-6)^2 - 4(1)(p) > 0$
$36 - 4p > 0$
$-4p > -36$
$p < 9$
$\{p | p < 9, p \in \text{ real numbers}\}$

149. $x^2 - 2x + p = 0$
$a = 1, b = -2, c = p$
$b^2 - 4ac < 0$
$(-2)^2 - 4(1)(p) < 0$
$4 - 4p < 0$
$-4p < -4$
$p > 1$
$\{p | p > 1, p \in \text{ real numbers}\}$

151. Using the quadratic formula with $a = 1$, $b = i$, and $c = 2$,
$x = \dfrac{-i \pm \sqrt{i^2 - 4(1)(2)}}{2(1)}$
$= \dfrac{-i \pm \sqrt{-1 - 8}}{2}$
$= \dfrac{-i \pm \sqrt{-9}}{2}$
$= \dfrac{-i \pm 3i}{2}$
$x = \dfrac{-i + 3i}{2} \qquad x = \dfrac{-i - 3i}{2}$
$\quad = \dfrac{2i}{2} \qquad\qquad = \dfrac{-4i}{2}$
$\quad = i \qquad\qquad\quad = -2i$
The values of x are $i, -2i$.

Chapter 8: Quadratic Equations And Inequalities

153. Strategy
To find the time it takes for the ball to hit the ground, use the value for height ($h = 0$) and solve for t.

Solution
$$h = -16t^2 + 70t + 4$$
$$0 = -16t^2 + 70t + 4$$
$$-4 = -16t^2 + 70t$$
$$-\frac{1}{16}(-4) = -\frac{1}{16}(16t^2 + 70t)$$
$$\frac{1}{4} = t^2 - \frac{35}{8}t$$

Complete the square.
$$\frac{1}{4} + \frac{1225}{256} = t^2 - \frac{35}{8}t + \frac{1225}{256}$$
$$\frac{1289}{256} = \left(t - \frac{35}{16}\right)^2$$
$$\sqrt{\frac{1289}{256}} = \sqrt{\left(t - \frac{35}{16}\right)^2}$$
$$\pm\sqrt{\frac{1289}{256}} = t - \frac{35}{16}$$

$$t - \frac{35}{16} = \sqrt{\frac{1289}{256}} \qquad t - \frac{35}{16} = -\sqrt{\frac{1289}{256}}$$
$$t = \frac{35}{16} + \sqrt{\frac{1289}{256}} \qquad t = \frac{35}{16} - \sqrt{\frac{1289}{256}}$$
$$t \approx 4.431 \qquad t \approx -0.0564$$

The solution $t = -0.0564$ is not possible because it represents a time before the ball is thrown.
The ball takes about 4.43 seconds to hit the ground.

155. a.
$$x^2 - s_a x - s_j s_s f = 0$$
$$x^2 - 0.97x - (0.34)(0.97)(0.24) = 0$$
$$x^2 - 0.97x - 0.079152 = 0$$
$$x = \frac{0.97 \pm \sqrt{(-0.97)^2 - 4(1)(-0.079152)}}{2(1)}$$
$$x \approx 1.05 \text{ and } x \approx -0.08$$
The larger of the two roots is 1.05.
$1.05 > 1$
The model predicts that the population will increase.

b.
$$x^2 - s_a x - s_j s_s f = 0$$
$$x^2 - 0.94x - (0.11)(0.71)(0.24) = 0$$
$$x^2 - 0.94x - 0.018744 = 0$$
$$x = \frac{0.94 \pm \sqrt{(-0.94)^2 - 4(1)(-0.018744)}}{2(1)}$$
$$x \approx 0.96 \text{ and } x \approx -0.02$$
The larger of the two roots is 0.96.
$0.96 < 1$
The model predicts that the population will decrease.

157.
$$ax^2 + bx + c = 0$$
$$ax^2 + bx = -c$$
$$4a^2x^2 + 4abx = -4ac$$
$$4a^2x^2 + 4abx + b^2 = b^2 - 4ac$$
$$(2ax + b)^2 = b^2 - 4ac$$
$$2ax + b = \pm\sqrt{b^2 - 4ac}$$
$$2ax = -b \pm \sqrt{b^2 - 4ac}$$
$$x = \frac{-b \pm \sqrt{b^2 - 4ac}}{2a}$$

Section 8.3

Concept Review 8.3

1. Always true

3. Always true

Objective 8.3.1 Exercises

1. x^4

3. An equation is quadratic in form if it can be written as $au^2 + bu + c = 0$.

5.
$$x^4 - 13x^2 + 36 = 0$$
$$(x^2)^2 - 13(x^2) + 36 = 0$$
$$u^2 - 13u + 36 = 0$$
$$(u - 4)(u - 9) = 0$$
$$u - 4 = 0 \qquad u - 9 = 0$$
$$u = 4 \qquad u = 9$$
Replace u by x^2.
$$x^2 = 4 \qquad x^2 = 9$$
$$\sqrt{x^2} = \sqrt{4} \qquad \sqrt{x^2} = \sqrt{9}$$
$$x = \pm 2 \qquad x = \pm 3$$
The solutions are 2, –2, 3, and –3.

7.
$$z^4 - 6z^2 + 8 = 0$$
$$(z^2)^2 - 6(z^2) + 8 = 0$$
$$u^2 - 6u + 8 = 0$$
$$(u - 4)(u - 2) = 0$$
$$u - 4 = 0 \qquad u - 2 = 0$$
$$u = 4 \qquad u = 2$$
Replace u by z^2.
$$z^2 = 4 \qquad z^2 = 2$$
$$\sqrt{z^2} = \sqrt{4} \qquad \sqrt{z^2} = \sqrt{2}$$
$$z = \pm 2 \qquad z = \pm\sqrt{2}$$
The solutions are 2, –2, $\sqrt{2}$, and $-\sqrt{2}$.

9.
$$p - 3p^{1/2} + 2 = 0$$
$$(p^{1/2})^2 - 3(p^{1/2}) + 2 = 0$$
$$u^2 - 3u + 2 = 0$$
$$(u-1)(u-2) = 0$$
$$u - 1 = 0 \quad u - 2 = 0$$
$$u = 1 \quad u = 2$$
Replace u by $p^{1/2}$.
$$p^{1/2} = 1 \quad p^{1/2} = 2$$
$$(p^{1/2})^2 = 1^2 \quad (p^{1/2})^2 = 2^2$$
$$p = 1 \quad p = 4$$
The solutions are 1 and 4.

11.
$$x - x^{1/2} - 12 = 0$$
$$(x^{1/2})^2 - (x^{1/2}) - 12 = 0$$
$$u^2 - u - 12 = 0$$
$$(u+3)(u-4) = 0$$
$$u + 3 = 0 \quad u - 4 = 0$$
$$u = -3 \quad u = 4$$
Replace u by $x^{1/2}$.
$$x^{1/2} = -3 \quad x^{1/2} = 4$$
$$(x^{1/2})^2 = (-3)^2 \quad (x^{1/2})^2 = 4^2$$
$$x = 9 \quad x = 16$$
16 checks as a solution.
9 does not check as a solution.
The solution is 16.

13.
$$z^4 + 3z^2 - 4 = 0$$
$$(z^2)^2 + 3(z^2) - 4 = 0$$
$$u^2 + 3u - 4 = 0$$
$$(u+4)(u-1) = 0$$
$$u + 4 = 0 \quad u - 1 = 0$$
$$u = -4 \quad u = 1$$
Replace u by z^2.
$$z^2 = -4 \quad z^2 = 1$$
$$\sqrt{z^2} = \sqrt{-4} \quad \sqrt{z^2} = \sqrt{1}$$
$$z = \pm 2i \quad z = \pm 1$$
The solutions are $2i$, $-2i$, 1, and -1.

15.
$$x^4 + 12x^2 - 64 = 0$$
$$(x^2)^2 + 12(x^2) - 64 = 0$$
$$u^2 + 12u - 64 = 0$$
$$(u+16)(u-4) = 0$$
$$u + 16 = 0 \quad u - 4 = 0$$
$$u = -16 \quad u = 4$$
Replace u by x^2.
$$x^2 = -16 \quad x^2 = 4$$
$$\sqrt{x^2} = \sqrt{-16} \quad \sqrt{x^2} = \sqrt{4}$$
$$x = \pm 4i \quad x = \pm 2$$
The solutions are $4i$, $-4i$, 2, and -2.

17.
$$p + 2p^{1/2} - 24 = 0$$
$$(p^{1/2})^2 + 2(p^{1/2}) - 24 = 0$$
$$u^2 + 2u - 24 = 0$$
$$(u+6)(u-4) = 0$$
$$u + 6 = 0 \quad u - 4 = 0$$
$$u = -6 \quad u = 4$$
Replace u by $p^{1/2}$.
$$p^{1/2} = -6 \quad p^{1/2} = 4$$
$$(p^{1/2})^2 = (-6)^2 \quad (p^{1/2})^2 = 4^2$$
$$p = 36 \quad p = 16$$
16 checks as a solution.
36 does not check as a solution.
The solution is 16.

19.
$$y^{2/3} - 9y^{1/3} + 8 = 0$$
$$(y^{1/3})^2 - 9(y^{1/3}) + 8 = 0$$
$$u^2 - 9u + 8 = 0$$
$$(u-1)(u-8) = 0$$
$$u - 1 = 0 \quad u - 8 = 0$$
$$u = 1 \quad u = 8$$
Replace u by $y^{1/3}$.
$$y^{1/3} = 1 \quad y^{1/3} = 8$$
$$(y^{1/3})^3 = 1^3 \quad (y^{1/3})^3 = 8^3$$
$$y = 1 \quad y = 512$$
The solutions are 1 and 512.

21.
$$x^6 - 9x^3 + 8 = 0$$
$$(x^3)^2 - 9(x^3) + 8 = 0$$
$$u^2 - 9u + 8 = 0$$
$$(u-8)(u-1) = 0$$
$$u - 8 = 0 \quad u - 1 = 0$$
Replace u by x^3.
$$x^3 - 8 = 0 \qquad x^3 - 1 = 0$$
$$(x-2)(x^2 + 2x + 4) = 0 \quad (x-1)(x^2 + x + 1) = 0$$

$$x - 2 = 0 \qquad x^2 + 2x + 4 = 0$$
$$x = 2 \qquad x = \frac{-2 \pm \sqrt{2^2 - 4(1)(4)}}{2(1)}$$
$$= \frac{-2 \pm \sqrt{-12}}{2}$$
$$= \frac{-2 \pm 2i\sqrt{3}}{2}$$
$$= -1 \pm i\sqrt{3}$$

$$x - 1 = 0 \qquad x^2 + x + 1 = 0$$
$$x = 1 \qquad x = \frac{-1 \pm \sqrt{1^2 - 4(1)(1)}}{2(1)}$$
$$= \frac{-1 \pm \sqrt{-3}}{2}$$
$$= \frac{-1 \pm i\sqrt{3}}{2}$$

The solutions are 2, 1, $-1 + i\sqrt{3}, -1 - i\sqrt{3}, -\frac{1}{2} + \frac{\sqrt{3}}{2}i$, and $-\frac{1}{2} - \frac{\sqrt{3}}{2}i$.

23.
$$z^8 - 17z^4 + 16 = 0$$
$$(z^4)^2 - 17(z^4) + 16 = 0$$
$$u^2 - 17u + 16 = 0$$
$$(u-16)(u-1) = 0$$
$$u - 16 = 0 \quad u - 1 = 0$$
Replace u by z^4.
$$z^4 - 16 = 0 \qquad z^4 - 1 = 0$$
$$(z^2)^2 - 16 = 0 \quad (z^2)^2 - 1 = 0$$
$$v^2 - 16 = 0 \qquad v^2 - 1 = 0$$

$$(v+4)(v-4) = 0 \qquad (v+1)(v-1) = 0$$
$$v + 4 = 0 \quad v - 4 = 0 \qquad v + 1 = 0$$
$$v = -4 \quad v = 4 \qquad v = -1$$
$$v - 1 = 0$$
$$v = 1$$

Replace v by z^2.
$$z^2 = -4 \qquad z^2 = 4 \qquad z^2 = -1$$
$$\sqrt{z^2} = \sqrt{-4} \quad \sqrt{z^2} = \sqrt{4} \quad \sqrt{z^2} = \sqrt{-1}$$
$$z = \pm 2i \qquad z = \pm 2 \qquad z = \pm i$$
$$z^2 = 1$$
$$\sqrt{z^2} = 1$$
$$z = \pm 1$$
The solutions are $-2, 2, 2i, -2i, -1, 1, i$, and $-i$.

25.
$$p^{2/3} + 2p^{1/3} - 8 = 0$$
$$(p^{1/3})^2 + 2(p^{1/3}) - 8 = 0$$
$$u^2 + 2u - 8 = 0$$
$$(u+4)(u-2) = 0$$
$$u + 4 = 0 \qquad u - 2 = 0$$
$$u = -4 \qquad u = 2$$
Replace u by $p^{1/3}$.
$$p^{1/3} = -4 \qquad p^{1/3} = 2$$
$$(p^{1/3})^3 = (-4)^3 \qquad (p^{1/3})^3 = 2^3$$
$$p = -64 \qquad p = 8$$
The solutions are -64 and 8.

27.
$$2x - 3x^{1/2} + 1 = 0$$
$$2(x^{1/2})^2 - 3(x^{1/2}) + 1 = 0$$
$$2u^2 - 3u + 1 = 0$$
$$(2u-1)(u-1) = 0$$
$$2u - 1 = 0 \qquad u - 1 = 0$$
$$2u = 1 \qquad u = 1$$
$$u = \frac{1}{2}$$
Replace u by $x^{1/2}$.
$$x^{1/2} = \frac{1}{2} \qquad x^{1/2} = 1$$
$$(x^{1/2})^2 = \left(\frac{1}{2}\right)^2 \qquad (x^{1/2})^2 = 1^2$$
$$x = \frac{1}{4} \qquad x = 1$$

The solutions are $\frac{1}{4}$ and 1.

29. **a** and **d** because both of these equations are in a quadratic form similar to that of the given equation.

Objective 8.3.2 Exercises

31. $\sqrt{a+1}$; 7; square

33. 38, 39, 40, 44, 45, 46, 48, 49, 52

35.
$$\sqrt{x+1} + x = 5$$
$$\sqrt{x+1} = 5 - x$$
$$(\sqrt{x+1})^2 = (5-x)^2$$
$$x + 1 = 25 - 10x + x^2$$
$$0 = 24 - 11x + x^2$$
$$0 = (3-x)(8-x)$$
$$3 - x = 0 \qquad 8 - x = 0$$
$$3 = x \qquad 8 = x$$
3 checks as a solution.
8 does not check as a solution.
The solution is 3.

37.
$$x = \sqrt{x} + 6$$
$$x - 6 = \sqrt{x}$$
$$(x-6)^2 = (\sqrt{x})^2$$
$$x^2 - 12x + 36 = x$$
$$x^2 - 13x + 36 = 0$$
$$(x-4)(x-9) = 0$$
$$x - 4 = 0 \qquad x - 9 = 0$$
$$x = 4 \qquad x = 9$$
9 checks as a solution.
4 does not check as a solution.
The solution is 9.

39.
$$\sqrt{3w+3} = w + 1$$
$$(\sqrt{3w+3})^2 = (w+1)^2$$
$$3w + 3 = w^2 + 2w + 1$$
$$0 = w^2 - w - 2$$
$$0 = (w-2)(w+1)$$
$$w - 2 = 0 \qquad w + 1 = 0$$
$$w = 2 \qquad w = -1$$
2 and –1 check as solutions.
The solutions are 2 and –1.

41.
$$\sqrt{4y+1} - y = 1$$
$$\sqrt{4y+1} = y + 1$$
$$(\sqrt{4y+1})^2 = (y+1)^2$$
$$4y + 1 = y^2 + 2y + 1$$
$$0 = y^2 - 2y$$
$$0 = y(y-2)$$
$$y = 0 \qquad y - 2 = 0$$
$$\qquad\qquad y = 2$$
0 and 2 check as solutions.
The solutions are 0 and 2.

43.
$$\sqrt{10x+5} - 2x = 1$$
$$\sqrt{10x+5} = 2x + 1$$
$$(\sqrt{10x+5})^2 = (2x+1)^2$$
$$10x + 5 = 4x^2 + 4x + 1$$
$$0 = 4x^2 - 6x - 4$$
$$0 = 2(2x^2 - 3x - 2)$$
$$0 = 2(2x+1)(x-2)$$
$$2x + 1 = 0 \qquad x - 2 = 0$$
$$2x = -1 \qquad x = 2$$
$$x = -\frac{1}{2}$$
$-\frac{1}{2}$ and 2 check as solutions.
The solutions are $-\frac{1}{2}$ and 2.

45.
$$\sqrt{p+11} = 1 - p$$
$$(\sqrt{p+11})^2 = (1-p)^2$$
$$p + 11 = 1 - 2p + p^2$$
$$0 = -10 - 3p + p^2$$
$$0 = p^2 - 3p - 10$$
$$0 = (p-5)(p+2)$$
$$p - 5 = 0 \qquad p + 2 = 0$$
$$p = 5 \qquad p = -2$$
–2 checks as a solution.
5 does not check as a solution.
The solution is –2.

47.
$$\sqrt{x-1} - \sqrt{x} = -1$$
$$\sqrt{x-1} = \sqrt{x} - 1$$
$$(\sqrt{x-1})^2 = (\sqrt{x} - 1)^2$$
$$x - 1 = x - 2\sqrt{x} + 1$$
$$2\sqrt{x} = 2$$
$$\sqrt{x} = 1$$
$$(\sqrt{x})^2 = 1^2$$
$$x = 1$$
1 checks as a solution.
The solution is 1.

49.
$$\sqrt{2x-1} = 1 - \sqrt{x-1}$$
$$(\sqrt{2x-1})^2 = (1-\sqrt{x-1})^2$$
$$2x-1 = 1 - 2\sqrt{x-1} + x - 1$$
$$2\sqrt{x-1} = -x+1$$
$$(2\sqrt{x-1})^2 = (-x+1)^2$$
$$4(x-1) = x^2 - 2x + 1$$
$$4x - 4 = x^2 - 2x + 1$$
$$0 = x^2 - 6x + 5$$
$$0 = (x-5)(x-1)$$
$$x - 5 = 0 \qquad x - 1 = 0$$
$$x = 5 \qquad x = 1$$
1 checks as a solution.
5 does not check as a solution.
The solution is 1.

51.
$$\sqrt{t+3} + \sqrt{2t+7} = 1$$
$$\sqrt{2t+7} = 1 - \sqrt{t+3}$$
$$(\sqrt{2t+7})^2 = (1 - \sqrt{t+3})^2$$
$$2t + 7 = 1 - 2\sqrt{t+3} + t + 3$$
$$t + 3 = -2\sqrt{t+3}$$
$$(t+3)^2 = (-2\sqrt{t+3})^2$$
$$t^2 + 6t + 9 = 4(t+3)$$
$$t^2 + 6t + 9 = 4t + 12$$
$$t^2 + 2t - 3 = 0$$
$$(t+3)(t-1) = 0$$
$$t + 3 = 0 \qquad t - 1 = 0$$
$$t = -3 \qquad t = 1$$
–3 checks as a solution.
1 does not check as a solution.
The solution is –3.

Objective 8.3.3 Exercises

53.
a. y; $y-4$; $y(y-4)$

b. $4y - 16$

c. $y^2 - 6y$

d. $4y - 16$; $y^2 - 6y$

55.
$$x = \frac{10}{x-9}$$
$$(x-9)x = (x-9)\frac{10}{x-9}$$
$$x^2 - 9x = 10$$
$$x^2 - 9x - 10 = 0$$
$$(x-10)(x+1) = 0$$
$$x - 10 = 0 \qquad x + 1 = 0$$
$$x = 10 \qquad x = -1$$
The solutions are 10 and –1.

57.
$$\frac{t}{t+1} = \frac{-2}{t-1}$$
$$(t-1)(t+1)\frac{t}{t+1} = (t-1)(t+1)\frac{-2}{t-1}$$
$$(t-1)t = (t+1)(-2)$$
$$t^2 - t = -2t - 2$$
$$t^2 + t + 2 = 0$$
$$t = \frac{-b \pm \sqrt{b^2 - 4ac}}{2a}$$
$$= \frac{-1 \pm \sqrt{1^2 - 4(1)(2)}}{2(1)}$$
$$= \frac{-1 \pm \sqrt{1-8}}{2} = \frac{-1 \pm \sqrt{-7}}{2} = \frac{-1 \pm i\sqrt{7}}{2}$$
The solutions are $-\frac{1}{2} + \frac{\sqrt{7}}{2}i$ and $-\frac{1}{2} - \frac{\sqrt{7}}{2}i$.

59.
$$\frac{y-1}{y+2} + y = 1$$
$$(y+2)\left(\frac{y-1}{y+2} + y\right) = (y+2)1$$
$$(y+2)\frac{y-1}{y+2} + (y+2)y = y+2$$
$$y - 1 + y^2 + 2y = y + 2$$
$$y^2 + 3y - 1 = y + 2$$
$$y^2 + 2y - 3 = 0$$
$$(y+3)(y-1) = 0$$
$$y + 3 = 0 \qquad y - 1 = 0$$
$$y = -3 \qquad y = 1$$
The solutions are –3 and 1.

61.
$$\frac{3r+2}{r+2} - 2r = 1$$
$$(r+2)\left(\frac{3r+2}{r+2} - 2r\right) = (r+2)1$$
$$(r+2)\frac{3r+2}{r+2} - (r+2)2r = r + 2$$
$$3r + 2 - 2r^2 - 4r = r + 2$$
$$-2r^2 - r + 2 = r + 2$$
$$-2r^2 - 2r = 0$$
$$-2r(r+1) = 0$$
$$-2r = 0 \qquad r + 1 = 0$$
$$r = 0 \qquad r = -1$$
The solutions are 0 and –1.

63.
$$\frac{2}{2x+1} + \frac{1}{x} = 3$$
$$x(2x+1)\left(\frac{2}{2x+1} + \frac{1}{x}\right) = x(2x+1)3$$
$$x(2x+1)\frac{2}{2x+1} + x(2x+1)\frac{1}{x} = 3x(2x+1)$$
$$2x + 2x + 1 = 6x^2 + 3x$$
$$4x + 1 = 6x^2 + 3x$$
$$0 = 6x^2 - x - 1$$
$$0 = (2x-1)(3x+1)$$

$2x - 1 = 0 \qquad 3x + 1 = 0$
$2x = 1 \qquad\quad 3x = -1$
$x = \dfrac{1}{2} \qquad\quad x = -\dfrac{1}{3}$

The solutions are $\dfrac{1}{2}$ and $-\dfrac{1}{3}$.

65.
$$\frac{16}{z-2} + \frac{16}{z+2} = 6$$
$$(z-2)(z+2)\left(\frac{16}{z-2} + \frac{16}{z+2}\right) = (z-2)(z+2)6$$
$$(z-2)(z+2)\frac{16}{z-2} + (z-2)(z+2)\frac{16}{z+2} = (z^2-4)6$$
$$(z+2)16 + (z-2)16 = 6z^2 - 24$$
$$16z + 32 + 16z - 32 = 6z^2 - 24$$
$$32z = 6z^2 - 24$$
$$0 = 6z^2 - 32z - 24$$
$$0 = 2(3z^2 - 16z - 12)$$
$$0 = 2(3z+2)(z-6)$$

$3z + 2 = 0 \qquad z - 6 = 0$
$3z = -2 \qquad\quad z = 6$
$z = -\dfrac{2}{3}$

The solutions are $-\dfrac{2}{3}$ and 6.

67.

$$\frac{t}{t-2}+\frac{2}{t-1}=4$$

$$(t-2)(t-1)\left(\frac{t}{t-2}+\frac{2}{t-1}\right)=(t-2)(t-1)4$$

$$(t-2)(t-1)\frac{t}{t-2}+(t-2)(t-1)\frac{2}{t-1}=(t^2-3t+2)4$$

$$(t-1)t+(t-2)2=4t^2-12t+8$$

$$t^2-t+2t-4=4t^2-12t+8$$

$$t^2+t-4=4t^2-12t+8$$

$$0=3t^2-13t+12$$

$$0=(3t-4)(t-3)$$

$3t-4=0 \qquad t-3=0$
$3t=4 \qquad t=3$
$t=\frac{4}{3}$

The solutions are $\frac{4}{3}$ and 3.

69.

$$\frac{5}{2p-1}+\frac{4}{p+1}=2$$

$$(2p-1)(p+1)\left(\frac{5}{2p-1}+\frac{4}{p+1}\right)=2(p-1)(p+1)2$$

$$(2p-1)(p+1)\frac{5}{2p-1}+(2p-1)(p+1)\frac{4}{p+1}=(2p^2+p-1)2$$

$$(p+1)5+(2p-1)4=4p^2+2p-2$$

$$5p+5+8p-4=4p^2+2p-2$$

$$13p+1=4p^2+2p-2$$

$$0=4p^2-11p-3$$

$$0=(4p+1)(p-3)$$

$4p+1=0 \qquad p-3=0$
$4p=-1 \qquad p=3$
$p=-\frac{1}{4}$

The solutions are $-\frac{1}{4}$ and 3.

71.
$$\frac{2v}{v+2} + \frac{3}{v+4} = 1$$
$$(v+2)(v+4)\left(\frac{2v}{v+2} + \frac{3}{v+4}\right) = (v+2)(v+4)1$$
$$(v+2)(v+4)\frac{2v}{v+2} + (v+2)(v+4)\frac{3}{v+4} = v^2 + 6v + 8$$
$$(v+4)2v + (v+2)3 = v^2 + 6v + 8$$
$$2v^2 + 8v + 3v + 6 = v^2 + 6v + 8$$
$$2v^2 + 11v + 6 = v^2 + 6v + 8$$
$$v^2 + 5v - 2 = 0$$
$$v = \frac{-b \pm \sqrt{b^2 - 4ac}}{2a} = \frac{-5 \pm \sqrt{5^2 - 4(1)(-2)}}{2(1)} = \frac{-5 \pm \sqrt{25 + 8}}{2} = \frac{-5 \pm \sqrt{33}}{2}$$

The solutions are $\frac{-5 + \sqrt{33}}{2}$ and $\frac{-5 - \sqrt{33}}{2}$.

Applying Concepts 8.3

73.
$$\frac{x^2}{4} + \frac{x}{2} = 6$$
$$4\left(\frac{x^2}{4} + \frac{x}{2}\right) = 4(6)$$
$$x^2 + 2x = 24$$
$$x^2 + 2x - 24 = 0$$
$$(x+6)(x-4) = 0$$
$$x + 6 = 0 \qquad x - 4 = 0$$
$$x = -6 \qquad x = 4$$
The solutions are −6 and 4.

75.
$$\frac{x+2}{3} + \frac{2}{x-2} = 3$$
$$3(x-2)\left(\frac{x+2}{3} + \frac{2}{x-2}\right) = 3(x-2)3$$
$$(x-2)(x+2) + 3 \cdot 2 = 9(x-2)$$
$$x^2 - 4 + 6 = 9x - 18$$
$$x^2 + 2 = 9x - 18$$
$$x^2 - 9x + 20 = 0$$
$$(x-4)(x-5) = 0$$
$$x - 4 = 0 \qquad x - 5 = 0$$
$$x = 4 \qquad x = 5$$
The solutions are 4 and 5.

77.
$$\frac{x^4}{3} - \frac{8x^2}{3} = 3$$
$$3\left(\frac{x^4}{3} - \frac{8x^2}{3}\right) = 3(3)$$
$$x^4 - 8x^2 = 9$$
$$x^4 - 8x^2 - 9 = 0$$
$$(x^2)^2 - 8(x^2) - 9 = 0$$
$$u^2 - 8u - 9 = 0$$
$$(u-9)(u+1) = 0$$
$$u - 9 = 0 \qquad u + 1 = 0$$
$$u = 9 \qquad u = -1$$
Replace u by x^2.
$$x^2 = 9 \qquad x^2 = -1$$
$$\sqrt{x^2} = \sqrt{9} \qquad \sqrt{x^2} = \sqrt{-1}$$
$$x = \pm 3 \qquad x = \pm i$$
The solutions are 3, −3, i, and −i.

79.
$$\frac{x^4}{8}+\frac{x^2}{4}=3$$
$$8\left(\frac{x^4}{8}+\frac{x^2}{4}\right)=8(3)$$
$$x^4+2x^2=24$$
$$x^4+2x^2-24=0$$
$$(x^2)^2+2(x^2)-24=0$$
$$u^2+2u-24=0$$
$$(u+6)(u-4)=0$$
$$u+6=0 \quad u-4=0$$
$$u=-6 \quad u=4$$

Replace u by x^2.
$$x^2=-6 \quad x^2=4$$
$$\sqrt{x^2}=\sqrt{-6} \quad \sqrt{x^2}=\sqrt{4}$$
$$x=\pm i\sqrt{6} \quad x=\pm 2$$

The solutions are 2, –2, $i\sqrt{6}$, and $-i\sqrt{6}$.

81.
$$\sqrt{x^4+4}=2x$$
$$(\sqrt{x^4+4})^2=(2x)^2$$
$$x^4+4=4x^2$$
$$x^4-4x^2+4=0$$
$$(x^2)^2-4(x^2)+4=0$$
$$u^2-4u+4=0$$
$$(u-2)(u-2)=0$$
$$u-2=0 \quad u-2=0$$
$$u=2 \quad u=2$$

Replace u by x^2.
$$x^2=2$$
$$\sqrt{x^2}=\sqrt{2}$$
$$x=\pm\sqrt{2}$$

$\sqrt{2}$ checks as a solution.
$-\sqrt{2}$ does not check as a solution.
The solution is $\sqrt{2}$.

83.
$$(\sqrt{x}+3)^2-4\sqrt{x}-17=0$$
$$(\sqrt{x}+3)^2-4\sqrt{x}-12-5=0$$
$$(\sqrt{x}+3)^2-4(\sqrt{x}+3)-5=0$$

Let $u=\sqrt{x}+3$.
$$u^2-4u-5=0$$
$$(u-5)(u+1)=0$$
$$u-5=0 \quad u+1=0$$
$$u=5 \quad u=-1$$

Replace u by $\sqrt{x}+3$.
$$\sqrt{x}+3=5 \quad \sqrt{x}+3=-1$$
$$\sqrt{x}=2 \quad \sqrt{x}=-4$$
$$(\sqrt{x})^2=2^2 \quad (\sqrt{x})^2=(-4)^2$$
$$x=4 \quad x=16$$

16 does not check in the original equation.
The solution is 4.

Section 8.4

Concept Review 8.4

1. Sometimes true
If $x=2$, $x^2=2^2$ and $(x+2)^2=4^2$.
These are the squares of two consecutive even integers.

3. Always true

Objective 8.4.1 Exercises

1. a. $2w-4$

b. $w(2w-4)$

c. $w(2w-4)$; 96

3. False. If it takes 2 h for both pipes to fill the tank, the larger pipe working alone must take longer than 2 h, not less than 2 h.

5. Strategy
This is a geometry problem.
The width of the rectangle: x
The length of the rectangle: $2x+8$
The area of the rectangle is 640 ft^2. Use the equation for the area of a rectangle $(A=L\cdot W)$.

Solution
$$A=L\cdot W$$
$$640=(2x+8)x$$
$$640=2x^2+8x$$
$$0=2x^2+8x-640$$
$$0=2(x^2+4x-320)$$
$$0=2(x+20)(x-16)$$
$$x+20=0 \quad x-16=0$$
$$x=-20 \quad x=16$$

Since the width of the rectangle cannot be negative, –20 cannot be a solution.
$2x+8=2(16)+8=32+8=40$
The width of the rectangle is 16 ft.
The length of the rectangle is 40 ft.

270 Chapter 8: Quadratic Equations And Inequalities

7. Strategy
This is a geometry problem.
The width of the rectangle: x
The length of the rectangle: $3x - 2$
The area of the rectangle is 65 ft². Use the equation for the area of a rectangle ($A = L \cdot W$).

Solution
$A = L \cdot W$
$65 = (3x - 2)x$
$65 = 3x^2 - 2x$
$0 = 3x^2 - 2x - 65$
$0 = (3x + 13)(x - 5)$

$3x + 13 = 0 \qquad x - 5 = 0$
$x = -\dfrac{13}{3} \qquad\quad x = 5$

Since the width cannot be negative, $-\dfrac{13}{3}$ cannot be a solution.
$3x + 2 = 3(5) - 2 = 15 - 2 = 13$
The length is of the rectangle is 13 ft.
The width of the rectangle is 5 ft.

9. Strategy
This is a geometry problem.
The side of the square: x
The length of the side of the square that is folded up: $x - 20$
The height of the box: 10
The volume of the box is 49,000 cm³. Use the equation for the volume ($A = L \cdot W \cdot H$).

Solution
$A = L \cdot W \cdot H$
$49,000 = (x - 20)(x - 20)10$
$4900 = x^2 - 40x + 400$
$0 = x^2 - 40x - 4500$
$0 = (x - 90)(x + 50)$

$x - 90 = 0 \qquad x + 50 = 0$
$x = 90 \qquad\quad x = -50$

Since the width cannot be negative, -50 cannot be a solution.
The cardboard needs to be 90 cm by 90 cm.

11. Strategy
This is a uniform motion problem.
Rate of truck on return trip: r

	Distance	Rate	Time
With load	550	$r - 5$	$\dfrac{550}{r-5}$
Without load	550	r	$\dfrac{550}{r}$

The total time of the trip was 21 h.

Solution
$\dfrac{550}{r-5} + \dfrac{550}{r} = 21$

$r(r-5)\left(\dfrac{550}{r-5} + \dfrac{550}{r}\right) = r(r-5)21$

$550r + 550(r-5) = (r^2 - 5r)21$
$550r + 550r - 2750 = 21r^2 - 105r$
$1100r - 2750 = 21r^2 - 105r$
$0 = 21r^2 - 1205 + 2750$
$0 = (21r - 50)(r - 55)$

$21r - 50 = 0 \qquad r - 55 = 0$
$21r = 0 \qquad\qquad r = 55$
$r = \dfrac{50}{21}$

$r - 5 = \dfrac{50}{21} = \dfrac{50 - 105}{21} = -\dfrac{55}{21}$ or
$r - 5 = 55 - 5 = 50$

$\dfrac{50}{21}$ cannot be a solution because then the rate of the trip with the load would be negative. The rate of the truck on the return trip was 55 mph.

13. Strategy
To find the time for a projectile to return to Earth, substitute the values for height ($s = 0$) and initial velocity ($v_0 = 200$ ft/s) and solve for t.

Solution
$s = v_0 t - 16t^2$
$0 = 200t - 16t^2$
$0 = 8t(25 - 2t)$

$8t = 0 \qquad 25 - 2t = 0$
$t = 0 \qquad\quad -2t = -25$
$\qquad\qquad\quad t = 12.5$

The solution $t = 0$ is not appropriate because the projectile has not yet left Earth. The rocket takes 12.5 s to return to Earth.

15. Strategy
To find the maximum speed, substitute for distance ($d = 150$) and solve for v.

Solution
$$d = 0.019v^2 + 0.69v$$
$$150 = 0.019v^2 + 0.69v$$
$$0 = 0.019v^2 + 0.69v - 150$$
$$v = \frac{-b \pm \sqrt{b^2 - 4ac}}{2a}$$
$$= \frac{-0.69 \pm \sqrt{(0.69)^2 - 4(0.019)(-150)}}{2(0.019)}$$
$$= \frac{-0.69 \pm \sqrt{11.8761}}{0.038}$$
$$= 72.5 \text{ or } -108.85$$

Since the speed cannot be negative, -108.85 cannot be a solution.
The maximum speed a driver can be going and still be able to stop within 150 m is 72.5 km/h.

17. Strategy
To find when the rocket will be 300 ft above the ground, substitute for height ($h = 300$) and solve for t.

Solution
$$h = -16t^2 + 200t$$
$$300 = -16t^2 + 200t$$
$$0 = -16t^2 + 200t - 300$$
$$t = \frac{-b \pm \sqrt{b^2 - 4ac}}{2a}$$
$$t = \frac{-200 \pm \sqrt{(200)^2 - 4(-16)(-300)}}{2(-16)}$$
$$t = \frac{-200 \pm \sqrt{20{,}800}}{-32}$$
$$t = 1.74 \text{ or } 10.76$$

The rocket will be 300 ft above the ground 1.74 s and 10.76 s after the launch.

19. Strategy
This is a work problem.
Time for the smaller pipe to fill the tank: t
Time for the larger pipe to fill the tank: $t - 6$

	Rate	Time	Part
Smaller pipe	$\frac{1}{t}$	4	$\frac{4}{t}$
Larger pipe	$\frac{1}{t-6}$	4	$\frac{4}{t-6}$

The sum of the parts of the task completed must equal 1.
$$\frac{4}{t} + \frac{4}{t-6} = 1$$

Solution
$$\frac{4}{t} + \frac{4}{t-6} = 1$$
$$t(t-6)\left(\frac{4}{t} + \frac{4}{t-6}\right) = t(t-6)1$$
$$(t-6)4 + 4t = t^2 - 6t$$
$$4t - 24 + 4t = t^2 - 6t$$
$$8t - 24 = t^2 - 6t$$
$$0 = t^2 - 14t + 24$$
$$0 = (t-12)(t-2)$$
$$t - 12 = 0 \qquad t - 2 = 0$$
$$t = 12 \qquad t = 2$$
$$t - 6 = 12 - 6 = 6 \qquad t - 6 = 2 - 6 = -4$$

The solution -4 is not possible, since time cannot be a negative number.
It would take the larger pipe 6 min to fill the tank.
It would take the smaller pipe 12 min to fill the tank.

21. Strategy
This is a distance-rate problem.
Rate of the cruise ship for the first 40 mi: r
Rate of the cruise ship for the next 60 mi: $r + 5$

	Distance	Rate	Time
First 40 mi	40	r	$\frac{40}{r}$
Next 60 mi	60	$r + 5$	$\frac{60}{r+5}$

The total time of travel was 8 h.
$$\frac{40}{r} + \frac{60}{r+5} = 8$$

Solution
$$\frac{40}{r} + \frac{60}{r+5} = 8$$
$$r(r+5)\left(\frac{40}{r} + \frac{60}{r+5}\right) = r(r+5)8$$
$$(r+5)40 + 60r = 8r(r+5)$$
$$40r + 200 + 60r = 8r^2 + 40r$$
$$100r + 200 = 8r^2 + 40r$$
$$0 = 8r^2 - 60r - 200$$
$$0 = 4(2r^2 - 15r - 50)$$
$$0 = 4(2r+5)(r-10)$$
$$2r + 5 = 0 \qquad r - 10 = 0$$
$$2r = -5 \qquad r = 10$$
$$r = -\frac{5}{2}$$

The solution $-\frac{5}{2}$ is not possible, since rate cannot be a negative number. The rate of the cruise ship for the first 40 mi was 10 mph.

272 Chapter 8: Quadratic Equations And Inequalities

23. Strategy
This is a distance-rate problem.
Rate of the wind: w

	Distance	Rate	Time
With wind	240	$100 + w$	$\dfrac{240}{100 + w}$
Against wind	240	$100 - w$	$\dfrac{240}{100 - w}$

The time with the wind is 1 h less than the time against the wind.

$$\dfrac{240}{100 + w} = \dfrac{240}{100 - w} - 1$$

Solution

$$\dfrac{240}{100 + w} = \dfrac{240}{100 - w} - 1$$

$$(100 + w)(100 - w)\left(\dfrac{240}{100 + w}\right) = (100 + w)(100 - w)\left(\dfrac{240}{100 - w} - 1\right)$$

$$240(100 - w) = 240(100 + w) - (100 + w)(100 - w)$$

$$24{,}000 - 240w = 24{,}000 + 240w - 10{,}000 + w^2$$

$$0 = w^2 + 480w - 10{,}000$$

$$0 = (w + 500)(w - 20)$$

$w + 500 = 0 \qquad w - 20 = 0$
$w = -500 \qquad w = 20$

The solution −500 is not possible, since rate cannot be a negative number.
The rate of the wind is 20 mph.

25. Strategy
This is a distance-rate problem.
Rate of crew in calm water: x

	Distance	Rate	Time
With current	16	$x + 2$	$\dfrac{16}{x + 2}$
Against current	16	$x - 2$	$\dfrac{16}{x - 2}$

The total trip took 6 h.

$$\dfrac{16}{x + 2} + \dfrac{16}{x - 2} = 6$$

Solution

$$\dfrac{16}{x + 2} + \dfrac{16}{x - 2} = 6$$

$$(x + 2)(x - 2)\left(\dfrac{16}{x + 2} + \dfrac{16}{x - 2}\right) = (x + 2)(x - 2)6$$

$$16(x - 2) + 16(x + 2) = 6(x^2 - 4)$$

$$16x - 32 + 16x + 32 = 6x^2 - 24$$

$$32x = 6x^2 - 24$$

$$0 = 6x^2 - 32x - 24$$

$$0 = 2(3x^2 - 16x - 12)$$

$$0 = 2(3x + 2)(x - 6)$$

$3x + 2 = 0 \qquad x - 6 = 0$
$x = -\dfrac{2}{3} \qquad x = 6$

The solution $x = -\dfrac{2}{3}$ is not possible, since rate cannot be a negative number.
The rate of the crew in calm water is 6 mph.

Applying Concepts 8.4

27. **Strategy**
The numerator of the fraction: n
The denominator of the fraction: $n + 3$
The fraction: $\dfrac{n}{n+3}$
Four times the reciprocal of the fraction:
$4 \cdot \dfrac{n+3}{n} = \dfrac{4(n+3)}{n}$
The sum of the fraction and four times its reciprocal is $\dfrac{17}{2}$.

Solution
$$\dfrac{n}{n+3} + \dfrac{4(n+3)}{n} = \dfrac{17}{2}$$
$$\dfrac{n}{n+3} + \dfrac{4n+12}{n} = \dfrac{17}{2}$$
$$2n(n+3)\left(\dfrac{n}{n+3} + \dfrac{4n+12}{n}\right) = 2n(n+3)\left(\dfrac{17}{2}\right)$$
$$2n^2 + 2(n+3)(4n+12) = n(n+3)(17)$$
$$2n^2 + 2(4n^2 + 24n + 36) = 17n(n+3)$$
$$2n^2 + 8n^2 + 48n + 72 = 17n^2 + 51n$$
$$10n^2 + 48n + 72 = 17n^2 + 51n$$
$$-7n^2 - 3n + 72 = 0$$
$$7n^2 + 3n - 72 = 0$$
$$(7n + 24)(n - 3) = 0$$
$7n + 24 = 0 \qquad n - 3 = 0$
$7n = -24 \qquad n = 3$
$n = -\dfrac{24}{7}$

The numerator cannot be a fraction, so $-\dfrac{24}{7}$ cannot be a solution.
$\dfrac{n}{n+3} = \dfrac{3}{3+3} = \dfrac{3}{6}$
The fraction is $\dfrac{3}{6}$.

29. **Strategy**
This is a geometry problem.
Width of rectangular piece of cardboard: w
Length of rectangular piece of cardboard: $w + 8$
Width of open box: $w - 2(2) = w - 4$
Length of open box: $w + 8 - 2(2) = w + 4$
Height of open box: 2
Use the equation $V = lwh$.

Solution
$V = lwh$
$256 = (w+4)(w-4)(2)$
$256 = (w^2 - 16)(2)$
$256 = 2w^2 - 32$
$0 = 2w^2 - 288$
$0 = 2(w^2 - 144)$
$0 = (w+12)(w-12)$
$w + 12 = 0 \qquad w - 12 = 0$
$w = -12 \qquad w = 12$
The width cannot be negative, so -12 is not a solution.
$w - 4 = 12 - 4 = 8$
$w + 4 = 12 + 4 = 16$
The width is 8 cm.
The length is 16 cm.
The height is 2 cm.

31. To find when the depth will be 10 cm, substitute 10 for d and solve for t.
$d = 0.0034t^2 - 0.52518t + 20$
$10 = 0.0034t^2 - 0.52518t + 20$
$0 = 0.0034t^2 - 0.52518t + 10$
$$t = \dfrac{-b \pm \sqrt{b^2 - 4ac}}{2a}$$
$$t = \dfrac{-(-0.52518) \pm \sqrt{(0.0034)^2 - 4(0.0034)(10)}}{2(0.0034)}$$
$$t = \dfrac{0.52518 \pm \sqrt{0.139814}}{0.0068}$$
$t \approx 132.2$ or $t \approx 22.2$
The depth will first reach 10 cm in 22.2 s.

33. **Strategy**
This is a geometry problem.
Use the Pythagorean formula $(a^2 + b^2 = c^2)$, with the legs being $a = 1.5$, $b = 3.5$, and the hypotenuse being $c = x + 1.5$.

Solution
$a^2 + b^2 = c^2$
$(1.5)^2 + (3.5)^2 = (x+1.5)^2$
$14.5 = (x+1.5)^2$
$\pm 3.8 \approx x + 1.5$
$3.8 = x + 1.5 \qquad -3.8 = x + 1.5$
$2.3 = x \qquad -5.3 = x$
The solution $x = -5.3$ is not possible since distance cannot be negative.
The bottom of the scoop of ice cream is 2.3 in. from the bottom of the cone.

Section 8.5

Concept Review 8.5

1. Sometimes true
   The end points of $x^2 - 4 \leq 0$ are included in the solution set.

Objective 8.5.1 Exercises

1. 3 is not a possible element of the solution set because the value 3 makes the denominator 0.

3. positive; negative; negative

5. $(x-4)(x+2) > 0$
   $\{x \mid x < -2 \text{ or } x > 4\}$

7. $x^2 - 3x + 2 \geq 0$
   $(x-1)(x-2) \geq 0$
   $\{x \mid x \leq 1 \text{ or } x \geq 2\}$

9. $x^2 - x - 12 < 0$
   $(x+3)(x-4) < 0$
   $\{x \mid -3 < x < 4\}$

11. $(x-1)(x+2)(x-3) < 0$
    $\{x \mid x < -2 \text{ or } 1 < x < 3\}$

13. $\dfrac{x-4}{x+2} > 0$
    $\{x \mid x < -2 \text{ or } x > 4\}$

15. $\dfrac{x-3}{x+1} \leq 0$
    $\{x \mid -1 < x \leq 3\}$

17. $\dfrac{(x-1)(x+2)}{x-3} \leq 0$
    $\{x \mid x \leq -2 \text{ or } 1 \leq x < 3\}$

19. **a** and **d**

21. $x^2 - 16 > 0$
    $(x-4)(x+4) > 0$
    $\{x \mid x < -4 \text{ or } x > 4\}$

23. $x^2 - 4x + 4 > 0$
    $(x-2)(x-2) > 0$
    $\{x \mid x < 2 \text{ or } x > 2\}$

25. $x^2 - 9x \leq 36$
    $x^2 - 9x - 36 \leq 0$
    $(x+3)(x-12) \leq 0$
    $\{x \mid -3 \leq x \leq 12\}$

27. $2x^2 - 5x + 2 \geq 0$
    $(2x-1)(x-2) \geq 0$
    $\left\{x \mid x \leq \dfrac{1}{2} \text{ or } x \geq 2\right\}$

29. $4x^2 - 8x + 3 < 0$

$(2x-1)(2x-3) < 0$

$\left\{x \mid \dfrac{1}{2} < x < \dfrac{3}{2}\right\}$

31. $(x-6)(x+3)(x-2) \leq 0$

$\{x \mid x \leq -3 \text{ or } 2 \leq x \leq 6\}$

33. $(2x-1)(x-4)(2x+3) > 0$

$\left\{x \mid -\dfrac{3}{2} < x < \dfrac{1}{2} \text{ or } x > 4\right\}$

35. $x^3 + 3x^2 - x - 3 \leq 0$

$x^2(x+3) - 1(x+3) \leq 0$

$(x+3)(x^2-1) \leq 0$

$(x+3)(x+1)(x-1) \leq 0$

$\{x \mid x \leq -3 \text{ or } -1 \leq x \leq 1\}$

37. $x^3 - x^2 - 4x + 4 \geq 0$

$x^2(x-1) - 4(x-1) \geq 0$

$(x-1)(x^2-4) \geq 0$

$(x-1)(x+2)(x-2) \geq 0$

$\{x \mid -2 \leq x \leq 1 \text{ or } x \geq 2\}$

39. $\dfrac{3x}{x-2} > 1$

$\dfrac{3x}{x-2} - 1 > 0$

$\dfrac{3x}{x-2} - \dfrac{x-2}{x-2} > 0$

$\dfrac{2x+2}{x-2} > 0$

$\{x \mid x < -1 \text{ or } x > 2\}$

41. $\dfrac{2}{x+1} \geq 2$

$\dfrac{2}{x+1} - 2 \geq 0$

$\dfrac{2}{x+1} - \dfrac{2x+2}{x+1} \geq 0$

$\dfrac{-2x}{x+1} \geq 0$

$\{x \mid -1 < x \leq 0\}$

43. $\dfrac{x}{(x-1)(x+2)} \geq 0$

$\{x \mid -2 < x \leq 0 \text{ or } x > 1\}$

45. $\dfrac{1}{x} < 2$

$\dfrac{1}{x} - 2 < 0$

$\dfrac{1}{x} - \dfrac{2x}{x} < 0$

$\dfrac{1-2x}{x} < 0$

$\left\{x \mid x < 0 \text{ or } x > \dfrac{1}{2}\right\}$

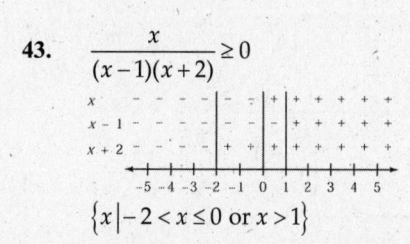

Applying Concepts 8.5

47. $(x+2)(x-3)(x+1)(x+4) > 0$

$\{x \mid x < -4 \text{ or } -2 < x < -1 \text{ or } x > 3\}$

49. $(x^2+2x-8)(x^2-2x-3) < 0$

$(x+4)(x-2)(x-3)(x+1) < 0$

$\{x \mid -4 < x < -1 \text{ or } 2 < x < 3\}$

276 Chapter 8: Quadratic Equations And Inequalities

51. $(x^2+1)(x^2-3x+2) > 0$
    $(x^2+1)(x-2)(x-1) > 0$

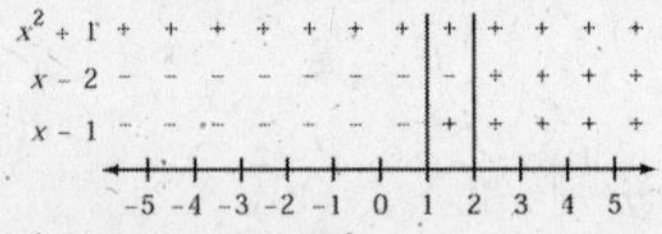

	-5	-4	-3	-2	-1	0	1	2	3	4	5
x^2+1	+	+	+	+	+	+	+	+	+	+	+
$x-2$	-	-	-	-	-	-	-	-	+	+	+
$x-1$	-	-	-	-	-	-	-	+	+	+	+

    $\{x \mid x < 1 \text{ or } x > 2\}$

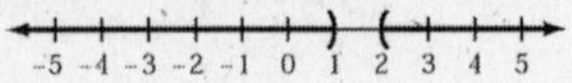

53. $\dfrac{x^2(3-x)(2x+1)}{(x+4)(x+2)} \geq 0$

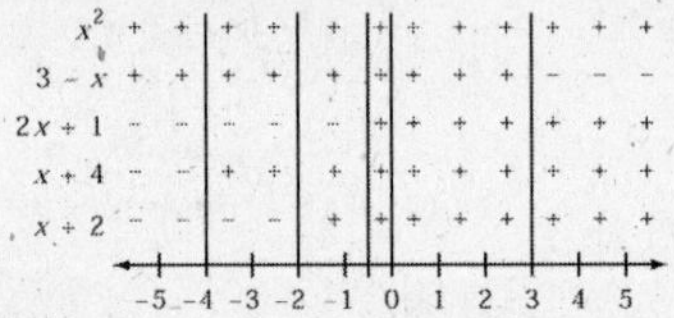

	-5	-4	-3	-2	-1	0	1	2	3	4	5
x^2	+	+	+	+	+	0	+	+	+	+	+
$3-x$	+	+	+	+	+	+	+	+	0	-	-
$2x+1$	-	-	-	-	-	-	+	+	+	+	+
$x+4$	-	0	+	+	+	+	+	+	+	+	+
$x+2$	-	-	-	0	+	+	+	+	+	+	+

    $\left\{x \mid -4 < x < -2 \text{ or } -\dfrac{1}{2} \leq x \leq 3\right\}$

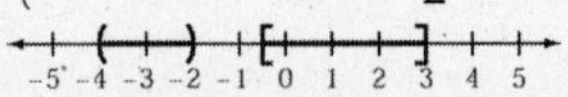

55. $3x - \dfrac{1}{x} \leq 2$

    $3x - \dfrac{1}{x} - 2 \leq 0$

    $\dfrac{3x^2 - 1 - 2x}{x} \leq 0$

    $\dfrac{3x^2 - 2x - 1}{x} \leq 0$

    $\dfrac{(3x+1)(x-1)}{x} \leq 0$

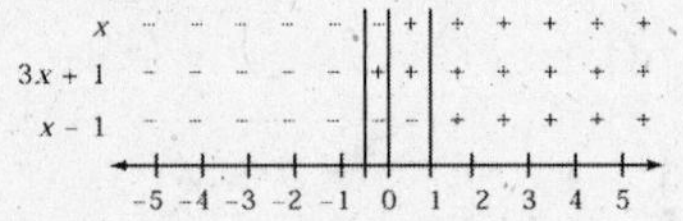

	-5	-4	-3	-2	-1	0	1	2	3	4	5
x	-	-	-	-	-	0	+	+	+	+	+
$3x+1$	-	-	-	-	-	-	+	+	+	+	+
$x-1$	-	-	-	-	-	-	0	+	+	+	+

    $\left\{x \mid x \leq -\dfrac{1}{3} \text{ or } 0 < x \leq 1\right\}$

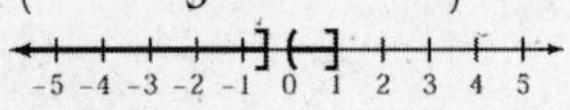

Section 8.6

Concept Review 8.6

1. Sometimes true
   If the vertex is on an axis, the axis of symmetry will intersect the origin.

3. Sometimes true
   A parabola may have one, two or no x-intercepts.

5. Always true

Objective 8.6.1 Exercises

1. A quadratic function is a function of the form $f(x) = ax^2 + bx + c$, $a \neq 0$.

3. The x-coordinate of the vertex is -5.

5. The axis of symmetry is $x = 7$.

7. a. 3; 0; 1
   b. parabola; up
   c. 0; 3; 0
   d. 0; –1; 0; –1

9. $a = 1, b = 0$
   $-\dfrac{b}{2a} = -\dfrac{0}{2(1)} = 0$
   $y = 0^2 = 0$
   Vertex: (0, 0)

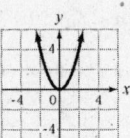

   Axis of symmetry: $x = 0$

11. $a = 1, b = 0$
    $-\dfrac{b}{2a} = -\dfrac{0}{2(1)} = 0$
    $y = 0^2 - 2 = -2$
    Vertex: (0, –2)

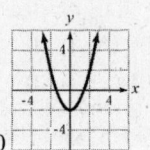

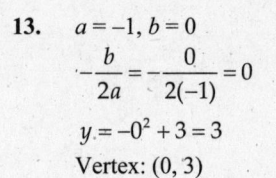

    Axis of symmetry: $x = 0$

13. $a = -1, b = 0$
    $-\dfrac{b}{2a} = -\dfrac{0}{2(-1)} = 0$
    $y = -0^2 + 3 = 3$
    Vertex: (0, 3)

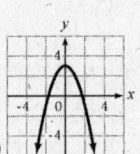

    Axis of symmetry: $x = 0$

15. $a = \dfrac{1}{2}, b = 0$
    $-\dfrac{b}{2a} = -\dfrac{0}{2\left(\frac{1}{2}\right)} = 0$
    $y = \dfrac{1}{2}(0)^2 = 0$
    Vertex: (0, 0)

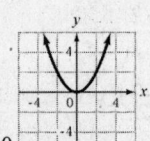

    Axis of symmetry: $x = 0$

Copyright © Houghton Mifflin Company. All rights reserved.

Chapter 8: Quadratic Equations And Inequalities   277

17. $a = 2, b = 0$
$-\dfrac{b}{2a} = -\dfrac{0}{2(2)} = 0$
$y = 2(0)^2 - 1 = -1$
Vertex: $(0, -1)$

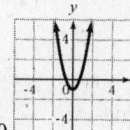

Axis of symmetry: $x = 0$

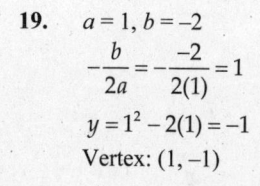

19. $a = 1, b = -2$
$-\dfrac{b}{2a} = -\dfrac{-2}{2(1)} = 1$
$y = 1^2 - 2(1) = -1$
Vertex: $(1, -1)$

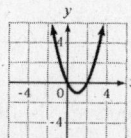

Axis of symmetry: $x = 1$

21. $a = -2, b = 4$
$-\dfrac{b}{2a} = -\dfrac{4}{2(-2)} = 1$
$y = -2(1)^2 + 4(1) = 2$
Vertex: $(1, 2)$

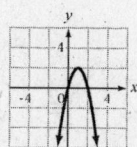

Axis of symmetry: $x = 1$

23. $a = 1, b = -1$
$-\dfrac{b}{2a} = -\dfrac{-1}{2(1)} = \dfrac{1}{2}$
$y = \left(\dfrac{1}{2}\right)^2 - \dfrac{1}{2} - 2 = -\dfrac{9}{4}$
Vertex: $\left(\dfrac{1}{2}, -\dfrac{9}{4}\right)$

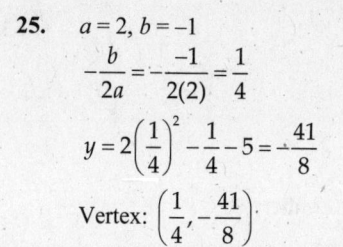

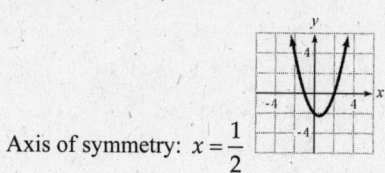

Axis of symmetry: $x = \dfrac{1}{2}$

25. $a = 2, b = -1$
$-\dfrac{b}{2a} = -\dfrac{-1}{2(2)} = \dfrac{1}{4}$
$y = 2\left(\dfrac{1}{4}\right)^2 - \dfrac{1}{4} - 5 = -\dfrac{41}{8}$
Vertex: $\left(\dfrac{1}{4}, -\dfrac{41}{8}\right)$

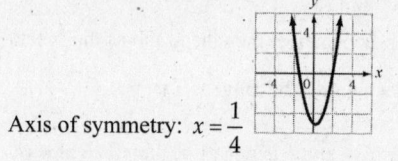

Axis of symmetry: $x = \dfrac{1}{4}$

27.

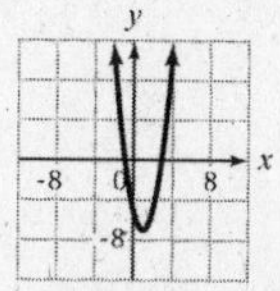

Domain: $\{x \mid x \in \text{real numbers}\}$
Range: $\{y \mid y \geq -7\}$

29.

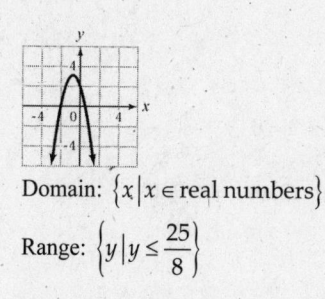

Domain: $\{x \mid x \in \text{real numbers}\}$
Range: $\left\{y \mid y \leq \dfrac{25}{8}\right\}$

31.

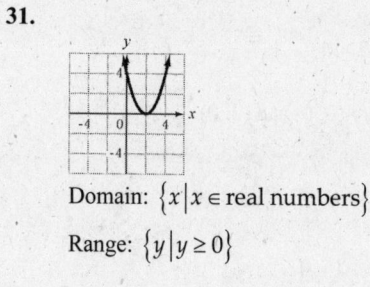

Domain: $\{x \mid x \in \text{real numbers}\}$
Range: $\{y \mid y \geq 0\}$

33.

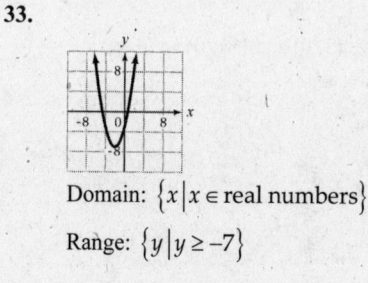

Domain: $\{x \mid x \in \text{real numbers}\}$
Range: $\{y \mid y \geq -7\}$

Copyright © Houghton Mifflin Company. All rights reserved.

35.

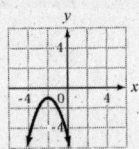

Domain: $\{x \mid x \in \text{real numbers}\}$

Range: $\{y \mid y \leq -1\}$

37. False. If a is negative, then the graph of the function opens downward. The range is $\left\{y \mid y \leq f\left(-\dfrac{b}{2a}\right)\right\}$

39. If a is positive and b is negative, then the value of $-\dfrac{b}{2a}$ is positive. The axis of symmetry, $x = -\dfrac{b}{2a}$, is to the right of the y-axis.

Objective 8.6.2 Exercises

41. 3; 0; 3; 0; x

43. To find the x-intercepts for the graph of the quadratic function $f(x) = ax^2 + bx + c$, solve the equation $0 = ax^2 + bx + c$ for x.

45. $y = x^2 - 4$
$0 = x^2 - 4$
$0 = (x-2)(x+2)$
$x - 2 = 0 \quad x + 2 = 0$
$x = 2 \quad\quad x = -2$
The x-intercepts are $(2, 0)$ and $(-2, 0)$.

47. $y = 2x^2 - 4x$
$0 = 2x^2 - 4x$
$0 = 2x(x-2)$
$2x = 0 \quad x - 2 = 0$
$x = 0 \quad\quad x = 2$
The x-intercepts are $(0, 0)$ and $(2, 0)$.

49. $y = x^2 - x - 2$
$0 = x^2 - x - 2$
$0 = (x-2)(x+1)$
$x - 2 = 0 \quad x + 1 = 0$
$x = 2 \quad\quad x = -1$
The x-intercepts are $(2, 0)$ and $(-1, 0)$.

51. $y = 2x^2 - 5x - 3$
$0 = 2x^2 - 5x - 3$
$0 = (2x+1)(x-3)$
$2x + 1 = 0 \quad x - 3 = 0$
$2x = -1 \quad\quad x = 3$
$x = -\dfrac{1}{2}$
The x-intercepts are $\left(-\dfrac{1}{2}, 0\right)$ and $(3, 0)$.

53. $y = 3x^2 - 19x - 14$
$0 = 3x^2 - 19x - 14$
$0 = (3x+2)(x-7)$
$3x + 2 = 0 \quad x - 7 = 0$
$3x = -2 \quad\quad x = 7$
$x = -\dfrac{2}{3}$
The x-intercepts are $\left(-\dfrac{2}{3}, 0\right)$ and $(7, 0)$.

55. $y = 3x^2 - 19x + 20$
$0 = 3x^2 - 19x + 20$
$0 = (3x-4)(x-5)$
$3x - 4 = 0 \quad x - 5 = 0$
$3x = 4 \quad\quad x = 5$
$x = \dfrac{4}{3}$
The x-intercepts are $\left(\dfrac{4}{3}, 0\right)$ and $(5, 0)$.

57. $y = 9x^2 - 12x + 4$
$0 = 9x^2 - 12x + 4$
$0 = (3x-2)(3x-2)$
$3x - 2 = 0 \quad 3x - 2 = 0$
$3x = 2 \quad\quad 3x = 2$
$x = \dfrac{2}{3} \quad\quad x = \dfrac{2}{3}$
The x-intercept is $\left(\dfrac{2}{3}, 0\right)$.

59. $y = 9x^2 - 2$
$0 = 9x^2 - 2$
$2 = 9x^2$
$\dfrac{2}{9} = x^2$
$\sqrt{\dfrac{2}{9}} = \sqrt{x^2}$
$\pm \dfrac{\sqrt{2}}{3} = x$
The x-intercepts are $\left(\dfrac{\sqrt{2}}{3}, 0\right)$ and $\left(-\dfrac{\sqrt{2}}{3}, 0\right)$.

61. $y = 4x^2 - 4x - 15$
$0 = 4x^2 - 4x - 15$
$0 = (2x+3)(2x-5)$
$2x + 3 = 0 \quad (2x-5) = 0$
$2x = -3 \quad\quad 2x = 5$
$x = -\dfrac{3}{2} \quad\quad x = \dfrac{5}{2}$
The x-intercepts are $\left(-\dfrac{3}{2}, 0\right)$ and $\left(\dfrac{5}{2}, 0\right)$.

63. $y = x^2 + 4x - 3$
$0 = x^2 + 4x - 3$
$a = 1, b = 4, c = -3$
$x = \dfrac{-b \pm \sqrt{b^2 - 4ac}}{2a}$
$= \dfrac{-4 \pm \sqrt{4^2 - 4(1)(-3)}}{2(1)}$
$= \dfrac{-4 \pm \sqrt{16 + 12}}{2}$
$= \dfrac{-4 \pm \sqrt{28}}{2}$
$= \dfrac{-4 \pm 2\sqrt{7}}{2}$
$= -2 \pm \sqrt{7}$

The x-intercepts are $(-2 + \sqrt{7}, 0)$ and $(-2 - \sqrt{7}, 0)$.

65. $y = -x^2 - 4x - 5$
$0 = -x^2 - 4x - 5$
$a = -1, b = -4, c = -5$
$x = \dfrac{-b \pm \sqrt{b^2 - 4ac}}{2a}$
$= \dfrac{-(-4) \pm \sqrt{(-4)^2 - 4(-1)(-5)}}{2(-1)}$
$= \dfrac{4 \pm \sqrt{16 - 20}}{-2}$
$= \dfrac{4 \pm \sqrt{-4}}{-2}$
$= \dfrac{4 \pm 2i}{-2}$
$= -2 \pm i$

The equation has no real solutions.
The parabola has no x-intercepts.

67. $y = -x^2 - 2x + 1$
$0 = -x^2 - 2x + 1$
$0 = x^2 + 2x - 1$
$a = 1, b = 2, c = -1$
$x = \dfrac{-b \pm \sqrt{b^2 - 4ac}}{2a}$
$= \dfrac{-2 \pm \sqrt{2^2 - 4(1)(-1)}}{2(1)}$
$= \dfrac{-2 \pm \sqrt{4 + 4}}{2}$
$= \dfrac{-2 \pm \sqrt{8}}{2}$
$= \dfrac{-2 \pm 2\sqrt{2}}{2}$
$= -1 \pm \sqrt{2}$

The x-intercepts are $(-1 + \sqrt{2}, 0)$ and $(-1 - \sqrt{2}, 0)$.

69. $f(x) = x^2 + 3x + 2$
$0 = x^2 + 3x + 2$
$0 = (x + 2)(x + 1)$
$x + 2 = 0 \qquad x + 1 = 0$
$x = -2 \qquad x = -1$
The zeros are $-2, -1$.

71. $f(x) = -x^2 + 4x - 5$
$0 = -x^2 + 4x - 5$
$a = -1, b = 4, c = -5$
$x = \dfrac{-b \pm \sqrt{b^2 - 4ac}}{2a}$
$= \dfrac{-4 \pm \sqrt{(4)^2 - 4(-1)(-5)}}{2(-1)}$
$= \dfrac{-4 \pm \sqrt{16 - 20}}{-2}$
$= \dfrac{-4 \pm \sqrt{-4}}{-2}$
$= \dfrac{-4 \pm 2i}{-2}$
$= 2 \pm i$

The zeros are $2 + i, 2 - i$.

73. $f(x) = 2x^2 - 3x$
$0 = 2x^2 - 3x$
$0 = x(2x - 3)$
$x = 0 \qquad 2x - 3 = 0$
$\qquad\qquad 2x = 3$
$\qquad\qquad x = \dfrac{3}{2}$

The zeros are 0 and $\dfrac{3}{2}$.

75. $f(x) = 2x^2 - 4$
$0 = 2x^2 - 4$
$4 = 2x^2$
$2 = x^2$
$\sqrt{2} = \sqrt{x^2}$
$\pm\sqrt{2} = x$

The zeros are $\sqrt{2}$ and $-\sqrt{2}$.

77. $f(x) = 2x^2 + 3x + 2$
$0 = 2x^2 + 3x + 2$
$a = 2, b = 3, c = 2$
$x = \dfrac{-b \pm \sqrt{b^2 - 4ac}}{2a}$
$= \dfrac{-3 \pm \sqrt{3^2 - 4(2)(2)}}{2(2)}$
$= \dfrac{-3 \pm \sqrt{9 - 16}}{4}$
$= \dfrac{-3 \pm \sqrt{-7}}{4}$
$= \dfrac{-3 \pm i\sqrt{7}}{4}$

The zeros are $-\dfrac{3}{4} + \dfrac{\sqrt{7}}{4}i$ and $-\dfrac{3}{4} - \dfrac{\sqrt{7}}{4}i$.

79. $f(x) = -3x^2 + 4x - 1$
$0 = -3x^2 + 4x - 1$
$0 = (-3x + 1)(x - 1)$
$-3x + 1 = 0 \qquad x - 1 = 0$
$-3x = -1 \qquad x = 1$
$x = \dfrac{1}{3}$

The zeros are $\dfrac{1}{3}$ and 1.

81. If the zeros of the function are −1 and 3, then the x-intercepts are (−1, 0) and (3, 0).

83.

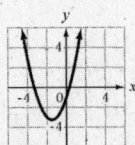

To the nearest tenth, the zeros of $f(x) = x^2 + 3x - 1$ are −3.3 and 0.3.

85.

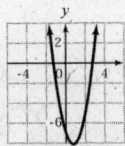

To the nearest tenth, the zeros of $f(x) = 2x^2 - 3x - 7$ are −1.3 and 2.8.

87.

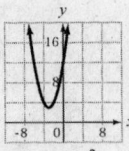

$f(x) = x^2 + 6x + 12$ has no x-intercepts.

89. $f(x) = x^2 - 10x - 5$
$2 = x^2 - 10x - 5$
$0 = x^2 - 10x - 7$
$a = 1, b = -10, c = -7$
$x = \dfrac{-b \pm \sqrt{b^2 - 4ac}}{2a}$
$x = \dfrac{10 \pm \sqrt{(-10)^2 - 4(1)(-7)}}{2(1)}$
$= \dfrac{10 \pm 8\sqrt{2}}{2} = 5 \pm 4\sqrt{2}$

91. $f(x) = x^2 - 2x - 10$
$7 = x^2 - 2x - 10$
$0 = x^2 - 2x - 17$
$a = 1, b = -2, c = -17$
$x = \dfrac{-b \pm \sqrt{b^2 - 4ac}}{2a}$
$x = \dfrac{2 \pm \sqrt{(-2)^2 - 4(1)(-17)}}{2(1)}$
$= \dfrac{2 \pm \sqrt{72}}{2} = 1 \pm 3\sqrt{2}$

93. $f(x) = x^2 - 2x + 11$
$2 = x^2 - 2x + 11$
$0 = x^2 - 2x + 9$
$a = 1, b = -2, c = 9$
$x = \dfrac{-b \pm \sqrt{b^2 - 4ac}}{2a}$
$x = \dfrac{2 \pm \sqrt{(-2)^2 - 4(1)(9)}}{2(1)}$
$= \dfrac{2 \pm \sqrt{-32}}{2}$

The solutions of $0 = x^2 - 2x + 9$ are complex numbers.

95. $y = 2x^2 + x + 1$
$a = 2, b = 1, c = 1$
$b^2 - 4ac$
$1^2 - 4(2)(1) = 1 - 8 = -7$
$-7 < 0$
Since the discriminant is less than zero, the parabola has no x-intercepts.

97. $y = -x^2 - x + 3$
$a = -1, b = -1, c = 3$
$b^2 - 4ac$
$(-1)^2 - 4(-1)(3) = 1 + 12 = 13$
$13 > 0$
Since the discriminant is greater than zero, the parabola has two x-intercepts.

99. $y = x^2 - 8x + 16$
$a = 1, b = -8, c = 16$
$b^2 - 4ac$
$(-8)^2 - 4(1)(16) = 64 - 64 = 0$
Since the discriminant is equal to zero, the parabola has one x-intercept.

101. $y = -3x^2 - x - 2$
$a = -3, b = -1, c = -2$
$b^2 - 4ac$
$(-1)^2 - 4(-3)(-2) = 1 - 24 = -23$
$-23 < 0$
Since the discriminant is less than zero, the parabola has no x-intercepts.

103. $y = 4x^2 - x - 2$
$a = 4, b = -1, c = -2$
$b^2 - 4ac$
$(-1)^2 - 4(4)(-2) = 1 + 32 = 33$
$33 > 0$
Since the discriminant is greater than zero, the parabola has two x-intercepts.

105. $y = -2x^2 - x - 5$
$a = -2, b = -1, c = -5$
$b^2 - 4ac$
$(-1)^2 - 4(-2)(-5) = 1 - 40 = -39$
$-39 < 0$
Since the discriminant is less than zero, the parabola has no x-intercepts.

107. $y = x^2 + 8x + 16$
$a = 1, b = 8, c = 16$
$b^2 - 4ac$
$8^2 - 4(1)(16) = 64 - 64 = 0$
Since the discriminant is equal to zero, the parabola has one x-intercept.

109. $y = x^2 + x - 3$
$a = 1, b = 1, c = -3$
$b^2 - 4ac$
$1^2 - 4(1)(-3) = 1 + 12 = 13$
$13 > 0$
Since the discriminant is greater than zero, the parabola has two x-intercepts.

111. True

113. True

Applying Concepts 8.6

115. $y = x^2 + 2x + k$
$1 = (-3)^2 + 2(-3) + k$
$1 = 9 - 6 + k$
$1 = 3 + k$
$-2 = k$
The value of k is –2.

117. $y = 3x^2 + kx - 6$
$4 = 3(-2)^2 + k(-2) - 6$
$4 = 3(4) - 2k - 6$
$4 = 12 - 2k - 6$
$4 = 6 - 2k$
$0 = 2 - 2k$
$2k = 2$
$k = 1$
The value of k is 1.

119. $f(x) = x^3 + x^2 + 4x + 4$
$0 = x^3 + x^2 + 4x + 4$
$0 = x^2(x + 1) + 4(x + 1)$
$0 = (x + 1)(x^2 + 4)$
$x + 1 = 0 \quad x^2 + 4 = 0$
$x = -1 \quad x^2 = -4$
$\sqrt{x^2} = \sqrt{-4}$
$|x| = 2i$
$x = \pm 2i$

121. $f(x) = x^4 + 3x^2 - 4$
$0 = x^4 + 3x^2 - 4$
$0 = (x^2 - 1)(x^2 + 4)$
$x^2 - 1 = 0 \quad x^2 + 4 = 0$
$x^2 = 1 \quad x^2 = -4$
$\sqrt{x^2} = \sqrt{1} \quad \sqrt{x^2} = \sqrt{-4}$
$|x| = 1 \quad |x| = 2i$
$x = \pm 1 \quad x = \pm 2i$

123. If the vertex is on the x-axis, then the value of y is zero.
Find the value of the x-coordinate of the vertex:
$x = \dfrac{-b}{2a}$
$= \dfrac{-(-8)}{2(1)}$
$= 4$
$y = x^2 - 8x + k$
$0 = (4)^2 - 8(4) + k$
$0 = 16 - 32 + k$
$0 = -16 + k$
$16 = k$
The value of k is 16.

125. Substitute 5 for y in the equation and solve for x.
$$5 = 3x^2 - 2x - 1$$
$$0 = 3x^2 - 2x - 6$$
$$a = 3, b = -2, c = -6$$
$$x = \frac{-b \pm \sqrt{b^2 - 4ac}}{2a}$$
$$= \frac{-(-2) \pm \sqrt{(-2)^2 - 4(3)(-6)}}{2(3)}$$
$$= \frac{2 \pm \sqrt{4 + 72}}{6}$$
$$= \frac{2 \pm \sqrt{76}}{6}$$
$$= \frac{2 \pm 2\sqrt{19}}{6}$$
$$= \frac{1 \pm \sqrt{19}}{3}$$

$\frac{1 - \sqrt{19}}{3}$ is not a possible solution because $\left(\frac{1 - \sqrt{19}}{3}, 5\right)$ is in quadrant II.

The solution is $\frac{1 + \sqrt{19}}{3}$.

127. a. The function $f(x) = x^2$ has a minimum vertex (0, 0).
The graph is E.

b. The function $f(x) = x^2 - 1$ has a minimum vertex (−1, 0).
The graph is F.

c. The function $f(x) = -x^2$ has a maximum vertex (0, 0).
The graph is C.

d. The function $f(x) = -x^2 + 3$ has a maximum vertex (3, 0).
The graph is A.

e. The function $f(x) = x^2 + 3x - 2$ has a minimum vertex (−1.5, −4.25).
The graph is B.

f. The function $f(x) = -x^2 - 3x + 2$ has a maximum vertex (−1.5, 4.25).
The graph is D.

129. $F(v) = 0.151v^2$
$F(15) = 0.151(15)^2$
$= 33.975$
The drag force is 33.975 lb.

131. $y = x^2 - 4x + 7$
$= (x^2 - 4x) + 7$
$= (x^2 - 4x + 4) - 4 + 7$
$= (x - 2)^2 + 3$
$x - 2 = 0$
$x = 2$
$y = (x - 2)^2 + 3$
$= (2 - 2)^2 + 3$
$= 3$
The vertex is (2, 3).

133. $y = x^2 + x + 2$
$= (x^2 + x) + 2$
$= \left(x^2 + x + \frac{1}{4}\right) - \frac{1}{4} + 2$
$= \left(x + \frac{1}{2}\right)^2 + \frac{7}{4}$

$x + \frac{1}{2} = 0 \qquad y = \left(x + \frac{1}{2}\right)^2 + \frac{7}{4}$
$x = -\frac{1}{2} \qquad = \left(-\frac{1}{2} + \frac{1}{2}\right)^2 + \frac{7}{4}$
$\qquad = \frac{7}{4}$

The vertex is $\left(-\frac{1}{2}, \frac{7}{4}\right)$.

135. $y = a(x - h)^2 + k$
Vertex (1, 2): $h = 1, k = 2$
$P(2, 5): x = 2, y = 5$
Substitute into the equation:
$5 = a(2 - 1)^2 + 2$
$5 = a(1)^2 + 2$
$5 = a + 2$
$3 = a$
$y = 3(x - 1)^2 + 2$
$y = 3(x^2 - 2x + 1) + 2$
$y = 3x^2 - 6x + 3 + 2$
$y = 3x^2 - 6x + 5$

Section 8.7
Concept Review 8.7

1. Sometimes true
If the parabola opens down, it has a maximum value. The range of the parabola would be from negative infinity to the maximum value. Thus, there would be no minimum value.

3. Always true

5. Sometimes true
If a parabola opens down and the parabola is below the x-axis, the maximum value of the parabola is a negative number.

Objective 8.7.1 Exercises

1. The minimum value or the maximum value of a quadratic function is the value of the function at the vertex of the graph of the function; it is the y-coordinate of the vertex. A quadratic function has a minimum value when $a > 0$ in $f(x) = ax^2 + bx + c$. A quadratic function has a maximum value when $a < 0$ in $f(x) = ax^2 + bx + c$.

3. Because $a = \underline{-8}$, the graph of f opens <u>down</u>. The vertex is the highest point on the graph, so its y-coordinate is the maximum value of $f(x)$.

5. a. The function $f(x) = -x^2 + 6x - 1$ has a maximum since $a = -1 < 0$.

   b. The function $f(x) = 2x^2 - 4$ has a minimum since $a = 2 > 0$.

   c. The function $f(x) = -5x^2 + x$ has a maximum since $a = -5 < 0$.

7. $f(x) = x^2 - 2x + 3$
$$x = -\frac{b}{2a} = -\frac{-2}{2(1)} = 1$$
$$f(x) = x^2 - 2x + 3$$
$$f(1) = (1)^2 - 2(1) + 3 = 2$$
Since a is positive, the function has a minimum value. The minimum value of the function is 2.

9. $f(x) = -2x^2 + 4x - 3$
$$x = -\frac{b}{2a} = -\frac{4}{2(-2)} = 1$$
$$f(x) = -2x^2 + 4x - 3$$
$$f(1) = -2(1)^2 + 4(1) - 3$$
$$= -2 + 4 - 3 = -1$$
Since a is negative, the function has a maximum value. The maximum value of the function is -1.

11. $f(x) = -2x^2 - 3x + 4$
$$x = -\frac{b}{2a} = -\frac{-3}{2(-2)} = -\frac{3}{4}$$
$$f(x) = -2x^2 - 3x + 4$$
$$f\left(-\frac{3}{4}\right) = -2\left(-\frac{3}{4}\right)^2 - 3\left(-\frac{3}{4}\right) + 4$$
$$= -\frac{9}{8} + \frac{9}{4} + 4 = \frac{41}{8}$$
Since a is negative, the function has a maximum value. The maximum value of the function is $\frac{41}{8}$.

13. $f(x) = 2x^2 + 3x - 8$
$$x = -\frac{b}{2a} = -\frac{3}{2(2)} = -\frac{3}{4}$$
$$f(x) = 2x^2 + 3x - 8$$
$$f\left(-\frac{3}{4}\right) = 2\left(-\frac{3}{4}\right)^2 + 3\left(-\frac{3}{4}\right) - 8$$
$$= \frac{9}{8} - \frac{9}{4} - 8$$
$$= -\frac{73}{8}$$
Since a is positive, the function has a minimum value. The minimum value of the function is $-\frac{73}{8}$.

15. $f(x) = -3x^2 + x - 6$
$$x = -\frac{b}{2a} = -\frac{1}{2(-3)} = \frac{1}{6}$$
$$f(x) = -3x^2 + x - 6$$
$$f\left(\frac{1}{6}\right) = -3\left(\frac{1}{6}\right)^2 + \left(\frac{1}{6}\right) - 6 = -\frac{1}{12} + \frac{1}{6} - 6 = -\frac{71}{12}$$
Since a is negative, the function has a maximum value. The maximum value of the function is $-\frac{71}{12}$.

17. $f(x) = x^2 - 5x + 3$
$$x = -\frac{b}{2a} = -\frac{-5}{2(1)} = \frac{5}{2}$$
$$f(x) = x^2 - 5x + 3$$
$$f\left(\frac{5}{2}\right) = \left(\frac{5}{2}\right)^2 - 5\left(\frac{5}{2}\right) + 3$$
$$= \frac{25}{4} - \frac{25}{2} + 3$$
$$= -\frac{13}{4}$$
Since a is positive, the function has a minimum value. The minimum value of the function is $-\frac{13}{4}$.

19. Strategy
To find the highest minimum value, find the x-coordinate of the vertex for each parabola and evaluate the parabola at that point.

Solution
For a, $y = x^2 - 2x - 3$,
$$\frac{-b}{2a} = \frac{-(-2)}{2(1)} = \frac{2}{2} = 1$$
$$y = (1)^2 - 2(1) - 3 = -4$$

For b, $y = x^2 - 10x + 20$,
$$\frac{-b}{2a} = \frac{-(-10)}{2(1)} = \frac{10}{2} = 5$$
$$y = (5)^2 - 10(5) + 20 = -5$$

For c, $y = 3x^2 - 6$,
$$\frac{-b}{2a} = \frac{0}{2(3)} = 0$$
$$y = 3(0)^2 - 6 = -6$$
a has the highest minimum value.

21. Since the parabola open up, it has a minimum value. The minimum value of the function is 7.

23. True

Objective 8.7.2 Exercises

25. To find out after how many seconds the ball reaches its maximum height, find the t-coordinate of the vertex of the graph of h: $t = -\frac{b}{2a} = -\frac{30}{2(-5)} = 3$. The ball reaches its maximum height after 3 s.

27. Strategy
To find the time it takes the rock to reach its maximum height, find the t-coordinate of the vertex. To find the maximum height, evaluate the function at the t-coordinate of the vertex.

Solution
$$t = -\frac{b}{2a} = -\frac{64}{2(-16)} = 2$$
The rock reaches its maximum height in 2 s.
$$s(t) = -16t^2 + 64t + 50$$
$$s(2) = -16(2)^2 + 64(2) + 50$$
$$= -64 + 128 + 50 = 114$$
The maximum height is 114 ft.

29. Strategy
To find the number of tickets that will give the maximum profit, find the x-coordinate of the vertex and evaluate the function at that point.

Solution
$$x = -\frac{b}{2a} = -\frac{40}{2(-0.25)} = 80$$
$$P(x) = 40x - 0.25x^2$$
$$P(80) = 40(80) - 0.25(80)^2$$
$$= 3200 - 1600$$
$$= 1600$$
The operator can expect a maximum profit of $1600.

31. Strategy
To find the number of days for the least amount of algae, find the t-coordinate of the vertex.

Solution
$$t = -\frac{b}{2a} = -\frac{-400}{2(40)} = 5$$
The pool will have the least amount of algae 5 days after treatment.

33. Strategy
To find the point where the thickness will be a minimum, find the x-coordinate of the vertex. To find the minimum thickness, evaluate the function at the x-coordinate of the vertex.

Solution
$$h(x) = 0.000379x^2 - 0.0758x + 24$$
$$x = -\frac{b}{2a} = -\frac{-0.0758}{2(0.000379)} = 100$$
The thickness is a minimum 100 inches from the edge.
$$h(x) = 0.000379x^2 - 0.0758x + 24$$
$$h(100) = 0.000379(100)^2 - 0.0758(100) + 24$$
$$= 3.79 - 7.58 + 24$$
$$= 20.21$$
The minimum thickness is 20.21 in.

35. **Strategy**
 To find how far up the water will land, evaluate the function at 40 ft.

 Solution
 $$s(x) = -\frac{1}{30}x^2 + 2x + 5$$
 $$s(40) = -\frac{1}{30}(40)^2 + 2(40) + 5 = 31\frac{2}{3}$$
 The water will land at $31\frac{2}{3}$ ft.

37. **Strategy**
 The first number: x
 The second number: $20 - x$
 Find the number that will maximize the product of the two numbers.

 Solution
 $$x(20 - x) = 0$$
 $$-x^2 + 20x = 0$$
 $$x = -\frac{b}{2a} = -\frac{20}{2(-1)} = 10$$
 $$20 - x = 20 - 10 = 10$$
 The numbers are 10 and 10.

39. **Strategy**
 The length of fencing is 200 ft.
 $200 = 2W + L$
 $200 - 2W = L$
 The area is $LW = (200 - 2W)W = 200W - 2W^2$
 To find the width that will maximize the area, find the W-coordinate of the vertex.
 To find the length, replace W in $200 - 2W$ by the W-coordinate of the vertex and evaluate.

 Solution
 $$A = 200W - 2W^2$$
 $$W = -\frac{b}{2a} = -\frac{200}{2(-2)} = 50$$
 $$L = 200 - 2(W)$$
 $$= 200 - 2(50)$$
 $$= 100$$
 To maximize the area, the length is 100 ft and the width is 50 ft.

Applying Concepts 8.7

41. The minimum value of the function $f(x) = x^4 - 2x^2 + 4$ is 3.0.

43. The maximum value of the function $f(x) = -x^6 + x^4 - x^3 + x$ is 0.5.

45. When the highest power is an even exponent and the sign of the leading coefficient of a polynomial function is positive, the function will have a minimum value. When the highest power is an even exponent and the sign of the leading coefficient of a polynomial function is negative, the function will have a maximum value.

47. Here are some highlights. For more information see *The History of Mathematics: An Introduction* (1985) by David Burton or *A History of Mathematics: An Introduction* (1993) by Victor Katz.
 The Babylonians had solved quadratic equations by 2000 B.C. The solution of cubic equations did not come until much later. Around 1100 A.D., Omar Khayyam gave the first complete *geometric* solutions of method that relied heavily on Euclid's *Elements*.
 Scipione del Ferro (1465-1526) was the first mathematician to give an algebraic solution of the cubic equation. He solved the special case $x^3 + cx = d$. Academic promotion during del Ferr's time was frequently based on public challenges in which mathematicians would challenge each other to solve certain problems.
 Having the secret of the solution of $x^3 + cx = d$ guaranteed del Ferro's continued success at winning these competitions and, consequently, retaining his professorship.
 In the sixteenth century, Ludovico Ferrari solved a fourth-degree equation by reducing it to third-degree equation.
 Paolo Ruffini (1765-1822) was the first to prove that the quintic equation could not be solved by explicit formulas similar to quadratic, cubic, and quartic equations. His proof was generally sound but contained some flaws. Neils Henrik Abel (1802-1829) gave a more rigorous proof in 1824. It was Evariste Galois (1812-1832) who first showed that no polynomial equation of degree greater than 4 could be solved by a general formula.

CHAPTER 8 REVIEW EXERCISES

1. $$2x^2 - 3x = 0$$
 $$x(2x - 3) = 0$$
 $$x = 0 \qquad 2x - 3 = 0$$
 $$\qquad\qquad 2x = 3$$
 $$\qquad\qquad x = \frac{3}{2}$$
 The solutions are 0 and $\frac{3}{2}$.

2.
$$6x^2 + 9xc = 6c^2$$
$$6x^2 + 9xc - 6c^2 = 0$$
$$3(2x^2 + 3cx - 2c^2) = 0$$
$$3(2x - c)(x + 2c) = 0$$
$$2x - c = 0 \quad x + 2c = 0$$
$$2x = c \quad x = -2c$$
$$x = \frac{c}{2}$$

The solutions are $\frac{c}{2}$ and $-2c$.

3.
$$x^2 = 48$$
$$\sqrt{x^2} = \sqrt{48}$$
$$x = \pm\sqrt{48} = \pm 4\sqrt{3}$$

The solutions are $4\sqrt{3}$ and $-4\sqrt{3}$.

4.
$$\left(x + \frac{1}{2}\right)^2 + 4 = 0$$
$$\left(x + \frac{1}{2}\right)^2 = -4$$
$$\sqrt{\left(x + \frac{1}{2}\right)^2} = \sqrt{-4}$$
$$x + \frac{1}{2} = \pm\sqrt{-4} = \pm 2i$$
$$x + \frac{1}{2} = 2i \quad x + \frac{1}{2} = -2i$$
$$x = -\frac{1}{2} + 2i \quad x = -\frac{1}{2} - 2i$$

The solutions are $-\frac{1}{2} + 2i$ and $-\frac{1}{2} - 2i$.

5.
$$-\frac{b}{2a} = -\frac{-7}{2(1)} = \frac{7}{2}$$
$$f(x) = x^2 - 7x + 8$$
$$f\left(\frac{7}{2}\right) = \left(\frac{7}{2}\right)^2 - 7\left(\frac{7}{2}\right) + 8$$
$$= \frac{49}{4} - \frac{49}{2} + 8$$
$$= -\frac{17}{4}$$

The minimum value of the function is $-\frac{17}{4}$.

6.
$$-\frac{b}{2a} = -\frac{4}{2(-2)} = 1$$
$$f(x) = -2x^2 + 4x + 1$$
$$f(1) = -2(1)^2 + 4(1) + 1$$
$$= -2 + 4 + 1$$
$$= 3$$

The maximum value of the function is 3.

7.
$$(x - r_1)(x - r_2) = 0$$
$$\left(x - \frac{1}{3}\right)[x - (-3)] = 0$$
$$\left(x - \frac{1}{3}\right)(x + 3) = 0$$
$$x^2 + \frac{8}{3}x - 1 = 0$$
$$3\left(x^2 + \frac{8}{3}x - 1\right) = 3 \cdot 0$$
$$3x^2 + 8x - 3 = 0$$

8.
$$2x^2 + 9x = 5$$
$$2x^2 + 9x - 5 = 0$$
$$(2x - 1)(x + 5) = 0$$
$$2x - 1 = 0 \quad x + 5 = 0$$
$$2x = 1 \quad x = -5$$
$$x = \frac{1}{2}$$

The solutions are $\frac{1}{2}$ and -5.

9.
$$2(x + 1)^2 - 36 = 0$$
$$2(x + 1)^2 = 36$$
$$(x + 1)^2 = 18$$
$$\sqrt{(x + 1)^2} = \sqrt{18}$$
$$x + 1 = \pm 3\sqrt{2}$$
$$x + 1 = 3\sqrt{2} \quad x + 1 = -3\sqrt{2}$$
$$x = -1 + 3\sqrt{2} \quad x = -1 - 3\sqrt{2}$$

The solutions are $-1 + 3\sqrt{2}$ and $-1 - 3\sqrt{2}$.

10.
$$x^2 + 6x + 10 = 0$$
$$a = 1, b = 6, c = 10$$
$$x = \frac{-b \pm \sqrt{b^2 - 4ac}}{2a}$$
$$= \frac{-6 \pm \sqrt{6^2 - 4(1)(10)}}{2(1)}$$
$$= \frac{-6 \pm \sqrt{36 - 40}}{2}$$
$$= \frac{-6 \pm \sqrt{-4}}{2}$$
$$= \frac{-6 \pm 2i}{2}$$
$$= -3 \pm i$$

The solutions are $-3 + i$ and $-3 - i$.

Chapter 8: Quadratic Equations And Inequalities   **287**

11.
$$\frac{2}{x-4}+3=\frac{x}{2x-3}$$
$$(x-4)(2x-3)\left(\frac{2}{x-4}+3\right)=(x-4)(2x-3)\frac{x}{2x-3}$$
$$(2x-3)2+3(x-4)(2x-3)=(x-4)x$$
$$4x-6+3(2x^2-11x+12)=x^2-4x$$
$$4x-6+6x^2-33x+36=x^2-4x$$
$$6x^2-29x+30=x^2-4x$$
$$5x^2-25x+30=0$$
$$5(x^2-5x+6)=0$$
$$5(x-2)(x-3)=0$$
$$x-2=0 \quad x-3=0$$
$$x=2 \quad x=3$$
The solutions are 2 and 3.

12.
$$x^4-6x^2+8=0$$
$$(x^2)^2-6(x^2)+8=0$$
$$u^2-6u+8=0$$
$$(u-2)(u-4)=0$$
$$u-2=0 \quad u-4=0$$
$$u=2 \quad u=4$$
Replace u by x^2.
$$x^2=2 \quad x^2=4$$
$$\sqrt{x^2}=\sqrt{2} \quad \sqrt{x^2}=\sqrt{4}$$
$$x=\pm\sqrt{2} \quad x=\pm 2$$
The solutions are $\sqrt{2}$, $-\sqrt{2}$, 2, and -2

13.
$$\sqrt{2x-1}+\sqrt{2x}=3$$
$$\sqrt{2x-1}=3-\sqrt{2x}$$
$$\left(\sqrt{2x-1}\right)^2=\left(3-\sqrt{2x}\right)^2$$
$$2x-1=9-6\sqrt{2x}+2x$$
$$-10=-6\sqrt{2x}$$
$$\frac{-10}{-6}=\frac{-6\sqrt{2x}}{-6}$$
$$\frac{5}{3}=\sqrt{2x}$$
$$\left(\frac{5}{3}\right)^2=\left(\sqrt{2x}\right)^2$$
$$\frac{25}{9}=2x$$
$$\frac{1}{2}\left(\frac{25}{9}\right)=\frac{1}{2}(2x)$$
$$\frac{25}{18}=x$$
The solution is $\frac{25}{18}$.

14.
$$2x^{2/3}+3x^{1/3}-2=0$$
$$2(x^{1/3})^2+3(x^{1/3})-2=0$$
$$2u^2+3u-2=0$$
$$(2u-1)(u+2)=0$$
$$2u-1=0 \quad u+2=0$$
$$2u=1 \quad u=-2$$
$$u=\frac{1}{2}$$
Replace u by $x^{1/3}$.
$$x^{1/3}=\frac{1}{2} \quad\quad x^{1/3}=-2$$
$$(x^{1/3})^3=\left(\frac{1}{2}\right)^3 \quad (x^{1/3})^3=(-2)^3$$
$$x=\frac{1}{8} \quad\quad x=-8$$
The solutions are $\frac{1}{8}$ and -8.

15.
$$\sqrt{3x-2}+4=3x$$
$$\sqrt{3x-2}=3x-4$$
$$\left(\sqrt{3x-2}\right)^2=(3x-4)^2$$
$$3x-2=9x^2-24x+16$$
$$0=9x^2-27x+18$$
$$0=9(x^2-3x+2)$$
$$0=9(x-2)(x-1)$$
$$x-2=0 \quad x-1=0$$
$$x=2 \quad x=1$$
1 does not check as a solution.
The solution is 2.

16.
$$x^2-6x-2=0$$
$$a=1, b=-6, c=-2$$
$$x=\frac{-b\pm\sqrt{b^2-4ac}}{2a}$$
$$=\frac{6\pm\sqrt{(-6)^2-4(1)(-2)}}{2(1)}$$
$$=\frac{6\pm\sqrt{36+8}}{2}$$
$$=\frac{6\pm\sqrt{44}}{2}$$
$$=\frac{6\pm 2\sqrt{11}}{2}$$
$$=3\pm\sqrt{11}$$
The solutions are $3+\sqrt{11}$ and $3-\sqrt{11}$.

17.
$$\frac{2x}{x-4} + \frac{6}{x+1} = 11$$
$$(x-4)(x+1)\left(\frac{2x}{x-4} + \frac{6}{x+1}\right) = (x-4)(x+1)11$$
$$(x-4)(x+1)\frac{2x}{x-4} + (x-4)(x+1)\frac{6}{x+1} = 11(x-4)(x+1)$$
$$2x(x+1) + 6(x-4) = 11(x^2 - 3x - 4)$$
$$2x^2 + 2x + 6x - 24 = 11x^2 - 33x - 44$$
$$2x^2 + 8x - 24 = 11x^2 - 33x - 44$$
$$0 = 9x^2 - 41x - 20$$
$$0 = (9x + 4)(x - 5)$$

$9x + 4 = 0 \qquad x - 5 = 0$
$9x = -4 \qquad x = 5$
$x = -\frac{4}{9}$

The solutions are $-\frac{4}{9}$ and 5.

18. $\quad 2x^2 - 2x = 1$
$2x^2 - 2x - 1 = 0$
$a = 2,\ b = -2,\ c = -1$
$x = \dfrac{-b \pm \sqrt{b^2 - 4ac}}{2a}$
$= \dfrac{2 \pm \sqrt{(-2)^2 - 4(2)(-1)}}{2(2)}$
$= \dfrac{2 \pm \sqrt{4 + 8}}{4}$
$= \dfrac{2 \pm \sqrt{12}}{4}$
$= \dfrac{2 \pm 2\sqrt{3}}{4}$
$= \dfrac{1 \pm \sqrt{3}}{2}$

The solutions are $\dfrac{1+\sqrt{3}}{2}$ and $\dfrac{1-\sqrt{3}}{2}$.

19. $\quad 2x = 4 - 3\sqrt{x-1}$
$2x - 4 = -3\sqrt{x-1}$
$(2x - 4)^2 = \left(-3\sqrt{x-1}\right)^2$
$4x^2 - 16x + 16 = 9(x - 1)$
$4x^2 - 16x + 16 = 9x - 9$
$4x^2 - 25x + 25 = 0$
$(4x - 5)(x - 5) = 0$
$4x - 5 = 0 \qquad x - 5 = 0$
$4x = 5 \qquad x = 5$
$x = \dfrac{5}{4}$

5 does not check as a solution. The solution is $\dfrac{5}{4}$.

20. $\quad 3x = \dfrac{9}{x-2}$
$3x(x-2) = \dfrac{9}{x-2}(x-2)$
$3x^2 - 6x = 9$
$3x^2 - 6x - 9 = 0$
$3(x^2 - 2x - 3) = 0$
$3(x - 3)(x + 1) = 0$
$x - 3 = 0 \qquad x + 1 = 0$
$x = 3 \qquad x = -1$
The solutions are 3 and -1.

21.
$$\frac{3x+7}{x+2} + x = 3$$
$$(x+2)\left(\frac{3x+7}{x+2} + x\right) = (x+2)3$$
$$3x + 7 + x(x+2) = 3x + 6$$
$$3x + 7 + x^2 + 2x = 3x + 6$$
$$x^2 + 5x + 7 = 3x + 6$$
$$x^2 + 2x + 1 = 0$$
$$(x+1)^2 = 0$$
$$\sqrt{(x+1)^2} = \sqrt{0}$$
$$x + 1 = 0$$
$$x = -1$$
The solution is -1.

22.
$$\frac{x-2}{2x+3} - \frac{x-4}{x} = 2$$
$$x(2x+3)\left(\frac{x-2}{2x+3} - \frac{x-4}{x}\right) = x(2x+3)2$$
$$x(x-2) - (2x+3)(x-4) = 2x(2x+3)$$
$$x^2 - 2x - (2x^2 - 5x - 12) = 4x^2 + 6x$$
$$x^2 - 2x - 2x^2 + 5x + 12 = 4x^2 + 6x$$
$$-x^2 + 3x + 12 = 4x^2 + 6x$$
$$0 = 5x^2 + 3x - 12$$
$$a = 5, b = 3, c = -12$$
$$x = \frac{-b \pm \sqrt{b^2 - 4ac}}{2a}$$
$$= \frac{-3 \pm \sqrt{3^2 - 4(5)(-12)}}{2(5)}$$
$$= \frac{-3 \pm \sqrt{9 + 240}}{10}$$
$$= \frac{-3 \pm \sqrt{249}}{10}$$
The solutions are $\frac{-3 + \sqrt{249}}{10}$ and $\frac{-3 - \sqrt{249}}{10}$.

23.
$$1 - \frac{x+4}{2-x} = \frac{x-3}{x+2}$$
$$(x+2)(2-x)\left(1 - \frac{x+4}{2-x}\right) = (x+2)(2-x)\frac{x-3}{x+2}$$
$$(x+2)(2-x) - (x+2)(x+4) = (2-x)(x-3)$$
$$4 - x^2 - (x^2 + 6x + 8) = -x^2 + 5x - 6$$
$$4 - x^2 - x^2 - 6x - 8 = -x^2 + 5x - 6$$
$$-2x^2 - 6x - 4 = -x^2 + 5x - 6$$
$$0 = x^2 + 11x - 2$$
$$a = 1, b = 11, c = -2$$
$$x = \frac{-b \pm \sqrt{b^2 - 4ac}}{2a}$$
$$= \frac{-11 \pm \sqrt{11^2 - 4(1)(-2)}}{2(1)}$$
$$= \frac{-11 \pm \sqrt{121 + 8}}{2}$$
$$= \frac{-11 \pm \sqrt{129}}{2}$$
The solutions are $\frac{-11 + \sqrt{129}}{2}$ and $\frac{-11 - \sqrt{129}}{2}$.

24. $y = -x^2 + 6x - 5$
The axis of symmetry is the line with equation $x = -\frac{b}{2a}$.
$$a = -1, b = 6$$
$$x = -\frac{6}{2(-1)} = -\frac{6}{-2} = 3$$
The axis of symmetry is $x = 3$.

25. $y = -x^2 + 3x - 2$
The x-coordinate of the vertex is $-\frac{b}{2a}$.
$$a = -1, b = 3$$
$$-\frac{b}{2a} = -\frac{3}{2(-1)} = -\frac{3}{-2} = \frac{3}{2}$$
$$y = -x^2 + 3x - 2$$
$$= -\left(\frac{3}{2}\right)^2 + 3\left(\frac{3}{2}\right) - 2$$
$$= \frac{1}{4}$$
The vertex is $\left(\frac{3}{2}, \frac{1}{4}\right)$

26. $f(x) = x^2 - 8x - 4$
$1 = x^2 - 8x - 4$
$0 = x^2 - 8x - 5$
$a = 1, b = -8, c = -5$
$x = \dfrac{-b \pm \sqrt{b^2 - 4ac}}{2a}$
$= \dfrac{8 \pm \sqrt{(-8)^2 - 4(1)(-5)}}{2(1)}$
$= \dfrac{8 \pm \sqrt{84}}{2}$
$= \dfrac{8 \pm 2\sqrt{21}}{2} = 4 \pm \sqrt{21}$

27. $y = 3x^2 - 2x - 4$
$a = 3, b = -2, c = -4$
$b^2 - 4ac$
$(-2)^2 - 4(3)(-4) = 4 + 48 = 52$
$50 > 0$
Since the discriminant is greater than zero, the parabola has two x-intercepts.

28. $y = 4x^2 + 12x + 4$
$0 = 4x^2 + 12x + 4$
$0 = 4(x^2 + 3x + 1)$
$a = 1, b = 3, c = 1$
$x = \dfrac{-b \pm \sqrt{b^2 - 4ac}}{2a}$
$= \dfrac{-3 \pm \sqrt{3^2 - 4(1)(1)}}{2(1)}$
$= \dfrac{-3 \pm \sqrt{9 - 4}}{2}$
$= \dfrac{-3 \pm \sqrt{5}}{2}$
The x-intercepts are $\left(\dfrac{-3 + \sqrt{5}}{2}, 0\right)$ and $\left(\dfrac{-3 - \sqrt{5}}{2}, 0\right)$.

29. $y = -2x^2 - 3x + 2$
$0 = -2x^2 - 3x + 2$
$0 = (-2x + 1)(x + 2)$
$-2x + 1 = 0 \qquad x + 2 = 0$
$-2x = -1 \qquad x = -2$
$x = \dfrac{1}{2}$
The x-intercepts are $\left(\dfrac{1}{2}, 0\right)$ and $(-2, 0)$.

30. $f(x) = 3x^2 + 2x + 2$
$0 = 3x^2 + 2x + 2$
$a = 3, b = 2, c = -2$
$x = \dfrac{-b \pm \sqrt{b^2 - 4ac}}{2a}$
$= \dfrac{-2 \pm \sqrt{2^2 - 4(3)(2)}}{2(3)}$
$= \dfrac{-2 \pm \sqrt{4 - 24}}{6}$
$= \dfrac{-2 \pm \sqrt{-20}}{6}$
$= \dfrac{-2 \pm 2i\sqrt{5}}{6}$
$= \dfrac{-1 \pm i\sqrt{5}}{3}$
The zeros are $-\dfrac{1}{3} + \dfrac{\sqrt{5}}{3}i$ and $-\dfrac{1}{3} - \dfrac{\sqrt{5}}{3}i$.

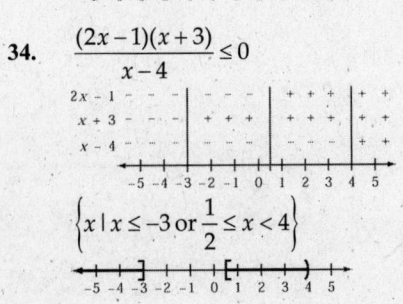

31. $(x + 3)(2x - 5) < 0$
$\left\{x \mid -3 < x < \dfrac{5}{2}\right\}$

32. $(x - 2)(x + 4)(2x + 3) \leq 0$
$\left\{x \mid x \leq -4 \text{ or } -\dfrac{3}{2} \leq x \leq 2\right\}$

33. $\dfrac{x - 2}{2x - 3} \geq 0$
$\left\{x \mid x < \dfrac{3}{2} \text{ or } x \geq 2\right\}$

34. $\dfrac{(2x - 1)(x + 3)}{x - 4} \leq 0$
$\left\{x \mid x \leq -3 \text{ or } \dfrac{1}{2} \leq x < 4\right\}$

35. $-\dfrac{b}{2a} = -\dfrac{2}{2(1)} = -1$

$y = (-1)^2 + 2(-1) - 4 = -5$

The vertex is (−1, −5).
The axis of symmetry is $x = -1$.

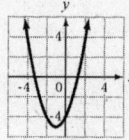

The domain is $\{x \mid x \in \text{real numbers}\}$
The range is $\{y \mid y \geq -5\}$.

36. $-\dfrac{b}{2a} = -\dfrac{-2}{2(1)} = 1$

$y = 1^2 - 2(1) + 3 = 2$

The vertex is (1, 2).
The axis of symmetry is $x = 1$.

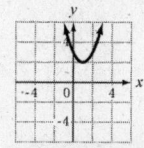

37. **Strategy**
This is an integer problem.
The first integer: x
The second consecutive even integer: $x + 2$
The third consecutive even integer: $x + 4$
The sum of the squares of the three consecutive integers is 56.
$x^2 + (x+2)^2 + (x+4)^2 = 56$

Solution
$x^2 + (x+2)^2 + (x+4)^2 = 56$
$x^2 + x^2 + 4x + 4 + x^2 + 8x + 16 = 56$
$3x^2 + 12x - 36 = 0$
$3(x^2 + 4 - 12) = 0$
$3(x+6)(x-2) = 0$
$x + 6 = 0 \quad x - 2 = 0$
$x = -6 \quad\quad x = 2$
$x = 2, x + 2 = 4, x + 4 = 6$
$x = -6, x + 2 = -4, x + 4 = -2$
The integers are 2, 4, and 6 or −6, −4, and −2.

38. **Strategy**
This is a geometry problem.
The width of the rectangle: x
The length of the rectangle: $2x + 2$
The area of the rectangle is 60 cm^2.
Use the equation for the area of the rectangle ($A = L \cdot W$).

Solution
$A = L \cdot W$
$60 = x(2x + 2)$
$60 = 2x^2 + 2x$
$0 = 2x^2 + 2x - 60$
$0 = 2(x^2 + x - 30)$
$0 = 2(x+6)(x-5)$
$x + 6 = 0 \quad x - 5 = 0$
$x = -6 \quad\quad x = 5$
Since the width cannot be negative, −6 cannot be a solution.
$2x + 2 = 2(5) + 2 = 10 + 2 = 12$
The width of the rectangle is 5 cm.
The length of the rectangle is 12 cm.

39. **Strategy**
This is a work problem.
Time for new computer to print payroll: x
Time for older computer to print payroll: $x + 12$

	Rate	Time	Part
New computer	$\dfrac{1}{x}$	8	$\dfrac{8}{x}$
Older computer	$\dfrac{1}{x+12}$	8	$\dfrac{8}{x+12}$

The sum of the parts of the task completed must be 1.

$\dfrac{8}{x} + \dfrac{8}{x+12} = 1$

Solution

$\dfrac{8}{x} + \dfrac{8}{x+12} = 1$

$x(x+12)\left(\dfrac{8}{x} + \dfrac{8}{x+12}\right) = x(x+12)(1)$

$8(x+12) + 8x = x(x+12)$
$8x + 96 + 8x = x^2 + 12x$
$16x + 96 = x^2 + 12x$
$0 = x^2 - 4x - 96$
$0 = (x-12)(x+8)$
$x - 12 = 0 \quad x + 8 = 0$
$x = 12 \quad\quad x = -8$
The solution −8 is not possible, since time cannot be a negative number. Working alone, the new computer can print the payroll in 12 min.

40. Strategy
This is a distance-rate problem.
Rate of the first car: r
Rate of the second car: $r + 10$

	Distance	Rate	Time
First car	200	r	$\dfrac{200}{r}$
Second car	200	$r + 10$	$\dfrac{200}{r+10}$

The second car's time is one hour less than the time of the first car.
$$\frac{200}{r+10} = \frac{200}{r} - 1$$

Solution
$$\frac{200}{r+10} = \frac{200}{r} - 1$$
$$r(r+10)\left(\frac{200}{r+10}\right) = r(r+10)\left(\frac{200}{r} - 1\right)$$
$$200r = 200(r+10) - r(r+10)$$
$$200r = 200r + 2000 - r^2 - 10r$$
$$r^2 + 10r - 2000 = 0$$
$$(r+50)(r-40) = 0$$
$r + 50 = 0 \qquad r - 40 = 0$
$r = -50 \qquad r = 40$

The solution −50 is not possible, since rate cannot be a negative number.
$r + 10 = 40 + 10 = 50$
The rate of the first car is 40 mph.
The rate of the second car is 50 mph.

CHAPTER 8 TEST

1.
$$2x^2 + x = 6$$
$$2x^2 + x - 6 = 0$$
$$(2x-3)(x+2) = 0$$
$2x - 3 = 0 \qquad x + 2 = 0$
$2x = 3 \qquad x = -2$
$x = \dfrac{3}{2}$

The solutions are $\dfrac{3}{2}$ and -2.

2.
$$12x^2 + 7x - 12 = 0$$
$$(3x+4)(4x-3) = 0$$
$3x + 4 = 0 \qquad 4x - 3 = 0$
$3x = -4 \qquad 4x = 3$
$x = -\dfrac{4}{3} \qquad x = \dfrac{3}{4}$

The solutions are $-\dfrac{4}{3}$ and $\dfrac{3}{4}$.

3. $f(x) = -x^2 + 8x - 7$
$$x = -\frac{b}{2a} = -\frac{8}{2(-1)} = 4$$
$$f(4) = -4^2 + 8(4) - 7 = 9.$$
The maximum value of the function is 9.

4.
$$(x - r_1)(x - r_2) = 0$$
$$\left[x - \left(-\frac{1}{3}\right)\right](x - 3) = 0$$
$$\left(x + \frac{1}{3}\right)(x - 3) = 0$$
$$x^2 - 3x + \frac{1}{3}x - 1 = 0$$
$$x^2 - \frac{8}{3}x - 1 = 0$$
$$3\left(x^2 - \frac{8}{3}x - 1\right) = 3 \cdot 0$$
$$3x^2 - 8x - 3 = 0$$

5. $2(x+3)^2 - 36 = 0$
$$2(x+3)^2 = 36$$
$$(x+3)^2 = 18$$
$$\sqrt{(x+3)^2} = \sqrt{18}$$
$$x + 3 = \pm 3\sqrt{2}$$
$$x = -3 \pm 3\sqrt{2}$$
The solutions are $-3 + 3\sqrt{2}$ and $-3 - 3\sqrt{2}$.

6. $x^2 + 4x - 1 = 0$
$$x^2 + 4x = 1$$
Complete the square.
$$x^2 + 4x + 4 = 1 + 4$$
$$(x+2)^2 = 5$$
$$\sqrt{(x+2)^2} = \sqrt{5}$$
$$x + 2 = \pm\sqrt{5}$$
$$x = -2 \pm \sqrt{5}$$
The solutions are $-2 + \sqrt{5}$ and $-2 - \sqrt{5}$.

7. $g(x) = x^2 + 3x - 8$
$0 = x^2 + 3x - 8$
$a = 1, b = 3, c = -8$
$x = \dfrac{-b \pm \sqrt{b^2 - 4ac}}{2a}$
$= \dfrac{-3 \pm \sqrt{3^2 - 4(1)(-8)}}{2(1)}$
$= \dfrac{-3 \pm \sqrt{9 + 32}}{2}$
$= \dfrac{-3 \pm \sqrt{41}}{2}$

The zeros of $g(x)$ are $\dfrac{-3 + \sqrt{41}}{2}$ and $\dfrac{-3 - \sqrt{41}}{2}$.

8. $3x^2 - x + 8 = 0$
$a = 3, b = -1, c = 8$
$x = \dfrac{-b \pm \sqrt{b^2 - 4ac}}{2a}$
$= \dfrac{-(-1) \pm \sqrt{(-1)^2 - 4(3)(8)}}{2(3)}$
$= \dfrac{1 \pm \sqrt{1 - 96}}{6}$
$= \dfrac{1 \pm \sqrt{-95}}{6}$
$= \dfrac{1 \pm i\sqrt{95}}{6}$

The solutions are $= \dfrac{1}{6} + \dfrac{\sqrt{95}}{6}i$ and $= \dfrac{1}{6} - \dfrac{\sqrt{95}}{6}i$.

9. $\dfrac{2x}{x-1} + \dfrac{3}{x+2} = 1$
$(x-1)(x+2)\left(\dfrac{2x}{x-1} + \dfrac{3}{x+2}\right) = (x-1)(x+2)1$
$2x(x+2) + 3(x-1) = x^2 + x - 2$
$2x^2 + 4x + 3x - 3 = x^2 + x - 2$
$2x^2 + 7x - 3 = x^2 + x - 2$
$x^2 + 6x - 1 = 0$
$a = 1, b = 6, c = -1$
$x = \dfrac{-b \pm \sqrt{b^2 - 4ac}}{2a}$
$= \dfrac{-6 \pm \sqrt{6^2 - 4(1)(-1)}}{2(1)}$
$= \dfrac{-6 \pm \sqrt{36 + 4}}{2}$
$= \dfrac{-6 \pm \sqrt{40}}{2}$
$= \dfrac{-6 \pm 2\sqrt{10}}{2}$
$= -3 \pm \sqrt{10}$

The solutions are $-3 + \sqrt{10}$ and $-3 - \sqrt{10}$.

10. $2x + 7x^{1/2} - 4 = 0$
$2\left(x^{1/2}\right)^2 + 7x^{1/2} - 4 = 0$
$2u^2 + 7u - 4 = 0$
$(2u - 1)(u + 4) = 0$
$2u - 1 = 0 \qquad u + 4 = 0$
$2u = 1 \qquad u = -4$
$u = \dfrac{1}{2}$

Replace u by $x^{1/2}$.

$x^{1/2} = \dfrac{1}{2} \qquad\qquad x^{1/2} = -4$
$\left(x^{1/2}\right)^2 = \left(\dfrac{1}{2}\right)^2 \qquad \left(x^{1/2}\right)^2 = (-4)^2$
$x = \dfrac{1}{4} \qquad\qquad x = 16$

16 does not check as a solution.
$\dfrac{1}{4}$ does check as a solution.

The solution is $\dfrac{1}{4}$.

11. $x^4 - 11x^2 + 18 = 0$
$\left(x^2\right)^2 - 11x^2 + 18 = 0$
$u^2 - 11u + 18 = 0$
$(u - 9)(u - 2) = 0$
$u - 9 = 0 \qquad u - 2 = 0$
$u = 9 \qquad u = 2$

Replace u by x^2.
$x^2 = 9 \qquad x^2 = 2$
$\sqrt{x^2} = \sqrt{9} \qquad \sqrt{x^2} = \sqrt{2}$
$x = \pm 3 \qquad x = \pm\sqrt{2}$

The solutions are 3, –3, $\sqrt{2}$, $-\sqrt{2}$.

12. $\sqrt{2x+1} + 5 = 2x$
$\sqrt{2x+1} = 2x - 5$
$\left(\sqrt{2x+1}\right)^2 = (2x-5)^2$
$2x + 1 = 4x^2 - 20x + 25$
$0 = 4x^2 - 22x + 24$
$0 = 2\left(2x^2 - 11x + 12\right)$
$0 = 2(2x - 3)(x - 4)$
$2x - 3 = 0 \qquad x - 4 = 0$
$2x = 3 \qquad x = 4$
$x = \dfrac{3}{2}$

$\dfrac{3}{2}$ does not check as a solution.
4 does check as a solution.
The solution is 4.

Chapter 8: Quadratic Equations And Inequalities

13. $f(x) = x^2 - 4x - 2$
$4 = x^2 - 4x - 2$
$0 = x^2 - 4x - 6$
$a = 1, b = -4, c = -6$
$x = \dfrac{-b \pm \sqrt{b^2 - 4ac}}{2a}$
$= \dfrac{4 \pm \sqrt{(-4)^2 - 4(1)(-6)}}{2(1)}$
$= \dfrac{4 \pm \sqrt{40}}{2}$
$= \dfrac{4 \pm 2\sqrt{10}}{2}$
$= 2 \pm \sqrt{10}$

14. $b^2 - 4ac = 2^2 - 4(3)(-4)$
$= 4 + 48$
$= 52$
The discriminant is positive.
The parabola has two x-intercepts.

15. $y = 2x^2 + 5x - 12$
$0 = 2x^2 + 5x - 12$
$0 = (2x - 3)(x + 4)$
$2x - 3 = 0 \qquad x + 4 = 0$
$2x = 3 \qquad\qquad x = -4$
$x = \dfrac{3}{2}$

The x-intercepts of the parabola are $\left(\dfrac{3}{2}, 0\right)$ and $(-4, 0)$.

16. $y = 2x^2 + 6x + 3$
The equation of the axis of symmetry is $x = -\dfrac{b}{2a}$.
$a = 2, b = 6$
$x = -\dfrac{b}{2a} = -\dfrac{6}{2(2)} = -\dfrac{6}{4} = -\dfrac{3}{2}$
The axis of symmetry is $x = -\dfrac{3}{2}$.

17. $-\dfrac{b}{2a} = -\dfrac{1}{2(\frac{1}{2})} = -1$
$y = \dfrac{1}{2}(-1)^2 - 1 - 4 = -4.5$
The vertex is $(-1, -4.5)$.
The axis of symmetry is $x = -1$.

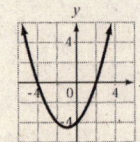

The domain is $\{x \mid x \in \text{real numbers}\}$.
The range is $\{y \mid y \geq -4.5\}$.

18. $\dfrac{2x - 3}{x + 4} \leq 0$

$\{x \mid -4 < x \leq \dfrac{3}{2}\}$

19. **Strategy**
This is a geometry problem.
The height of the triangle: x
The base of the triangle: $3x + 3$
The area of the triangle is 30 ft^2. Use the formula for the area of a triangle $\left(A = \dfrac{1}{2}bh\right)$.

Solution
$A = \dfrac{1}{2}bh$
$30 = \dfrac{1}{2}(3x + 3)x$
$60 = 3x^2 + 3x$
$0 = 3x^2 + 3x - 60$
$0 = 3(x^2 + x - 20)$
$0 = 3(x + 5)(x - 4)$
$x + 5 = 0 \qquad x - 4 = 0$
$x = -5 \qquad\qquad x = 4$
The solution -5 is not possible, since height cannot be a negative number.
$x = 4, 3x + 3 = 3(4) + 3 = 15$
The height of the triangle is 4 ft.
The base of the triangle is 15 ft.

20. **Strategy**
This is a distance-rate problem.
The rate of the canoe in calm water: x
The total traveling time is 4 h.
$\dfrac{6}{x+2} + \dfrac{6}{x-2} = 4$

Solution
$\dfrac{6}{x+2} + \dfrac{6}{x-2} = 4$
$(x+2)(x-2)\left(\dfrac{6}{x+2} + \dfrac{6}{x-2}\right) = (x+2)(x-2)4$
$6(x - 2) + 6(x + 2) = 4(x^2 - 4)$
$6x - 12 + 6x + 12 = 4x^2 - 16$
$12x = 4x^2 - 16$
$0 = 4x^2 - 12x - 16$
$0 = 4(x^2 - 3x - 4)$
$0 = 4(x - 4)(x + 1)$
$x - 4 = 0 \qquad x + 1 = 0$
$x = 4 \qquad\qquad x = -1$
The solution -1 is not possible, since the rate cannot be negative.
The rate of the canoe in calm water is 4 mph.

CUMULATIVE REVIEW EXERCISES

1. $2a^2 - b^2 \div c^2$
   $2(3)^2 - (-4)^2 \div (-2)^2$
   $= 2(9) - 16 \div 4$
   $= 18 - 16 \div 4$
   $= 18 - 4$
   $= 14$

2. $\dfrac{2x-3}{4} - \dfrac{x+4}{6} = \dfrac{3x-2}{8}$
   $24\left(\dfrac{2x-3}{4} - \dfrac{x+4}{6}\right) = 24\left(\dfrac{3x-2}{8}\right)$
   $6(2x-3) - 4(x+4) = 3(3x-2)$
   $12x - 18 - 4x - 16 = 9x - 6$
   $8x - 34 = 9x - 6$
   $-x - 34 = -6$
   $-x = 28$
   $x = -28$
   The solution is -28.

3. $P_1(3, -4), P_2(-1, 2)$
   $m = \dfrac{y_2 - y_1}{x_2 - x_1} = \dfrac{2-(-4)}{-1-3} = \dfrac{2+4}{-4} = \dfrac{6}{-4} = -\dfrac{3}{2}$

4. $x - y = 1$
   $y = x - 1$
   $m = 1 \quad (x_1, y_1) = (1, 2)$
   $y - y_1 = m(x - x_1)$
   $y - 2 = 1(x - 1)$
   $y - 2 = x - 1$
   $y = x + 1$

5. $-3x^3y + 6x^2y^2 - 9xy^3$
   $-3xy(x^2 - 2xy + 3y^2)$

6. $6x^2 - 7x - 20 = (2x - 5)(3x + 4)$

7. $a^n x + a^n y - 2x - 2y = a^n(x+y) - 2(x+y)$
   $\qquad\qquad\qquad\qquad\; = (x+y)(a^n - 2)$

8. $\begin{array}{r} x^2 - 3x - 4 \\ 3x-4\overline{)3x^3 - 13x^2 + 10} \\ \underline{3x^3 - 4x^2} \\ -9x^2 \\ \underline{-9x^2 + 12x} \\ -12x + 10 \\ \underline{-12x + 16} \\ -6 \end{array}$

   $(3x^3 - 13x^2 + 10) \div (3x-4) = x^2 - 3x - 4 - \dfrac{6}{3x-4}$

9. $\dfrac{x^2 + 2x + 1}{8x^2 + 8x} \cdot \dfrac{4x^3 - 4x^2}{x^2 - 1}$
   $= \dfrac{(x+1)(x+1)}{8x(x+1)} \cdot \dfrac{4x^2(x-1)}{(x+1)(x-1)}$
   $= \dfrac{(x+1)(x+1)4x^2(x-1)}{8x(x+1)(x+1)(x-1)}$
   $= \dfrac{x}{2}$

10. Distance between points is $\sqrt{(x_2 - x_1)^2 + (y_2 - y_1)^2}$
    Distance $= \sqrt{[2-(-2)]^2 + (5-3)^2}$
    $= \sqrt{4^2 + 2^2}$
    $= \sqrt{20}$
    $= \sqrt{4 \cdot 5}$
    $= 2\sqrt{5}$
    The distance between the points is $2\sqrt{5}$

11. $S = \dfrac{n}{2}(a+b)$
    $2S = 2\dfrac{n}{2}(a+b)$
    $2S = n(a+b)$
    $2S = an + bn$
    $2S - an = bn$
    $\dfrac{2S - an}{n} = b$

12. $-2i(7 - 4i) = -14i + 8i^2$
    $= -14i + 8(-1)$
    $= -8 - 14i$

13. $a^{-1/2}\left(a^{1/2} - a^{3/2}\right) = a^{-1/2+1/2} - a^{-1/2+3/2}$
    $= a^0 - a^1$
    $= 1 - a$

14. $\dfrac{\sqrt[3]{8x^4y^5}}{\sqrt[3]{16xy^6}} = \sqrt[3]{\dfrac{8x^4y^5}{16xy^6}}$
    $= \sqrt[3]{\dfrac{x^3}{2y}}$
    $= \sqrt[3]{\dfrac{x^3}{2y}} \sqrt[3]{\dfrac{4y^2}{4y^2}}$
    $= \sqrt[3]{\dfrac{4y^2 x^3}{8y^3}}$
    $= \sqrt[3]{\dfrac{x^3(4y^2)}{8y^3}}$
    $= \dfrac{x\sqrt[3]{4y^2}}{2y}$

15.
$$\frac{x}{x+2} - \frac{4x}{x+3} = 1$$
$$(x+2)(x+3)\left(\frac{x}{x+2} - \frac{4x}{x+3}\right) = (x+2)(x+3)1$$
$$x(x+3) - 4x(x+2) = (x+2)(x+3)$$
$$x^2 + 3x - 4x^2 - 8x = x^2 + 5x + 6$$
$$-3x^2 - 5x = x^2 + 5x + 6$$
$$0 = 4x^2 + 10x + 6$$
$$0 = 2(2x^2 + 5x + 3)$$
$$0 = 2(2x+3)(x+1)$$

$2x + 3 = 0 \quad\quad x + 1 = 0$
$2x = -3 \quad\quad x = -1$
$x = -\frac{3}{2}$

The solutions are $-\frac{3}{2}$ and -1.

16.
$$\frac{x}{2x+3} - \frac{3}{4x^2 - 9} = \frac{x}{2x-3}$$
$$\frac{x}{2x+3} - \frac{3}{(2x+3)(2x-3)} = \frac{x}{2x-3}$$
$$(2x+3)(2x-3)\left[\frac{x}{2x+3} - \frac{3}{(2x+3)(2x-3)}\right] = (2x+3)(2x-3)\frac{x}{2x-3}$$
$$(2x-3)x - 3 = (2x+3)x$$
$$2x^2 - 3x - 3 = 2x^2 + 3x$$
$$-3 = 6x$$
$$-\frac{1}{2} = x$$

Check: $\dfrac{-\frac{1}{2}}{2\left(-\frac{1}{2}\right)+3} - \dfrac{3}{4\left(-\frac{1}{2}\right)^2 - 9} = \dfrac{-\frac{1}{2}}{2\left(-\frac{1}{2}\right) - 3}$

$\dfrac{-\frac{1}{2}}{2} - \dfrac{3}{-8} = \dfrac{-\frac{1}{2}}{-4}$

$\dfrac{1}{8} = \dfrac{1}{8}$

The solution is $x = -\frac{1}{2}$.

17.
$$x^4 - 6x^2 + 8 = 0$$
$$(x^2)^2 - 6x^2 + 8 = 0$$
$$u^2 - 6u + 8 = 0$$
$$(u-4)(u-2) = 0$$

$u - 4 = 0 \quad\quad u - 2 = 0$
$u = 4 \quad\quad u = 2$

Replace u by x^2
$x^2 = 4 \quad\quad x^2 = 2$
$\sqrt{x^2} = \sqrt{4} \quad\quad \sqrt{x^2} = \sqrt{2}$
$x = \pm 2 \quad\quad x = \pm\sqrt{2}$

The solutions are $2, -2, \sqrt{2}$, and $-\sqrt{2}$.

18.
$$\sqrt{3x+1} - 1 = x$$
$$\sqrt{3x+1} = x + 1$$
$$\left(\sqrt{3x+1}\right)^2 = (x+1)^2$$
$$3x + 1 = x^2 + 2x + 1$$
$$0 = x^2 - x$$
$$0 = x(x-1)$$

$x = 0 \quad\quad x - 1 = 0$
$\quad\quad\quad\quad\quad x = 1$

0 and 1 both check as solutions. The solutions are 0 and 1.

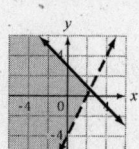

19. $|3x-2|<8$
$-8<3x-2<8$
$-8+2<3x-2+2<8+2$
$-6<3x<10$
$\frac{1}{3}\cdot(-6)<\frac{1}{3}\cdot(3x)<\frac{1}{3}\cdot 10$
$-2<x<\frac{10}{3}$
$\left\{x\mid -2<x<\frac{10}{3}\right\}$

20. $6x-5y=15$
$6x-5(0)=15$
$6x=15$
$x=\frac{15}{6}=\frac{5}{2}$

The x-intercept is $\left(\frac{5}{2},0\right)$.

$6x-5y=15$
$6(0)-5y=15$
$-5y=15$
$y=-3$

The y-intercept is $(0,-3)$.

21. Solve each inequality.

$x+y\leq 3$ $\quad\quad$ $2x-y<4$
$y\leq 3-x$ $\quad\quad$ $-y<4-2x$
$\quad\quad\quad\quad\quad\quad\quad\quad$ $y>-4+2x$

22. $x+y+z=2$
$-x+2y-3z=-9$
$x-2y-2z=-1$

$D=\begin{vmatrix}1 & 1 & 1\\ -1 & 2 & -3\\ 1 & -2 & -2\end{vmatrix}=\begin{vmatrix}2 & -3\\ -2 & -2\end{vmatrix}-\begin{vmatrix}-1 & -3\\ 1 & -2\end{vmatrix}+\begin{vmatrix}-1 & 2\\ 1 & -2\end{vmatrix}$

$=(-4-6)-[2-(-3)]+(2-2)=-10-5+0=-15$

$D_x=\begin{vmatrix}2 & 1 & 1\\ -9 & 2 & -3\\ -1 & -2 & -2\end{vmatrix}=2\begin{vmatrix}2 & -3\\ -2 & -2\end{vmatrix}-\begin{vmatrix}-9 & -3\\ -1 & -2\end{vmatrix}+\begin{vmatrix}-9 & 2\\ -1 & -2\end{vmatrix}$

$=2(-4-6)-(18-3)+[18-(-2)]=-20-15+20=-15$

$D_y=\begin{vmatrix}1 & 2 & 1\\ -1 & -9 & -3\\ 1 & -1 & -2\end{vmatrix}=\begin{vmatrix}-9 & -3\\ -1 & -2\end{vmatrix}-2\begin{vmatrix}-1 & -3\\ 1 & -2\end{vmatrix}+\begin{vmatrix}-1 & -9\\ 1 & -1\end{vmatrix}$

$=(18-3)-2[2-(-3)]+[1-(-9)]=15-10+10=15$

$D_z=\begin{vmatrix}1 & 1 & 2\\ -1 & 2 & -9\\ 1 & -2 & -1\end{vmatrix}=\begin{vmatrix}2 & -9\\ -2 & -1\end{vmatrix}-\begin{vmatrix}-1 & -9\\ 1 & -1\end{vmatrix}+2\begin{vmatrix}-1 & 2\\ 1 & -2\end{vmatrix}$

$=(-2-18)-[1-(-9)]+2(2-2)=-20-10+0=-30$

$x=\frac{D_x}{D}=\frac{-15}{-15}=1$
$y=\frac{D_y}{D}=\frac{15}{-15}=-1$
$z=\frac{D_z}{D}=\frac{-30}{-15}=2$

The solution is $(1,-1,2)$.

23. $f(x)=\frac{2x-3}{x^2-1}$

$f(-2)=\frac{2(-2)-3}{(-2)^2-1}=-\frac{7}{3}$

24. $f(x)=\frac{x-2}{x^2-2x-15}$

$f(x)=\frac{x-2}{(x-5)(x+3)}$

$0=(x-5)(x+3)$
$x-5=0\quad\quad x+3=0$
$x=5\quad\quad\quad x=-3$

The domain of $f(x)$ is $\{x\mid x\neq 5\text{ and }x\neq -3\}$

25. $x^3+x^2-6x<0$
$x(x^2+x-6)<0$
$x(x+3)(x-2)<0$

$\{x\mid x<-3\text{ or }0<x<2\}$

298 Chapter 8: Quadratic Equations And Inequalities

26. $\dfrac{(x-1)(x-5)}{x+3} \geq 0$

```
x - 1   - - - - - - - | + + + + | +
x - 5   - - - - - - - | - - - - | +
x + 3   - - | + + + + | + + + + | +
       -5 -4 -3 -2 -1 0 1 2 3 4 5
```

$\{x \mid -3 < x \leq 1 \text{ or } x \geq 5\}$

```
    -5 -4 -3 -2 -1 0 1 2 3 4 5
```

27. Between the age of 2 and 8,

average rate of change $= \dfrac{42-7}{8.1-0.8}$

$= \dfrac{35}{7.3}$

≈ 4.8

The average rate of change in the height of the trees to the diameter of the trees is 4.8 ft/in.

28. Strategy
Let p represent the length of the piston rod, T the tolerance, and m the given length.
Solve the absolute value inequality $|m - p| \leq T$ for m.

Solution

$|m - p| \leq T$

$\left| m - 9\dfrac{3}{8} \right| \leq \dfrac{1}{64}$

$-\dfrac{1}{64} \leq m - 9\dfrac{3}{8} \leq \dfrac{1}{64}$

$-\dfrac{1}{64} + 9\dfrac{3}{8} \leq m \leq \dfrac{1}{64} + 9\dfrac{3}{8}$

$9\dfrac{23}{64} \leq m \leq 9\dfrac{25}{64}$

The lower limit is $9\dfrac{23}{64}$ in.

The upper limit is $9\dfrac{25}{64}$ in.

29. $A = \dfrac{1}{2} b \cdot h$

$= \dfrac{1}{2}(x+8)(2x-4)$

$= \dfrac{1}{2}(2x^2 + 12x - 32)$

$= (x^2 + 6x - 16) \text{ ft}^2$

30. $m = \dfrac{y_2 - y_1}{x_2 - x_1}$

$= \dfrac{0 - 250{,}000}{30 - 0}$

$= \dfrac{-250{,}000}{30}$

$= -\dfrac{25{,}000}{3}$

The slope represents the amount in dollars that the building depreciates $8333.33 each year.

CHAPTER 9: FUNCTIONS AND RELATIONS

CHAPTER 9 PREP TEST

1. $|x+4| = |-6+4| = |-2| = 2$ [1.3.2]

2. $y = (-2)^2 + 2(-2) + 1$ [1.3.2]
   $= -4 - 4 + 1$
   $= -7$

3. $f(4) = (-4)^2 - 3(-4) + 2$ [3.2.1]
   $= 16 + 12 + 2$
   $= 30$

4. $p(2 + h) = (2 + h)^2 - 5$ [3.2.1]
   $= 4 + 4h + h^2 - 5$
   $= h^2 + 4h - 1$

5. $x = 2y + 4$ [3.3.2]
   $2y = x - 4$
   $y = \frac{1}{2}x - 2$

6. D: $\{-2, 3, 4, 6\}$; R: $\{4, 5, 6\}$; Yes   [3.2.1]

7. 8   [3.2.1]

8. [3.2.2]

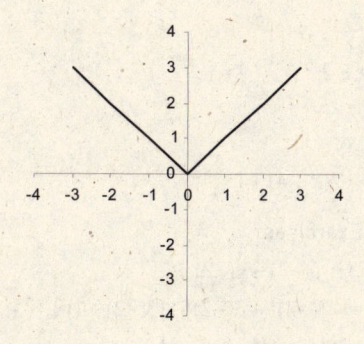

9. [3.2.2]

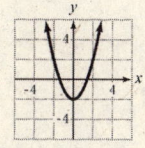

Section 9.1

Concept Review 9.1

1. Always true

3. Always true

5. Never true
   If c_1 and c_2 are positive constants, the graph of $y = f(x + c_1) + c_2$ is the graph of $y = f(x)$ translated to the left c_1 units and shifted upward c_2 units.

Objective 9.1.1 Exercises

1. down

3. left

5.

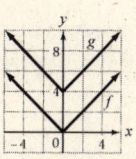

7.

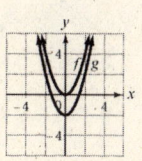

9.

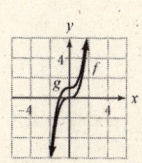

11.

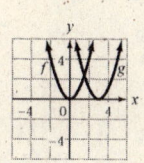

13.

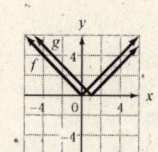

15.

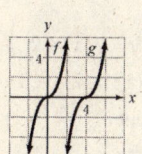

17.

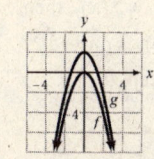

300 Chapter 9: Functions And Relations

19.

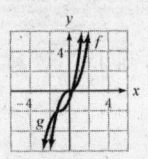

21.

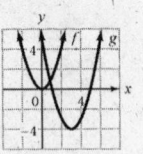

23.

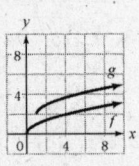

25.

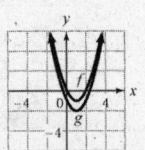

27.

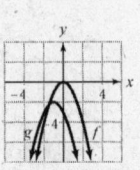

29.

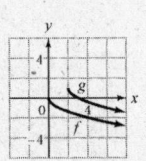

31.
  a. If (0, 7) is the y-intercept of $y = f(x) - 2$, then the y-intercept of $y = f(x)$ is moved upward 2 units. The y-intercept of $y = f(x)$ is (0, 9).

  b. If (5, 0) is the x-intercept of $y = f(x)$, then the x-intercept of $y = f(x - 3)$ is moved 2 units to the right. The x-intercept of $y = f(x - 3)$ is (8, 0).

Applying Concepts 9.1

33.

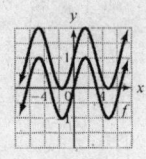

35.

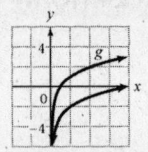

37.

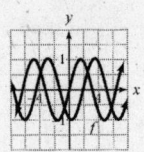

39.

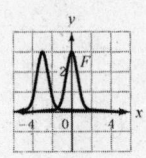

41.

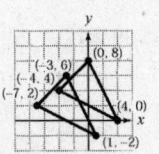

Section 9.2

Concept Review 9.2

1. Always true

3. Never true
   $(f \circ g)(x) = 2(x + 4) + 1$

Objective 9.2.1 Exercises

1. $(f - g)(-2) = f(-2) - g(-2)$
   $= [(-2)^2 + 2(-2)] - [3(-2) - 1]$
   $= [4 + -4] - [-6 - 1]$
   $= 0 - -7$
   $= 7$

3. $(f - g)(2) = f(2) - g(2)$
   $= [2(2)^2 - 3] - [-2(2) + 4]$
   $= 5 - 0$
   $= 5$
   $(f - g)(2) = 5$

5. $(f + g)(0) = f(0) + g(0)$
   $= [2(0)^2 - 3] + [-2(0) + 4]$
   $= -3 + 4$
   $= 1$
   $(f + g)(0) = 1$

7. $(f \cdot g)(2) = f(2) \cdot g(2)$
$= [2(2)^2 - 3] \cdot [-2(2) + 4]$
$= 5 \cdot 0$
$= 0$
$(f \cdot g)(2) = 0$

9. $\left(\dfrac{f}{g}\right)(4) = \dfrac{f(4)}{g(4)}$
$= \dfrac{2(4)^2 - 3}{-2(4) + 4}$
$= -\dfrac{29}{4}$
$\dfrac{f}{g}(4) = -\dfrac{29}{4}$

11. $\left(\dfrac{g}{f}\right)(-3) = \dfrac{g(-3)}{f(-3)}$
$= \dfrac{-2(-3) + 4}{2(-3)^2 - 3}$
$= \dfrac{10}{15}$
$= \dfrac{2}{3}$
$\left(\dfrac{g}{f}\right)(-3) = \dfrac{2}{3}$

13. $(f + g)(1) = f(1) + g(1)$
$= [2(1)^2 + 3(1) - 1] + [2(1) - 4]$
$= [2 + 3 - 1] + [2 - 4]$
$= 4 - 2$
$= 2$
$(f + g)(1) = 2$

15. $(f - g)(4) = f(4) - g(4)$
$= [2(4)^2 + 3(4) - 1] - [2(4) - 4]$
$= [32 + 12 - 1] - [8 - 4]$
$= 43 - 4$
$= 39$
$(f - g)(4) = 39$

17. $(f \cdot g)(1) = f(1) \cdot g(1)$
$= [2(1)^2 + 3(1) - 1] \cdot [2(1) - 4]$
$= [2 + 3 - 1] \cdot [2 - 4]$
$= 4(-2)$
$= -8$
$(f \cdot g)(1) = -8$

19. $\left(\dfrac{f}{g}\right)(-3) = \dfrac{f(-3)}{g(-3)}$
$= \dfrac{2(-3)^2 + 3(-3) - 1}{2(-3) - 4}$
$= \dfrac{18 - 9 - 1}{-6 - 4}$
$= \dfrac{8}{-10}$
$= -\dfrac{4}{5}$
$\left(\dfrac{f}{g}\right)(-3) = -\dfrac{4}{5}$

21. $(f - g)(2) = f(2) - g(2)$
$= [2^2 + 3(2) - 5] - [2^3 - 2(2) + 3]$
$= [4 + 6 - 5] - [8 - 4 + 3]$
$= 5 - 7$
$= -2$
$(f - g)(-2) = -2$

23. $\left(\dfrac{f}{g}\right)(-2) = \dfrac{f(-2)}{g(-2)}$
$= \dfrac{(-2)^2 + 3(-2) - 5}{(-2)^3 - 2(-2) + 3}$
$= \dfrac{4 - 6 - 5}{-8 + 4 + 3}$
$= \dfrac{-7}{-1} = 7$
$\left(\dfrac{f}{g}\right)(-2) = 7$

25. Undefined because $f(-a) = \sqrt{-a}$ is not defined.

27. Defined

Objective 9.2.2 Exercises

29. $f(g(x)) = f(4x)$
$= (4x)^2 - 2(4x)$
$= 16x^2 - 8x$

31. The expression $(f \circ g)(x)$ means to evaluate the function f at $g(x)$.

33. $f(x) = 2x - 3$
$f(0) = 2(0) - 3 = 0 - 3 = -3$
$g(x) = 4x - 1$
$g[f(0)] = g(-3) = 4(-3) - 1 = -12 - 1 = -13$
$g[f(0)] = -13$

35. $f(x) = 2x - 3$
$f(-2) = 2(-2) - 3 = -4 - 3 = -7$
$g(x) = 4x - 1$
$g[f(-2)] = g(-7) = 4(-7) - 1 = -28 - 1 = -29$
$g[f(-2)] = -29$

37. $f(x) = 2x - 3$
$g(x) = 4x - 1$
$g[f(x)] = g(2x - 3)$
$= 4(2x - 3) - 1$
$= 8x - 12 - 1$
$= 8x - 13$
$g[f(x)] = 8x - 13$

39. $g(x) = x^2 + 3$
$g(0) = 0^2 + 3 = 3$
$h(x) = x - 2$
$h[g(0)] = h(3) = 3 - 2 = 1$
$h[g(0)] = 1$

41. $g(x) = x^2 + 3$
$g(-2) = (-2)^2 + 3 = 4 + 3 = 7$
$h(x) = x - 2$
$h[g(-2)] = h(7) = 7 - 2 = 5$
$h[g(-2)] = 5$

43. $g(x) = x^2 + 3$
$h(x) = x - 2$
$h[g(x)] = h(x^2 + 3) = x^2 + 3 - 2 = x^2 + 1$
$h[g(x)] = x^2 + 1$

45. $f(x) = x^2 + x + 1$
$f(0) = 0^2 + 0 + 1 = 0 + 0 + 1 = 1$
$h(x) = 3x + 2$
$h[f(0)] = h(1) = 3(1) + 2 = 3 + 2 = 5$
$h[f(0)] = 5$

47. $f(x) = x^2 + x + 1$
$f(-2) = (-2)^2 - 2 + 1 = 4 - 2 + 1 = 3$
$h(x) = 3x + 2$
$h[f(-2)] = h(3) = 3(3) + 2 = 9 + 2 = 11$
$h[f(-2)] = 11$

49. $f(x) = x^2 + x + 1$
$h(x) = 3x + 2$
$h[f(x)] = h(x^2 + x + 1) = 3(x^2 + x + 1) + 2$
$= 3x^2 + 3x + 3 + 2$
$= 3x^2 + 3x + 5$
$h[f(x)] = 3x^2 + 3x + 5$

51. $g(x) = x^3$
$g(-1) = (-1)^3 = -1$
$f(x) = x - 2$
$f[g(-1)] = f(-1) = -1 - 2 = -3$
$f[g(-1)] = -3$

53. $f(x) = x - 2$
$f(-1) = -1 - 2 = -3$
$g(x) = x^3$
$g[f(-1)] = g(-3) = (-3)^3 = -27$
$g[f(-1)] = -27$

55. $f(x) = x - 2,\ g(x) = x^3$
$g[f(x)] = g(x - 2) = (x - 2)^3$
$g[f(x)] = x^3 - 6x^2 - 12x - 8$

57. $f(x) = x^2 \quad g(x) = 3x - 3$
$(f \circ g)(x) = f(g(x))$
$= f(3x - 3)$
$= (3x - 3)^2$
$= 9x^2 - 18x + 9$
No

59. $f(x) = 3x \quad g(x) = x^2 - 1$
$(f \circ g)(x) = f(g(x))$
$= f(x^2 - 1)$
$= 3(x^2 - 1)$
$= 3x^2 - 3$
Yes

Applying Concepts 9.2

61. $g(3 + h) - g(3) = (3 + h)^2 - 1 - [(3)^2 - 1]$
$= 9 + 6h + h^2 - 1 - 8$
$g(3 + h) - g(3) = h^2 + 6h$

63. $\dfrac{g(1 + h) - g(1)}{h} = \dfrac{[(1 + h)^2 - 1] - [(1)^2 - 1]}{h}$
$= \dfrac{1 + 2h + h^2 - 1 - 0}{h}$
$= \dfrac{2h + h^2}{h}$
$\dfrac{g(1 - h) - g(1)}{h} = 2 + h$

65.
$$\frac{g(a+h)-g(a)}{h} = \frac{[(a+h)^2-1]-(a^2-1)}{h}$$
$$= \frac{a^2+2ah+h^2-1-a^2+1}{h}$$
$$= \frac{2ah+h^2}{h}$$
$$\frac{g(a+h)-g(a)}{h} = 2a+h$$

67.
$f(x) = 2x$
$f(1) = 2 \cdot 1 = 2$
$h(x) = x - 2$
$h(2) = 2 - 2 = 0$
$g(x) = 3x - 1$
$g(0) = 3 \cdot 0 - 1 = -1$
$g(h[f(1)]) = -1$

69.
$g(x) = 3x - 1$
$g(0) = 3 \cdot 0 - 1 = -1$
$h(x) = x - 2$
$h(-1) = -1 - 2 = -3$
$f(x) = 2x$
$f(-3) = 2(-3) = -6$
$f(h[g(0)]) = -6$

71.
$h(x) = x - 2$
$f(x-2) = 2(x-2) = 2x - 4$
$g(2x-4) = 3(2x-4) - 1 = 6x - 12 - 1 = 6x - 13$

73. $f(g(2)) = f(0) = -2$

75. $f(g(-2)) = f(0) = -2$

77. $g(f(0)) = g(-2) = 0$

79. $g(f(-4)) = g(-4) = 6$

Section 9.3

Concept Review 9.3

1. Always true

3. Never true
The function $\{(2, 3), (4, 5), (6, 3)\}$ is not a 1-1 function. The function does not have an inverse.

5. Sometimes true
If a function is a 1-1 function, then the inverse is a function.

Objective 9.3.1 Exercises

1. A function is a set of ordered pairs in which no two ordered pairs that have the same first coordinate have different second coordinates. This means that given any x, there is only one y that can be paired with that x. A 1-1 function satisfies the additional condition that given any y, there is only one x that can be paired with the given y.

3. −2; 4

5. The graph represents a 1-1 function.

7. The graph is not a 1-1 function. It fails the horizontal-line test.

9. The graph is a 1-1 function.

11. The graph is not a 1-1 function. It fails the horizontal- and vertical-line tests.

13. The graph is not a 1-1 function. It fails the horizontal-line test.

15. The graph is not a 1-1 function. It fails the horizontal-line test.

17. Yes

19. No, its graph fails the horizontal line test.

Objective 9.3.2 Exercises

21. −3; 2

23. The inverse of $\{(1, 0), (2, 3), (3, 8), (4, 15)\}$ is $\{(0, 1), (3, 2), (8, 3), (15, 4)\}$.

25. $\{(3, 5), (-3, -5), (2, 5), (-2, -5)\}$ has no inverse because the numbers 5 and −5 would be paired with more than one member of the range.

27.
$f(x) = 4x - 8$
$y = 4x - 8$
$x = 4y - 8$
$x + 8 = 4y$
$\frac{1}{4}x + 2 = y$

The inverse function is $f^{-1}(x) = \frac{1}{4}x + 2$.

29. $f(x) = x^2 - 1$ is not a 1-1 function. Therefore, it has no inverse.

31.
$f(x) = x - 5$
$y = x - 5$
$x = y - 5$
$x + 5 = y$

The inverse function is $f^{-1}(x) = x + 5$.

33. $f(x) = \dfrac{1}{3}x + 2$

$y = \dfrac{1}{3}x + 2$

$x = \dfrac{1}{3}y + 2$

$x - 2 = \dfrac{1}{3}y$

$3x - 6 = y$

The inverse function is $f^{-1}(x) = 3x - 6$.

35. $f(x) = -3x - 9$

$y = -3x - 9$

$x = -3y - 9$

$3y = -x - 9$

$y = -\dfrac{1}{3}x - 3$

The inverse function is $f^{-1}(x) = -\dfrac{1}{3}x - 3$.

37. $f(x) = \dfrac{2}{3}x + 4$

$y = \dfrac{2}{3}x + 4$

$x = \dfrac{2}{3}y + 4$

$x - 4 = \dfrac{2}{3}y$

$\dfrac{3}{2}(x - 4) = y$

$\dfrac{3}{2}x - 6 = y$

The inverse function is $f^{-1}(x) = \dfrac{3}{2}x - 6$.

39. $f(x) = -\dfrac{1}{3}x + 1$

$y = -\dfrac{1}{3}x + 1$

$x = -\dfrac{1}{3}y + 1$

$x - 1 = -\dfrac{1}{3}y$

$-3(x - 1) = y$

$-3x + 3 = y$

The inverse function is $f^{-1}(x) = -3x + 3$.

41. $f(x) = 2x - 5$

$y = 2x - 5$

$x = 2y - 5$

$x + 5 = 2y$

$\dfrac{1}{2}x + \dfrac{5}{2} = y$

The inverse function is $f^{-1}(x) = \dfrac{1}{2}x + \dfrac{5}{2}$.

43. $f(x) = x^2 + 3$ is not a 1-1 function. Therefore, it has no inverse.

45. $f(x) = 3x - 5$

$y = 3x - 5$

$x = 3y - 5$

$x + 5 = 3y$

$\dfrac{1}{3}x + \dfrac{5}{3} = y$

$f^{-1}(x) = \dfrac{1}{3}x + \dfrac{5}{3}$

$f^{-1}(0) = \dfrac{1}{3}(0) + \dfrac{5}{3}$

$f^{-1}(0) = 0 + \dfrac{5}{3}$

$f^{-1}(0) = \dfrac{5}{3}$

47. $f(x) = 3x - 5$

From exercise 54,

$f^{-1}(x) = \dfrac{1}{3}x + \dfrac{5}{3}$

$f^{-1}(4) = \dfrac{1}{3}(4) + \dfrac{5}{3}$

$f^{-1}(4) = \dfrac{4}{3} + \dfrac{5}{3}$

$f^{-1}(4) = 3$

49. The graph passes the vertical line test. So the graph is a function. The graph passes the horizontal line test. The graph has an inverse.

51. $f(g(x)) = f\left(\dfrac{x}{4}\right)$

$= 4\left(\dfrac{x}{4}\right)$

$= x$

$g(f(x)) = g(4x)$

$= \dfrac{4x}{4}$

$= x$

The functions are inverses of each other.

53. $f(h(x)) = f\left(\dfrac{1}{3x}\right)$
$= 3\left(\dfrac{1}{3x}\right)$
$= \dfrac{1}{x}$
$h(f(x)) = h(3x)$
$= \dfrac{1}{3 \cdot 3x}$
$= \dfrac{1}{9x}$

The functions are not inverses of each other.

55. $g(f(x)) = g\left(\dfrac{1}{3}x - \dfrac{2}{3}\right)$
$= 3\left(\dfrac{1}{3}x - \dfrac{2}{3}\right) + 2$
$= x - 2 + 2$
$= x$
$f(g(x)) = f(3x + 2)$
$= \dfrac{1}{3}(3x + 2) - \dfrac{2}{3}$
$= x + \dfrac{2}{3} - \dfrac{2}{3}$
$= x$

The functions are inverses of each other.

57. $f(g(x)) = f(2x + 3)$
$= \dfrac{1}{2}(2x + 3) - \dfrac{3}{2}$
$= x + \dfrac{3}{2} - \dfrac{3}{2}$
$= x$
$g(f(x)) = g\left(\dfrac{1}{2}x - \dfrac{3}{2}\right)$
$= 2\left(\dfrac{1}{2}x - \dfrac{3}{2}\right) + 3$
$= x - 3 + 3$
$= x$

The functions are inverses of each other.

59. The domain of the inverse of f^{-1} is the range of f.

61. For any linear function f and its inverse f^{-1}, $f[f^{-1}(3)] = 3$.

63. If f is a 1-1 function and $f(0) = 5$, then $f^{-1}(5) = 0$.

65. If f is a 1-1 function and $f(2) = 9$ then $f^{-1}(9) = 2$.

67. If $f(c) = a$, then $f^{-1}(a) = c$.

69. If $f(c) = a$, then $f^{-1}(a) = c$. Therefore, $f(f^{-1}(a)) = f(c) = a$.

71. $f[g(x)] = f(5x + 4)$
$= 3(5x + 4) + 2$
$= 15x + 14$
$g[f(x)] = g(3x + 2)$
$= 5(3x + 2) + 4$
$= 15x + 14$

No, these functions are not inverses because to be inverses we must have $f(g(x)) = g(f(x)) = x$.

Applying Concepts 9.3

73.

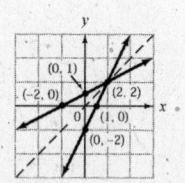

75.

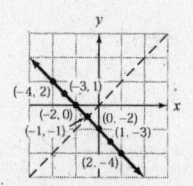

77.

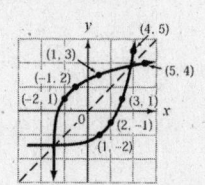

79. No, the inverse of the grading scale is not a function because each grade is paired with more than one score.

81.

a. $f(x) = 2x - 4$
$y = 2x - 4$
$x = 2y - 4$
$x + 4 = 2y$
$\dfrac{1}{2}x + 2 = y$
$f^{-1}(x) = \dfrac{1}{2}x + 2$

$f^{-1}(46204424) = \dfrac{1}{2}(46204424) + 2 = 23102214$

b. $f^{-1}(44205830) = \dfrac{1}{2}(44205830) + 2 = 22102917$
22 10 29 17 = MATH

83.
$f(x) = 3x$
$y = 3x$
$x = 3y$
$\frac{1}{3}x = y$
$f^{-1}(x) = \frac{1}{3}x$

The inverse function converts feet into yards.

85.
$f(x) = 4x$
$y = 4x$
$x = 4y$
$\frac{1}{4}x = y$
$f^{-1}(x) = \frac{1}{4}x$

The inverse function gives the length of a side of a square given its perimeter.

CHAPTER 9 REVIEW EXERCISES

1.

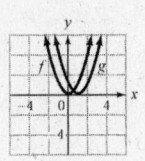

2.

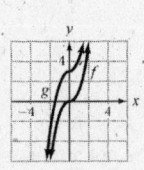

3.

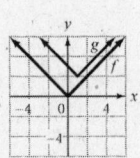

4.

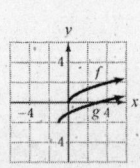

5. $(f+g)(2) = f(2) + g(2)$
$= [2^2 + 2(2) - 3] + [2^2 - 2]$
$= [4 + 4 - 3] + [4 - 2]$
$= 5 + 2$
$= 7$
$(f+g)(2) = 7$

6. $(f-g)(-4) = f(-4) - g(-4)$
$= [(-4)^2 + 2(-4) - 3] - [(-4)^2 - 2]$
$= [16 - 8 - 3] - [16 - 2]$
$= 5 - 14$
$= -9$
$(f-g)(-4) = -9$

7. $(f \cdot g)(-4) = f(-4) \cdot g(-4)$
$= [(-4)^2 + 2(-4) - 3] \cdot [(-4)^2 - 2]$
$= [16 - 8 - 3] \cdot [16 - 2]$
$= 5 \cdot 14$
$= 70$
$(f \cdot g)(-4) = 70$

8. $\left(\frac{f}{g}\right)(3) = \frac{f(3)}{g(3)}$
$= \frac{3^2 + 2(3) - 3}{3^2 - 2}$
$= \frac{9 + 6 - 3}{9 - 2}$
$= \frac{12}{7}$
$\left(\frac{f}{g}\right)(3) = \frac{12}{7}$

9. $f(x) = 3x^2 - 4, g(x) = 2x + 1$
$f[g(x)] = f(2x+1)$
$= 3(2x+1)^2 - 4$
$= 3(4x^2 + 4x + 1) - 4$
$= 12x^2 + 12x + 3 - 4$
$f[g(x)] = 12x^2 + 12x - 1$

10. $f(x) = x^2 + 4, g(x) = 4x - 1$
$g(0) = 4(0) - 1 = 0 - 1 = -1$
$f(-1) = (-1)^2 + 4 = 1 + 4 = 5$
$f[g(0)] = 5$

11. $f(x) = 6x + 8, g(x) = 4x + 2$
$f(-1) = 6(-1) + 8 = -6 + 8 = 2$
$g(2) = 4(2) + 2 = 8 + 2 = 10$
$g[f(-1)] = 10$

12. $f(x) = 2x^2 + x - 5, g(x) = 3x - 1$
$g[f(x)] = g(2x^2 + x - 5)$
$= 3(2x^2 + x - 5) - 1$
$= 6x^2 + 3x - 15 - 1$
$g[f(x)] = 6x^2 + 3x - 16$

13. No

14. The inverse of {(−2, 1), (2, 3), (5, −4), (7, 9)} is
{(1, −2), (3, 2), (−4, 5), (9, 7)}

15. A vertical line will intersect the graph at no more than one point. The graph is the graph of a function. A horizontal line will intersect the graph at more than one point. The graph is not the graph of a 1-1 function.

16. A vertical line will intersect the graph at no more than one point. The graph is the graph of a function. A horizontal line will intersect the graph at no more than one point. The graph is the graph of a 1-1 function.

17.
$$f(x) = \frac{1}{2}x + 8$$
$$y = \frac{1}{2}x + 8$$
$$x = \frac{1}{2}y + 8$$
$$x - 8 = \frac{1}{2}y$$
$$2(x - 8) = 2 \cdot \frac{1}{2}y$$
$$2x - 16 = y$$
The inverse function is $f^{-1}(x) = 2x - 16$.

18.
$$f(x) = -6x + 4$$
$$y = -6x + 4$$
$$x = -6y + 4$$
$$x - 4 = -6y$$
$$-\frac{1}{6}x + \frac{2}{3} = y$$
The inverse function is $f^{-1}(x) = -\frac{1}{6}x + \frac{2}{3}$.

19.
$$f(x) = \frac{2}{3}x - 12$$
$$y = \frac{2}{3}x - 12$$
$$x = \frac{2}{3}y - 12$$
$$x + 12 = \frac{2}{3}y$$
$$\frac{3}{2}(x + 12) = y$$
$$\frac{3}{2}x + 18 = y$$
The inverse function is $f^{-1}(x) = \frac{3}{2}x + 18$.

20.
$$f(g(x)) = f(-4x + 5) = -\frac{1}{4}(-4x + 5) + \frac{5}{4}$$
$$= x - \frac{5}{4} + \frac{5}{4} = x$$
$$g(f(x)) = g\left(-\frac{1}{4}x + \frac{5}{4}\right) = -4\left(-\frac{1}{4}x + \frac{5}{4}\right) + 5$$
$$= x - 5 + 5 = x$$
The functions are inverses of each other.

CHAPTER 9 TEST

1. A vertical line intersects the graph at more than one point. The graph is not a function.

2. $(f - g)(2) = f(2) - g(2)$
$= [2^2 + 2(2) - 3] - [2^3 - 1]$
$= [4 + 4 - 3] - [8 - 1]$
$= 5 - 7$
$= -2$
$(f - g)(2) = -2$

3. $(f \cdot g)(-3) = f(-3) \cdot g(-3)$
$= [(-3)^3 + 1][2(-3) - 3]$
$= [-27 + 1][-9]$
$= 234$
$(f \cdot g)(-3) = 234$

4.
$$\left(\frac{f}{g}\right)(-2) = \frac{f(-2)}{g(-2)}$$
$$= \frac{4(-2) - 5}{(-2)^2 + 3(-2) + 4}$$
$$= \frac{-8 - 5}{4 - 6 + 4}$$
$$= -\frac{13}{2}$$
$$\frac{f}{g}(-2) = \frac{-13}{2} = -\frac{13}{2}$$

5. $(f - g)(-4) = f(-4) - g(-4)$
$= [(-4)^2 + 4] - [2(-4)^2 + 2(-4) + 1]$
$= [16 + 4] - [32 - 8 + 1]$
$= 20 - 25$
$= -5$
$(f - g)(-4) = -5$

6. $g(x) = \dfrac{x}{x+1}$

    $g(3) = \dfrac{3}{3+1} = \dfrac{3}{4}$

    $f(x) = 4x + 2$

    $f\left(\dfrac{3}{4}\right) = 4\left(\dfrac{3}{4}\right) + 2$

    $f[g(3)] = 3 + 2 = 5$

7.

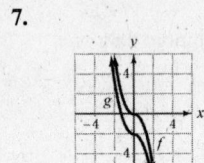

8.

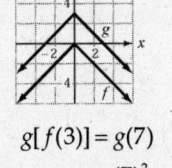

9. $g[f(3)] = g(7)$

    $= (7)^2 - 2$

    $= 47$

10.

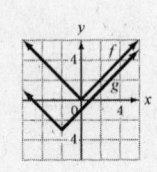

11.

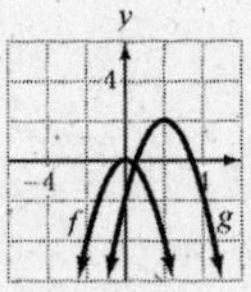

12. $f(x) = \dfrac{1}{4}x - 4$

    $y = \dfrac{1}{4}x - 4$

    $x = \dfrac{1}{4}y - 4$

    $x + 4 = \dfrac{1}{4}y$

    $4x + 16 = y$

    The inverse of the function is $f^{-1}(x) = 4x + 16$.

13. The inverse of the function of {(2, 6), (3, 5), (4, 4), (5, 3)} is {(6, 2), (5, 3), (4, 4), (3, 5)}.

14. $f[g(x)] = \dfrac{1}{2}(2x - 4) + 2$

    $= x - 2 + 2$

    $= x$

    $g[f(x)] = 2\left(\dfrac{1}{2}x + 2\right) - 4$

    $= x + 4 - 4$

    $= x$

    The functions are inverses of each other.

15. $f[g(x)] = f(x - 1)$

    $f[g(x)] = 2(x - 1)^2 - 7$

    $= 2(x^2 - 2x + 1) - 7$

    $= 2x^2 - 4x + 2 - 7$

    $= 2x^2 - 4x - 5$

16. $f(x) = \dfrac{1}{2}x - 3$

    $y = \dfrac{1}{2}x - 3$

    $x = \dfrac{1}{2}y - 3$

    $2x = y - 6$

    $y = 2x + 6$

    $f^{-1}(x) = 2x + 6$

17. $f[g(x)] = f\left(\dfrac{3}{2}x - 3\right)$

    $= \dfrac{2}{3}\left(\dfrac{3}{2}x - 3\right) + 3$

    $= x - 2 + 3$

    $= x + 1$

    $g[f(x)] = g\left(\dfrac{2}{3}x + 3\right)$

    $= \dfrac{3}{2}\left(\dfrac{2}{3}x + 3\right) - 3$

    $= x + \dfrac{9}{2} - 3$

    $= x + \dfrac{3}{2}$

    The functions are not inverses of each other.

18. The graph does not represent a 1-1 function. It fails the horizontal- and vertical-line tests.

CUMULATIVE REVIEW EXERCISES

1.
$$-3a + \left|\frac{3b-ab}{3b-c}\right| = -3(2) + \left|\frac{3(2)-2(2)}{3(2)-(-2)}\right|$$
$$= -6 + \left|\frac{6-4}{6+2}\right|$$
$$= -6 + \left|\frac{2}{8}\right|$$
$$= -6 + \left|\frac{1}{4}\right|$$
$$= -6 + \frac{1}{4} = -\frac{23}{4}$$

2. [number line with open parenthesis at -3, shaded right]

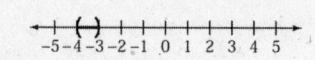

3.
$$\frac{3x-1}{6} - \frac{5-x}{4} = \frac{5}{6}$$
$$12\left(\frac{3x-1}{6} - \frac{5-x}{4}\right) = 12\left(\frac{5}{6}\right)$$
$$2(3x-1) - 3(5-x) = 2(5)$$
$$6x - 2 - 15 + 3x = 10$$
$$9x - 17 = 10$$
$$9x = 27$$
$$x = 3$$

4. $4x - 2 < -10$ or $3x - 1 > 8$
$4x < -8$ $3x > 9$
$x < -2$ $x > 3$
$\{x|x < -2\}$ $\{x|x > 3\}$
$\{x|x < -2\} \cup \{x|x > 3\} = \{x|x < -2 \text{ or } x > 3\}$

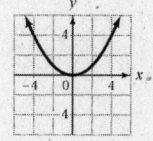

5. Vertex: $(0, 0)$
Axis of symmetry: $x = 0$
[graph of parabola]

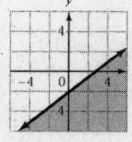

6. $3x - 4y \geq 8$
$-4y \geq -3x + 8$
$y \leq \frac{3}{4}x - 2$
[graph of inequality]

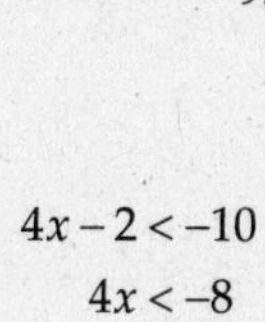

7. $|8 - 2x| \geq 0$
$8 - 2x \leq 0$ or $8 - 2x \geq 0$
$8 \leq 2x$ $8 \geq 2x$
$4 \leq x$ $4 \geq x$
$\{x|x \geq 4\}$ $\{x|x \leq 4\}$
$\{x|x \geq 4\} \cup \{x|x \leq 4\} = \{x|x \in \text{real numbers}\}$

8.
$$\left(\frac{3a^3b}{2a}\right)^2 \left(\frac{a^2}{-3b^2}\right)^3 = \left(\frac{3a^2b}{2}\right)^2 \left(\frac{a^2}{-3b^2}\right)^3$$
$$= \left(\frac{3^2 a^4 b^2}{2^2}\right)\left(\frac{a^6}{(-3)^3 b^6}\right)$$
$$= \left(\frac{9a^4b^2}{4}\right)\left(\frac{a^6}{-27b^6}\right)$$
$$= \frac{9a^4b^2 a^6}{4(-27)b^6}$$
$$= \frac{9a^{10}b^2}{-108b^6}$$
$$= -\frac{a^{10}}{12b^4}$$

9.
$\quad\quad 2x^2 + 4x - 1$
$\quad\quad\quad\quad x - 4$
$\overline{\quad -8x^2 - 16x + 4\quad}$
$\quad 2x^3 + 4x^2 - x$
$\overline{\quad 2x^3 - 4x^2 - 17x + 4\quad}$

10. $a^4 - 2a^2 - 8 = (a^2)^2 - 2(a^2) - 8$
$= (a^2 - 4)(a^2 + 2)$
$= (a+2)(a-2)(a^2+2)$

11. $x^3y + x^2y^2 - 6xy^3 = xy(x^2 + xy - 6y^2)$
$= xy(x+3y)(x-2y)$

12. $(b+2)(b-5) = 2b + 14$
$b^2 - 3b - 10 = 2b + 14$
$b^2 - 5b - 24 = 0$
$(b-8)(b+3) = 0$
$b - 8 = 0 \quad b + 3 = 0$
$b = 8 \quad\quad b = -3$
The solutions are 8 and -3.

13. $\quad x^2 - 2x > 15$
$x^2 - 2x - 15 > 0$
$(x-5)(x+3) > 0$ 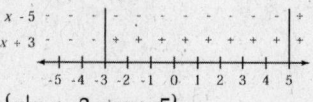
[sign chart]
$\{x|x < -3 \text{ or } x > 5\}$

Chapter 9: Functions And Relations

14. $\dfrac{x^2 + 4x - 5}{2x^2 - 3x + 1} - \dfrac{x}{2x - 1} = \dfrac{(x+5)(x-1)}{(2x-1)(x-1)} - \dfrac{x}{2x-1}$

$= \dfrac{x+5}{2x-1} - \dfrac{x}{2x-1}$

$= \dfrac{x+5-x}{2x-1}$

$= \dfrac{5}{2x-1}$

15.

$\dfrac{5}{x^2 + 7x + 12} = \dfrac{9}{x+4} - \dfrac{2}{x+3}$

$(x+4)(x+3) \dfrac{5}{(x+4)(x+3)} = (x+4)(x+3)\left[\dfrac{9}{x+4} - \dfrac{2}{x+3}\right]$

$5 = (x+3)9 - (x+4)2$

$5 = 9x + 27 - 2x - 8$

$5 = 7x + 19$

$-14 = 7x$

$-2 = x$

The solution is -2.

16. $\dfrac{4 - 6i}{2i} = \dfrac{4 - 6i}{2i} \cdot \dfrac{i}{i}$

$= \dfrac{4i - 6i^2}{2i^2}$

$= \dfrac{4i + 6}{-2}$

$= -3 - 2i$

17. $m = \dfrac{y_2 - y_1}{x_2 - x_1} = \dfrac{-6 - 4}{2 - (-3)} = \dfrac{-10}{5} = -2$

$y - y_1 = m(x - x_1)$

$y - 4 = -2[x - (-3)]$

$y - 4 = -2(x + 3)$

$y - 4 = -2x - 6$

$y = -2x - 2$

18. The product of the slopes of perpendicular lines is -1.

$2x - 3y = 6$ $\qquad m_1 \cdot m_2 = -1$

$-3y = -2x + 6$ $\qquad \dfrac{2}{3} \cdot m_2 = -1$

$y = \dfrac{2}{3}x - 2$ $\qquad m_2 = -\dfrac{3}{2}$

$y - y_1 = m(x - x_1)$

$y - 1 = -\dfrac{3}{2}[x - (-3)]$

$y - 1 = -\dfrac{3}{2}(x + 3)$

$y - 1 = -\dfrac{3}{2}x - \dfrac{9}{2}$

$y = -\dfrac{3}{2}x - \dfrac{7}{2}$

19.
$$3x^2 = 3x - 1$$
$$3x^2 - 3x + 1 = 0$$
$$a = 3, b = -3, c = 1$$
$$x = \frac{-b \pm \sqrt{b^2 - 4ac}}{2a}$$
$$= \frac{-(-3) \pm \sqrt{(-3)^2 - 4(3)(1)}}{2(3)}$$
$$= \frac{3 \pm \sqrt{9 - 12}}{6}$$
$$= \frac{3 \pm \sqrt{-3}}{6}$$
$$= \frac{3 \pm i\sqrt{3}}{6}$$
$$= \frac{1}{2} \pm \frac{\sqrt{3}}{6}i$$

The solutions are $\frac{1}{2} + \frac{\sqrt{3}}{6}i$ and $\frac{1}{2} - \frac{\sqrt{3}}{6}i$.

20.
$$\sqrt{8x+1} = 2x - 1$$
$$\left(\sqrt{8x+1}\right)^2 = (2x-1)^2$$
$$8x + 1 = 4x^2 - 4x + 1$$
$$0 = 4x^2 - 12x$$
$$0 = 4x(x - 3)$$
$$4x = 0 \quad x - 3 = 0$$
$$x = 0 \quad x = 3$$

Check:
$$\sqrt{8x+1} = 2x - 1$$

$\sqrt{8(0)+1}$	$2(0)-1$
$\sqrt{1}$	-1
1	$\neq -1$

$$\sqrt{8x+1} = 2x - 1$$

$\sqrt{8(3)+1}$	$2(3)-1$
$\sqrt{24+1}$	$6-1$
$\sqrt{25}$	5
5	$= 5$

The solution is 3.

21. $f(x) = 2x^2 - 3$
$$a = 2, b = 0, c = -3$$
$$x = -\frac{b}{2a} = \frac{-0}{2 \cdot 2} = 0$$
$$f(0) = 2(0)^2 - 3 = -3$$
The minimum value of the function is -3.

22. $f(x) = |3x - 4|$;
domain = {0, 1, 2, 3}
$f(x) = |3x - 4|$
$f(0) = |3(0) - 4| = |0 - 4| = |-4| = 4$
$f(1) = |3(1) - 4| = |3 - 4| = |-1| = 1$
$f(2) = |3(2) - 4| = |6 - 4| = |2| = 2$
$f(3) = |3(3) - 4| = |9 - 4| = |5| = 5$
The range is {1, 2, 4, 5}.

23. {(−3, 0), (−2, 0), (−1, 1), (0, 1)}
Each member of the domain is paired with only one member of the range. The set of ordered pairs is a function.

24.
$$\sqrt[3]{5x-2} = 2$$
$$\left(\sqrt[3]{5x-2}\right)^3 = 2^3$$
$$5x - 2 = 8$$
$$5x = 10$$
$$x = 2$$
The solution is 2.

25. $h(x) = \frac{1}{2}x + 4$
$$h(2) = \frac{1}{2}(2) + 4 = 1 + 4 = 5$$
$$g(x) = 3x - 5$$
$$g(5) = 3(5) - 5 = 15 - 5 = 10$$
$$g(h(2)) = 10$$

26. $f(x) = -3x + 9$
$$y = -3x + 9$$
$$x = -3y + 9$$
$$3y = -x + 9$$
$$y = -\frac{1}{3}x + 3$$
The inverse function is $f^{-1}(x) = -\frac{1}{3}x + 3$.

27. Strategy
Cost per pounds of the mixture: x

	Amount	Cost	Value
$4.50 tea	30	4.50	4.50(30)
$3.60 tea	45	3.60	3.60(45)
Mixture	75	x	$75x$

The sum of the values before mixing equals the value after mixing.

Solution
$$4.50(30) + 3.60(45) = 75x$$
$$135 + 162 = 75x$$
$$297 = 75x$$
$$3.96 = x$$
The cost per pound of the mixture is $3.96.

312 Chapter 9: Functions And Relations

28. Strategy
Pounds of 80% copper alloy: x

	Amount	Percent	Quantity
80%	x	0.80	$0.80x$
20%	50	0.20	$0.20(50)$
40%	$50 + x$	0.40	$0.40(50 + x)$

The sum of the quantities before mixing is equal to the quantity after mixing.

Solution
$$0.80x + 0.20(50) = 0.40(50 + x)$$
$$0.80x + 10 = 20 + 0.40x$$
$$0.40x + 10 = 20$$
$$0.40x = 10$$
$$x = 25$$
25 lb of the 80% copper alloy must be used.

29. Strategy
To find the additional amount of insecticide, write and solve a proportion, using x to represent the additional amount of insecticide. Then, $x + 6$ is the total amount of insecticide.

Solution
$$\frac{6}{16} = \frac{x+6}{28}$$
$$\frac{3}{8} = \frac{x+6}{28}$$
$$\frac{3}{8} \cdot 56 = \frac{x+6}{28} \cdot 56$$
$$21 = (x+6)2$$
$$21 = 2x + 12$$
$$9 = 2x$$
$$4.5 = x$$
An additional 4.5 oz of insecticide are required.

30. Strategy
This is a work problem.
Time for the smaller pipe to fill the tank: t
Time for the larger pipe to fill the tank: $t - 8$

	Rate	Time	Part
Smaller pipe	$\frac{1}{t}$	3	$\frac{3}{t}$
Larger pipe	$\frac{1}{t-8}$	3	$\frac{3}{t-8}$

The sum of the parts of the task completed must equal 1.

Solution
$$\frac{3}{t} + \frac{3}{t-8} = 1$$
$$t(t-8)\left(\frac{3}{t} + \frac{3}{t-8}\right) = t(t-8)$$
$$(t-8)3 + 3t = t^2 - 8t$$
$$3t - 24 + 3t = t^2 - 8t$$
$$6t - 24 = t^2 - 8t$$
$$0 = t^2 - 14t + 24$$
$$= (t-2)(t-12)$$
$$t - 2 = 0 \quad t - 12 = 0$$
$$t = 2 \quad t = 12$$
The solution 2 is not possible since the time for the larger pipe would then be a negative number.
$t - 8 = 12 - 8 = 4$
It would take the larger pipe 4 minutes to fill the tank.

31. Strategy
To find the distance:
Write the basic direct variation equation, replace the variables by the given values, and solve for k.
Write the direct variation equation, replacing k by its value. Substitute 40 for f and solve for d.

Solution
$$d = kf \qquad d = \frac{3}{5}f$$
$$30 = k(50) \qquad = \frac{3}{5}(40)$$
$$\frac{3}{5} = k \qquad = 24$$
A force of 40 lb will stretch the string 24 in.

32. Strategy
To find the average annual rate of change, divide the difference in energy generated by the change in years.

Solution
For 2003 to 2009,
$$\frac{50.20 - 11.19}{2009 - 2003} = \frac{39.01}{6} \approx 6.50$$

The average annual rate of change for 2003 to 2009 is 6.50 gigawatts per year.

33. Strategy
To find the frequency:
Write the basic inverse variation equation, replace the variables by the given values, and solve for k.
Write the inverse variation equation, replacing k by its value. Substitute 1.5 for L and solve for f.

Solution
$$f = \frac{k}{L} \qquad f = \frac{120}{L}$$
$$60 = \frac{k}{2} \qquad = \frac{120}{1.5}$$
$$120 = k \qquad = 80$$
The frequency is 80 vibrations per minute.

CHAPTER 10: EXPONENTIAL AND LOGARITHMIC FUNCTIONS

CHAPTER 10 PREP TEST

1. $3^{-2} = \dfrac{1}{9}$ [5.1.2]

2. $\left(\dfrac{1}{2}\right)^{-4} = 2^4 = 16$ [5.1.2]

3. $\dfrac{1}{8} = 2^{-3}$ [5.1.2]

4. $f(-1) = (-1)^4 + (-1)^3$ [3.2.1]
   $= 1 - 1$
   $= 0$
   $f(3) = (3)^4 + (3)^3$
   $= 81 + 27$
   $= 108$

5. $3x + 7 = x - 5$ [2.1.1]
   $2x = -12$
   $x = -6$

6. $16 = x^2 - 6x$ [8.1.1]
   $0 = x^2 - 6x - 16$
   $0 = (x-8)(x+2)$
   $x - 8 = 0 \quad x + 2 = 0$
   $x = 8 \qquad x = -2$
   The solutions are -2 and 8.

7. $5000(1+0.04)^6 = 5000(1.04)^6$ [1.3.2]
   $= 5000(1.265319)$
   $= 6326.60$

8. $f(x) = x^2 - 1$ [8.6.1]

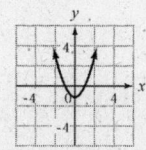

Section 10.1
Concept Review 10.1

1. Never true
   The domain is $\{x \mid x \text{ is a real number}\}$.

3. Never true
   The graph of $f(x) = b^x$ passes through the point $(0, 1)$.

5. Never true
   The range of $f(x) = b^x$ is $\{f(x) \mid f(x) > 0\}$. The function does not have x-intercepts.

Objective 10.1.1 Exercises

1. An exponential function with base b is defined by $f(x) = b^x$, $b > 0$, $b \neq 1$, and x is any real number.

3. c cannot be the base since $-5 < 0$.

5. 1; 0; 1

7. $f(x) = 3^x$
   a. $f(2) = 3^2 = 9$
   b. $f(0) = 3^0 = 1$
   c. $f(-2) = 3^{-2} = \dfrac{1}{3^2} = \dfrac{1}{9}$

9. $g(x) = 2^{x+1}$
   a. $g(3) = 2^{3+1} = 2^4 = 16$
   b. $g(1) = 2^{1+1} = 2^2 = 4$
   c. $g(-3) = 2^{-3+1} = 2^{-2} = \dfrac{1}{2^2} = \dfrac{1}{4}$

11. $P(x) = \left(\dfrac{1}{2}\right)^{2x}$
    a. $P(0) = \left(\dfrac{1}{2}\right)^{2 \cdot 0} = \left(\dfrac{1}{2}\right)^0 = 1$
    b. $P\left(\dfrac{3}{2}\right) = \left(\dfrac{1}{2}\right)^{2 \cdot \frac{3}{2}} = \left(\dfrac{1}{2}\right)^3 = \dfrac{1}{8}$
    c. $P(-2) = \left(\dfrac{1}{2}\right)^{2(-2)} = \left(\dfrac{1}{2}\right)^{-4} = 2^4 = 16$

13. $G(x) = e^{x/2}$
    a. $G(4) = e^{4/2} = e^2 \approx 7.3891$
    b. $G(-2) = e^{-2/2} = e^{-1} = \dfrac{1}{e} \approx 0.3679$
    c. $G\left(\dfrac{1}{2}\right) = e^{\frac{1}{2}/2} = e^{1/4} = e^{0.25} \approx 1.2840$

15. $H(r) = e^{-r+3}$
    a. $H(-1) = e^{-(-1)+3} = e^{1+3} = e^4 \approx 54.5982$
    b. $H(3) = e^{-3+3} = e^0 = 1$
    c. $H(5) = e^{-5+3} = e^{-2} = \dfrac{1}{e^2} \approx 0.1353$

Chapter 10: Exponential And Logarithmic Functions

17. $F(x) = 2^{x^2}$

  a. $F(2) = 2^{2^2} = 2^4 = 16$

  b. $F(-2) = 2^{(-2)^2} = 2^4 = 16$

  c. $F\left(\dfrac{3}{4}\right) = 2^{\left(\frac{3}{4}\right)^2} = 2^{\frac{9}{16}} = \sqrt[16]{2^9} = \sqrt[16]{512} \approx 1.4768$

19. $f(x) = e^{-x^2/2}$

  a. $f(-2) = e^{-(-2)^2/2}$
  $= e^{-4/2}$
  $= e^{-2}$
  $= \dfrac{1}{e^2}$
  ≈ 0.1353

  b. $f(2) = e^{-(2)^2/2} = e^{-4/2} = e^{-2} = \dfrac{1}{e^2} \approx 0.1353$

  c. $f(-3) = e^{-(-3)^2/2} = e^{-9/2} = \dfrac{1}{e^{9/2}} \approx 0.0111$

21. $\left(\dfrac{1}{2}\right)^{-3}$

23. Less than

Objective 10.1.2 Exercises

25. -3; 3^3; $\dfrac{1}{27}$; -3; $\dfrac{1}{27}$

27.

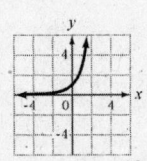

29.

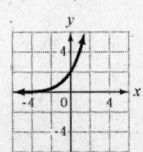

31.

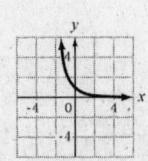

33.

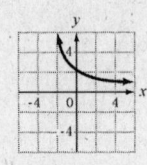

35.

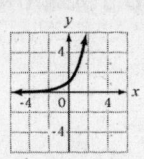

37.

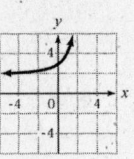

39. b and d have the same graph since $\left(\dfrac{1}{3}\right)^x = 3^{-x}$.

41. The graphs intersect at (0, 1).

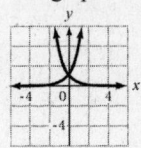

43. There is no x-intercept. The y-intercept is (0, 1).

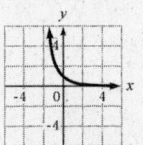

45. Greater than

47. $g(x) = b^x - a$

49.

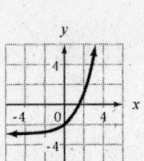

The zero of f is 1.6.

51.

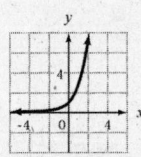

The value of x for which $f(x) = 3$ is 1.1.

53.

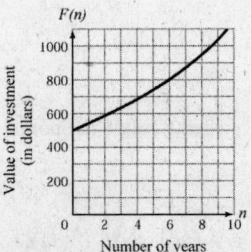

In approximately 9 years the investment will be worth $1000.

55.

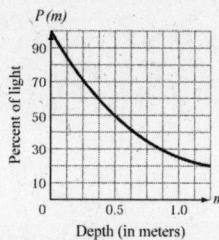

50% of the light reaches 0.5 meter below the surface of the ocean.

Applying Concepts 10.1

57.

a.

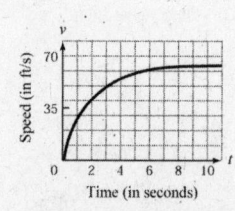

b. At $t = 4$ seconds after the object is dropped, it will be falling at a speed of 55.3 feet per second.

59. The graphs of g and h are exactly the same as the graph of f except for their position on the coordinate grid. The graph of g is shifted 2 units down; the graph of h is shifted two units up. The graphs of all three functions are shown here.

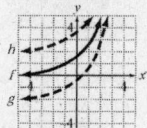

Section 10.2

Concept Review 10.2

1. Always true

3. Never true

   $\dfrac{\log x}{\log y}$ is in simplest form.

5. Always true

7. Always true

Objective 10.2.1 Exercises

1.
   a. A common logarithm is a logarithm with base 10.
   b. $\log 4z$

3. base b; x

5. $5^2 = 25$ is equivalent to $\log_5 25 = 2$.

7. $4^{-2} = \dfrac{1}{16}$ is equivalent to $\log_4\left(\dfrac{1}{16}\right) = -2$.

9. $10^y = x$ is equivalent to $\log_{10} x = y$.

11. $a^x = w$ is equivalent to $\log_a w = x$.

13. $\log_3 9 = 2$ is equivalent to $3^2 = 9$.

15. $\log 0.01 = -2$ is equivalent to $10^{-2} = 0.01$.

17. $\ln x = y$ is equivalent to $e^y = x$.

19. $\log_b u = v$ is equivalent to $b^v = u$.

21. $u = v$

23. $\log_3 81 = x$
    $3^x = 81$
    $3^x = 3^4$
    $x = 4$
    $\log_3 81 = 4$

25. $\log_2 128 = x$
    $2^x = 128$
    $2^x = 2^7$
    $x = 7$
    $\log_2 128 = 7$

27. $\log 100 = x$
    $10^x = 100$
    $10^x = 10^2$
    $x = 2$
    $\log 100 = 2$

29. $\ln e^3 = x$
    $3 \ln e = x$
    $3(1) = x$
    $x = 3$
    $\ln e^3 = 3$

31. $\log_8 1 = x$
    $8^x = 1$
    $x = 0$
    $\log_8 1 = 0$

33. $\log_5 625 = x$
    $5^x = 625$
    $5^x = 5^4$
    $x = 4$
    $\log_5 625 = 4$

35. $\log_3 x = 2$
    $3^2 = x$
    $9 = x$

37. $\log_4 x = 3$
$4^3 = x$
$64 = x$

39. $\log_7 x = -1$
$7^{-1} = x$
$\dfrac{1}{7} = x$

41. $\log_6 x = 0$
$6^0 = x$
$1 = x$

43. $\log x = 2.5$
$10^{2.5} = x$
$316.23 \approx x$

45. $\log x = -1.75$
$10^{-1.75} = x$
$0.02 \approx x$

47. $\ln x = 2$
$e^2 = x$
$7.39 \approx x$

49. $\ln x = -\dfrac{1}{2}$
$e^{-1/2} = x$
$0.61 \approx x$

51. Greater than

Objective 10.2.2 Exercises

53. Answers may vary. For example, the log of a product is equal to the sum of the logs: $\log_b(xy) = \log_b x + \log_b y$.

55. $r \log_b x$

57. $\log_8(xz) = \log_8 x + \log_8 z$

59. $\log_3 x^5 = 5\log_3 x$

61. $\ln\left(\dfrac{r}{s}\right) = \ln r - \ln s$

63. $\log_3(x^2 y^6) = \log_3 x^2 + \log_3 y^6$
$= 2\log_3 x + 6\log_3 y$

65. $\log_7\left(\dfrac{u^3}{v^4}\right) = \log_7 u^3 - \log_7 v^4$
$= 3\log_7 u - 4\log_7 v$

67. $\log_2(rs)^2 = 2\log_2(rs)$
$= 2(\log_2 r + \log_2 s)$
$= 2\log_2 r + 2\log_2 s$

69. $\log_9 x^2 yz = \log_9 x^2 + \log_9 y + \log_9 z$
$= 2\log_9 x + \log_9 y + \log_9 z$

71. $\ln\left(\dfrac{xy^2}{z^4}\right) = \ln(xy^2) - \ln z^4$
$= \ln x + \ln y^2 - \ln z^4$
$= \ln x + 2\ln y - 4\ln z$

73. $\log_8\left(\dfrac{x^2}{yz^2}\right) = \log_8 x^2 - \log_8(yz^2)$
$= \log_8 x^2 - (\log_8 y + \log_8 z^2)$
$= \log_8 x^2 - \log_8 y - \log_8 z^2$
$= 2\log_8 x - \log_8 y - 2\log_8 z$

75. $\log_7 \sqrt{xy} = \log_7(xy)^{1/2}$
$= \dfrac{1}{2}\log_7(xy)$
$= \dfrac{1}{2}(\log_7 x + \log_7 y)$
$= \dfrac{1}{2}\log_7 x + \dfrac{1}{2}\log_7 y$

77. $\log_2 \sqrt{\dfrac{x}{y}} = \log_2\left(\dfrac{x}{y}\right)^{1/2}$
$= \dfrac{1}{2}\log_2\left(\dfrac{x}{y}\right)$
$= \dfrac{1}{2}(\log_2 x - \log_2 y)$
$= \dfrac{1}{2}\log_2 x - \dfrac{1}{2}\log_2 y$

79. $\ln \sqrt{x^3 y} = \ln(x^3 y)^{1/2}$
$= \dfrac{1}{2}\ln(x^3 y)$
$= \dfrac{1}{2}(\ln x^3 + \ln y)$
$= \dfrac{1}{2}(3\ln x + \ln y)$
$= \dfrac{3}{2}\ln x + \dfrac{1}{2}\ln y$

81. $\log_7 \sqrt{\dfrac{x^3}{y}} = \log_7\left(\dfrac{x^3}{y}\right)^{1/2}$
$= \dfrac{1}{2}\log_7\left(\dfrac{x^3}{y}\right)$
$= \dfrac{1}{2}(\log_7 x^3 - \log_7 y)$
$= \dfrac{1}{2}(3\log_7 x - \log_7 y)$
$= \dfrac{3}{2}\log_7 x - \dfrac{1}{2}\log_7 y$

83. $\log_3 x^3 - \log_3 y = \log_3\left(\dfrac{x^3}{y}\right)$

85. $\log_8 x^4 + \log_8 y^2 = \log_8(x^4 y^2)$

87. $3\ln x = \ln x^3$

89. $3\log_5 x + 4\log_5 y = \log_5 x^3 + \log_5 y^4$
$= \log_5(x^3 y^4)$

91. $-2\log_4 x = \log_4(x^{-2}) = \log_4\left(\dfrac{1}{x^2}\right)$

93. $2\log_3 x - \log_3 y + 2\log_3 z$
$= \log_3 x^2 - \log_3 y + \log_3 z^2$
$= \log_3\left(\dfrac{x^2}{y}\right) + \log_3 z^2$
$= \log_3\left(\dfrac{x^2 z^2}{y}\right)$

95. $\log_b x - (2\log_b y + \log_b z)$
$= \log_b x - (\log_b y^2 + \log_b z)$
$= \log_b x - \log_b y^2 z$
$= \log_b\left(\dfrac{x}{y^2 z}\right)$

97. $2(\ln x + \ln y) = 2\ln(xy) = \ln(xy)^2 = \ln(x^2 y^2)$

99. $\dfrac{1}{2}(\log_6 x - \log_6 y) = \dfrac{1}{2}\log_6\left(\dfrac{x}{y}\right)$
$= \log_6\left(\dfrac{x}{y}\right)^{1/2}$
$= \log_6\sqrt{\dfrac{x}{y}}$

101. $2(\log_4 s - 2\log_4 t + \log_4 r)$
$= 2\left(\log_4\dfrac{s}{t^2} + \log_4 r\right)$
$= 2\log_4\left(\dfrac{sr}{t^2}\right)$
$= \log_4\left(\dfrac{sr}{t^2}\right)^2$
$= \log_4\dfrac{s^2 r^2}{t^4}$

103. $\log_5 x - 2(\log_5 y + \log_5 z) = \log_5 x - 2\log_5(yz)$
$= \log_5 x - \log_5(yz)^2$
$= \log_5 x - \log_5 y^2 z^2$
$= \log_5\dfrac{x}{y^2 z^2}$

105. $3\ln t - 2(\ln r - \ln v) = \ln t^3 - 2\ln\left(\dfrac{r}{v}\right)$
$= \ln t^3 - \ln\left(\dfrac{r}{v}\right)^2$
$= \ln t^3 - \ln\left(\dfrac{r^2}{v^2}\right)$
$= \ln\dfrac{t^3}{\frac{r^2}{v^2}}$
$= \ln\dfrac{t^3 v^2}{r^2}$

107. $\dfrac{1}{2}(3\log_4 x - 2\log_4 y + \log_4 z)$
$= \dfrac{1}{2}(\log_4 x^3 - \log_4 y^2 + \log_4 z)$
$= \dfrac{1}{2}\left(\log_4\dfrac{x^3}{y^2} + \log_4 z\right)$
$= \dfrac{1}{2}\log_4\left(\dfrac{x^3 z}{y^2}\right)$
$= \log_4\left(\dfrac{x^3 z}{y^2}\right)^{1/2}$
$= \log_4\sqrt{\dfrac{x^3 z}{y^2}}$

109. False. $\log_b 10 + \log_b 5 = \log_b 50$

111. True

113. $\ln 4 \approx 1.3863$

115. $\ln\left(\dfrac{17}{6}\right) = \ln 17 - \ln 6 \approx 1.0415$

117. $\log_8 6 = \dfrac{\log 6}{\log 8} \approx 0.8617$

119. $\log_5 30 = \dfrac{\log 30}{\log 5} \approx 2.1133$

121. $\log_3(0.5) = \dfrac{\log(0.5)}{\log 3} \approx -0.6309$

123. $\log_7(1.7) = \dfrac{\log 1.7}{\log 7} \approx 0.2727$

125. $\log_5 15 = \dfrac{\log 15}{\log 5} \approx 1.6826$

127. $\log_{12} 120 = \dfrac{\log 120}{\log 12} \approx 1.9266$

129. $\log_3(3x - 2) = \dfrac{\log(3x - 2)}{\log 3}$

131. $\log_8(4-9x) = \dfrac{\log(4-9x)}{\log 8}$

133. $5\log_9(6x+7) = 5\dfrac{\log(67x+7)}{\log 9}$
$ = \dfrac{5}{\log 9}\log(6x+7)$

135. $\log_2(x+5) = \dfrac{\ln(x+5)}{\ln 2}$

137. $\log_3(x^2+9) = \dfrac{\ln(x^2+9)}{\ln 3}$

139. $7\log_8(10x-7) = 7\dfrac{\ln(10x-7)}{\ln 8}$
$ = \dfrac{7}{\ln 8}\ln(10x-7)$

Applying Concepts 10.2

141. $\log_8 x = 3\log_8 2$
$\log_8 x = \log_8 2^3$
$\log_8 x = \log_8 8$
$x = 8$
The solution is 8.

143. $\log_4 x = \log_4 2 + \log_4 3$
$\log_4 x = \log_4(2\cdot 3)$
$\log_4 x = \log_4 6$
$x = 6$
The solution is 6.

145. $\log_6 x = 3\log_6 2 - \log_6 4$
$\log_6 x = \log_6 2^3 - \log_6 4$
$\log_6 x = \log_6 8 - \log_6 4$
$\log_6 x = \log_6\left(\dfrac{8}{4}\right)$
$\log_6 x = \log_6 2$
$x = 2$
The solution is 2.

147. $\log x = \dfrac{1}{3}\log 27$
$\log x = \log 27^{1/3}$
$\log x = \log \sqrt[3]{27}$
$\log x = \log 3$
$x = 3$
The solution is 3.

149. $a^c = b$

151. $S(t) = 8\log_5(6t+2)$
$S(2) = 8\log_5(6\cdot 2+2)$
$ = 8\log_5 14$
$ = 8\dfrac{\log 14}{\log 5}$
$ \approx 13.12$

153. $G(x) = -5\log_7(2x+19)$
$G(-3) = -5\log_7[2(-3)+19]$
$ = -5\log_7 13$
$ = -5\dfrac{\log 13}{\log 7}$
$ \approx -6.59$

155. $\ln(\ln x) = 1$
$e^1 = \ln x$
$e = \ln x$
$e^e = x$
$15.1543 \approx x$
The solution is 15.1543.

Section 10.3
Concept Review 10.3

1. Never true
The domain of $f(x) = \log_b x$ is $\{x|x>0\}$.

3. Always true

Objective 10.3.1 Exercises

1. Yes. Answers may vary. For example, it passes both the vertical line test and the horizontal line test.

3. They are the same graph.

5. y; exponential; $3^y + 1$; y; x

7. $f(x) = \log_4 x$
$y = \log_4 x$
$y = \log_4 x$ is equivalent to $x = 4^y$.

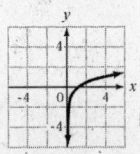

9. $f(x) = \log_3(2x-1)$
$y = \log_3(2x-1)$
$y = \log_3(2x-1)$ is equivalent to
$(2x-1) = 3^y,\ 2x = 3^y+1,\ \text{or}\ x = \dfrac{1}{2}(3^y+1)$.

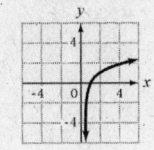

11. $f(x) = 3\log_2 x$

$y = 3\log_2 x$

$\dfrac{y}{3} = \log_2 x$

$\dfrac{y}{3} = \log_2 x$ is equivalent to $x = 2^{y/3}$.

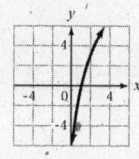

13. $f(x) = -\log_2 x$

$y = -\log_2 x$

$-y = \log_2 x$

$-y = \log_2 x$ is equivalent to $x = 2^{-y}$.

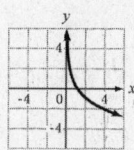

15. $f(x) = \log_2(x-1)$

$y = \log_2(x-1)$

$y = \log_2(x-1)$ is equivalent to $(x-1) = 2^y$, or $x = 2^y + 1$.

17. $f(x) = -\log_2(x-1)$

$y = -\log_2(x-1)$

$-y = \log_2(x-1)$

$-y = \log_2(x-1)$ is equivalent to $(x-1) = 2^{-y}$, or $x = 2^{-y} + 1$.

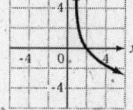

19. $f(x) = \log_2 x - 3$

$y = \log_2 x - 3$

$y = \dfrac{\log x}{\log 2} - 3$

$y = \dfrac{\log x}{0.3010} - 3$

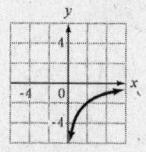

21. $f(x) = -\log_2 x + 2$

$y = -\log_2 x + 2$

$y = -\dfrac{\log x}{\log 2} + 2$

$y = -\dfrac{\log x}{0.3010} + 2$

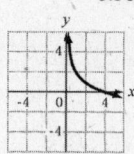

23. $f(x) = x - \log_2(1-x)$

$y = x - \log_2(1-x)$

$y = x - \dfrac{\log(1-x)}{\log 2}$

$y = x - \dfrac{\log(1-x)}{0.3010}$

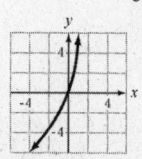

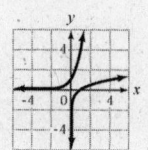

25.

27.

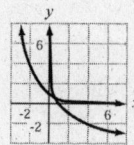

29. $f(x) = \log_3 3x = \log_3 3 + \log_3 x = 1 + \log_3 x$

$f(x) = \log_3 3x = \dfrac{\log 3x}{\log 3}$ by the Change-of-Base Formula.

b and **d** will have the same graph as the graph of the given function.

Applying Concepts 10.3

31.

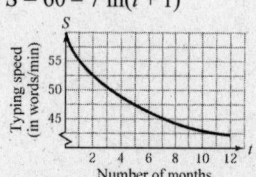

a. $S = 60 - 7\ln(t+1)$

b. After 4 months without typing to practice, a typist's proficiency decreases 49 words per minute.

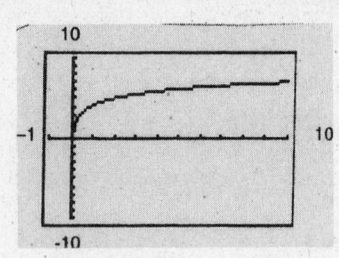

320 Chapter 10: Exponential And Logarithmic Functions

33.
a.

b. $y = 6$ when $t \approx 4$.
The energy production from renewable sources will first exceed 6 quadrillion BTU approximately 4 years after 2002, in 2006.

c. In 2010, $t = 8$. When $t = 8$, $y \approx 6.7$.
In 2010, the energy production from renewable energy sources will be approximately 6.7 quadrillion BTU.

35. quadratic

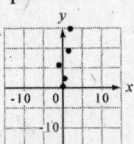

37. linear

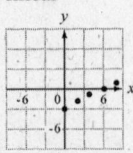

39. exponential

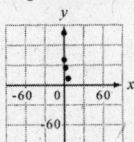

41. logarithmic

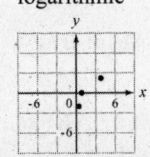

43. This is similar to the previous exercise. The graphs of $f(x) = e^{\ln x}$ and $g(x) = \ln e^x$ are shown below. The function f is defined for $x > 0$, whereas g is defined for all x

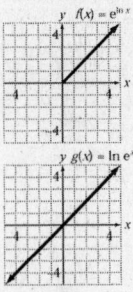

Section 10.4
Concept Review 10.4
1. Always true

3. Never true
The logarithm of a negative number is not defined.

5. Always true

Objective 10.4.1 Exercises

1. An exponential equation is one in which a variable occurs in an exponent.

3. 3; 2; $4x$; $3x - 1$; $4x$; -1

5. $5^{4x-1} = 5^{x+2}$
$4x - 1 = x + 2$
$3x - 1 = 2$
$3x = 3$
$x = 1$
The solution is 1.

7. $8^{x-4} = 8^{5x+8}$
$x - 4 = 5x + 8$
$-4x - 4 = 8$
$-4x = 12$
$x = -3$
The solution is -3.

9. $5^x = 6$
$\log 5^x = \log 6$
$x \log 5 = \log 6$
$x = \dfrac{\log 6}{\log 5}$
$x \approx 1.1133$
The solution is 1.1133.

11.
$$12^x = 6$$
$$\log 12^x = \log 6$$
$$x \log 12 = \log 6$$
$$x = \frac{\log 6}{\log 12}$$
$$x \approx 0.7211$$
The solution is 0.7211.

13.
$$\left(\frac{1}{2}\right)^x = 3$$
$$\log\left(\frac{1}{2}\right)^x = \log 3$$
$$x \log\left(\frac{1}{2}\right) = \log 3$$
$$x = \frac{\log 3}{\log \frac{1}{2}}$$
$$x = \frac{\log 3}{\log 0.5}$$
$$x \approx -1.5850$$
The solution is -1.5850.

15.
$$1.5^x = 2$$
$$\log 1.5^x = \log 2$$
$$x \log 1.5 = \log 2$$
$$x = \frac{\log 2}{\log 1.5}$$
$$x \approx 1.7095$$
The solution is 1.7095.

17.
$$10^x = 21$$
$$\log 10^x = \log 21$$
$$x \log 10 = \log 21$$
$$x = \frac{\log 21}{\log 10}$$
$$x \approx 1.3222$$
The solution is 1.3222.

19.
$$2^{-x} = 7$$
$$\log 2^{-x} = \log 7$$
$$-x \log 2 = \log 7$$
$$-x = \frac{\log 7}{\log 2}$$
$$x = -\frac{\log 7}{\log 2}$$
$$x \approx -2.8074$$
The solution is -2.8074.

21.
$$2^{x-1} = 6$$
$$\log 2^{x-1} = \log 6$$
$$(x-1)\log 2 = \log 6$$
$$x - 1 = \frac{\log 6}{\log 2}$$
$$x = \frac{\log 6}{\log 1} + 1$$
$$x \approx 3.5850$$
The solution is 3.5850.

23.
$$3^{2x-1} = 4$$
$$\log 3^{2x-1} = \log 4$$
$$(2x-1)\log 3 = \log 4$$
$$2x - 1 = \frac{\log 4}{\log 3}$$
$$2x = \frac{\log 4}{\log 3} + 1$$
$$x = \frac{1}{2}\left(\frac{\log 4}{\log 3} + 1\right)$$
$$x \approx 1.1309$$
The solution is 1.1309.

25.
$$9^x = 3^{x+1}$$
$$3^{2x} = 3^{x+1}$$
$$2x = x + 1$$
$$x = 1$$
The solution is 1.

27.
$$8^{x+2} = 16^x$$
$$(2^3)^{x+2} = 2^{4x}$$
$$2^{3x+6} = 2^{4x}$$
$$3x + 6 = 4x$$
$$6 = x$$
The solution is 6.

29.
$$5^{x^2} = 21$$
$$\log 5^{x^2} = \log 21$$
$$x^2 \log 5 = \log 21$$
$$x^2 = \frac{\log 21}{\log 5}$$
$$x = \pm\sqrt{\frac{\log 21}{\log 5}}$$
$$x = \pm 1.3754$$
The solutions are 1.3754 and -1.3754.

31.
$$2^{4x-2} = 20$$
$$\log 2^{4x-2} = \log 20$$
$$(4x-2)\log 2 = \log 20$$
$$4x - 2 = \frac{\log 20}{\log 2}$$
$$4x = \frac{\log 20}{\log 2} + 2$$
$$x = \frac{1}{4}\left(\frac{\log 20}{\log 2} + 2\right)$$
$$x = 1.5805$$
The solution is 1.5805.

33.
$$3^{-x+2} = 18$$
$$\log 3^{-x+2} = \log 18$$
$$(-x+2)\log 3 = \log 18$$
$$-x + 2 = \frac{\log 18}{\log 3}$$
$$-x = \frac{\log 18}{\log 3} - 2$$
$$x = -\frac{\log 18}{\log 3} + 2$$
$$x \approx -0.6309$$
The solution is −0.6309.

35.
$$4^{2x} = 100$$
$$\log 4^{2x} = \log 100$$
$$2x \log 4 = 2$$
$$2x = \frac{2}{\log 4}$$
$$x = \frac{1}{2}\left(\frac{2}{\log 4}\right) = \frac{1}{\log 4}$$
$$x \approx 1.6610$$
The solution is 1.6610.

37.
$$2.5^{-x} = 4$$
$$\log 2.5^{-x} = \log 4$$
$$-x \log 2.5 = \log 4$$
$$-x = \frac{\log 4}{\log 2.5}$$
$$x = -\frac{\log 4}{\log 2.5}$$
$$x \approx -1.5129$$
The solution is −1.5129.

39.
$$0.25^x = 0.125$$
$$\log 0.25^x = \log 0.125$$
$$x \log 0.25 = \log 0.125$$
$$x = \frac{\log 0.125}{\log 0.25}$$
$$x = 1.5$$
The solution is 1.5.

41.

a.
$$9^{x-a} = 81^x$$
$$9^{x-a} = (9^2)^x$$
$$9^{x-a} = 9^{2x}$$
$$x - a = 2x$$
$$-a = x$$

b.
$$5^{2x-2a} = 25^{2x}$$
$$5^{2x-2a} = (5^2)^{2x}$$
$$5^{2x-2a} = 5^{4x}$$
$$2x - 2a = 4x$$
$$-2a = 2x$$
$$-a = x$$

c.
$$49^{(x-a)/4} = 7^x$$
$$(7^2)^{(x-a)/4} = 7^x$$
$$7^{(x-a)/2} = 7^x$$
$$\frac{x-a}{2} = x$$
$$x - a = 2x$$
$$-a = x$$

d.
$$9^x = 81^{x-a}$$
$$9^x = (9^2)^{x-a}$$
$$9^x = 9^{2(x-a)}$$
$$x = 2x - 2a$$
$$2a = x$$

The equations with the same solution are **a**, **b**, and **c**.

43.
$$3^x = 2$$
$$3^x - 2 = 0$$

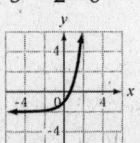

The solution is 0.63.

45. $2^x = 2x + 4$
$2^x - 2x - 4 = 0$

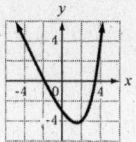

The solutions are -1.86 and 3.44.

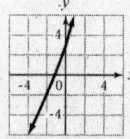

47. $e^x = -2x - 2$
$e^x + 2x + 2 = 0$

The solution is -1.16.

Objective 10.4.2 Exercises

49. A logarithmic equation is an equation in which one or more of the terms is a logarithmic expression.

51. exponential form

53.
  a. Exercise 52
  b. Exercise 51
  c. Exercise 51
  d. Exercise 52

55. $\log_3(x+1) = 2$
Rewrite in exponential form.
$3^2 = x + 1$
$9 = x + 1$
$8 = x$
The solution is 8.

57. $\log_2(2x - 3) = 3$
Rewrite in exponential form.
$2^3 = 2x - 3$
$8 = 2x - 3$
$11 = 2x$
$\frac{11}{2} = x$
The solution is $\frac{11}{2}$.

59. $\log_2(x^2 + 2x) = 3$
Rewrite in exponential form.
$2^3 = x^2 + 2x$
$8 = x^2 + 2x$
$0 = x^2 + 2x - 8$
$0 = (x+4)(x-2)$
$x + 4 = 0 \quad x - 2 = 0$
$x = -4 \quad x = 2$
The solutions are -4 and 2.

61. $\log_5 \frac{2x}{x-1} = 1$
Rewrite in exponential form.
$5^1 = \frac{2x}{x-1}$
$(x-1)5 = (x-1)\frac{2x}{x-1}$
$5x - 5 = 2x$
$3x - 5 = 0$
$3x = 5$
$x = \frac{5}{3}$
The solution is $\frac{5}{3}$.

63. $\log_7 x = \log_7(1-x)$
Use the fact that if $\log_b u = \log_b v$, then $u = v$.
$x = 1 - x$
$2x = 1$
$x = \frac{1}{2}$
The solution is $\frac{1}{2}$.

65. $\frac{2}{3}\log x = 6$
$\log x^{2/3} = 6$
Rewrite in exponential form.
$10^6 = x^{2/3}$
$(10^6)^{3/2} = (x^{2/3})^{3/2}$
$10^9 = x$
The solution is 1,000,000,000.

67. $\log_2(x-3) + \log_2(x+4) = 3$
$\log_2(x-3)(x+4) = 3$
Rewrite in exponential form.
$(x-3)(x+4) = 2^3$
$x^2 + x - 12 = 8$
$x^2 + x - 20 = 0$
$(x-4)(x+5) = 0$
$x - 4 = 0 \quad x + 5 = 0$
$x = 4 \quad x = -5$
−5 does not check as a solution. The solution is 4.

69. $\log_3 x + \log_3(x-1) = \log_3 6$
$\log_3 x(x-1) = \log_3 6$
Use the fact that if $\log_b u = \log_b v$, then $u = v$.
$x(x-1) = 6$
$x^2 - x = 6$
$x^2 - x - 6 = 0$
$(x+2)(x-3) = 0$
$x + 2 = 0 \quad x - 3 = 0$
$x = -2 \quad x = 3$
−2 does not check as a solution. The solution is 3.

71. $\log_2(8x) - \log_2(x^2 - 1) = \log_2 3$
$\log_2\left(\dfrac{8x}{x^2-1}\right) = \log_2 3$
Use the fact that if $\log_b u = \log_b v$, then $u = v$.
$\dfrac{8x}{x^2-1} = 3$
$(x^2-1)\dfrac{8x}{x^2-1} = (x^2-1)3$
$8x = 3x^2 - 3$
$0 = 3x^2 - 8x - 3$
$0 = (3x+1)(x-3)$
$3x + 1 = 0 \quad x - 3 = 0$
$3x = -1 \quad x = 3$
$x = -\dfrac{1}{3}$
$-\dfrac{1}{3}$ does not check as a solution. The solution is 3.

73. $\log_9 x + \log_9(2x-3) = \log_9 2$
$\log_9 x(2x-3) = \log_9 2$
Use the fact that if $\log_b u = \log_b v$, then $u = v$.
$x(2x-3) = 2$
$2x^2 - 3x = 2$
$2x^2 - 3x - 2 = 0$
$(2x+1)(x-2) = 0$
$2x + 1 = 0 \quad x - 2 = 0$
$2x = -1 \quad x = 2$
$x = -\dfrac{1}{2}$
$-\dfrac{1}{2}$ does not check as a solution. The solution is 2.

75. $\log_8(6x) = \log_8 2 + \log_8(x-4)$
$\log_8(6x) = \log_8 2(x-4)$
Use the fact that if $\log_b u = \log_b v$, then $u = v$.
$6x = 2(x-4)$
$6x = 2x - 8$
$4x = -8$
$x = -2$
−2 does not check as a solution. The equation has no solution.

77. $\log_9(7x) = \log_9 2 + \log_9(x^2 - 2)$
$\log_9(7x) = \log_9 2(x^2 - 2)$
Use the fact that if $\log_b u = \log_b v$, then $u = v$.
$7x = 2(x^2 - 2)$
$7x = 2x^2 - 4$
$0 = 2x^2 - 7x - 4$
$0 = (2x+1)(x-4)$
$2x + 1 = 0 \quad x - 4 = 0$
$2x = -1 \quad x = 4$
$x = -\dfrac{1}{2}$
$-\dfrac{1}{2}$ does not check as a solution. The solution is 4.

79. $\log(x^2+3) - \log(x+1) = \log 5$

$$\log\left(\frac{x^2+3}{x+1}\right) = \log 5$$

Use the fact that if $\log_b u = \log_b v$, then $u = v$.

$$\frac{x^2+3}{x+1} = 5$$

$$(x+1)\left(\frac{x^2+3}{x+1}\right) = (x+1)5$$

$$x^2 + 3 = 5x + 5$$

$$x^2 - 5x - 2 = 0$$

$$x = \frac{-(-5) \pm \sqrt{(-5)^2 - 4(1)(-2)}}{2(1)}$$

$$= \frac{5 \pm \sqrt{25+8}}{2}$$

$$= \frac{5 \pm \sqrt{33}}{2}$$

The solutions are $\frac{5+\sqrt{33}}{2}$ and $\frac{5-\sqrt{33}}{2}$.

81. $\log x = -x + 2$

$\log x + x - 2 = 0$

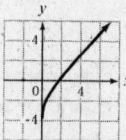

The solution is 1.76.

83. $\log(2x-1) = -x + 3$

$\log(2x-1) + x - 3 = 0$

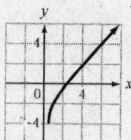

The solution is 2.42.

85. $\ln(x+2) = x^2 - 3$

$\ln(x+2) - x^2 + 3 = 0$

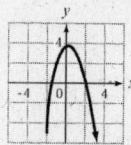

The solutions are -1.51 and 2.10.

Applying Concepts 10.4

87. $8^{x/2} = 6$

$\log 8^{x/2} = \log 6$

$\frac{x}{2} \log 8 = \log 6$

$\frac{x}{2} = \frac{\log 6}{\log 8}$

$x = \frac{2 \log 6}{\log 8}$

$x = 1.7233$

The solution is 1.7233.

89. $5^{3x/2} = 7$

$\log 5^{3x/2} = \log 7$

$\frac{3x}{2} \log 5 = \log 7$

$\frac{3x}{2} = \frac{\log 7}{\log 5}$

$x = \frac{2 \log 7}{3 \log 5}$

$x = 0.8060$

The solution is 0.8060.

91. $1.2^{(x/2)-1} = 1.4$

$\log 1.2^{(x/2)-1} = \log 1.4$

$\left(\frac{x}{2} - 1\right)(\log 1.2) = \log 1.4$

$\frac{x}{2} - 1 = \frac{\log 1.4}{\log 1.2}$

$\frac{x}{2} = \frac{\log 1.4}{\log 1.2} + 1$

$x = 2\left[\frac{\log 1.4}{\log 1.2} + 1\right]$

$x = 5.6910$

The solution is 5.6910.

93. $4^x = 7$

$\log 4^x = \log 7$

$x \log 4 = \log 7$

$x = \frac{\log 7}{\log 4}$

$x \approx 1.4036775$

$2^{(6x+3)} = 2^{(6 \cdot 1.4036775 + 3)}$

$= 2^{11.422065}$

$= 2744$

Objective 10.5.1 Exercises

1. Exponential decay is an example of an exponential equation in which the value of the dependent variable decreases exponentially as the value of the independent variable increases.

3. 4000; 3000; 0.0025; unknown

5. **Strategy**
   To find the value of the investment, solve the compound interest formula for P. Use $A = 1000$, $n = 8$, and $i = \dfrac{8\%}{4} = \dfrac{0.08}{4} = 0.02$.

   Solution
   $P = A(1+i)^n$
   $P = 1000(1+0.02)^8$
   $P = 1000(1.02)^8$
   $P \approx 1172$
   The value of the investment after 2 years is $1172.

7. **Strategy**
   To find the number of years, solve the compound interest formula for y. Use $P = 15{,}000$, $A = 5{,}000$, $n = 12y$, and $i = \dfrac{6\%}{12} = \dfrac{0.06}{12} = 0.005$.

   Solution
   $P = A(1+i)^n$
   $15{,}000 = 5{,}000(1+0.005)^{12y}$
   $3 = (1.005)^{12y}$
   $\log 3 = \log(1.005)^{12y}$
   $\log 3 = 12y \log(1.005)$
   $\dfrac{\log 3}{12 \log(1.005)} = y$
   $18 \approx y$
   In approximately 18 years the investment will be worth $15,000.

9. a. **Strategy**
   To find the level after 3 h, solve for A in the exponential decay equation. Use $A_0 = 30$, $t = 3$, and $k = 6$.

   Solution
   $A = A_0 \left(\dfrac{1}{2}\right)^{t/k}$
   $A = 30 \left(\dfrac{1}{2}\right)^{3/6}$
   $A = 21.2$
   After 3 h, the level will be 21.2 mg.

   b. **Strategy**
   To find the time, solve for t in the exponential decay equation. Use $k = 6$, $A_0 = 30$, and $A = 20$.

   Solution
   $A = A_0 \left(\dfrac{1}{2}\right)^{t/k}$
   $20 = 30 \left(\dfrac{1}{2}\right)^{t/6}$
   $0.\overline{6} = \left(\dfrac{1}{2}\right)^{t/6}$
   $0.\overline{6} = (0.5)^{t/6}$
   $\log(0.\overline{6}) = \log(0.5)^{t/6}$
   $\log(0.\overline{6}) = \dfrac{t}{6} \log(0.5)$
   $\dfrac{\log(0.\overline{6})}{\log(0.5)} = \dfrac{t}{6}$
   $\dfrac{6 \log(0.\overline{6})}{\log(0.5)} = t$
   $3.510 = t$
   It will take 3.5 hours for the injection to decay to 20 mg.

11. **Strategy**
    To find the half-life, solve for k in the exponential decay equation. Use $A_0 = 25$, $A = 18.95$, and $t = 1$.

    Solution
    $A = A_0 \left(\dfrac{1}{2}\right)^{t/k}$
    $18.95 = 25 \left(\dfrac{1}{2}\right)^{1/k}$
    $0.758 = \left(\dfrac{1}{2}\right)^{1/k}$
    $0.758 = (0.5)^{1/k}$
    $\log(0.758) = \log(0.5)^{1/k}$
    $\log(0.758) = \dfrac{1}{k} \log(0.5)$
    $k \log(0.758) = \log(0.5)$
    $k = \dfrac{\log(0.5)}{\log(0.758)}$
    $k \approx 2.5$
    The half-life is 2.5 years.

13. 2 years

15.
   a. **Strategy**
   To find the pressure at 40 km, solve for P in the equation. Use $h = 40$.

   Solution
   $P(40) = 10.13e^{-0.116(40)}$
   ≈ 0.098
   The pressure is approximately 0.098 newtons/cm^2.

   b. **Strategy**
   To find the pressure at Earth's surface, solve for P in the equation. Use $h = 0$.

   Solution
   $P(0) = 10.13e^{-0.116(0)}$
   $= 10.13$
   The pressure is 10.13 newtons/cm^2.

   c. The pressure decreases as you rise above Earth's surface.

17. **Strategy**
   To find the pH, replace H^+ with its given value and solve for pH.

   Solution
   $\text{pH} = -\log(3.97 \times 10^{-7})$
   $= -(\log 3.97 + \log 10^{-7})$
   $= -[0.5988 + (-7)]$
   $= 6.4012$
   The pH of the milk is 6.4.

19. **Strategy**
   To find the thickness, solve the equation for d. Use $P = 75\% = 0.75$ and $k = 0.05$.

   Solution
   $\log P = -kd$
   $\log(0.75) = -(0.05)d$
   $\dfrac{\log(0.75)}{-0.05} = d$
   $2.4987 \approx d$
   The depth must be 2.5 m.

21. **Strategy**
   To find the number of decibels, replace I with its given value in the equation and solve for D.

   Solution
   $D = 10(\log I + 16)$
   $= 10[\log(3.2 \times 10^{-10}) + 16]$
   $= 10[\log 3.2 + \log 10^{-10} + 16]$
   $= 10[0.5051 + (-10) + 16]$
   $= 10(6.5051)$
   $= 65.051$
   The number of decibels is 65.

23. **Strategy**
   To get the mass of the bacteria colony, multiply the population by the mass of one bacteria. To find how many hours, replace y in the equation with the given value and solve for t.

   Solution
   $y = (6.7 \times 10^{-15})2^{t/3}$
   $5.98 \times 10^{27} = (6.7 \times 10^{-15})2^{t/3}$
   $\dfrac{5.98 \times 10^{27}}{6.7 \times 10^{-15}} = 2^{t/3}$
   $\log\left(\dfrac{5.98 \times 10^{27}}{6.7 \times 10^{-15}}\right) = \log(2^{t/3})$
   $\log\left(\dfrac{5.98 \times 10^{27}}{6.7 \times 10^{-15}}\right) = \dfrac{t}{3}\log(2)$
   $\dfrac{\log\left(\dfrac{5.98 \times 10^{27}}{6.7 \times 10^{-15}}\right)}{\log(2)} = \dfrac{t}{3}$
   $3 \cdot \dfrac{\log\left(\dfrac{5.98 \times 10^{27}}{6.7 \times 10^{-15}}\right)}{\log(2)} = t$
   $418 \approx t$
   It would take 418 h for the mass of the colony to equal the mass of Earth. This is more time than predicted by the novel.

25. Strategy
To find the thickness needed, solve the given equation for x. Use $I = 0.25I_0$ and $k = 3.2$.

Solution
$$I = I_0 e^{-kx}$$
$$0.25I_0 = I_0 e^{-3.2x}$$
$$0.25 = e^{-3.2x}$$
$$\ln 0.25 = \ln(e^{-3.2x})$$
$$\ln 0.25 = -3.2x$$
$$\frac{\ln 0.25}{-3.2} = x$$
$$0.4 = x$$
Use a piece of copper that is 0.4 cm thick.

27. Strategy
To find the Richter scale magnitude, replace I with its given value in the equation and solve for M.

Solution
$$M = \log \frac{I}{I_0}$$
$$= \log \frac{3{,}162{,}277{,}000 I_0}{I_0}$$
$$\approx 9.5$$
The earthquake had a magnitude of 9.5.

29. Strategy
To find the intensity, replace M with its given value in the equation and solve for I.

Solution
$$M = \log \frac{I}{I_0}$$
$$8.5 = \log \frac{I}{I_0}$$
$$10^{8.5} = \frac{I}{I_0}$$
$$316{,}227{,}766 I_0 = I$$
The intensity was $316{,}228{,}000 I_0$.

31. $M = \log A + 3 \log 8t - 2.92$
$M = \log 23 + 3 \log[8(24)] - 2.92$
$M \approx 1.36173 + 6.84990 - 2.92$
$M \approx 5.3$
The magnitude is approximately 5.3.

33. $M = \log A + 3 \log 8t - 2.92$
$M = \log 28 + 3 \log[8(28)] - 2.92$
$M \approx 1.44716 + 7.05074 - 2.92$
$M \approx 5.6$
The magnitude is approximately 5.6.

Applying Concepts 10.5

35. Strategy
To find the number of years, write the compound interest formula using $1 for A, $2 for P, and 5% for i. Solve for n.

Solution
$$P = A(1+i)^n$$
$$2 = 1(1+0.05)^n$$
$$2 = (1.05)^n$$
$$\log 2 = \log(1.05)^n$$
$$\log 2 = n(\log 1.05)$$
$$\frac{\log 2}{\log 1.05} = n$$
$$14.2067 = n$$
The price will double in 14 years.

37. Strategy
To find the value of the investment, solve the continuous compounding formula for P. Use $A = 2500$, $n = 5$, and $r = 0.05$.

Solution
$$P = Ae^{rt}$$
$$P = 2500 e^{0.05(5)}$$
$$P = 2500 e^{0.25}$$
$$P = 2500(1.284)$$
$$P = 3210.06$$
The investment has a value of $3210.06 after 5 years.

39.

a. $8\% \div 12 = 0.08 \div 12 \approx 0.00667$

$$y = A(1+i)^x + B$$

$$y = \frac{Pi - M}{i}(1+i)^x + \frac{M}{i}$$

$$90{,}000 = \frac{100{,}000(0.00667) - 733.76}{0.00667}(1.00667)^x + \frac{733.76}{0.00667}$$

$$90{,}000 \approx -10{,}009(1.00667)^x + 110{,}009$$

$$10{,}009(1.00667)^x = 20{,}009$$

$$(1.00667)^x \approx 1.9991$$

$$\log(1.0067)^x \doteq \log 1.991$$

$$x \log 1.0067 = \log 1.991$$

$$x = \frac{\log 1.991}{\log 1.00667} \approx 104$$

104 months are required to reduce the loan amount to $90,000.

b. $$50{,}000 = -10{,}009(1.00667)^x + 110{,}009$$

$$10{,}009(1.00667)^x = 60{,}009$$

$$(1.00667)^x \approx 5.9955$$

$$x \approx \frac{\log 5.9955}{\log 1.00667} \approx 269$$

269 months are required to reduce the loan amount to $50,000.

c. $I = Mx + A(1+i)^x + B - P$

$$I = Mx + \frac{Pi - M}{i}(1+i)^x + \frac{M}{i} - P$$

$$I = 733.76x - 10{,}009(1.00667)^x + 110{,}009 - 100{,}000$$

$$I = 733.76x - 10{,}009(1.00667)^x + 10{,}009$$

Using a graphing utility, I is 100,000 when $x \approx 163$. The total interest paid exceeds $100,000 in month 163.

41. Carbon-14 (the 14 means there are 14 neutrons) occurs naturally as a result of cosmic rays (alpha particles) passing through the atmosphere and producing neutrons that convert nitrogen-14 into carbon-14. The carbon-14 atom continuously decays back to nitrogen-14 in 5730 years (the half-life of carbon-14).
W.F. Libby and others used this information to determine the age of archeological objects. As a result of his investigations into carbon dating, Libby was awarded the Nobel Prize in chemistry in 1960. Carbon dating is effective for dating organic material (material containing carbon, such as wood, bones, and cloth). Because the half-life of carbon-14 is small by geological standards, it is not useful for dating objects that are more than 25,000 years old.
Rubidium dating is another method of determining the age of an object. This element is used to date noncarbon objects such as rocks. Rubidium-87 is an isotope of rubidium and decays to strontium-86, which is stable. The half-life of rubidium-87 is 4.86×10^{10} years.
The uranium-thorium dating method is useful for dating rocks that do not contain rubidium. This method has been used only to date skeletal remains in which trace amounts of uranium were found. Uranium-238 decays to uranium-234 with a half-life of 4.5×10^9 years. Uranium-234 then decays to thorium-230 in 248,000 years. Recently, uranium-thorium dating was applied to a skeleton found in Del Mar, California. Carbon dating had determined the skeleton to be 48,000 years old. However, the uranium-thorium method showed the skeleton to be 11,000 years old. One reason that has been proposed to explain the discrepancy between the two dates is that carbon dating is prone to inaccuracies because the ratios of carbon-12 to carbon-14 present in the environment have changed over time.

CHAPTER 10 REVIEW EXERCISES

1. $\log_4 16 = x$
$4^x = 16$
$4^x = 4^2$
$x = 2$
$\log_4 16 = 2$

2. $\frac{1}{2}(\log_3 x - \log_3 y) = \frac{1}{2}\left(\log_3 \frac{x}{y}\right)$
$= \log_3 \left(\frac{x}{y}\right)^{1/2}$
$= \log_3 \sqrt{\frac{x}{y}}$

3. $f(x) = e^{x-2}$
$f(2) = e^{2-2}$
$f(2) = e^0$
$f(2) = 1$

4. $8^x = 2^{x-6}$
$(2^3)^x = 2^{x-6}$
$2^{3x} = 2^{x-6}$
$3x = x - 6$
$2x = -6$
$x = -3$
The solution is -3.

5. $f(x) = \left(\frac{2}{3}\right)^x$
$f(0) = \left(\frac{2}{3}\right)^0$
$f(0) = 1$

6. $\log_3 x = -2$
$3^{-2} = x$
$\frac{1}{3^2} = x$
$\frac{1}{9} = x$
The solution is $\frac{1}{9}$.

7. $2^5 = 32$ is equivalent to $\log_2 32 = 5$.

8. $\log x + \log(x-4) = \log 12$
$\log x(x-4) = \log 12$
$x(x-4) = 12$
$x^2 - 4x = 12$
$x^2 - 4x - 12 = 0$
$(x-6)(x+2) = 0$
$x - 6 = 0 \quad x + 2 = 0$
$x = 6 \quad x = -2$
-2 does not check as a solution. 6 checks as a solution. The solution is 6.

9. $\log_6 \sqrt{xy^3} = \log_6 \sqrt{x}\sqrt{y^3}$
$= \log_6 \sqrt{x} + \log_6 \sqrt{y^3}$
$= \log_6 x^{1/2} + \log_6 y^{3/2}$
$= \frac{1}{2}\log_6 x + \frac{3}{2}\log_6 y$

10. $4^{5x-2} = 4^{3x+2}$
$5x - 2 = 3x + 2$
$2x - 2 = 2$
$2x = 4$
$x = 2$
The solution is 2.

11. $3^{7x+1} = 3^{4x-5}$
$7x + 1 = 4x - 5$
$3x + 1 = -5$
$3x = -6$
$x = -2$
The solution is -2.

12. $f(x) = 3^{x+1}$
$f(-2) = 3^{-2+1} = 3^{-1} = \frac{1}{3}$

13. $\log_2 16 = x$
$2^x = 16$
$2^x = 2^4$
$x = 4$
The solution is 4.

14. $\log_6 2x = \log_6 2 + \log_6 (3x-4)$
$\log_6 2x = \log_6 2(3x-4)$
$\log_6 2x = \log_6 (6x-8)$
$2x = 6x - 8$
$-4x = -8$
$x = 2$
The solution is 2.

15. $\log_2 5 = \dfrac{\log 5}{\log 2} = 2.3219$

16. $\log_6 22 = \dfrac{\log 22}{\log 6} = 1.7251$

17. $\quad 4^x = 8^{x-1}$
$(2^2)^x = (2^3)^{x-1}$
$2^{2x} = 2^{3x-3}$
$2x = 3x - 3$
$-x = -3$
$x = 3$
The solution is 3.

18. $f(x) = \left(\dfrac{1}{4}\right)^x$
$f(-1) = \left(\dfrac{1}{4}\right)^{-1} = 4$

19. $\log_5 \sqrt{\dfrac{x}{y}} = \log_5 \dfrac{\sqrt{x}}{\sqrt{y}}$
$= \log_5 \sqrt{x} - \log_5 \sqrt{y}$
$= \log_5 x^{1/2} - \log_5 y^{1/2}$
$= \dfrac{1}{2} \log_5 x - \dfrac{1}{2} \log_5 y$

20. $\log_5 \dfrac{7x+2}{3x} = 1$
$5^1 = \dfrac{7x+2}{3x}$
$5 = \dfrac{7x+2}{3x}$
$15x = 7x + 2$
$8x = 2$
$x = \dfrac{1}{4}$
The solution is $\dfrac{1}{4}$.

21. $\log_5 x = 3$
$5^3 = x$
$125 = x$
The solution is 125.

22. $\log x + \log(2x+3) = \log 2$
$\log x(2x+3) = \log 2$
$x(2x+3) = 2$
$2x^2 + 3x = 2$
$2x^2 + 3x - 2 = 0$
$(2x-1)(x+2) = 0$
$2x - 1 = 0 \quad x + 2 = 0$
$2x = 1 \quad\quad x = -2$
$x = \dfrac{1}{2}$
-2 does not check as a solution. $\dfrac{1}{2}$ checks as a solution. The solution is $\dfrac{1}{2}$.

23. $3\log_b x - 5\log_b y = \log_b x^3 - \log_b y^5 = \log_b \dfrac{x^3}{y^5}$

24. $f(x) = 2^{-x-1}$
$f(-3) = 2^{-(-3)-1}$
$f(-3) = 2^{3-1}$
$f(-3) = 2^2 = 4$

25. $\log_3 19 = \dfrac{\log 19}{\log 3} = 2.6801$

26. $\quad 3^{x+2} = 5$
$\log 3^{x+2} = \log 5$
$(x+2)\log 3 = \log 5$
$x + 2 = \dfrac{\log 5}{\log 3}$
$x = \dfrac{\log 5}{\log 3} - 2$
$x = -0.5350$
The solution is -0.5350.

27.

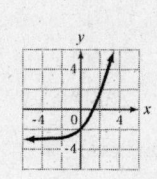

28.

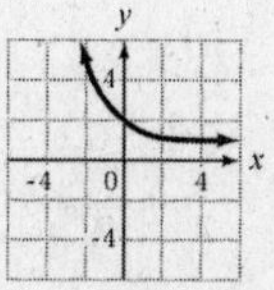

29.

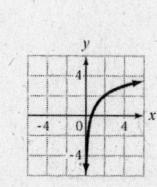

30.

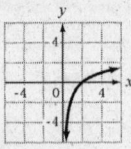

31. 1.0

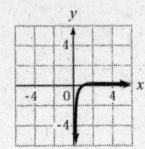

32. **Strategy**
To find the half-life, solve for k in the exponential decay equation. Use $A_0 = 10$, $A = 9$, and $t = 5$.

Solution
$$A = A_0\left(\frac{1}{2}\right)^{t/k}$$
$$9 = 10\left(\frac{1}{2}\right)^{5/k}$$
$$0.9 = \left(\frac{1}{2}\right)^{5/k}$$
$$0.9 = (0.5)^{5/k}$$
$$\log(0.9) = \log(0.5)^{5/k}$$
$$\log(0.9) = \frac{5}{k}\log(0.5)$$
$$k = \frac{5\log 0.5}{\log 0.9} = 32.89$$
The half-life is 33 h.

33. **Strategy**
To find the thickness, solve the equation $\log P = -0.5d$ for d. Use $P = 50\% = 0.5$.

Solution
$$\log P = -0.5d$$
$$\log 0.5 = -0.5d$$
$$\frac{\log 0.5}{-0.5} = d$$
$$0.602 = d$$
The material must be 0.602 cm thick.

CHAPTER 10 TEST

1. $f(x) = \left(\frac{3}{4}\right)^x$
$f(0) = \left(\frac{3}{4}\right)^0 = 1$

2. $f(x) = 4^{x-1}$
$f(-2) = 4^{-2-1}$
$f(-2) = 4^{-3}$
$f(-2) = \frac{1}{4^3} = \frac{1}{64}$

3. $\log_4 64 = x$
$4^x = 64$
$4^x = 4^3$
$x = 3$

4. $\log_4 x = -2$
$4^{-2} = x$
$\frac{1}{4^2} = x$
$\frac{1}{16} = x$
The solution is $\frac{1}{16}$.

5. $\log_6 \sqrt[3]{x^2 y^5} = \log_6 (x^2 y^5)^{1/3}$
$= \frac{1}{3}\log_6(x^2 y^5)$
$= \frac{1}{3}(\log_6 x^2 + \log_6 y^5)$
$= \frac{1}{3}(2\log_6 x + 5\log_6 y)$
$= \frac{2}{3}\log_6 x + \frac{5}{3}\log_6 y$

6. $\frac{1}{2}(\log_5 x - \log_5 y) = \frac{1}{2}\left(\log_5 \frac{x}{y}\right)$
$= \log_5 \left(\frac{x}{y}\right)^{1/2}$
$= \log_5 \sqrt{\frac{x}{y}}$

7. $\log_6 x + \log_6(x-1) = 1$
$\log_6 x(x-1) = 1$
$x(x-1) = 6^1$
$x^2 - x = 6$
$x^2 - x - 6 = 0$
$(x-3)(x+2) = 0$
$x - 3 = 0 \quad x + 2 = 0$
$x = 3 \quad x = -2$
-2 does not check as a solution. 3 checks as a solution. The solution is 3.

8. $f(x) = 3^{x+1}$
$f(-2) = 3^{-2+1}$
$f(-2) = 3^{-1} = \frac{1}{3}$

9.
$$3^x = 17$$
$$\log 3^x = \log 17$$
$$x \log 3 = \log 17$$
$$x = \frac{\log 17}{\log 3}$$
$$x = 2.5789$$

10.
$$\log_2 x + 3 = \log_2(x^2 - 20)$$
$$3 = \log_2(x^2 - 20) - \log_2 x$$
$$3 = \log_2 \frac{(x^2 - 20)}{x}$$
$$2^3 = \frac{x^2 - 20}{x}$$
$$8 = \frac{x^2 - 20}{x}$$
$$8x = x^2 - 20$$
$$0 = x^2 - 8x - 20$$
$$0 = (x-10)(x+2)$$
$$x - 10 = 0 \quad x + 2 = 0$$
$$x = 10 \quad x = -2$$
−2 does not check as a solution. 10 checks as a solution.
The solution is 10.

11.
$$5^{6x-2} = 5^{3x+7}$$
$$6x - 2 = 3x + 7$$
$$3x - 2 = 7$$
$$3x = 9$$
$$x = 3$$
The solution is 3.

12.
$$4^x = 2^{3x+4}$$
$$(2^2)^x = 2^{3x+4}$$
$$2^{2x} = 2^{3x+4}$$
$$2x = 3x + 4$$
$$-x = 4$$
$$x = -4$$
The solution is −4.

13.
$$\log(2x+1) + \log x = \log 6$$
$$\log x(2x+1) = \log 6$$
$$x(2x+1) = 6$$
$$2x^2 + x = 6$$
$$2x^2 + x - 6 = 0$$
$$(2x-3)(x+2) = 0$$
$$2x - 3 = 0 \quad x + 2 = 0$$
$$2x = 3 \quad x = -2$$
$$x = \frac{3}{2}$$
−2 does not check as a solution. $\frac{3}{2}$ checks as a solution.

The solution is $\frac{3}{2}$.

14.

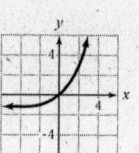

15.

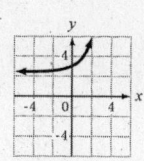

16.

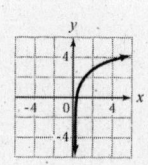

17.

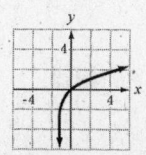

18. 1.6

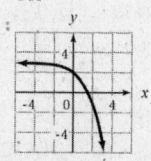

19. Strategy
To find the value, solve the compound interest formula for P. Use $A = 10,000$,
$$i = \frac{7.5\%}{12} = \frac{0.075}{12} = 0.00625, \; n = 72.$$

Solution
$$P = A(1+i)^n$$
$$P = 10,000(1 + 0.00625)^{72}$$
$$\approx 15,661$$
The value is $15,661.

334 Chapter 10: Exponential And Logarithmic Functions

20. Strategy
To find the half-life, solve for k in the exponential decay equation. Use $A_0 = 40$, $A = 30$, and $t = 10$.

Solution
$$A = A_0 \left(\frac{1}{2}\right)^{t/k}$$
$$30 = 40\left(\frac{1}{2}\right)^{10/k}$$
$$0.75 = \left(\frac{1}{2}\right)^{10/k}$$
$$0.75 = (0.5)^{10/k}$$
$$\log 0.75 = \log(0.5)^{10/k}$$
$$\log(0.75) = \frac{10}{k}\log(0.5)$$
$$k = \frac{10\log 0.5}{\log 0.75} \approx 24.09$$

The half-life is 24 h.

CUMULATIVE REVIEW EXERCISES

1. $4 - 2[x - 3(2 - 3x) - 4x] = 2x$
$4 - 2[x - 6 + 9x - 4x] = 2x$
$4 - 2[6x - 6] = 2x$
$4 - 12x + 12 = 2x$
$-12x + 16 = 2x$
$-14x = -16$
$x = \frac{8}{7}$

The solution is $\frac{8}{7}$.

2. $S = 2WH + 2WL + 2LH$
$S - 2WH = 2WL + 2LH$
$S - 2WH = L(2W + 2H)$
$\frac{S - 2WH}{2W + 2H} = L$

3. $|2x - 5| \leq 3$
$-3 \leq 2x - 5 \leq 3$
$-3 + 5 \leq 2x - 5 + 5 \leq 3 + 5$
$2 \leq 2x \leq 8$
$1 \leq x \leq 4$
$\{x | 1 \leq x \leq 4\}$

4. $4x^{2n} + 7x^n + 3 = (4x^n + 3)(x^n + 1)$

5. $x^2 + 4x - 5 \leq 0$
$(x + 5)(x - 1) \leq 0$

$\{x | -5 \leq x \leq 1\}$

6. $\frac{1 - \frac{5}{x} + \frac{6}{x^2}}{1 + \frac{1}{x} - \frac{6}{x^2}} = \frac{1 - \frac{5}{x} + \frac{6}{x^2}}{1 + \frac{1}{x} - \frac{6}{x^2}} \cdot \frac{x^2}{x^2}$
$= \frac{x^2 - 5x + 6}{x^2 + x - 6}$
$= \frac{(x-2)(x-3)}{(x+3)(x-2)}$
$= \frac{x-3}{x+3}$

7. $\frac{\sqrt{xy}}{\sqrt{x} - \sqrt{y}} = \frac{\sqrt{xy}}{\sqrt{x} - \sqrt{y}} \cdot \frac{\sqrt{x} + \sqrt{y}}{\sqrt{x} + \sqrt{y}}$
$= \frac{\sqrt{x^2y} + \sqrt{xy^2}}{(\sqrt{x})^2 - (\sqrt{y})^2}$
$= \frac{x\sqrt{y} + y\sqrt{x}}{x - y}$

8. $y\sqrt{18x^5y^4} - x\sqrt{98x^3y^6}$
$= y\sqrt{2 \cdot 3^2 x^5 y^4} - x\sqrt{2 \cdot 7^2 x^3 y^6}$
$= y\sqrt{3^2 x^4 y^4 (2x)} - x\sqrt{7^2 x^2 y^6 (2x)}$
$= y\sqrt{3^2 x^4 y^4}\sqrt{2x} - x\sqrt{7^2 x^2 y^6}\sqrt{2x}$
$= y \cdot 3x^2 y^2 \sqrt{2x} - x \cdot 7xy^3 \sqrt{2x}$
$= 3x^2 y^3 \sqrt{2x} - 7x^2 y^3 \sqrt{2x}$
$= -4x^2 y^3 \sqrt{2x}$

9. $\frac{i}{2-i} = \frac{i}{2-i} \cdot \frac{2+i}{2+i}$
$= \frac{2i + i^2}{4 - i^2}$
$= \frac{2i - 1}{4 + 1}$
$= \frac{-1 + 2i}{5}$
$= -\frac{1}{5} + \frac{2}{5}i$

10. $2x - y = 5$
$-y = -2x + 5$
$y = 2x - 5$
$m = 2 \quad (x_1, y_1) = (2, -2)$
$y - y_1 = m(x - x_1)$
$y - (-2) = 2(x - 2)$
$y + 2 = 2x - 4$
$y = 2x - 6$

The equation of the line is $y = 2x - 6$.

Copyright © Houghton Mifflin Company. All rights reserved.

11. $(x - r_1)(x - r_2) = 0$

$\left(x - \dfrac{1}{3}\right)(x - (-3)) = 0$

$\left(x - \dfrac{1}{3}\right)(x + 3) = 0$

$x^2 + \dfrac{8}{3}x - 1 = 0$

$3\left(x^2 + \dfrac{8}{3}x - 1\right) = 3(0)$

$3x^2 + 8x - 3 = 0$

12. $x^2 - 4x - 6 = 0$

$x^2 - 4x = 6$

$x^2 - 4x + 4 = 6 + 4$

$(x - 2)^2 = 10$

$\sqrt{(x - 2)^2} = \sqrt{10}$

$x - 2 = \pm\sqrt{10}$

$x = 2 \pm \sqrt{10}$

The solutions are is $2 + \sqrt{10}$ and $2 - \sqrt{10}$.

13. $f(x) = x^2 - 3x - 4$

$f(-1) = (-1)^2 - 3(-1) - 4 = 1 + 3 - 4 = 0$

$f(0) = 0^2 - 3(0) - 4 = -4$

$f(1) = 1^2 - 3(1) - 4 = 1 - 3 - 4 = -6$

$f(2) = 2^2 - 3(2) - 4 = 4 - 6 - 4 = -6$

$f(3) = 3^2 - 3(3) - 4 = 9 - 9 - 4 = -4$

The range is $\{-6, -4, 0\}$.

14. $g(x) = 2x - 3$

$g(0) = 2(0) - 3 = -3$

$f(x) = x^2 + 2x + 1$

$f(-3) = (-3)^2 + 2(-3) + 1$

$= 9 - 6 + 1$

$= 4$

$f[g(0)] = 4$

15. (1) $\quad 3x - y + z = 3$

(2) $\quad x + y + 4z = 7$

(3) $\quad 3x - 2y + 3z = 8$

Eliminate y.

Add equation (1) and (2).

$3x - y + z = 3$

$x + y + 4z = 7$

(4) $4x + 5z = 10$

Multiply equation (2) by 2 and add to equation (3).

$2(x + y + 4z) = 7(2)$

$3x - 2y + 3z = 8$

$2x + 2y + 8z = 14$

$3x - 2y + 3z = 8$

(5) $5x + 11z = 22$

Multiply equation (4) by −5. Multiply equation (5) by 4 and add.

$-5(4x + 5z) = -5(10)$

$4(5x + 11z) = 4(22)$

$-20x - 25z = -50$

$20x + 44z = 88$

$19z = 38$

$z = 2$

Substitute 2 for z in equation (5).

$5x + 11(2) = 22$

$5x + 22 = 22$

$5x = 0$

$x = 0$

Substitute 0 for x and 2 for z in equation (2).

$x + y + 4z = 7$

$0 + y + 4(2) = 7$

$y + 8 = 7$

$y = -1$

The solution is $(0, -1, 2)$.

16. (1) $y = -2x - 3$

(2) $y = 2x - 1$

Solve by the substitution method.

$y = 2x - 1$

$-2x - 3 = 2x - 1$

$-4x - 3 = -1$

$-4x = 2$

$x = -\dfrac{1}{2}$

Substitute into equation (2).

$y = 2x - 1$

$y = 2\left(-\dfrac{1}{2}\right) - 1$

$y = -1 - 1$

$y = -2$

The solution is $\left(-\dfrac{1}{2}, -2\right)$.

17. $f(x) = 3^{-x+1}$

$f(-4) = 3^{-(-4)+1} = 3^{4+1} = 3^5 = 243$

336 Chapter 10: Exponential And Logarithmic Functions

18. $\log_4 x = 3$
    $4^3 = x$
    $64 = x$
    The solution is 64.

19. $2^{3x+2} = 4^{x+5}$
    $2^{3x+2} = (2^2)^{x+5}$
    $2^{3x+2} = 2^{2x+10}$
    $3x + 2 = 2x + 10$
    $x + 2 = 10$
    $x = 8$
    The solution is 8.

20. $\log x + \log(3x + 2) = \log 5$
    $\log x(3x + 2) = \log 5$
    $x(3x + 2) = 5$
    $3x^2 + 2x = 5$
    $3x^2 + 2x - 5 = 0$
    $(3x + 5)(x - 1) = 0$
    $3x + 5 = 0 \quad x - 1 = 0$
    $3x = -5 \quad x = 1$
    $x = -\dfrac{5}{3}$

    $-\dfrac{5}{3}$ does not check as a solution. 1 does check as a solution. The solution is 1.

21. [number line graph]

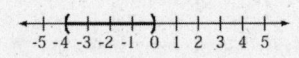

22. $\dfrac{x+2}{x-1} \geq 0$ 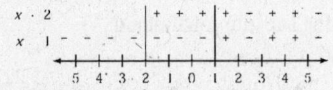
    [sign chart]
    $\{x \mid x \leq -2 \text{ or } x > 1\}$
    [number line graph]

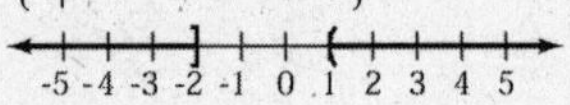

23. $y = -x^2 - 2x + 3$
    $-\dfrac{b}{2a} = -\dfrac{(-2)}{2(-1)} = -1$
    $y = -(-1)^2 - 2(-1) + 3 = -1 + 2 + 3 = 4$
    Vertex: $(-1, 4)$
    Axis of symmetry: $x = -1$
    [parabola graph]

24. [V-shaped graph]

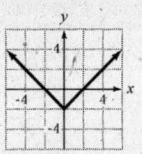

25. [exponential decay graph]

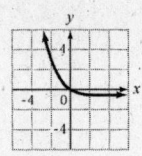

26. [logarithmic graph]

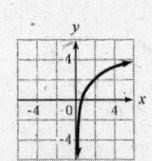

27. **Strategy**
    Pounds of 25% alloy: x
    Pounds of 50% alloy: $2000 - x$

	Amount	Percent	Quantity
25% alloy	x	0.25	$0.25x$
50% alloy	$2000 - x$	0.50	$0.50(2000 - x)$
40% alloy	2000	0.40	$0.40(2000)$

    The sum of the quantities before mixing equals the quantity after mixing.

    Solution
    $0.25x + 0.50(2000 - x) = 0.40(2000)$
    $0.25x + 1000 - 0.50x = 800$
    $-0.25x + 1000 = 800$
    $-0.25x = -200$
    $x = 800$
    $2000 - x = 2000 - 800 = 1200$
    800 lb of the alloy containing 25% tin and 1200 lb of the alloy containing 50% tin were used.

28. **Strategy**
    To find the amount, write and solve an inequality using x to represent the amount of sales.

    Solution
    $500 + 0.08x \geq 3000$
    $0.08x \geq 2500$
    $x \geq 31{,}250$

    To earn $3000 or more a month, the sales executive must sell $31,250 or more.

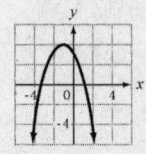

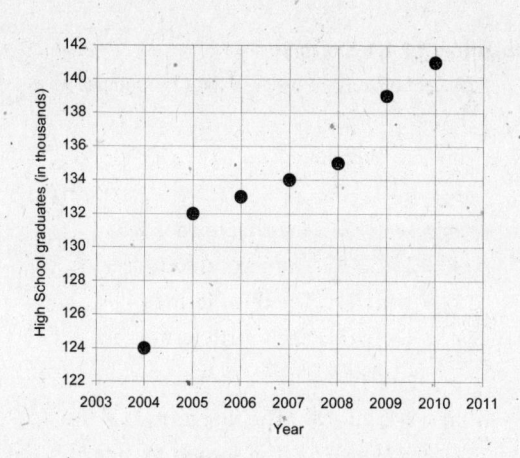

29.

[Scatter plot: High School graduates (in thousands) vs Year, showing points approximately: 2004→124, 2005→132, 2006→133, 2007→134, 2008→135, 2009→139, 2010→141]

30. Strategy
Time to print the checks when both printers are operating:

	Rate	Time	Part
Old printer	$\frac{1}{30}$	t	$\frac{t}{30}$
New printer	$\frac{1}{10}$	t	$\frac{t}{10}$

The sum of the parts of the task completed by each printer equals 1.

Solution
$$\frac{t}{30} + \frac{t}{10} = 1$$
$$30\left(\frac{t}{30} + \frac{t}{10}\right) = 30(1)$$
$$t + 3t = 30$$
$$4t = 30$$
$$t = 7.5$$

When both printers are operating, it will take 7.5 min to print the checks.

31. Strategy
To find the pressure:
Write the basic inverse variation equation, replace the variables by the given values, and solve for k.
Write the inverse variation equation, replacing k by its value. Substitute 25 for V and solve for P.

Solution
$$P = \frac{k}{V}$$
$$50 = \frac{k}{250}$$
$$12{,}500 = k$$

$$P = \frac{12{,}500}{V}$$
$$= \frac{12{,}500}{25}$$
$$= 500$$

When the volume is 25 ft³, the pressure is 500 lb/in².

32. Strategy
Cost per yard of nylon carpet: n
Cost per yard of wool carpet: w
First purchase:

	Amount	Unit cost	Value
Nylon carpet	45	n	$45n$
Wool carpet	30	w	$30w$

Second purchase:

	Amount	Unit cost	Value
Nylon carpet	25	n	$25n$
Wool carpet	80	w	$80w$

The total of the first purchase was $2340.
The total of the second purchase was $2820.
$45n + 30w = 2340$
$25n + 80w = 2820$

Solution
$45n + 30w = 2340$
$25n + 80w = 2820$

$25(45n + 30w) = 25(2340)$
$-45(25n + 80w) = -45(2820)$
$1125n + 750w = 58{,}500$
$-1125n - 3600w = -126{,}900$
$-2850w = -68{,}400$
$w = 24$

The cost per yard of the wool carpet is $24.

33. Strategy
To find the value of the investment, solve the compound interest formula for P. Use $A = 10{,}000$, $i = \frac{9\%}{12} = \frac{0.09}{12} = 0.0075$, and $n = 12 \cdot 5 = 60$.

Solution
$P = A(1+i)^n$
$= 10{,}000(1 + 0.0075)^{60}$
$= 10{,}000(1.0075)^{60}$
$\approx 15{,}657$

The value of the investment after 5 years is $15,657.

CHAPTER 11: SEQUENCES AND SERIES

CHAPTER 11 PREP TEST

1. $[3(1)-2]+[3(2)-2]+[3(3)-2]$ [1.2.4]
   $=[3-2]+[6-2]+[9-2]$
   $=1+4+7$
   $=12$

2. $f(n)=\dfrac{n}{n+2}$ [3.2.1]
   $f(6)=\dfrac{6}{6+2}$
   $=\dfrac{6}{8}=\dfrac{3}{4}$

3. $a_1+(n-1)d$ [1.3.2]
   $2+(5-1)4$
   $=2+(4)4$
   $=2+16=18$

4. $a_1 r^{n-1}$ [1.3.2]
   $=-3(-2)^{6-1}$
   $=-3(-2)^5$
   $=-3(-32)=96$

5. $\dfrac{a_1(1-r^n)}{1-r}$ [1.3.2]
   $=\dfrac{-2(1-(-4)^5)}{1-(-4)}$
   $=\dfrac{-2(1+1024)}{5}$
   $=\dfrac{-2(1025)}{5}=-410$

6. $\dfrac{\frac{4}{10}}{1-\frac{1}{10}}=\dfrac{\frac{4}{10}}{\frac{9}{10}}=\dfrac{4}{10}\cdot\dfrac{10}{9}=\dfrac{4}{9}$ [1.2.4]

7. $(x+y)^2=(x+y)(x+y)$ [5.3.3]
   $=x^2+2xy+y^2$

8. $(x+y)^3=(x+y)(x+y)(x+y)$ [5.3.2, 5.3.3]
   $=(x^2+2xy+y^2)(x+y)$
   $=x^3+3x^2y+3xy^2+y^3$

Section 11.1

Concept Review 11.1

1. Always true
3. Always true

Objective 11.1.1 Exercises

1. A sequence is an ordered list of numbers.
3. 8
5. first; second; nth
7. $a_n=n+1$
   $a_1=1+1=2$ The first term is 2.
   $a_2=2+1=3$ The second term is 3.
   $a_3=3+1=4$ The third term is 4.
   $a_4=4+1=5$ The fourth term is 5.

9. $a_n=2n+1$
   $a_1=2(1)+1=3$ The first term is 3.
   $a_2=2(2)+1=5$ The second term is 5.
   $a_3=2(3)+1=7$ The third term is 7.
   $a_4=2(4)+1=9$ The fourth term is 9.

11. $a_n=2-2n$
    $a_1=2-2(1)=0$ The first term is 0.
    $a_2=2-2(2)=-2$ The second term is –2.
    $a_3=2-2(3)=-4$ The third term is –4.
    $a_4=2-2(4)=-6$ The fourth term is –6.

13. $a_n=2^n$
    $a_1=2^1=2$ The first term is 2.
    $a_2=2^2=4$ The second term is 4.
    $a_3=2^3=8$ The third term is 8.
    $a_4=2^4=16$ The fourth term is 16.

15. $a_n=n^2+1$
    $a_1=1^2+1=2$ The first term is 2.
    $a_2=2^2+1=5$ The second term is 5.
    $a_3=3^2+1=10$ The third term is 10.
    $a_4=4^2+1=17$ The fourth term is 17.

17. $a_n=\dfrac{n}{n^2+1}$
    $a_1=\dfrac{1}{1^2+1}=\dfrac{1}{2}$ The first term is $\dfrac{1}{2}$.
    $a_2=\dfrac{2}{2^2+1}=\dfrac{2}{5}$ The second term is $\dfrac{2}{5}$.
    $a_3=\dfrac{3}{3^2+1}=\dfrac{3}{10}$ The third term is $\dfrac{3}{10}$.
    $a_4=\dfrac{4}{4^2+1}=\dfrac{4}{17}$ The fourth term is $\dfrac{4}{17}$.

19. $a_n = n - \dfrac{1}{n}$

   $a_1 = 1 - \dfrac{1}{1} = 0$ The first term is 0.

   $a_2 = 2 - \dfrac{1}{2} = \dfrac{3}{2}$ The second term is $\dfrac{3}{2}$.

   $a_3 = 3 - \dfrac{1}{3} = \dfrac{8}{3}$ The third term is $\dfrac{8}{3}$.

   $a_4 = 4 - \dfrac{1}{4} = \dfrac{15}{4}$ The fourth term is $\dfrac{15}{4}$.

21. $a_n = (-1)^{n+1} n$

   $a_1 = (-1)^2 (1) = 1$ The first term is 1.
   $a_2 = (-1)^3 (2) = -2$ The second term is -2.
   $a_3 = (-1)^4 (3) = 3$ The third term is 3.
   $a_4 = (-1)^5 (4) = -4$ The fourth term is -4.

23. $a_n = \dfrac{(-1)^{n+1}}{n^2 + 1}$

   $a_1 = \dfrac{(-1)^2}{1^2 + 1} = \dfrac{1}{2}$ The first term is $\dfrac{1}{2}$.

   $a_2 = \dfrac{(-1)^3}{2^2 + 1} = -\dfrac{1}{5}$ The second term is $-\dfrac{1}{5}$.

   $a_3 = \dfrac{(-1)^4}{3^2 + 1} = \dfrac{1}{10}$ The third term is $\dfrac{1}{10}$.

   $a_4 = \dfrac{(-1)^5}{4^2 + 1} = -\dfrac{1}{17}$ The fourth term is $-\dfrac{1}{17}$.

25. $a_n = (-1)^n 2^n$

   $a_1 = (-1)^1 \cdot 2^1 = -2$ The first term is -2.
   $a_2 = (-1)^2 \cdot 2^2 = 4$ The second term is 4.
   $a_3 = (-1)^3 \cdot 2^3 = -8$ The third term is -8.
   $a_4 = (-1)^4 \cdot 2^4 = 16$ The fourth term is 16.

27. $a_n = 2\left(\dfrac{1}{3}\right)^{n+1}$

   $a_1 = 2\left(\dfrac{1}{3}\right)^2 = \dfrac{2}{9}$ The first term is $\dfrac{2}{9}$.

   $a_2 = 2\left(\dfrac{1}{3}\right)^3 = \dfrac{2}{27}$ The second term is $\dfrac{2}{27}$.

   $a_3 = 2\left(\dfrac{1}{3}\right)^4 = \dfrac{2}{81}$ The third term is $\dfrac{2}{81}$.

   $a_4 = 2\left(\dfrac{1}{3}\right)^5 = \dfrac{2}{243}$ The fourth term is $\dfrac{2}{243}$.

29. $a_n = 2n - 5$
   $a_{10} = 2(10) - 5 = 15$
   The tenth term is 15.

31. $a_n = \dfrac{n}{n+1}$

   $a_{12} = \dfrac{12}{12+1} = \dfrac{12}{13}$

   The twelfth term is $\dfrac{12}{13}$.

33. $a_n = (-1)^{n-1}(n-1)$
   $a_{25} = (-1)^{24}(25 - 1) = 24$
   The twenty-fifth term is 24.

35. $a_n = \left(\dfrac{2}{3}\right)^n$

   $a_5 = \left(\dfrac{2}{3}\right)^5 = \dfrac{32}{243}$

   The fifth term is $\dfrac{32}{243}$.

37. $a_n = (n+4)(n+1)$
   $a_7 = (7+4)(7+1) = (11)(8) = 88$
   The seventh term is 88.

39. $a_n = \dfrac{(-1)^{2n}}{n+4}$

   $a_{16} = \dfrac{(-1)^{32}}{16 + 4} = \dfrac{1}{20}$

   The sixteenth term is $\dfrac{1}{20}$.

41. $a_n = \dfrac{1}{3}n + n^2$

   $a_6 = \dfrac{1}{3}(6) + 6^2 = 2 + 36 = 38$

   The sixth term is 38.

43. False. If $b = -2$, $a_4 = -2(4) = -8$, which is less than b, not greater than it.

Objective 11.1.2 Exercises

45. A series is the sum of the terms of a sequence.

47. 1; 2; 3; 4; 4; 7; 10; 13; 34

49. $\displaystyle\sum_{n=1}^{7}(i+2) = (1+2)+(2+2)+(3+2)+(4+2)+(5+2)+(6+2)+(7+2)$
    $= 3+4+5+6+7+8+9$
    $= 42$

51. $\displaystyle\sum_{n=1}^{7} n = 1+2+3+4+5+6+7 = 28$

53. $\sum_{i=1}^{5}(i^2+1) = (1^2+1)+(2^2+1)+(3^2+1)+(4^2+1)+(5^2+1)$

$\quad\quad = 2+5+10+17+26$

$\quad\quad = 60$

55. $\sum_{n=1}^{4}\frac{1}{2n} = \frac{1}{2(1)}+\frac{1}{2(2)}+\frac{1}{2(3)}+\frac{1}{2(4)} = \frac{1}{2}+\frac{1}{4}+\frac{1}{6}+\frac{1}{8} = \frac{12+6+4+3}{24} = \frac{25}{24}$

57. $\sum_{n=2}^{4} 2^n = 2^2+2^3+2^4 = 4+8+16 = 28$

59. $\sum_{i=3}^{6}\frac{i+1}{i} = \frac{3+1}{3}+\frac{4+1}{4}+\frac{5+1}{5}+\frac{6+1}{6} = \frac{4}{3}+\frac{5}{4}+\frac{6}{5}+\frac{7}{6} = \frac{80+75+72+70}{60} = \frac{297}{60} = \frac{99}{20}$

61. $\sum_{i=1}^{5}\frac{1}{2i} = \frac{1}{2(1)}+\frac{1}{2(2)}+\frac{1}{2(3)}+\frac{1}{2(4)}+\frac{1}{2(5)} = \frac{1}{2}+\frac{1}{4}+\frac{1}{6}+\frac{1}{8}+\frac{1}{10} = \frac{60+30+20+15+12}{120} = \frac{137}{120}$

63. $\sum_{i=1}^{4}(-1)^{i-1}(i+1) = (-1)^0(1+1)+(-1)^1(2+1)+(-1)^2(3+1)+(-1)^3(4+1)$

$\quad\quad = 2+(-3)+4+(-5) = -2$

65. $\sum_{n=4}^{7}\frac{(-1)^{n-1}}{n-3} = \frac{(-1)^3}{4-3}+\frac{(-1)^4}{5-3}+\frac{(-1)^5}{6-3}+\frac{(-1)^6}{7-3} = -1+\frac{1}{2}-\frac{1}{3}+\frac{1}{4} = \frac{-12+6-4+3}{12} = -\frac{7}{12}$

67. $\sum_{n=1}^{4}\frac{2n}{x} = \frac{2(1)}{x}+\frac{2(2)}{x}+\frac{2(3)}{x}+\frac{2(4)}{x} = \frac{2}{x}+\frac{4}{x}+\frac{6}{x}+\frac{8}{x}$

69. $\sum_{i=1}^{4}\frac{x^i}{i+1} = \frac{x^1}{1+1}+\frac{x^2}{2+1}+\frac{x^3}{3+1}+\frac{x^4}{4+1} = \frac{x}{2}+\frac{x^2}{3}+\frac{x^3}{4}+\frac{x^4}{5}$

71. $\sum_{i=2}^{4}\frac{x^i}{2i-1} = \frac{x^2}{2(2)-1}+\frac{x^3}{2(3)-1}+\frac{x^4}{2(4)-1} = \frac{x^2}{3}+\frac{x^3}{5}+\frac{x^4}{7}$

73. $\sum_{n=1}^{4} x^{2n-1} = x^{2(1)-1}+x^{2(2)-1}+x^{2(3)-1}+x^{2(4)-1} = x+x^3+x^5+x^7$

75. c

Applying Concepts 11.1

77. The sequence of the odd natural numbers is expressed by the formula $a_n = 2n-1$.

79. The sequence of the negative odd integers is expressed by the formula $a_n = -2n+1$.

81. The sequence of the positive integers that are divisible by 4 is expressed by the formula $a_n = 4n$.

83. $\sum_{i=1}^{4} \log 2i$

$= \log[2(1)] + \log[2(2)] + \log[2(3)] + \log[2(4)]$

$= \log 2 + \log 4 + \log 6 + \log 8$

$= \log(2 \cdot 4 \cdot 6 \cdot 8)$

$= \log 384$

85. Strategy
Multiply 6 by the number of 6s in the ones, tens, and hundreds places. Then multiply each product by the place value. Add the products and find the hundreds digit.

Solution
$31 \times 6 = 186 \quad\quad 186(1) = 186$
$30 \times 6 = 180 \quad\quad 180(10) = 1,800$
$29 \times 6 = 174 \quad\quad 174(100) = \underline{17,400}$
$\quad\quad\quad\quad\quad\quad\quad\quad\quad = 19,386$

The hundreds digit is 3.

87. $a_1 = 1, a_2 = 1, a_n = a_{n-1} + a_{n-2}, n \geq 3$
$a_1 = 1$
$a_2 = 1$
$a_3 = a_{3-1} + a_{3-2} = a_2 + a_1 = 1 + 1 = 2$
$a_4 = a_{4-1} + a_{4-2} = a_3 + a_2 = 2 + 1 = 3$
$a_5 = a_{5-1} + a_{5-2} = a_4 + a_3 = 3 + 2 = 5$
The first four terms are 1, 2, 3, and 5.

89. $\dfrac{1}{1} + \dfrac{1}{2} + \dfrac{1}{3} + \ldots + \dfrac{1}{n} = \sum_{i=1}^{n} \dfrac{1}{i}$

Section 11.2
Concept Review 11.2

1. Sometimes true
The successive terms s of the series $\sum_{i=1}^{n}(3-i)$ decreases in value.

3. Sometimes true
The first term of –4, –2, 0, 2, … is a negative number.

Objective 11.2.1 Exercises

1. Only in an arithmetic sequence is the difference between any two consecutive terms constant.

3.
   a. 4; common
   b. nth; 50; 3; 4; 199

5. $d = a_2 - a_1 = 11 - 1 = 10$
$a_n = a_1 + (n-1)d$
$a_{15} = 1 + (15-1)(10) = 1 + 14(10) = 1 + 140$
$a_{15} = 141$

7. $d = a_2 - a_1 = -2 - (-6) = 4$
$a_n = a_1 + (n-1)d$
$a_{15} = -6 + (15-1)4 = -6 + (14)4 = -6 + 56$
$a_{15} = 50$

9. $d = a_2 - a_1 = 7 - 3 = 4$
$a_n = a_1 + (n-1)d$
$a_{18} = 3 + (18-1)4 = 3 + 17(4) = 3 + 68$
$a_{18} = 71$

11. $d = a_2 - a_1 = 0 - \left(-\dfrac{3}{4}\right) = \dfrac{3}{4}$
$a_n = a_1 + (n-1)d$
$a_{11} = -\dfrac{3}{4} + (11-1)\dfrac{3}{4}$
$= -\dfrac{3}{4} + 10\left(\dfrac{3}{4}\right)$
$= -\dfrac{3}{4} + \dfrac{30}{4}$
$a_{13} = \dfrac{27}{4}$

13. $d = a_2 - a_1 = \dfrac{5}{2} - 2 = \dfrac{1}{2}$
$a_n = a_1 + (n-1)d$
$a_{31} = 2 + (31-1)\dfrac{1}{2} = 2 + 30\left(\dfrac{1}{2}\right) = 2 + 15$
$a_{31} = 17$

15. $d = a_2 - a_1 = 5.75 - 6 = -0.25$
$a_n = a_1 + (n-1)d$
$a_{10} = 6 + (10-1)(-0.25) = 6 + 9(-0.25) = 6 - 2.25$
$a_{10} = 3.75$

17. $d = a_2 - a_1 = 2 - 1 = 1$
$a_n = a_1 + (n-1)d$
$a_n = 1 + (n-1)1$
$a_n = 1 + n - 1$
$a_n = n$

19. $d = a_2 - a_1 = 2 - 6 = -4$
$a_n = a_1 + (n-1)d$
$a_n = 6 + (n-1)(-4)$
$a_n = 6 - 4n + 4$
$a_n = -4n + 10$

21. $d = a_2 - a_1 = \dfrac{7}{2} - 2 = \dfrac{3}{2}$
$a_n = a_1 + (n-1)d$
$a_n = 2 + (n-1)\dfrac{3}{2}$
$a_n = 2 + \dfrac{3}{2}n - \dfrac{3}{2}$
$a_n = \dfrac{3}{2}n + \dfrac{1}{2}$
$a_n = \dfrac{3n+1}{2}$

23. $d = a_2 - a_1 = -13 - (-8) = -5$
$a_n = a_1 + (n-1)d$
$a_n = -8 + (n-1)(-5)$
$a_n = -8 - 5n + 5$
$a_n = -5n - 3$

25. $d = a_2 - a_1 = 16 - 26 = -10$
$a_n = a_1 + (n-1)d$
$a_n = 26 + (n-1)(-10)$
$a_n = 26 - 10n + 10$
$a_n = -10n + 36$

27. $d = a_2 - a_1 = 11 - 7 = 4$
$a_n = a_1 + (n-1)d$
$171 = 7 + (n-1)4$
$171 = 7 + 4n - 4$
$171 = 3 + 4n$
$168 = 4n$
$n = 42$
There are 42 terms in the sequence

29. $d = a_2 - a_1 = \frac{5}{3} - \frac{1}{3} = \frac{4}{3}$
$a_n = a_1 + (n-1)d$
$\frac{61}{3} = \frac{1}{3} + (n-1)\frac{4}{3}$
$\frac{61}{3} = \frac{1}{3} + \frac{4}{3}n - \frac{4}{3}$
$\frac{61}{3} = -1 + \frac{4}{3}n$
$\frac{64}{3} = \frac{4}{3}n$
$16 = n$
There are 16 terms in the sequence.

31. $d = a_2 - a_1 = 8 - 3 = 5$
$a_n = a_1 + (n-1)d$
$98 = 3 + (n-1)5$
$98 = 3 + 5n - 5$
$98 = 5n - 2$
$100 = 5n$
$20 = n$
There are 20 terms in the sequence.

33. $d = a_2 - a_1 = -3 - 1 = -4$
$a_n = a_1 + (n-1)d$
$-75 = 1 + (n-1)(-4)$
$-75 = 1 - 4n + 4$
$-75 = 5 - 4n$
$-80 = -4n$
$20 = n$
There are 20 terms in the sequence.

35. $d = a_2 - a_1 = \frac{13}{3} - \frac{7}{3} = 2$
$a_n = a_1 + (n-1)d$
$\frac{79}{3} = \frac{7}{3} + (n-1)2$
$\frac{79}{3} = \frac{7}{3} + 2n - 2$
$\frac{79}{3} = 2n + \frac{1}{3}$
$26 = 2n$
$13 = n$
There are 13 terms in the sequence.

37. $d = a_2 - a_1 = 2 - 3.5 = -1.5$
$a_n = a_1 + (n-1)d$
$-25 = 3.5 + (n-1)(-1.5)$
$-25 = 3.5 - 1.5n + 1.5$
$-25 = 5 - 1.5n$
$-30 = -1.5n$
$20 = n$
There are 20 terms in the sequence.

39. True

41. True

Objective 11.2.2 Exercises

43. 10; $(n-3)$; 1; -2; 10; 7; Arithmetic Series; 10; 10; -2; 7; 5; 5; 25

45. $d = a_2 - a_1 = 4 - 2 = 2$
$a_n = a_1 + (n-1)d$
$a_{25} = 2 + (25-1)2 = 2 + 24(2) = 50$
$S_n = \frac{n}{2}(a_1 + a_2)$
$S_{25} = \frac{25}{2}(2 + 50) = \frac{25}{2}(52) = 650$

47. $d = a_2 - a_1 = 20 - 25 = -5$
$a_n = a_1 + (n-1)d$
$a_{22} = 25 + (22-1)(-5)$
$= 25 + 21(-5)$
$= 25 - 105$
$= -80$
$S_n = \frac{n}{2}(a_1 + a_n)$
$S_{22} = \frac{22}{2}[25 + (-80)] = \frac{22}{2}(-55) = -605$

49. $d = a_2 - a_1 = \dfrac{11}{4} - 2 = \dfrac{3}{4}$

$a_n = a_1 + (n-1)d$

$a_{10} = 2 + (10-1)\dfrac{3}{4} = 2 + 9\left(\dfrac{3}{4}\right) = 2 + \dfrac{27}{4} = \dfrac{35}{4}$

$S_{10} = \dfrac{n}{2}(a_1 + a_n)$

$S_{10} = \dfrac{10}{2}\left(2 + \dfrac{35}{4}\right) = \dfrac{10}{2}\left(\dfrac{43}{4}\right) = \dfrac{215}{4}$

51. $a_i = 3i + 4$

$a_1 = 3(1) + 4 = 7$

$a_{15} = 3(15) + 4 = 49$

$S_i = \dfrac{i}{2}(a_1 + a_i)$

$S_{15} = \dfrac{15}{2}(7 + 49) = \dfrac{15}{2}(56) = 420$

53. $a_n = 1 - 4n$

$a_1 = 1 - 4(1) = -3$

$a_{10} = 1 - 4(10) = -39$

$S_n = \dfrac{n}{2}(a_1 + a_n)$

$S_{10} = \dfrac{10}{2}(-3 - 39)$

$= \dfrac{10}{2}(-42)$

$= -210$

55. $a_n = 5 - n$

$a_1 = 5 - 1 = 4$

$a_{10} = 5 - 10 = -5$

$S_n = \dfrac{n}{2}(a_1 + a_n)$

$S_{10} = \dfrac{10}{2}(4 - 5) = \dfrac{10}{2}(-1) = -5$

57. The sum will be negative.

Objective 11.2.3 Exercises

59. 8; 8; 3; 17; 80

61. Strategy
To find the number of weeks:
Write the arithmetic sequence.
Find the common difference of the arithmetic sequence.
Use the Formula for the nth Term of an Arithmetic Sequence to find the number of terms in the sequence.

Solution
12, 18, 24, ... 60
$d = a_2 - a_1 = 18 - 12 = 6$
$a_n = a_1 + (n-1)d$
$60 = 12 + (n-1)6$
$60 = 12 + 6n - 6$
$60 = 6 + 6n$
$54 = 6n$
$9 = n$

In 9 weeks the person will walk 60 min per day.

63. Strategy
To find the total number of seats:
Write the arithmetic sequence.
Find the common difference of the arithmetic sequence.
Use the Formula for the nth Term of an Arithmetic Sequence to find the sum of the sequence.
Use the Formula for the Sum of n Terms of an Arithmetic Sequence to find the sum of the sequence.

Solution
52, 58, 64, ...
$d = a_2 - a_1 = 58 - 52 = 6$
$a_n = a_1 + (n-1)d$
$a_{20} = 52 + (20-1)6 = 52 + 19(6) = 52 + 114 = 166$
$S_n = \dfrac{n}{2}(a_1 + a_n)$
$S_{20} = \dfrac{20}{2}(52 + 166) = 10(218) = 2180$

There are 2180 seats in the theater.

65. Strategy
To find the salary for the ninth month, use the Formula for the nth Term of an Arithmetic Sequence.
To find the total salary, use the Formula for the Sum of n Terms of an Arithmetic Sequence to find the sum of the sequence.

Solution
$a_n = a_1 + (n-1)d$
$a_{10} = 2200 + (10-1)150$
$= 2200 + (9)150$
$= 2200 + 1350$
$= 3550$

$S_n = \dfrac{n}{2}(a_1 + a_n)$
$S_{10} = \dfrac{10}{2}(2200 + 3550) = \dfrac{10}{2}(5750) = 28{,}750$

The salary for the tenth month is $3550.
The total salary for the ten-month period is $28,750.

67. Exercise 60

Applying Concepts 11.2

69. $d = a_2 - a_1 = 2 - (-3) = 5$

$a_n = a_1 + (n-1)d$
$= -3 + (n-1)5$
$= -3 + 5n - 5$
$= 5n - 8$

$S_n = \dfrac{n}{2}(a_1 + a_n)$

$116 = \dfrac{n}{2}(-3 + 5n - 8)$

$116 = \dfrac{n}{2}(5n - 11)$

$232 = n(5n - 11)$

$232 = 5n^2 - 11n$

$0 = 5n^2 - 11n - 232$

$0 = (5n + 29)(n - 8)$

$5n + 29 = 0 \qquad n - 8 = 0$

$5n = -29 \qquad n = 8$

$n = -\dfrac{29}{5} \qquad n = 8$

The number of terms must be a natural number, so $-\dfrac{29}{5}$ is not a solution.

8 terms must be added together.

71. Find d in the series in which $a_1 = 9$ and $a_6 = 29$.

$a_n = a_1 + (n-1)d$
$29 = 9 + (6-1)d$
$29 = 9 + 5d$
$20 = 5d$
$4 = d$

d is the same for the series in which $a_4 = 9$ and $a_9 = 29$.

$a_n = a_1 + (n-1)d$
$29 = a_1 + (9-1)4$
$29 = a_1 + 8(4)$
$29 = a_1 + 32$
$-3 = a_1$

The first term is -3.

73. Strategy
To find the sum of the angles:
Write the arithmetic sequence with third term 180°, fourth term 360°, and fifth term 540°.
Find the common difference of the arithmetic sequence.
Find the first term of the arithmetic sequence using the Formula for the nth Term of an Arithmetic Sequence.
Use the Formula for the nth Term of an Arithmetic Sequence to find the 12th term.

Solution
$a_1, a_2, 180, 360, 540, \ldots$

$d = a_4 - a_3 = 360 - 180 = 180$

$a_n = a_1 + (n-1)d$
$a_3 = a_1 + (3-1)180$
$180 = a_1 + 2(180)$
$180 = a_1 + 360$
$-180 = a_1$

$a_n = a_1 + (n-1)d$
$a_{12} = -180 + (12-1)180$
$= -180 + 11(180)$
$= 1800$

The sum of the angles in a dodecagon is 1800°.
The general term is
$a_n = -180 + (n-1)180$
$= -180 + 180n - 180$
$= 180n - 360$
$= 180(n - 2)$

The formula for the sum of the angles of an n-sided polygon is $180(n-2)$.

75. Strategy
To find a formula for the sequence of distances:
Find the first term of the sequence by finding the circumference of a circle whose radius is equal to the first term of the sequence in Exercise 60, and adding this to twice the length of the straight part of the track, 83.4.
The common difference of the sequence is the circumference of a circle whose radius is equal to the width of a lane, 1.22.
Use the Formula for the nth Term of an Arithmetic Sequence.

Solution
$a_1 = 2\pi(37.11) + 2(83.4)$
$a_1 = 399.85$

$d = 2\pi(1.22)$
$d = 7.66$

$a_n = a_1 + (n-1)d$
$a_n = 399.85 + (n-1)(7.66)$
$a_n = 7.66n + 392.19$

A formula for the sequence of distances around the track for each lane is $a_n = 7.66n + 392.19$. The farther a lane is from the innermost lane of the track, the longer the distance around the track. The starting positions must be staggered in order to ensure that each runner is running the same distance from the starting position to the finish line.

77. The first seven terms of the Fibonacci sequence are 1, 1, 2, 3, 5, 8, 13. In general, $a_1 = 1$, $a_2 = 1$, and $a_n = a_{n-2} + a_{n-1}$, $n \geq 3$. This sequence was discussed by Fibonacci in his book *Liver Abaci*, which was written around A.D. 1200. The sequence is the solution to a problem in the book that is now called the rabbit problem. If one pair of the rabbits can produce a pair of rabbits each month, and in the second month after birth the new pair can give birth to a pair of rabbits, which in turn can produce a pair of rabbits that breed with the same pattern, how many pairs of rabbits will there be in 12 months? Fibonacci sequences appear often. Daisies generally contain 21, 34, or 55 petals; from C to C on a piano keyboard there are 13 keys, 5 black and 8 white; the numbers of scales on a pinecone and on a pineapple are also terms of a Fibonacci sequence. The Fibonacci sequence is also related to the golden ratio, $\lim_{n \to \infty} \frac{F_{n+1}}{F_n} = \frac{\sqrt{5}+1}{2}$.

Section 11.3
Concept Review 11.3
1. Always true

3. Sometimes true
   If r is a negative number, the sum of a geometric series will oscillate.

5. Never true
   $\frac{a_{n+1}}{a_n}$ is not a constant ratio.

Objective 11.3.1 Exercises
1. An arithmetic sequence is one in which the *difference* between any two consecutive terms is a constant. A geometric sequence is one in which each successive term of the sequence is the same *nonzero constant multiple* of the preceding term.

3.
   a. 3; common ratio
   b. nth; 7; 2; 3; 7; 729; 1428

5. $r = \frac{a_2}{a_1} = \frac{8}{2} = 4$
   $a_n = a_1 r^{n-1}$
   $a_9 = 2(4)^{9-1} = 2(4)^8 = 2(65,536) = 131,072$

7. $r = \frac{a_2}{a_1} = \frac{-4}{6} = -\frac{2}{3}$
   $a_n = a_1 r^{n-1}$
   $a_7 = 6\left(-\frac{2}{3}\right)^{7-1} = 6\left(-\frac{2}{3}\right)^6 = 6\left(\frac{64}{729}\right) = \frac{128}{243}$

9. $r = \frac{a_2}{a_1} = \frac{\sqrt{2}}{1} = \sqrt{2}$
   $a_n = a_1 r^{n-1}$
   $a_9 = 1(\sqrt{2})^{9-1} = 1(\sqrt{2})^8 = 1(2^4) = 16$

   $a_n = a_1 r^{n-1}$
   $a_4 = 9r^{4-1}$
   $\frac{8}{3} = 9r^{4-1}$
   $\frac{8}{3} = 9r^3$
   $\frac{8}{27} = r^3$
   $\frac{2}{3} = r$

11. $a_n = a_1 r^{n-1}$
    $a_2 = 9\left(\frac{2}{3}\right)^{2-1} = 9\left(\frac{2}{3}\right) = 6$
    $a_3 = 9\left(\frac{2}{3}\right)^{3-1} = 9\left(\frac{2}{3}\right)^2 = 9\left(\frac{4}{9}\right) = 4$

    $a_n = a_1 r^{n-1}$
    $a_4 = 3r^{4-1}$
    $-\frac{8}{9} = 3r^{4-1}$
    $-\frac{8}{9} = 3r^3$
    $-\frac{8}{27} = r^3$
    $-\frac{2}{3} = r$

13. $a_n = a_1 r^{n-1}$
    $a_2 = 3\left(-\frac{2}{3}\right)^{2-1} = 3\left(-\frac{2}{3}\right) = -2$
    $a_3 = 3\left(-\frac{2}{3}\right)^{3-1} = 3\left(-\frac{2}{3}\right)^2 = 3\left(\frac{4}{9}\right) = \frac{4}{3}$

15. $a_n = a_1 r^{n-1}$
    $a_4 = (-3)r^{4-1}$
    $192 = (-3)r^{4-1}$
    $192 = (-3)r^3$
    $-64 = r^3$
    $-4 = r$
    $a_n = a_1 r^{n-1}$
    $a_2 = -3(-4)^{2-1} = -3(-4) = 12$
    $a_3 = -3(-4)^{3-1} = -3(-4)^2 = -3(16) = -48$

17. False. If $r < 0$, the terms of the sequence will alternate between negative and positive.

Chapter 11: Sequences And Series

Objective 11.3.2 Exercises

19. sum; number; first; common ratio

21. $r = \dfrac{a_2}{a_1} = \dfrac{6}{2} = 3$

$S_n = \dfrac{a_1(1-r^n)}{1-r}$

$S_7 = \dfrac{2(1-3^7)}{1-3}$

$= \dfrac{2(1-2187)}{-2}$

$= \dfrac{2(-2186)}{-2}$

$= 2186$

23. $r = \dfrac{a_2}{a_1} = \dfrac{9}{12} = \dfrac{3}{4}$

$S_n = \dfrac{a_1(1-r^n)}{1-r}$

$S_5 = \dfrac{12\left[1-\left(\frac{3}{4}\right)^5\right]}{1-\frac{3}{4}}$

$= \dfrac{12\left(1-\frac{243}{1024}\right)}{\frac{1}{4}}$

$= 48\left(\dfrac{781}{1024}\right)$

$= \dfrac{2343}{64}$

25. $a_i = (2)^i$

$a_1 = (2)^1 = 2$

$a_2 = (2)^2 = 4$

$r = \dfrac{a_2}{a_1} = \dfrac{4}{2} = 2$

$S_i = \dfrac{a_1(1-r^i)}{1-r}$

$S_5 = \dfrac{2(1-2^5)}{1-2} = \dfrac{2(1-32)}{-1} = -2(-31) = 62$

27. $a_i = \left(\dfrac{1}{3}\right)^i$

$a_1 = \left(\dfrac{1}{3}\right)^1 = \dfrac{1}{3}$

$a_2 = \left(\dfrac{1}{3}\right)^2 = \dfrac{1}{9}$

$r = \dfrac{a_2}{a_1} = \dfrac{1}{9} \div \dfrac{1}{3} = \dfrac{1}{9} \cdot \dfrac{3}{1} = \dfrac{1}{3}$

$S_i = \dfrac{a_1(1-r^i)}{1-r}$

$S_5 = \dfrac{\frac{1}{3}\left[1-\left(\frac{1}{3}\right)^5\right]}{1-\frac{1}{3}} = \dfrac{\frac{1}{3}\left(1-\frac{1}{243}\right)}{\frac{2}{3}} = \dfrac{1}{2}\left(\dfrac{242}{243}\right) = \dfrac{121}{243}$

29. $a_i = (4)^i$

$a_1 = 4^1 = 4$

$a_2 = 4^2 = 16$

$r = \dfrac{a_2}{a_1} = \dfrac{16}{4} = 4$

$S_i = \dfrac{a_1(1-r^i)}{1-r}$

$S_5 = \dfrac{4(1-4^5)}{1-4}$

$= \dfrac{4(1-1024)}{1-4}$

$= \dfrac{-4(1023)}{-3}$

$= 1364$

31. $a_i = (7)^i$

$a_1 = 7^1 = 7$

$a_2 = 7^2 = 49$

$r = \dfrac{a_2}{a_1} = \dfrac{49}{7} = 7$

$S_i = \dfrac{a_1(1-r^i)}{1-r}$

$S_4 = \dfrac{7(1-7^4)}{1-7}$

$= \dfrac{7(1-2401)}{-6}$

$= \dfrac{-7(2400)}{-6}$

$= 2800$

33. $a_i = \left(\dfrac{3}{4}\right)^i$

$a_1 = \left(\dfrac{3}{4}\right)^1 = \dfrac{3}{4}$

$a_2 = \left(\dfrac{3}{4}\right)^2 = \dfrac{9}{16}$

$r = \dfrac{a_2}{a_1} = \dfrac{9}{16} \div \dfrac{3}{4} = \dfrac{9}{16} \cdot \dfrac{4}{3} = \dfrac{3}{4}$

$S_i = \dfrac{a_1(1-r^i)}{1-r}$

$S_5 = \dfrac{\frac{3}{4}\left[1-\left(\frac{3}{4}\right)^5\right]}{1-\frac{3}{4}}$

$= \dfrac{\frac{3}{4}\left(1-\frac{243}{1024}\right)}{1-\frac{3}{4}}$

$= \dfrac{\frac{3}{4}\left(\frac{781}{1024}\right)}{\frac{1}{4}}$

$= 3\left(\dfrac{781}{1024}\right)$

$= \dfrac{2343}{1024}$

35. $a_i = \left(\dfrac{5}{3}\right)^i$

$a_1 = \left(\dfrac{5}{3}\right)^1 = \dfrac{5}{3}$

$a_2 = \left(\dfrac{5}{3}\right)^2 = \dfrac{25}{9}$

$r = \dfrac{a_2}{a_1} = \dfrac{25}{9} \div \dfrac{5}{3} = \dfrac{25}{9} \cdot \dfrac{3}{5} = \dfrac{5}{3}$

$S_i = \dfrac{a_1(1-r^i)}{1-r}$

$S_4 = \dfrac{\frac{5}{3}\left[1 - \left(\frac{5}{3}\right)^4\right]}{1 - \frac{5}{3}}$

$= \dfrac{\frac{5}{3}\left(1 - \frac{625}{81}\right)}{-\frac{2}{3}}$

$= \dfrac{-\frac{5}{3}\left(\frac{544}{81}\right)}{-\frac{2}{3}}$

$= \dfrac{5}{2}\left(\dfrac{544}{81}\right)$

$= \dfrac{1360}{81}$

37.
  a. No, because the series does not have a common ratio.
  b. Yes, because the series has a common ratio.

Objective 11.3.3 Exercises

39. -1; 1; 1

41. $r = \dfrac{a_2}{a_1} = \dfrac{2}{3}$

$S = \dfrac{a_1}{1-r} = \dfrac{3}{1-\frac{2}{3}} = \dfrac{3}{\frac{1}{3}} = 9$

43. $r = \dfrac{a_2}{a_1} = \dfrac{-4}{6} = -\dfrac{2}{3}$

$S = \dfrac{a_1}{1-r} = \dfrac{6}{1-\left(-\frac{2}{3}\right)} = \dfrac{6}{\frac{5}{3}} = \dfrac{18}{5}$

45. $r = \dfrac{a_2}{a_1} = \dfrac{\frac{7}{100}}{\frac{7}{10}} = \dfrac{1}{10}$

$S = \dfrac{a_1}{1-r} = \dfrac{\frac{7}{10}}{1-\frac{1}{10}} = \dfrac{\frac{7}{10}}{\frac{9}{10}} = \dfrac{7}{9}$

47.
  a. Yes, because $|r| < 1$.
  b. No, because $|r| > 1$.

49. $0.8\overline{8} = 0.8 + 0.08 + 0.008 + \ldots$

$= \dfrac{8}{10} + \dfrac{8}{100} + \dfrac{8}{1000} + \ldots$

$S = \dfrac{a_1}{1-r} = \dfrac{\frac{8}{10}}{1-\frac{1}{10}} = \dfrac{\frac{8}{10}}{\frac{9}{10}} = \dfrac{8}{9}$

An equivalent fraction is $\dfrac{8}{9}$.

51. $0.2\overline{2} = 0.2 + 0.02 + 0.002 + \ldots$

$= \dfrac{2}{10} + \dfrac{2}{100} + \dfrac{2}{1000} + \ldots$

$S = \dfrac{a_1}{1-r} = \dfrac{\frac{2}{10}}{1-\frac{1}{10}} = \dfrac{\frac{2}{10}}{\frac{9}{10}} = \dfrac{2}{9}$

An equivalent fraction $\dfrac{2}{9}$.

53. $0.45\overline{45} = 0.45 + 0.0045 + 0.000045 + \ldots$

$= \dfrac{45}{100} + \dfrac{45}{10,000} + \dfrac{45}{1,000,000} + \ldots$

$S = \dfrac{a_1}{1-r} = \dfrac{\frac{45}{100}}{1-\frac{1}{100}} = \dfrac{\frac{45}{100}}{\frac{99}{100}} = \dfrac{45}{99} = \dfrac{5}{11}$

An equivalent fraction is $\dfrac{5}{11}$.

55. $0.16\overline{6} = 0.1 + 0.06 + 0.006 + 0.0006 + \ldots$

$= \dfrac{1}{10} + \dfrac{6}{100} + \dfrac{6}{1000} + \dfrac{6}{10,000} + \ldots$

$S = \dfrac{a_1}{1-r} = \dfrac{\frac{6}{100}}{1-\frac{1}{10}} = \dfrac{\frac{6}{100}}{\frac{9}{10}} = \dfrac{6}{90} = \dfrac{1}{15}$

$0.16\overline{6} = \dfrac{1}{10} + \dfrac{1}{15} = \dfrac{5}{30} = \dfrac{1}{6}$

An equivalent fraction is $\dfrac{1}{6}$.

57. 700; 1400; geometric; 350; 2

Objective 11.3.4 Exercises

59. Strategy
Use the Formula for the nth Term of a Geometric Sequence.

Solution
$a_1 = 1, r = 3$
$a_n = a_1 r^{n-1}$
$a_n = 1 \cdot 3^{n-1}$
$a_n = 3^{n-1}$

Chapter 11: Sequences And Series

61. Strategy
To find the amount of radioactive material at the beginning of the seventh day, use the Formula for the nth Term of a Geometric Sequence.

Solution
$n = 7, a_1 = 500, r = \dfrac{1}{2}$

$a_n = a_1 r^{n-1}$

$a_7 = 500\left(\dfrac{1}{2}\right)^{7-1} = 500\left(\dfrac{1}{2}\right)^6 = 500\left(\dfrac{1}{64}\right) = 7.8125$

There will be 7.8125 mg of radioactive material in the sample at the beginning of the seventh day.

63. Strategy
To find the height of the ball on the fifth bounce, use the Formula for the nth Term of a Geometric Sequence. Let a_1 be the height of the ball after the first bounce.

Solution
$n = 5, a_1 = 80\% \text{ of } 8 = 6.4, r = 80\% = \dfrac{4}{5}$

$a_n = a_1 r^{n-1}$

$a_5 = 6.4\left(\dfrac{4}{5}\right)^{5-1} = 6.4\left(\dfrac{4}{5}\right)^4 = 6.4\left(\dfrac{256}{625}\right) \approx 2.6$

The ball bounces to a height of 2.6 ft on the fifth bounce.

65. Strategy
To find the value of the land in 15 years, use the Formula for the nth Term of a Geometric Sequence. Let a_1 be the value of the land after 1 year.

Solution
$n = 15, a_1 = 1.12(15,000) = 16,800,$
$r = 112\% = 1.12$

$a_n = a_1 r^{n-1}$

$a_{15} = 16,800(1.12)^{15-1}$

$\phantom{a_{15}} = 16,800(4.8871123)$

$\phantom{a_{15}} \approx 82,103.49$

The value of the land in 15 years will be $82,103.49.

67. Strategy
To find the value of the house in 30 years, use the Formula for the nth Term of a Geometric Sequence.

Solution
$n = 30, a_1 = 1.05(100,000) = 105,000,$
$r = 105\% = 1.05$

$a_n = a_1 r^{n-1}$

$a_{30} = 105,000(1.05)^{30-1}$

$\phantom{a_{30}} = (105,000)(4.1161356)$

$\phantom{a_{30}} = 432,194.24$

The value of the house in 30 years will be $432,194.24.

69. $4, -2, 1, \ldots$

$r = -\dfrac{1}{2}$

$a_4 = a_3 r = 1\left(-\dfrac{1}{2}\right) = -\dfrac{1}{2}$

The sequence is geometric. (G)

The next term is $-\dfrac{1}{2}$.

71. $5, 6.5, 8, \ldots$

$d = 1.5$

$a_4 = a_3 + d = 8 + 1.5 = 9.5$

The sequence is arithmetic. (A)
The next term is 9.5.

73. $1, 4, 9, 16, \ldots$

$a_n = n^2$

$a_5 = 5^2 = 25$

The sequence is neither arithmetic nor geometric. (N) The next term is 25.

75. $x^8, x^6, x^4, \ldots$

$r = x^{-2}$

$a_4 = a_3 r = x^4(x^{-2}) = x^2$

The sequence is geometric. (G)
The next term is x^2.

77. $\log x, 2\log x, 3\log x, \ldots$

$d = \log x$

$a_4 = a_3 + d = 3\log x + \log x = 4\log x$

The sequence is arithmetic. (A)
The next term is $4\log x$.

Applying Concepts 11.3

79. Find r for the geometric series in which $a_1 = 3$ and $a_4 = \frac{1}{9}$.

$$a_n = a_1 r^{n-1}$$
$$\frac{1}{9} = 3r^{4-1}$$
$$\frac{1}{9} = 3r^3$$
$$\frac{1}{27} = r^3$$
$$\sqrt[3]{\frac{1}{27}} = \sqrt[3]{r^3}$$
$$\frac{1}{3} = r$$

r is the same for the geometric series in which $a_3 = 3$ and $a_6 = \frac{1}{9}$.

$$a_n = a_1 r^{n-1}$$
$$\frac{1}{9} = a_1 \left(\frac{1}{3}\right)^{6-1}$$
$$\frac{1}{9} = a_1 \left(\frac{1}{3}\right)^5$$
$$\frac{1}{9} = a_1 \left(\frac{1}{243}\right)$$
$$27 = a_1$$

The first term is 27.

81. $a_n = 2^n$
$a_1 = 2^1 = 1$
$a_2 = 2^2 = 4$
$a_3 = 2^3 = 8$
$\vdots$
$a_n = 2^n$
$a_{n+1} = 2^{n+1}$

$b_n = \log a_n$
$b_1 = \log a_1 = \log 2$
$b_2 = \log a_2 = \log 4 = \log 2^2 = 2 \log 2$
$b_3 = \log a_3 = \log 8 = \log 2^3 = 3 \log 2$
$\vdots$
$b_n = \log a_n = \log 2^n = n \log 2$
$b_{n+1} = \log a_{n+1} = \log 2^{n+1} = (n+1) \log 2$
$b_{n+1} - b_n = (n+1) \log 2 - n \log 2$
$\qquad = n \log 2 + \log 2 - n \log 2 = \log 2$

The common difference is $\log 2$.

83. $a_n = 3n - 2$
$a_1 = 3(1) - 2 = 1$
$a_2 = 3(2) - 2 = 4$
$a_3 = 3(3) - 2 = 7$
$\vdots$
$a_n = 3n - 2$
$a_{n+1} = 3(n+1) - 2 = 3n + 3 - 2 = 3n + 1$
$b_n = 2^{a_n}$
$b_1 = 2^{a_1} = 2^1 = 2$
$b_2 = 2^{a_2} = 2^4 = 16$
$b_3 = 2^{a_3} = 2^7 = 128$
$\vdots$
$b_n = 2^{a_n} = 2^{3n-2}$
$b_{n+1} = 2^{a_{n+1}} = 2^{3n+1}$
$$\frac{b^{n+1}}{b^n} = \frac{2^{3n+1}}{2^{3n-2}} = 2^3 = 8$$

The common ratio is 8.

85.

a. $R_n = R_1(1.0075)^{n-1}$
$R_{27} = 66.29(1.0075)^{27-1} = 66.29(1.0075)^{26}$
≈ 80.50
$80.50 of the loan is repaid in the twenty-seventh payment.

b. $T = \sum_{k=1}^{n} R_1(1.0075)^{k-1}$
$T = \sum_{k=1}^{20} 66.29(1.0075)^{k-1}$
$T = 66.29(1.0075)^0 + 66.29(1.0075)^1 + 66.29(1.0075)^2 + \ldots + 66.29(1.0075)^{19}$
$a_1 = 66.29(1.0075)^0 = 66.25$
$S_{20} = \dfrac{66.29(1-1.0075^{20})}{1-1.0075} \approx 1424.65$
The total amount repaid after 20 payments is $1424.65.

c. $5000 - 1424.65 = 3575.35$
The unpaid amount repaid after 20 payments is $3575.35.

Section 11.4
Concept Review 11.4

1. Never true
 $0! \cdot 4! = 1 \cdot 24 = 24$

3. Never true
 The exponent on the fifth term is 4.

5. Never true
 There are $n + 1$ terms in the expansion of $(a+b)^n$.

Objective 11.4.1 Exercises

1. The factorial of a number is the product of all the natural numbers less than or equal to the number.

3. factorial

5. $n!$; $(n-r)!$

7. $3! = 3 \cdot 2 \cdot 1 = 6$

9. $8! = 8 \cdot 7 \cdot 6 \cdot 5 \cdot 4 \cdot 3 \cdot 2 \cdot 1 = 40{,}320$

11. $0! = 1$

13. $\dfrac{5!}{2!3!} = \dfrac{5 \cdot 4 \cdot 3 \cdot 2 \cdot 1}{(2 \cdot 1)(3 \cdot 2 \cdot 1)} = 10$

15. $\dfrac{6!}{6!0!} = \dfrac{6 \cdot 5 \cdot 4 \cdot 3 \cdot 2 \cdot 1}{(6 \cdot 5 \cdot 4 \cdot 3 \cdot 2 \cdot 1)(1)} = 1$

17. $\dfrac{9!}{6!3!} = \dfrac{9 \cdot 8 \cdot 7 \cdot 6 \cdot 5 \cdot 4 \cdot 3 \cdot 2 \cdot 1}{(6 \cdot 5 \cdot 4 \cdot 3 \cdot 2 \cdot 1)(3 \cdot 2 \cdot 1)} = 84$

19. $\binom{7}{2} = \dfrac{7!}{(7-2)!2!} = \dfrac{7!}{5!2!} = \dfrac{7 \cdot 6 \cdot 5 \cdot 4 \cdot 3 \cdot 2 \cdot 1}{(5 \cdot 4 \cdot 3 \cdot 2 \cdot 1)(2 \cdot 1)} = 21$

21. $\binom{10}{2} = \dfrac{10}{(10-2)!2!}$
$= \dfrac{10}{8!2!}$
$= \dfrac{10 \cdot 9 \cdot 8 \cdot 7 \cdot 6 \cdot 5 \cdot 4 \cdot 3 \cdot 2 \cdot 1}{(2 \cdot 1)(6 \cdot 5 \cdot 4 \cdot 3 \cdot 2 \cdot 1)}$
$= 45$

23. $\binom{9}{0} = \dfrac{9!}{(9-0)!0!}$
$= \dfrac{9!}{9!0!}$
$= \dfrac{9 \cdot 8 \cdot 7 \cdot 6 \cdot 5 \cdot 4 \cdot 3 \cdot 2 \cdot 1}{(9 \cdot 8 \cdot 7 \cdot 6 \cdot 5 \cdot 4 \cdot 3 \cdot 2 \cdot 1)(1)}$
$= 1$

25. $\binom{6}{3} = \dfrac{6!}{(6-3)!3!} = \dfrac{6!}{3!3!} = \dfrac{6 \cdot 5 \cdot 4 \cdot 3 \cdot 2 \cdot 1}{(3 \cdot 2 \cdot 1)(3 \cdot 2 \cdot 1)} = 20$

27. $\binom{11}{1} = \dfrac{11!}{(11-1)!1!}$
$= \dfrac{11!}{10!1!}$
$= \dfrac{11 \cdot 10 \cdot 9 \cdot 8 \cdot 7 \cdot 6 \cdot 5 \cdot 4 \cdot 3 \cdot 2 \cdot 1}{(10 \cdot 9 \cdot 8 \cdot 7 \cdot 6 \cdot 5 \cdot 4 \cdot 3 \cdot 2 \cdot 1)(1)}$
$= 11$

29. $\binom{4}{2} = \dfrac{4!}{(4-2)!2!} = \dfrac{4!}{2!2!} = \dfrac{4 \cdot 3 \cdot 2 \cdot 1}{(2 \cdot 1)(2 \cdot 1)} = 6$

31. False

33. x^5 ; 32

35. $(x+y)^4$
$$= \binom{4}{0}x^4 + \binom{4}{1}x^3y + \binom{4}{2}x^2y^2 + \binom{4}{3}xy^3 + \binom{4}{4}y^4$$
$$= x^4 + 4x^3y + 6x^2y^2 + 4xy^3 + y^4$$

37. $(x-y)^5 = \binom{5}{0}x^5 + \binom{5}{1}x^4(-y) + \binom{5}{2}x^3(-y)^2 + \binom{5}{3}x^2(-y)^3 + \binom{5}{4}x(-y)^4 + \binom{5}{5}(-y)^5$
$$= x^5 - 5x^4y + 10x^3y^2 - 10x^2y^3 + 5xy^4 - y^5$$

39. $(2m+1)^4 = \binom{4}{0}(2m)^4 + \binom{4}{1}(2m)^3(1) + \binom{4}{2}(2m)^2(1)^2 + \binom{4}{3}(2m)(1)^3 + \binom{4}{4}(1)^4$
$$= 1(16m^4) + 4(8m^3) + 6(4m^2) + 4(2m) + 1(1)$$
$$= 16m^4 + 32m^3 + 24m^2 + 8m + 1$$

41. $(2r-3)^5 = \binom{5}{0}(2r)^5 + \binom{5}{1}(2r)^4(-3) + \binom{5}{2}(2r)^3(-3)^2 + \binom{5}{3}(2r)^2(-3)^3 + \binom{5}{4}(2r)(-3)^4 + \binom{5}{5}(-3)^5$
$$= 1(32r^5) + 5(16r^4)(-3) + 10(8r^3)(9) + 10(4r^2)(-27) + 5(2r)(81) + 1(-243)$$
$$= 32r^5 - 240r^4 + 720r^3 - 1080r^2 + 810r - 243$$

43. $(a+b)^{10} = \binom{10}{0}a^{10} + \binom{10}{1}a^9b + \binom{10}{2}a^8b^2 + \ldots$
$$= a^{10} + 10a^9b + 45a^8b^2 + \ldots$$

45. $(a-b)^{11} = \binom{11}{0}a^{11} + \binom{11}{1}a^{10}(-b) + \binom{11}{2}a^9(-b)^2 + \ldots$
$$= (1)a^{11} + 11a^{10}(-b) + 55a^9b^2 + \ldots$$
$$= a^{11} - 11a^{10}b + 55a^9b^2 + \ldots$$

47. $(2x+y)^8 = \binom{8}{0}(2x)^8 + \binom{8}{1}(2x)^7 y + \binom{8}{2}(2x)^6 y^2 + \ldots$
$$= 1(256x^8) + 8(128x^7)y + 28(64x^6)y^2 + \ldots$$
$$= 256x^8 + 1024x^7y + 1792x^6y^2 + \ldots$$

49. $(4x-3y)^8 = \binom{8}{0}(4x)^8 + \binom{8}{1}(4x)^7(-3y) + \binom{8}{2}(4x)^6(-3y)^2 + \ldots$
$$= 1(65,536x^8) + 8(16,384x^7)(-3y) + 28(4096x^6)(9y^2) + \ldots$$
$$= 65,536x^8 - 393,216x^7y + 1,032,192x^6y^2 + \ldots$$

51. $\left(x+\dfrac{1}{x}\right)^7 = \binom{7}{0}x^7 + \binom{7}{1}x^6\left(\dfrac{1}{x}\right) + \binom{7}{2}x^5\left(\dfrac{1}{x}\right)^2 + \ldots$
$$= 1(x^7) + 7x^6\left(\dfrac{1}{x}\right) + 21x^5\left(\dfrac{1}{x^2}\right) + \ldots$$
$$= x^7 + 7x^5 + 21x^3 + \ldots$$

53. $(x^2+3)^5 = \binom{5}{0}(x^2)^5 + \binom{5}{1}(x^2)^4(3) = \binom{5}{2}(x^2)^3(3)^2 + \ldots$
$$= 1(x^{10}) + 5(x^8)(3) + 10(x^6)(9) + \ldots$$
$$= x^{10} + 15x^8 + 90x^6 + \ldots$$

352 Chapter 11: Sequences And Series

55. $n = 7, a = 2x, b = -1, r = 4$

$\binom{7}{4-1}(2x)^{7-4+1}(-1)^{4-1} = \binom{7}{3}(2x)^4(-1)^3 = 35(16x^4)(-1) = -560x^4$

57. $n = 6, a = x^2, b = -y^2, r = 2$

$\binom{6}{2-1}(x^2)^{6-2+1}(-y^2)^{2-1} = \binom{6}{1}(x^2)^5(-y^2) = 6x^{10}(-y^2) = -6x^{10}y^2$

59. $n = 9, a = y, b = -1, r = 5$

$\binom{9}{5-1}y^{9-5+1}(-1)^{5-1} = \binom{9}{4}y^5(-1)^4 = 126y^5(1) = 126y^5$

61. $n = 5, a = n, b = \dfrac{1}{n}, r = 2$

$\binom{5}{2-1}n^{5-2+1}\left(\dfrac{1}{n}\right)^{2-1} = \binom{5}{1}n^4\left(\dfrac{1}{n}\right) = 5n^4\left(\dfrac{1}{n}\right) = 5n^3$

63. $n = 5, a = \dfrac{x}{2}, b = 2, r = 1$

$\binom{5}{1-1}\left(\dfrac{x}{2}\right)^{5-1+1}(2)^{1-1} = \binom{5}{0}\left(\dfrac{x}{2}\right)^5(2)^0 = 1\left(\dfrac{x^5}{32}\right)(1) = \dfrac{x^5}{32}$

65. The constant term of the expanded form is b^n. If b is negative and n is odd, then the constant term is negative.

Applying Concepts 11.4

67.
$$\begin{array}{ccccccccccccccc}
 & & & & & & & 1 & & & & & & & \\
 & & & & & & 1 & & 1 & & & & & & \\
 & & & & & 1 & & 2 & & 1 & & & & & \\
 & & & & 1 & & 3 & & 3 & & 1 & & & & \\
 & & & 1 & & 4 & & 6 & & 4 & & 1 & & & \\
 & & 1 & & 5 & & 10 & & 10 & & 5 & & 1 & & \\
 & 1 & & 6 & & 15 & & 20 & & 15 & & 6 & & 1 & \\
1 & & 7 & & 21 & & 35 & & 35 & & 21 & & 7 & & 1
\end{array}$$

69. $\dfrac{n!}{(n-1)!} = \dfrac{n(n-1)!}{(n-1)!} = n$

71. $(x^{1/2} + 2)^4 = \binom{4}{0}(x^{1/2})^4 + \binom{4}{1}(x^{1/2})^3(2) + \binom{4}{2}(x^{1/2})^2(2^2) + \binom{4}{3}x^{1/2}(2^3) + \binom{4}{4}2^4$

$= x^2 + 4(2x^{3/2}) + 6(4x) + 4(8x^{1/2}) + 16$

$= x^2 + 8x^{3/2} + 24x + 32x^{1/2} + 16$

73. $(1+i)^6 = \binom{6}{0}1^6 + \binom{6}{1}1^5 i + \binom{6}{2}1^4 i^2 + \binom{6}{3}1^3 i^3 + \binom{6}{4}1^2 i^4 + \binom{6}{5}1 i^5 + \binom{6}{6}i^6$

$= 1 + 6i + 15i^2 + 20i^3 + 15i^4 + 6i^5 + i^6$

$= 1 + 6i - 15 - 20i + 15 + 6i - 1 = -8i$

75. $\dfrac{2 \cdot 4 \cdot 6 \cdot 8 \cdots (2n)}{2^n n!} = \dfrac{2(1) \cdot 2(2) \cdot 2(3) \cdot 2(4) \cdots 2(n)}{2^n n!}$

$= \dfrac{2^n (1 \cdot 2 \cdot 3 \cdots n)}{2^n n!}$

$= \dfrac{2^n n!}{2^n n!} = 1$

Copyright © Houghton Mifflin Company. All rights reserved.

77. Strategy

To find the coefficient of $a^4b^2c^3$ in the expansion of $(a+b+c)^9$, use the formula

$$\frac{n!}{r!k!(n-r-k)!}$$

where $n=9$, $r=4$, $k=2$

Solution

$$\frac{n!}{r!k!(n-r-k)!} = \frac{9!}{4!2!(9-4-2)!}$$
$$= \frac{9!}{4!2!3!} = \frac{9\cdot 8\cdot 7\cdot 6\cdot 5\cdot 4!}{4!2\cdot 6}$$
$$= \frac{9\cdot 8\cdot 7\cdot 6\cdot 5}{2\cdot 6} = 1260$$

The coefficient is 1260.

CHAPTER 11 REVIEW EXERCISES

1. $\sum_{i=1}^{4} 3x^i = 3x + 3x^2 + 3x^3 + 3x^4$

2. $d = a_2 - a_1 = -8 - (-5) = -3$
$a_n = a_1 + (n-1)d$
$-50 = -5 + (n-1)(-3)$
$-50 = -5 - 3n + 3$
$-50 = -2 - 3n$
$-48 = -3n$
$16 = n$
There are 16 terms.

3. $r = \frac{a_2}{a_1} = \frac{4\sqrt{2}}{4} = \sqrt{2}$
$a_n = a_1 r^{n-1}$
$a_7 = 4(\sqrt{2})^{7-1}$
$= 4(\sqrt{2})^6$
$= 4(8)$
$= 32$

4. $r = \frac{a_2}{a_1} = \frac{3}{4}$
$S = \frac{a_1}{1-r} = \frac{4}{1-\frac{3}{4}} = \frac{4}{\frac{1}{4}} = 16$

5. $\binom{9}{3} = \frac{9!}{(9-3)!3!}$
$= \frac{9!}{6!3!}$
$= \frac{9\cdot 8\cdot 7\cdot 6\cdot 5\cdot 4\cdot 3\cdot 2\cdot 1}{(6\cdot 5\cdot 4\cdot 3\cdot 2\cdot 1)(3\cdot 2\cdot 1)}$
$= 84$

6. $a_n = \frac{8}{n+2}$
$a_{14} = \frac{8}{14+2} = \frac{8}{16} = \frac{1}{2}$

7. $d = a_2 - a_1 = -4 - (-10) = 6$
$a_n = a_1 + (n-1)d$
$a_{10} = -10 + (10-1)6$
$= -10 + 9(6)$
$= -10 + 54$
$= 44$

8. $d = a_2 - a_1 = -19 - (-25) = 6$
$a_n = a_1 + (n-1)d$
$a_{18} = -25 + (18-1)6$
$= -25 + (17)6$
$= -25 + 102$
$= 77$

$S_n = \frac{n}{2}(a_1 + a_n)$
$S_{18} = \frac{18}{2}(-25 + 77) = 9(52) = 468$

9. $r = \frac{a_2}{a_1} = \frac{12}{-6} = -2$
$S_n = \frac{a_1(1-r^n)}{1-r}$
$S_5 = \frac{-6[1-(-2)^5]}{1-(-2)}$
$= \frac{-6[1-(-32)]}{3}$
$= \frac{-6(33)}{3}$
$= -66$

10. $\frac{8!}{4!4!} = \frac{8\cdot 7\cdot 6\cdot 5\cdot 4\cdot 3\cdot 2\cdot 1}{(4\cdot 3\cdot 2\cdot 1)(4\cdot 3\cdot 2\cdot 1)} = 70$

11. $n=9, a=3x, b=y, r=7$
$\binom{n}{r-1}a^{n-r+1}b^{r-1}$
$\binom{9}{7-1}(3x)^{9-7+1}y^{7-1} = \binom{9}{6}(3x)^3 y^6$
$= 84(27x^3 y^6)$
$= 2268x^3 y^6$

12. $\sum_{n=1}^{n}(3n+1)$
$= [3(1)+1]+[3(2)+1]+[3(3)+1]+[3(4)+1]$
$= 4+7+10+13 = 34$

13. $a_n = \frac{n+1}{n}$
$a_6 = \frac{6+1}{6} = \frac{7}{6}$

14. $d = a_2 - a_1 = 9 - 12 = -3$
$a_n = a_1 + (n-1)d$
$a_n = 12 + (n-1)(-3)$
$a_n = 12 - 3n + 3$
$a_n = -3n + 15$

15. $r = \dfrac{a_2}{a_1} = \dfrac{2}{6} = \dfrac{1}{3}$
$a_n = a_1 r^{n-1}$
$a_5 = 6\left(\dfrac{1}{3}\right)^{5-1} = 6\left(\dfrac{1}{3}\right)^4 = 6\left(\dfrac{1}{81}\right) = \dfrac{2}{27}$

16. $0.23\overline{3} = 0.02 + 0.03 + 0.003 + 0.0003 + \ldots$
$= \dfrac{2}{10} + \dfrac{3}{100} + \dfrac{3}{1000} + \dfrac{3}{10{,}000} + \ldots$
$S = \dfrac{a_1}{1-r} = \dfrac{\frac{3}{100}}{1-\frac{1}{10}} = \dfrac{\frac{3}{100}}{\frac{9}{10}} = \dfrac{1}{30}$
$0.23\overline{3} = \dfrac{2}{10} + \dfrac{1}{30} = \dfrac{7}{30}$

An equivalent fraction is $\dfrac{7}{30}$.

17. $d = a_2 - a_1 = -16 - (-13) = -3$
$a_n = a_1 + (n-1)d$
$a_{35} = -13 + (35-1)(-3)$
$= -13 + (34)(-3)$
$= -13 - 102$
$= -115$

18. $S_n = \dfrac{a_1(1-r^n)}{1-r}$
$S_6 = \dfrac{1\left[1-\left(\frac{3}{2}\right)^6\right]}{1-\frac{3}{2}} = \dfrac{1-\frac{729}{64}}{-\frac{1}{2}} = \dfrac{-\frac{665}{64}}{-\frac{1}{2}} = \dfrac{665}{32}$

19. $d = a_2 - a_1 = 12 - 5 = 7$
$a_n = a_1 + (n-1)d$
$= 5 + (21-1)7$
$= 5 + 20(7)$
$= 5 + 140 = 145$
$S_n = \dfrac{n}{2}(a_1 + a_n)$
$S_{21} = \dfrac{21}{2}(5 + 145) = \dfrac{21}{2}(150) = 1575$

20. $n = 7, a = x, b = -2y, r = 4$
$\binom{n}{r-1} a^{n-r+1} b^{r-1}$
$\binom{7}{4-1} x^{7-4+1}(-2y)^{4-1} = \binom{7}{3} x^4(-2y)^3$
$= 35x^4(-8y^3)$
$= -280x^4 y^3$

21. $d = a_2 - a_1 = 7 - 1 = 6$
$a_n = a_1 + (n-1)d$
$121 = 1 + (n-1)6$
$121 = 1 + 6n - 6$
$121 = 6n - 5$
$126 = 6n$
$21 = n$
There are 21 terms in the sequence.

22. $r = \dfrac{a_2}{a_1} = \dfrac{\frac{3}{4}}{\frac{3}{8}} = 2$
$a_n = a_1 r^{n-1}$
$a_8 = \dfrac{3}{8}(2)^{8-1} = \dfrac{3}{8}(2)^7 = \dfrac{3}{8}(128) = 48$

23. $\sum_{i=1}^{5} 2i = 2(1) + 2(2) + 2(3) + 2(4) + 2(5)$
$= 2 + 4 + 6 + 8 + 10 = 30$

24. $r = \dfrac{a_2}{a_1} = \dfrac{4}{1} = 4$
$S_n = \dfrac{a_1(1-r^n)}{1-r}$
$S_5 = \dfrac{1(1-4^5)}{1-4} = \dfrac{1(1-1024)}{-3} = \dfrac{-1023}{-3} = 341$

25. $5! = 5 \cdot 4 \cdot 3 \cdot 2 \cdot 1 = 120$

26. $n = 6, a = x, b = -4, r = 3$
$\binom{n}{r-1} a^{n-r+1} b^{r-1}$
$\binom{6}{3-1} x^{6-3+1}(-4)^{3-1} = \binom{6}{2} x^4(-4)^2$
$= 15x^4(16) = 240x^4$

27. $d = a_2 - a_1 = 3 - (-2) = 5$
$a_n = a_1 + (n-1)d$
$a_{30} = -2 + (30-1)5$
$= -2 + (29)(5)$
$= -2 + 145$
$= 143$

28. $d = a_2 - a_1 = 21 - 25 = -4$
$a_n = a_1 + (n-1)d$
$a_{25} = 25 + (25-1)(-4)$
$= 25 + (24)(-4)$
$= 25 - 96 = -71$
$S_n = \dfrac{n}{2}(a_1 + a_n)$
$S_{25} = \dfrac{25}{2}[25 + (-71)] = \dfrac{25}{2}(-46) = -575$

29. $a_n = \dfrac{(-1)^{2n-1} n}{n^2 + 2}$

$a_5 = \dfrac{(-1)^{2(5)-1} \cdot 5}{(5)^2 + 2} = \dfrac{(-1)^9 \cdot 5}{25 + 2} = \dfrac{-5}{27}$

The fifth term is $-\dfrac{5}{27}$.

30. $\sum_{i=1}^{4} 2x^{i-1} = 2x^{1-1} + 2x^{2-1} + 2x^{3-1} + 2x^{4-1}$

$= 2 + 2x + 2x^2 + 2x^3$

31. $0.\overline{23} = 0.23 + 0.0023 + 0.000023 + \ldots$

$= \dfrac{23}{100} + \dfrac{23}{10,000} + \dfrac{23}{1,000,000} + \ldots$

$S = \dfrac{a_1}{1-r} = \dfrac{\frac{23}{100}}{1 - \frac{1}{100}} = \dfrac{\frac{23}{100}}{\frac{99}{100}} = \dfrac{23}{99}$

An equivalent fraction is $\dfrac{23}{99}$.

32. $r = \dfrac{a_2}{a_1} = \dfrac{-1}{4} = -\dfrac{1}{4}$

$S = \dfrac{a_1}{1-r} = \dfrac{4}{1-\left(-\frac{1}{4}\right)} = \dfrac{4}{\frac{5}{4}} = \dfrac{16}{5}$

33. $a_n = 2(3)^n$

$a_1 = 2(3)^1 = 6$

$a_2 = 2(3)^2 = 18$

$r = \dfrac{a_2}{a_1} = \dfrac{18}{6} = 3$

$S_n = \dfrac{a_1(1-r^n)}{1-r}$

$S_5 = \dfrac{6(1-3^5)}{1-3} = \dfrac{6(1-243)}{-2} = -3(-242) = 726$

34. $n = 11, a = x, b = -2y, r = 8$

$\binom{n}{r-1} a^{n-r+1} b^{r-1}$

$\binom{11}{8-1}(x)^{11-8+1}(-2y)^{8-1} = \binom{11}{7} x^4 (-2y)^7$

$= 330 x^4 (-128 y^7)$

$= -42{,}240 x^4 y^7$

35. $a_n = \left(\dfrac{1}{2}\right)^n$

$a_1 = \left(\dfrac{1}{2}\right)^1 = \dfrac{1}{2}$

$a_2 = \left(\dfrac{1}{2}\right)^2 = \dfrac{1}{4}$

$r = \dfrac{a_2}{a_1} = \dfrac{\frac{1}{4}}{\frac{1}{2}} = \dfrac{1}{4} \cdot \dfrac{2}{1} = \dfrac{1}{2}$

$S_n = \dfrac{a_1(1-r^n)}{1-r}$

$= \dfrac{\frac{1}{2}\left(1-\left(\frac{1}{2}\right)^8\right)}{1-\frac{1}{2}}$

$= \dfrac{\frac{1}{2}\left(1-\frac{1}{256}\right)}{\frac{1}{2}}$

$= \dfrac{255}{256}$

≈ 0.996

36. $r = \dfrac{a_2}{a_1} = \dfrac{\frac{4}{3}}{2} = \dfrac{2}{3}$

$S = \dfrac{a_1}{1-r} = \dfrac{2}{1-\frac{2}{3}} = \dfrac{2}{\frac{1}{3}} = 6$

37. $0.6\overline{3} = 0.6 + 0.03 + 0.003 + \ldots$

$= \dfrac{6}{10} + \dfrac{3}{100} + \dfrac{3}{1000} + \ldots$

$S = \dfrac{a_1}{1-r} = \dfrac{\frac{3}{100}}{1-\frac{1}{10}} = \dfrac{\frac{3}{100}}{\frac{9}{10}} = \dfrac{3}{100} \cdot \dfrac{10}{9} = \dfrac{1}{30}$

$0.63\overline{3} = \dfrac{6}{10} + \dfrac{1}{30} = \dfrac{19}{30}$

38. $(x-3y^2)^5 = \binom{5}{0} x^5 + \binom{5}{1} x^4 (-3y^2)^1 + \binom{5}{2} x^3 (-3y^2)^2 + \binom{5}{3} x^2 (-3y^2)^3 + \binom{5}{4} x^1 (-3y^2)^4 + \binom{5}{5}(-3y^2)^5$

$= x^5 + 5x^4(-3y^2) + 10x^3(9y^4) + 10x^2(-27y^6) + 5x(81y^8) + 1(-243y^{10})$

$= x^5 - 15x^4 y^2 + 90x^3 y^4 - 270x^2 y^6 + 405xy^8 - 243y^{10}$

356 Chapter 11: Sequences And Series

39. $d = a_2 - a_1 = 2 - 8 = -6$
$a_n = a_1 + (n-1)d$
$-118 = 8 + (n-1)d$
$-118 = 8 - 6n + 6$
$-118 = 14 - 6n$
$-132 = -6n$
There are 22 terms in the sequence.

40. $\dfrac{12!}{5!8!} = \dfrac{12 \cdot 11 \cdot 10 \cdot 9 \cdot 8 \cdot 7 \cdot 6 \cdot 5 \cdot 4 \cdot 3 \cdot 2 \cdot 1}{(5 \cdot 4 \cdot 3 \cdot 2 \cdot 1)(8 \cdot 7 \cdot 6 \cdot 5 \cdot 4 \cdot 3 \cdot 2 \cdot 1)} = 99$

41. $\displaystyle\sum_{i=1}^{5} \dfrac{(2x)^i}{i} = \dfrac{(2x)^1}{1} + \dfrac{(2x)^2}{2} + \dfrac{(2x)^3}{3} + \dfrac{(2x)^4}{4} + \dfrac{(2x)^5}{5}$
$= 2x + \dfrac{4x^2}{2} + \dfrac{8x^3}{3} + \dfrac{16x^4}{4} + \dfrac{32x^5}{5}$
$= 2x + 2x^2 + \dfrac{8}{3}x^3 + 4x^4 + \dfrac{32}{5}x^5$

42. $\displaystyle\sum_{n=1}^{4} \dfrac{(-1)^{n-1} n}{n+1} = \dfrac{(-1)^{1-1} \cdot 1}{1+1} + \dfrac{(-1)^{2-1} \cdot 2}{2+1} + \dfrac{(-1)^{3-1} \cdot 3}{3+1} + \dfrac{(-1)^{4-1} \cdot 4}{4+1}$
$= \dfrac{1}{2} + \left(-\dfrac{2}{3}\right) + \dfrac{3}{4} + \left(-\dfrac{4}{5}\right) = \dfrac{30 - 40 + 45 - 48}{60}$
$= -\dfrac{13}{60}$

43. **Strategy**
To find the total salary for the nine-month period:
Write the arithmetic sequence.
Find the common difference of the arithmetic sequence.
Use the Formula for the nth Term of an Arithmetic Sequence to find the ninth term.
Use the Formula for the Sum of n Terms of an Arithmetic Sequence to find the sum of nine terms of the sequence.

Solution
$\$2400, \$2480, \$2560, \ldots$
$d = a_2 - a_1 = 2480 - 2400 = 80$
$a_n = a_1 + (n-1)d$
$a_9 = 2400 + (9-1)80 = 2400 + 640 = 3040$
$S_n = \dfrac{n}{2}(a_1 + a_n)$
$S_9 = \dfrac{9}{2}(2400 + 3040) = \$24,480$
The total salary for the nine-month period is $\$24,480$.

44. **Strategy**
To find the temperature of the spa after 8 hours, use the Formula for the nth Term of a Geometric Sequence.

Solution
$n = 8, a_1 = 102(0.95) = 96.9, r = 0.95$
$a_n = a_1 r^{n-1}$
$a_8 = 96.9(0.95)^7 \approx 67.7$
The temperature is $67.7°$ F.

CHAPTER 11 TEST

1. $a_n = \dfrac{6}{n+4}$
$a_{14} = \dfrac{6}{14+4} = \dfrac{6}{18} = \dfrac{1}{3}$
The fourteenth term is $\dfrac{1}{3}$.

2. $a_n = \dfrac{n-1}{n}$
$a_9 = \dfrac{9-1}{9} = \dfrac{8}{9}$
The ninth term is $\dfrac{8}{9}$.
$a_{10} = \dfrac{10-1}{10} = \dfrac{9}{10}$
The tenth term is $\dfrac{9}{10}$.

3. $\displaystyle\sum_{n=1}^{4}(2n+3)$
$= [2(1) + 3] + [2(2) + 3] + [2(3) + 3] + [2(4) + 3]$
$= 5 + 7 + 9 + 11$
$= 32$

4. $\displaystyle\sum_{i=1}^{4} 2x^{2i} = 2x^{2(1)} + 2x^{2(2)} + 2x^{2(3)} + 2x^{2(4)}$
$= 2x^2 + 2x^4 + 2x^6 + 2x^8$

5. $d = a_2 - a_1 = -16 - (-12) = -16 + 12 = -4$
$a_n = a_1 + (n-1)d$
$a_{28} = -12 + (28-1)(-4) = -12 + 27(-4) = -120$

6. $d = a_2 - a_1 = -1 - (-3) = -1 + 3 = 2$
   $a_n = a_1 + (n-1)d$
   $a_n = -3 + (n-1)(2) = -3 + 2n - 2$
   $a_n = 2n - 5$

7. $d = a_2 - a_1 = 3 - 7 = -4$
   $a_n = a_1 + (n-1)d$
   $-77 = 7 + (n-1)d$
   $-77 = 7 - 4n + 4$
   $-77 = 11 - 4n$
   $-88 = -4n$
   $22 = n$

8. $d = a_2 - a_1 = -33 - (-42) = -33 + 42 = 9$
   $a_n = a_1 + (n-1)d$
   $a_{15} = -42 + (15-1)9 = -42 + 14(9) = 84$
   $S_n = \frac{n}{2}(a_1 + a_n)$
   $S_{15} = \frac{15}{2}(-42 + 84) = 315$

9. $d = a_2 - a_1 = 2 - (-4) = 2 + 4 = 6$
   $a_n = a_1 + (n-1)d$
   $a_{24} = -4 + (24-1)6 = -4 + 23(6) = 134$
   $S_n = \frac{n}{2}(a_1 + a_n)$
   $S_{24} = \frac{24}{2}(-4 + 134) = 1560$

10. $\frac{10!}{5!5!} = \frac{10 \cdot 9 \cdot 8 \cdot 7 \cdot 6 \cdot 5 \cdot 4 \cdot 3 \cdot 2 \cdot 1}{(5 \cdot 4 \cdot 3 \cdot 2 \cdot 1)(5 \cdot 4 \cdot 3 \cdot 2 \cdot 1)} = 252$

11. $r = \frac{a_2}{a_1} = \frac{-4\sqrt{2}}{4} = -\sqrt{2}$
    $a_n = 4r^{n-1}$
    $a_{10} = 4(-\sqrt{2})^{10-1} = 4 \cdot (-\sqrt{2})^9 = -64\sqrt{2}$

12. $r = \frac{a_2}{a_1} = \frac{3}{5}$
    $a_n = a_1 r^{n-1}$
    $a_5 = 5 \cdot \left(\frac{3}{5}\right)^{5-1} = 5 \cdot \left(\frac{3}{5}\right)^4 = 5 \cdot \frac{81}{625} = \frac{81}{125}$

13. $r = \frac{a_2}{a_1} = \frac{3/4}{1} = \frac{3}{4}$
    $S_n = \frac{a_1(1-r^n)}{1-r}$
    $S_5 = \frac{1\left(1 - \left(\frac{3}{4}\right)^5\right)}{1 - \frac{3}{4}} = \frac{1 - \frac{243}{1024}}{\frac{1}{4}} = \frac{\frac{781}{1024}}{\frac{1}{4}} = \frac{781}{256}$

14. $r = \frac{a_2}{a_1} = \frac{10}{-5} = -2$
    $S_n = \frac{a_1(1-r^n)}{1-r}$
    $S_5 = \frac{-5(1-(-2)^5)}{1-(-2)}$
    $= \frac{-5(1-(-32))}{3}$
    $= \frac{-5(33)}{3}$
    $= -55$

15. $r = \frac{a_2}{a_1} = \frac{1}{2}$
    $S_n = \frac{a_1}{1-r} = \frac{2}{1-\frac{1}{2}} = \frac{2}{\frac{1}{2}} = 4$

16. $0.2\overline{3} = 0.2 + 0.03 + 0.003 + \ldots$
    $= \frac{2}{10} + \frac{3}{100} + \frac{3}{1000} + \ldots$
    $S = \frac{a_1}{1-r} = \frac{\frac{3}{100}}{1-\frac{1}{10}} = \frac{\frac{3}{100}}{\frac{9}{10}} = \frac{3}{90} = \frac{1}{30}$
    $0.2\overline{33} = \frac{2}{10} + \frac{1}{30} = \frac{7}{30}$
    An equivalent fraction is $\frac{7}{30}$.

17. $\binom{11}{4} = \frac{11!}{(11-4)!4!}$
    $= \frac{11!}{7!4!}$
    $= \frac{11 \cdot 10 \cdot 9 \cdot 8 \cdot 7 \cdot 6 \cdot 5 \cdot 4 \cdot 3 \cdot 2 \cdot 1}{(7 \cdot 6 \cdot 5 \cdot 4 \cdot 3 \cdot 2 \cdot 1)(4 \cdot 3 \cdot 2 \cdot 1)}$
    $= 330$

18. $n = 8, a = 3x, b = -y, r = 5$
    $\binom{n}{r-1}a^{n-r+1}b^{r-1}$
    $\binom{8}{5-1}(3x)^{8-5+1}(-y)^{5-1} = \binom{8}{4}(3x)^4(-y)^4$
    $= 70(81x^4)(y^4)$
    $= 5670x^4y^4$

358 Chapter 11: Sequences And Series

19. Strategy
To find how much material was in stock after the shipment on October 1:
Write the arithmetic sequence.
Find the common difference of the arithmetic sequence.
Use the Formula for the nth Term of an Arithmetic Sequence to find the tent

Solution
7500, 6950, 6400, …
$d = a_2 - a_1 = 6950 - 7500 = -550$
$a_n = a_1 + (n-1)d$
$a_{10} = 7500 + (10-1)(-550) = 2550$
The inventory after the October 1 shipment was 2550 yd.

20. Strategy
To find the amount of radioactive material at the beginning of the fifth day, use the Formula for the nth Term of a Geometric Sequence.

Solution
$n = 5, a_1 = 320, r = \dfrac{1}{2}$
$a_n = a_1 r^{n-1}$
$a_5 = 320\left(\dfrac{1}{2}\right)^{5-1} = 320\left(\dfrac{1}{2}\right)^4 = 20$
There will be 20 mg of radioactive material in the sample at the beginning of the fifth day.

CUMULATIVE REVIEW EXERCISES

1. $\dfrac{4x^2}{x^2 + x - 2} - \dfrac{3x-2}{x+2}$

$= \dfrac{4x^2}{(x+2)(x-1)} - \dfrac{3x-2}{x+2} \cdot \dfrac{x-1}{x-1}$

$= \dfrac{4x^2 - (3x^2 - 5x + 2)}{(x+2)(x-1)}$

$= \dfrac{x^2 + 5x - 2}{(x+2)(x-1)}$

2. $2x^6 + 16 = 2(x^6 + 8)$
$= 2((x^2)^3 + 2^3)$
$= 2(x^2 + 2)(x^4 - 2x^2 + 4)$

3. $\sqrt{2y}(\sqrt{8xy} - \sqrt{y}) = \sqrt{16xy^2} - \sqrt{2y^2}$
$= \sqrt{16y^2(x)} - \sqrt{y^2(2)}$
$= 4y\sqrt{x} - y\sqrt{2}$

4. $\left(\dfrac{x^{-\frac{3}{4}} x^{\frac{3}{2}}}{x^{-\frac{5}{2}}}\right)^{-8} = \dfrac{x^6 x^{-12}}{x^{20}} = \dfrac{x^{-6}}{x^{20}} = \dfrac{1}{x^{26}}$

5. $5 - \sqrt{x} = \sqrt{x+5}$
$(5 - \sqrt{x})^2 = (\sqrt{x+5})^2$
$25 - 10\sqrt{x} + x = x + 5$
$-10\sqrt{x} = -20$
$\sqrt{x} = 2$
$(\sqrt{x})^2 = 2^2$
$x = 4$
Check:

$5 - \sqrt{x} = \sqrt{x+5}$	
$5 - \sqrt{4}$	$\sqrt{4+5}$
$5 - 2$	$\sqrt{9}$
$3 = 3$	

The solution is 4.

6. $2x^2 - x + 7 = 0$
$a = 2, b = -1, c = 7$
$x = \dfrac{-b \pm \sqrt{b^2 - 4ac}}{2a}$
$= \dfrac{-(-1) \pm \sqrt{(-1)^2 - 4(2)(7)}}{2(2)}$
$= \dfrac{1 \pm \sqrt{1 - 56}}{4}$
$= \dfrac{1 \pm \sqrt{-55}}{4}$
$= \dfrac{1}{4} \pm \dfrac{\sqrt{55}}{4} i$

The solutions are $\dfrac{1}{4} + \dfrac{\sqrt{55}}{4}i$ and $\dfrac{1}{4} - \dfrac{\sqrt{55}}{4}i$.

7. (1) $3x - 3y = 2$
(2) $6x - 4y = 5$
Eliminate x.
Multiply equation (1) by -2 and add equation (2).
$(-2)(3x - 3y) = (-2)(2)$
$6x - 4y = 5$
$-6x + 6y = -4$
$6x - 4y = 5$
$2y = 1$
$y = \dfrac{1}{2}$

Substitute $\dfrac{1}{2}$ for y in equation (2).
$6x - 4\left(\dfrac{1}{2}\right) = 5$
$6x - 2 = 5$
$6x = 7$
$x = \dfrac{7}{6}$

The solution is $\left(\dfrac{7}{6}, \dfrac{1}{2}\right)$.

Copyright © Houghton Mifflin Company. All rights reserved.

8. $2x - 1 > 3$ or $1 - 3x > 7$
   $2x > 4$ $\quad -3x > 6$
   $x > 2$ $\quad x < -2$
   $\{x | x > 2\}$ $\quad \{x | x < -2\}$
   $\{x | x < -2 \text{ or } x > 2\}$

9. $\begin{vmatrix} -3 & 1 \\ 4 & 2 \end{vmatrix} = -3(2) - 4(1) = -6 - 4 = -10$

10. $\log_5 \sqrt{\dfrac{x}{y}} = \log_5 \left(\dfrac{x}{y}\right)^{\frac{1}{2}}$
    $= \dfrac{1}{2} \log_5 \left(\dfrac{x}{y}\right)$
    $= \dfrac{1}{2}(\log_5 x - \log_5 y)$
    $= \dfrac{1}{2} \log_5 x - \dfrac{1}{2} \log_5 y$

11. $4^x = 8^{x-1}$
    $(2^2)^x = (2^3)^{x-1}$
    $2^{2x} = 2^{3(x-1)}$
    $2x = 3(x-1)$
    $2x = 3x - 3$
    $-x = -3$
    $x = 3$

12. $a_n = n(n-1)$
    $a_5 = 5(5-1) = 5(4) = 20$ The fifth term is 20.
    $a_6 = 6(6-1) = 6(5) = 30$ The sixth term is 30.

13. $\sum_{n=1}^{7} (-1)^{n-1}(n+2) = (-1)^{1-1}(1+2) + (-1)^{2-1}(2+2) + (-1)^{3-1}(3+2) + (-1)^{4-1}(4+2) + (-1)^{5-1}(5+2)$
    $\qquad\qquad\qquad\qquad\qquad +(-1)^{6-1}(6+2) + (-1)^{7-1}(7+2)$
    $= (-1)^0(3) + (-1)^1(4) + (-1)^2(5) + (-1)^3(6) + (-1)^4(7) + (-1)^5(8) + (-1)^6(9)$
    $= 3 - 4 + 5 - 6 + 7 - 8 + 9 = 6$

14. (1) $\quad x + 2y + z = 3$
    (2) $\quad 2x - y + 2z = 6$
    (3) $\quad 3x + y - z = 5$
    To eliminate y add equations (2) and (3).
    (2) $\quad 2x - y + 2z = 6$
    (3) $\quad 3x + y - z = 5$
    (4) $\quad 5x + z = 11$
    To eliminate y from equations (1) and (2), multiply equation (2) by 2 and add it to equation (1)
    $x + 2y + z = 3$
    $2(2x - y + 2z) = 2(6)$

    $x + 2y + z = 3$
    $4x - 2y + 4z = 12$
    (5) $\quad 5x + 5z = 15$
    Eliminate x from equations (4) and (5) by multiplying equation (4) by -1 and adding to equation (5).
    $-1(5x + z) = (-1)(11)$
    $5x + 5z = 15$

    $-5x - z = -11$
    $5x + 5z = 15$

    $4z = 4$
    $z = 1$

    Solve for x by substituting 1 for z in equation (5).
    $5x + 5z = 15$
    $5x + 5(1) = 15$
    $5x + 5 = 15$
    $5x = 10$
    $x = 2$
    Solve for y by substituting 1 for z and 2 for x in equation (3).
    $3x + y - z = 5$
    $3(2) + y - 1 = 5$
    $6 + y - 1 = 5$
    $5 + y = 5$
    $y = 0$
    The solution is (2, 0, 1).

15. $\log_6 x = 3$
    $6^3 = x$
    $216 = x$
    The solution is 216.

Chapter 11: Sequences And Series

16. $(4x^3 - 3x + 5) \div (2x + 1)$

$$\begin{array}{r} 2x^2 - x - 1 \\ 2x+1 \overline{) 4x^3 + 0x^2 - 3x + 5} \\ \underline{4x^3 + 2x^2} \\ -2x^2 - 3x \\ \underline{-2x^2 - x} \\ -2x + 5 \\ \underline{-2x - 1} \\ 6 \end{array}$$

The solution is $2x^2 - x - 1 + \dfrac{6}{2x+1}$.

17. $g(x) = -3x + 4$
$g(1+h) = -3(1+h) + 4$
$ = -3 - 3h + 4 = -3h + 1$

18. $f(a) = \dfrac{a^3 - 1}{2a + 1}$

The domain is {0, 1, 2}.

$f(0) = \dfrac{0^3 - 1}{2(0) + 1} = \dfrac{-1}{1} = -1$

$f(1) = \dfrac{1^3 - 1}{2(1) + 1} = \dfrac{0}{3} = 0$

$f(2) = \dfrac{2^3 - 1}{2(2) + 1} = \dfrac{8 - 1}{5} = \dfrac{7}{5}$

The range is $\{-1, 0, \dfrac{7}{5}\}$.

19. $3x - 2y = -4$
$-2y = -3x - 4$
$y = \dfrac{3}{2}x + 2$

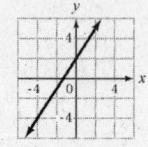

20. $2x - 3y < 9$
$-3y < -2x + 9$
$y > \dfrac{2}{3}x - 3$

21. Strategy
Time required for the older computer: t
Time required for the new computer: $t - 16$

	Rate	Time	Part
Older computer	$\dfrac{1}{t}$	15	$\dfrac{15}{t}$
New computer	$\dfrac{1}{t-16}$	15	$\dfrac{15}{t-16}$

The sum of the part of the task completed by the older computer and the part of the task completed by the new computer is 1.

$\dfrac{15}{t} + \dfrac{15}{t-16} = 1$

Solution

$\dfrac{15}{t} + \dfrac{15}{t-16} = 1$

$t(t-16)\left(\dfrac{15}{t} + \dfrac{15}{t-16}\right) = 1(t(t-16))$

$15(t - 16) + 15t = t^2 - 16t$

$15t - 240 + 15t = t^2 - 16t$

$0 = t^2 - 46t + 240$

$0 = (t - 40)(t - 6)$

$t - 40 = 0 \quad t - 6 = 0$
$t = 40 \quad\quad t = 6$

$t = 6$ does not check as a solution since $t - 16 = 6 - 16 = -10$.
$t - 16 = 40 - 16 = 24$

The new computer takes 24 min to complete the payroll.
The old computer takes 40 min to complete the payroll.

22. Strategy
Rate of boat in calm water: x
Rate of current: y

	Rate	Time	Distance
With current	$x + y$	2	$2(x+y)$
Against current	$x - y$	3	$3(x-y)$

The distance traveled with the current is 15 mi.
The distance traveled against the current is 15 mi.
$2(x + y) = 15$
$3(x - y) = 15$

Solution

$2(x+y) = 15 \quad \dfrac{1}{2} \cdot 2(x+y) = \dfrac{1}{2} \cdot 15$

$3(x-y) = 15 \quad \dfrac{1}{3} \cdot 3(x-y) = \dfrac{1}{3} \cdot 15$

$x + y = 7.5$
$x - y = 5$
$2x = 12.5$
$x = 6.25$
$x + y = 7.5$
$6.25 + y = 7.5$
$y = 1.25$

The rate of the boat in calm water is 6.25 mph.
The rate of the current is 1.25 mph.

23. To find the half-life, solve for k in the exponential decay equation.
Use $A_0 = 80, A = 55, t = 30$.

Solution

$$A = A_0 \left(\frac{1}{2}\right)^{t/k}$$

$$55 = 80\left(\frac{1}{2}\right)^{30/k}$$

$$0.6875 = (0.5)^{30/k}$$

$$\log 0.6875 = \frac{30}{k} \log (0.5)$$

$$k = \frac{30 \log 0.5}{\log 0.6875}$$

$$\approx 55.49$$

The half-life is approximately 55 days.

24. Strategy
To find the total number of seats in the 12 rows of the theater.
Write the arithmetic sequence.
Find the common difference of the arithmetic sequence.
Use the Formula for the nth Term of an Arithmetic Sequence to find the 12th term.
Use the Formula for the Sum of n Terms of an Arithmetic Sequence to find the sum of the 12 terms of the sequence.

Solution
62, 74, 86, ...

$$d = a_2 - a_1 = 74 - 62 = 12$$

$$a_n = a_1 + (n-1)d$$

$$a_{12} = 62 + (12-1)12 = 62 + 132 = 194$$

$$S_n = \frac{n}{2}(a_1 + a_n)$$

$$S_{12} = \frac{12}{2}(62 + 194) = 1536$$

The total number of seats in the theater is 1536.

25. Strategy
To find the height of the ball on the fifth bounce, use the Formula for the nth Term of a Geometric Sequence.

Solution
$n = 5, a_1 = 80\%$ of $10 = 8, r = 80\% = 0.8$

$$a_n = a_1 r^{n-1}$$

$$a_5 = 8(0.8)^{5-1}$$

$$= 8(0.4096) = 3.2768$$

The height of the ball on the fifth bounce is 3.3 ft.

CHAPTER 12: CONIC SECTIONS

CHAPTER 12 PREP TEST

1. $d = \sqrt{(4-(-2))^2 + (-1-3)^2}$ [3.1.2]
   $= \sqrt{6^2 + (-4)^2}$
   $= \sqrt{52}$
   ≈ 7.21
   The distance is approximately 7.21.

2. $x^2 - 8x + \left(\frac{1}{2}(-8)\right)^2 = x^2 - 8x + 16$ [8.2.1]
   $(x-4)^2$

3. $\frac{x^2}{16} + \frac{y^2}{9} = 1$ [8.1.3]
   For $y = 3$,
   $\frac{x^2}{16} + \frac{3^2}{9} = 1$
   $\frac{x^2}{16} + \frac{9}{9} = 1$
   $\frac{x^2}{16} + 1 = 1$
   $\frac{x^2}{16} = 0$
   $x^2 = 0$
   $x = 0$
   For $y = 0$,
   $\frac{x^2}{16} + \frac{0^2}{9} = 1$
   $x^2 = 16$
   $x = \pm 4$

4. $7x + 4y = 3$
   $y = x - 2$
   $7x + 4(x-2) = 3$
   $7x + 4x - 8 = 3$
   $11x = 11$
   $x = 1$
   $y = 1 - 2 = -1$
   The solution is $(1, -1)$.

5. (1) $4x - y = 9$ [4.2.1]
   (2) $2x + 3y = -13$
   Eliminate y.
   $3(4x - y) = 3(9)$
   $2x + 3y = -13$
   $12x - 3y = 27$
   $2x + 3y = -13$
   Add the equations.
   $14x = 14$
   $x = 1$
   Replace x in equation (2).
   $2(1) + 3y = -13$
   $3y = -15$
   $y = -5$
   The solution is $(1, -5)$.

6. $y = x^2 - 4x + 2$ [8.6.1]
   $a = 1, b = -4$
   $-\frac{b}{2a} = -\frac{-4}{2(1)} = 2$
   $y = (2)^2 - 4(2) + 2 = -2$
   Vertex: $(2, -2)$
   Axis of symmetry: $x = 2$

7. $f(x) = -2x^2 + 4x$ [8.6.1]

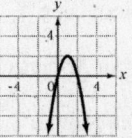

8. $5x - 2y > 10$ [3.7.1]
   $-2y > -5x + 10$
   $y < \frac{5}{2}x - 5$

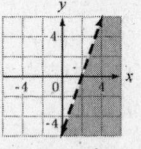

9. $x + 2y \le 4$ [4.5.1]
   $x - y \le 2$

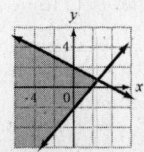

Section 12.1

Concept Review 12.1

1. Sometimes true
   The graph of the parabola $y = x^2 - 2$ is the graph of a function. The graph of the parabola $x = y^2 - 2$ is not the graph of a function.

3. Sometimes true
A parabola may have one intercept, two intercepts, or no intercepts.

5. Sometimes true
The axis of symmetry of the parabola $y = x^2 + 2x - 3$ is the line $x = -1$.

Objective 12.1.1 Exercises

1.
   a. Axis of symmetry is a vertical line.
   b. The parabola opens up since $3 > 0$.

3.
   a. Axis of symmetry is a horizontal line.
   b. The parabola opens right since $1 > 0$.

5.
   a. Axis of symmetry is a horizontal line.
   b. The parabola opens left since $-\frac{1}{2} < 0$.

7. y; positive; negative

9. $y = x^2 - 2x - 4$
$-\frac{b}{2a} = -\frac{-2}{2(1)} = 1$
$y = 1^2 - 2(1) - 4 = -5$
Vertex: $(1, -5)$
Axis of symmetry: $x = 1$

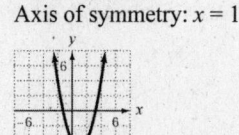

11. $y = -x^2 + 2x - 3$
$-\frac{b}{2a} = -\frac{2}{2(-1)} = 1$
$y = -(1)^2 + 2(1) - 3 = -2$
Vertex: $(1, -2)$
Axis of symmetry: $x = 1$

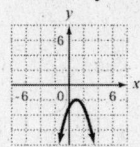

13. $x = y^2 + 6y + 5$
$-\frac{b}{2a} = -\frac{6}{2(1)} = -3$
$x = (-3)^2 + 6(-3) + 5 = -4$
Vertex: $(-4, -3)$
Axis of symmetry: $y = -3$

15. $y = 2x^2 - 4x + 1$
$-\frac{b}{2a} = -\frac{-4}{2(2)} = 1$
$y = 2(1)^2 - 4(1) + 1 = -1$
Vertex: $(1, -1)$
Axis of symmetry: $x = 1$

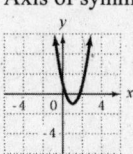

17. $y = x^2 - 5x + 4$
$-\frac{b}{2a} = -\frac{-5}{2(1)} = \frac{5}{2}$
$y = \left(\frac{5}{2}\right)^2 - 5\left(\frac{5}{2}\right) + 4 = -\frac{9}{4}$
Vertex: $\left(\frac{5}{2}, -\frac{9}{4}\right)$
Axis of symmetry: $x = \frac{5}{2}$

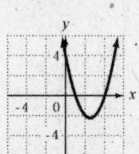

19. $x = y^2 - 2y - 5$
$-\frac{b}{2a} = -\frac{-2}{2(1)} = 1$
$x = 1^2 - 2(1) - 5 = -6$
Vertex: $(-6, 1)$
Axis of symmetry: $y = 1$

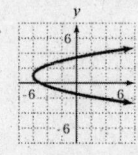

364 Chapter 12: Conic Sections

21. $y = -3x^2 - 9x$

$-\dfrac{b}{2a} = -\dfrac{-9}{2(-3)} = -\dfrac{3}{2}$

$y = -3\left(-\dfrac{3}{2}\right)^2 - 9\left(-\dfrac{3}{2}\right) = \dfrac{27}{4}$

Vertex: $\left(-\dfrac{3}{2}, \dfrac{27}{4}\right)$

Axis of symmetry: $x = -\dfrac{3}{2}$

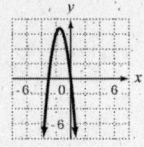

23. $x = -\dfrac{1}{2}y^2 + 4$

$-\dfrac{b}{2a} = -\dfrac{0}{2\left(-\frac{1}{2}\right)} = 0$

$x = -\dfrac{1}{2}(0)^2 + 4 = 4$

Vertex: $(4, 0)$
Axis of symmetry: $y = 0$

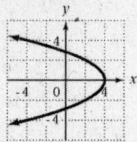

25. $x = \dfrac{1}{2}y^2 - y + 1$

$-\dfrac{b}{2a} = -\dfrac{-1}{2\left(\frac{1}{2}\right)} = 1$

$x = \dfrac{1}{2}(1)^2 - 1 + 1 = \dfrac{1}{2}$

Vertex: $\left(\dfrac{1}{2}, 1\right)$

Axis of symmetry: $y = 1$

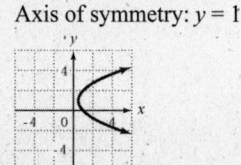

27. $y = \dfrac{1}{2}x^2 + 2x - 6$

$-\dfrac{b}{2a} = -\dfrac{2}{2\left(\frac{1}{2}\right)} = -2$

$y = \dfrac{1}{2}(-2)^2 + 2(-2) - 6 = -8$

Vertex: $(-2, -8)$
Axis of symmetry: $x = -2$

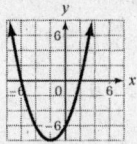

29. $x = ay^2 + by + c$; $a > 0$ and $c < 0$

31. $y = ax^2 + bx + c$; $a > 0$ and $c > 0$

Applying Concepts 12.1

33. $y = x^2 - 4x - 2$

$-\dfrac{b}{2a} = -\dfrac{-4}{2(1)} = \dfrac{4}{2} = 2$

$y = x^2 - 4x - 2$
$ = (2)^2 - 4(2) - 2$
$ = 4 - 8 - 2$
$ = -6$

a is positive. The minimum value of y is -6.
The domain is all real numbers.
The range is all real numbers greater than or equal to -6.

35. $x = y^2 + 6y - 5$

$-\dfrac{b}{2a} = -\dfrac{6}{2(1)} = -\dfrac{6}{2} = -3$

$x = y^2 + 6y - 5$
$ = (-3)^2 + 6(-3) - 5$
$ = 9 - 18 - 5$
$ = -14$

a is positive. The minimum value of x is -14.
The domain is all real numbers greater than or equal to -14. The range is all real numbers.

37. $p = \dfrac{1}{4a}$; $a = 2$

$p = \dfrac{1}{4(2)} = \dfrac{1}{8}$

The focus is $\left(0, \dfrac{1}{8}\right)$.

39.

a. The parabola passes through the point whose coordinates are (3.79, 100). Because the parabola opens to the right and its vertex is at the origin, the equation of the parabola is of the form $x = ay^2$.

$$x = ay^2$$
$$3.79 = a(100)^2 \quad \text{• Substitute (3.79, 100).}$$
$$0.000379 = a \quad \text{• Solve for } a.$$

The equation of the mirror is $x = 0.000379y^2$ or $x = \dfrac{1}{2639}y^2$.

b. The equation is valid over the interval $0 \le x \le 3.79$, or [0, 3.79].

41. The graph of $f(x) = ax^2$ opens upward and becomes narrower as a increases ($a > 0$). If $a > 0$, the graph has similar characteristics but opens downward. Some graphs of $f(x) = ax^2$ for various values of a are shown here.

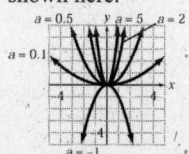

43. If the solutions of the quadratic equation $ax^2 + bx + c = 0$ are complex numbers, the graph of the corresponding equation $y = ax^2 + bx + c$ does not pass through the x-axis and has no x-intercepts. Since the values of $y = x^2 + 2x + 3$ are imaginary numbers when $y = 0$, the graph of this equation has no intercepts.

Section 12.2

Concept Review 12.2

1. Sometimes true
The center of the circle given by the equation $(x - 3)^2 + (y + 1)^2 = 4$ is (3, –1).

3. Always true

5. Never true
The square of the radius of a circle cannot be a negative value.

Objective 12.2.1 Exercises

1. The points on the circumference of a circle are all equidistant from the center of the circle. The common distance is the radius of the circle.

3. 1; –4; 6; 1; –4; 6

5.

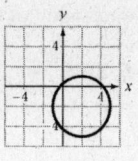

7.

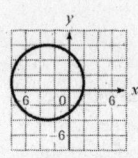

9.

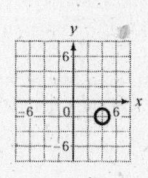

11.

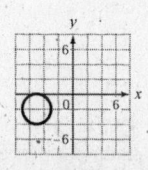

13. Quadrant IV

15.
$$(x - h)^2 + (y - k)^2 = r^2$$
$$(x - 2)^2 + [y - (-1)]^2 = 2^2$$
$$(x - 2)^2 + (y + 1)^2 = 4$$

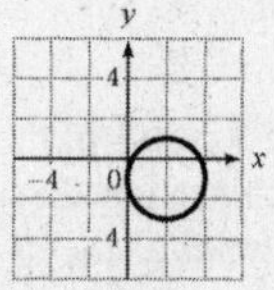

17. $(x_1, y_1) = (1, 2), (x_2, y_2) = (-1, 1)$
$$d = \sqrt{(x_2 - x_1)^2 + (y_2 - y_1)^2}$$
$$= \sqrt{(-1 - 1)^2 + (1 - 2)^2}$$
$$= \sqrt{(-2)^2 + (-1)^2}$$
$$= \sqrt{4 + 1}$$
$$= \sqrt{5}$$

$$(x - h)^2 + (y - k)^2 = r^2$$
$$[x - (-1)]^2 + (y - 1)^2 = (\sqrt{5})^2$$
$$(x + 1)^2 + (y - 1)^2 = 5$$

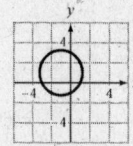

19. The endpoints of the diameter are $(-1, 4)$ and $(-5, 8)$. The center of the circle is the midpoint of the diameter.

$(x_1, y_1) = (-1, 4) \quad (x_2, y_2) = (-5, 8)$

$x_m = \dfrac{x_1 + x_2}{2} \qquad y_m = \dfrac{y_1 + y_2}{2}$

$= \dfrac{-1 + (-5)}{2} \qquad = \dfrac{4 + 8}{2}$

$= -3 \qquad\qquad = 6$

The center of the circle is $(-3, 6)$. The radius of the circle is the length of the segment connecting the center of the circle $(-3, 6)$ to an endpoint of the diameter (use either $(-1, 4)$ or $(-5, 8)$).

$r = \sqrt{(x_1 - x_m)^2 + (y_1 - y_m)^2}$

$r = \sqrt{(-1 - (-3))^2 + (4 - 6)^2}$

$r = \sqrt{4 + 4}$

$r = \sqrt{8}$

Write the equation of the circle with center $(-3, 6)$ and radius $\sqrt{8}$.

$(x + 3)^2 + (y - 6)^2 = 8$

21. The endpoints of the diameter are $(-4, 2)$ and $(0, 0)$. The center of the circle is the midpoint of the diameter.

$(x_1, y_1) = (-4, 2) \quad (x_2, y_2) = (0, 0)$

$x_m = \dfrac{x_1 + x_2}{2} \qquad y_m = \dfrac{y_1 + y_2}{2}$

$= \dfrac{-4 + 0}{2} \qquad = \dfrac{2 + 0}{2}$

$= -2 \qquad\qquad = 1$

The center of the circle is $(-2, 1)$. The radius of the circle is the length of the segment connecting the center of the circle $(-2, 1)$ to an endpoint of the diameter (use either $(-4, 2)$ or $(0, 0)$).

$r = \sqrt{(x_1 - x_m)^2 + (y_1 - y_m)^2}$

$r = \sqrt{(-4 - (-2))^2 + (2 - 1)^2}$

$r = \sqrt{4 + 1}$

$r = \sqrt{5}$

Write the equation of the circle with center $(-2, 1)$ and radius $\sqrt{5}$.

$(x + 2)^2 + (y - 1)^2 = 5$

Objective 12.2.2 Exercises

23.
  a. $8x$; $4y$; 5
  b. 16; 4; 16; 4; 16; 4
  c. $x + 4$; $y - 2$; 25

25. $x^2 + y^2 - 2x + 4y - 20 = 0$

$(x^2 - 2x) + (y^2 + 4y) = 20$

$(x^2 - 2x + 1) + (y^2 + 4y + 4) = 20 + 1 + 4$

$(x - 1)^2 + (y + 2)^2 = 25$

Center: $(1, -2)$
Radius: 5

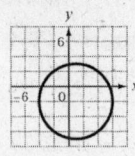

27. $x^2 + y^2 + 6x + 8y + 9 = 0$

$(x^2 + 6x) + (y^2 + 8y) = -9$

$(x^2 + 6x + 9) + (y^2 + 8y + 16) = -9 + 9 + 16$

$(x + 3)^2 + (y + 4)^2 = 16$

Center: $(-3, -4)$
Radius: 4

29. $x^2 + y^2 - x + 4y + \dfrac{13}{4} = 0$

$(x^2 - x) + (y^2 + 4y) = -\dfrac{13}{4}$

$\left(x^2 - x + \dfrac{1}{4}\right) + (y^2 + 4y + 4) = -\dfrac{13}{4} + \dfrac{1}{4} + 4$

$\left(x - \dfrac{1}{2}\right)^2 + (y + 2)^2 = 1$

Center: $\left(\dfrac{1}{2}, -2\right)$
Radius: 1

31. $x^2 + y^2 - 6x + 4y + 4 = 0$

$(x^2 - 6x) + (y^2 + 4y) = -4$

$(x^2 - 6x + 9) + (y^2 + 4y + 4) = -4 + 9 + 4$

$(x - 3)^2 + (y + 2)^2 = 9$

Center: $(3, -2)$
Radius: 3

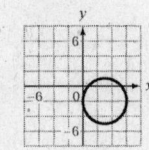

33. B
35. D

Applying Concepts 12.2

37. $(x_1, y_1) = (3, 0)$ $(x_2, y_2) = (0, 0)$
$r = \sqrt{(x_2 - x_1)^2 + (y_2 - y_1)^2}$
$= \sqrt{(0-3)^2 + (0-0)^2} = \sqrt{9} = 3$
$(x - h)^2 + (y - k)^2 = r^2$
$(x - 3)^2 + (y - 0)^2 = 3^2$
$(x - 3)^2 + y^2 = 9$

39. If the circle lies in quadrant II, has a radius of 1, and is tangent to both axes, then it must pass through the points (0, 1), (–1, 0), (–2, 1), and (–1, 2). The center must be (–1, 1).
$(x - h)^2 + (y - k)^2 = r^2$
$[x - (-1)]^2 + (y - 1)^2 = 1^2$
$(x + 1)^2 + (y - 1)^2 = 1$

41. Attempt to write the equation $x^2 + y^2 + 4x + 8y + 24 = 0$ in standard form.
$x^2 + y^2 + 4x + 8y + 24 = 0$
$(x^2 + 4x) + (y^2 + 8y) = -24$
$(x^2 + 4x + 4) + (y^2 + 8y + 16) = -24 + 4 + 16$
$(x + 2)^2 + (y + 4)^2 = -4$

This is not the equation of a circle because r^2 is negative ($r^2 = -4$), and the square of a real number cannot be negative.

43. The distance formula is used to derive the equation of a circle. If $C(h, k)$ are the coordinates of a fixed point in the plane and $P(x, y)$ is any other point in the plane, then the distance between C and P is $r = \sqrt{(x-h)^2 + (y-k)^2}$. Squaring each side of this equation gives the equation of a circle in standard form: $r^2 = (x-h)^2 + (y-k)^2$. Students might also note the connection between the *distance* formula and the definition of a circle as the set of all points in the plane that are a fixed *distance* from the center; hence the same derivation for the two formulas.

Section 12.3

Concept Review 12.3

1. Never true
A vertical line will intersect the graph of an ellipse at more than one point. By the vertical-line test, the graph of an ellipse is not the graph of a function.

3. Always true

5. Sometimes true
The hyperbola $\dfrac{y^2}{4} - \dfrac{x^2}{16} = 1$ has no *x*-intercepts.

Objective 12.3.1 Exercises

1. ellipse; the origin; $(a, 0)$; $(-a, 0)$; $(0, b)$; $(0, -b)$

3. *x*-intercepts: (2, 0) and (–2, 0)
y-intercepts: (0, 3) and (0, –3)

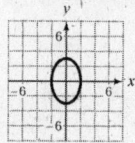

5. *x*-intercepts: (5, 0) and (–5, 0)
y-intercepts: (0, 3) and (0, –3)

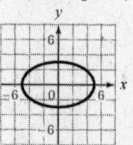

7. *x*-intercepts: (6, 0) and (–6, 0)
y-intercepts: (0, 4) and (0, –4)

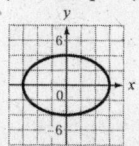

9. *x*-intercepts: (3, 0) and (–3, 0)
y-intercepts: (0, 5) and (0, –5)

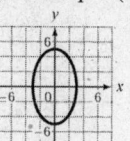

11. *x*-intercepts: (6, 0) and (–6, 0)
y-intercepts: (0, 3) and (0, –3)

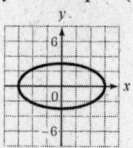

13. *x*-intercepts: $(2\sqrt{3}, 0)$ and $(-2\sqrt{3}, 0)$
y-intercepts: (0, 2) and (0, –2)

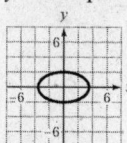

15. Less than

Objective 12.3.2 Exercises

17. If the equation is in standard form, the graph will be an ellipse if the terms are added. The graph will be a hyperbola if one term is subtracted from the other.

19. 25; 49

21. *y*; (0, 7); (0, –7)

368 Chapter 12: Conic Sections

23. Axis of symmetry: x-axis
Vertices: $(3, 0)$ and $(-3, 0)$
Asymptotes: $y = \frac{4}{3}x$ and $y = -\frac{4}{3}x$

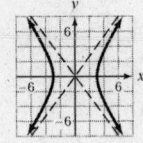

25. Axis of symmetry: y-axis
Vertices: $(0, 4)$ and $(0, -4)$
Asymptotes: $y = \frac{4}{3}x$ and $y = -\frac{4}{3}x$

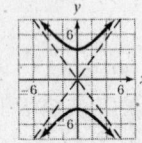

27. Axis of symmetry: x-axis
Vertices: $(2, 0)$ and $(-2, 0)$
Asymptotes: $y = \frac{5}{2}x$ and $y = -\frac{5}{2}x$

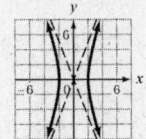

29. Axis of symmetry: y-axis
Vertices: $(0, 5)$ and $(0, -5)$
Asymptotes: $y = \frac{5}{3}x$ and $y = -\frac{5}{3}x$

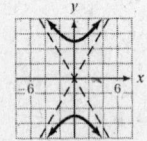

31. Axis of symmetry: x-axis
Vertices: $(5, 0)$ and $(-5, 0)$
Asymptotes: $y = \frac{4}{5}x$ and $y = -\frac{4}{5}x$

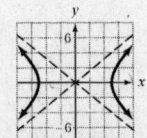

33. Axis of symmetry: y-axis
Vertices: $(0, 4)$ and $(0, -4)$
Asymptotes: $y = 2x$ and $y = -2x$

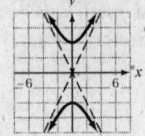

35. Axis of symmetry: x-axis
Vertices: $(5, 0)$ and $(-5, 0)$
Asymptotes: $y = \frac{3}{5}x$ and $y = -\frac{3}{5}x$

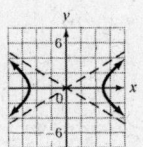

37. Axis of symmetry: y-axis
Vertices: $(0, 4)$ and $(0, -4)$
Asymptotes: $y = x$ and $y = -x$

Applying Concepts 12.3

39. The graph of $4x^2 + y^2 = 16$ is an ellipse.
$$4x^2 + y^2 = 16$$
$$\frac{4x^2 + y^2}{16} = \frac{16}{16}$$
$$\frac{4x^2}{16} + \frac{y^2}{16} = 1$$
$$\frac{x^2}{4} + \frac{y^2}{16} = 1$$
x-intercepts: $(2, 0)$ and $(-2, 0)$
y-intercepts: $(0, 4)$ and $(0, -4)$

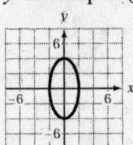

41. The graph of $y^2 - 4x^2 = 16$ is a hyperbola.
$$y^2 - 4x^2 = 16$$
$$\frac{y^2 - 4x^2}{16} = \frac{16}{16}$$
$$\frac{y^2}{16} - \frac{4x^2}{16} = 1$$
$$\frac{y^2}{16} - \frac{x^2}{4} = 1$$
Axis of symmetry: y-axis
Vertices: $(0, 4)$ and $(0, -4)$
Asymptotes: $y = 2x$ and $y = -2x$

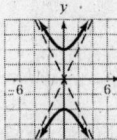

43. The graph of $9x^2 - 25y^2 = 225$ is a hyperbola.

$$9x^2 - 25y^2 = 225$$
$$\frac{9x^2 - 25y^2}{225} = \frac{225}{225}$$
$$\frac{9x^2}{225} - \frac{25y^2}{225} = 1$$
$$\frac{x^2}{25} - \frac{y^2}{9} = 1$$

Axis of symmetry: x-axis
Vertices: $(5, 0)$ and $(-5, 0)$
Asymptotes: $y = \frac{3}{5}x$ and $y = -\frac{3}{5}x$

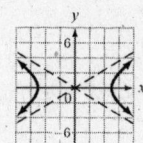

45.

a. $\frac{x^2}{a^2} + \frac{y^2}{b^2} = 1$

$\frac{x^2}{18^2} + \frac{y^2}{4.5^2} = 1$

$\frac{x^2}{324} + \frac{y^2}{20.25} = 1$

b. Major axis = 36 AU
36 AU ÷ 2 = 18 AU
18 AU = 18(92,960,000 mi)
= 1,673,280,000 mi
$\sqrt{a^2 - b^2} = \sqrt{324 - 20.25}$
$= \sqrt{303.75}$
≈ 17.42842506
17.42842506 AU
= 17.42842506(92,960,000 mi)
$\approx 1,620,146,393$ mi
1,673,280,000 + 1,620,146,393
$\approx 3,293,400,000$
The distance from the Sun to the point at the aphelion is about 3,293,400,000 mi.

c. 18 AU $- \sqrt{a^2 - b^2}$ AU
= 1,673,280,000 mi $-$ 1,620,146,393 mi
= 53,133,607 $\approx$ 53,100,000
The distance from the Sun to the point at the perihelion is about 53,100,000 mi.

47.

a. $\frac{x^2}{a^2} + \frac{y^2}{b^2} = 1$

$\frac{x^2}{1.52^2} + \frac{y^2}{1.495^2} = 1$

$\frac{x^2}{2.310} + \frac{y^2}{2.235} = 1$

b. Major axis = 3.04 AU
3.04 AU ÷ 2 = 1.52 AU
1.52 AU = 1.52(92,960,000 mi)
= 141,299,200 mi
$\sqrt{a^2 - b^2} = \sqrt{2.310 - 2.235}$
$= \sqrt{0.075}$
≈ 0.2738612788
0.2738612788 AU
= 0.2738612788(92,960,000 mi)
$\approx 25,488,144$ mi
141,299,200 + 25,458,144 = 168,757,344
$\approx 166,800,000$
The aphelion is about 166,800,000 mi.

c. 1.52 AU $- \sqrt{a^2 - b^2}$ AU
= 141,299,200 mi $-$ 25,458,144 mi
$\approx 115,800,000$
The perihelion is about 115,800,000 mi.

Section 12.4

Concept Review 12.4

1. Never true
Two ellipses with centers at the origin may intersect at four points or they may not intersect.

3. Sometimes true
A straight line may intersect a parabola at one point or two points, or the line may not intersect the parabola.

5. Always true

Objective 12.4.1 Exercises

1. Nonlinear systems of equations are systems in which one or more equations are not linear equations. If all equations are linear, the system is a linear system.

3. (1) $y = x^2 - x - 1$
(2) $y = 2x + 9$
Use the substitution method.
$y = x^2 - x - 1$
$2x + 9 = x^2 - x - 1$
$0 = x^2 - 3x - 10$
$0 = (x - 5)(x + 2)$
$x - 5 = 0 \quad x + 2 = 0$
$x = 5 \quad\quad x = -2$
Substitute into equation (2).
$y = 2x + 9 \quad\quad y = 2x + 9$
$y = 2(5) + 9 \quad y = 2(-2) = 9$
$y = 10 + 9 \quad\quad y = -4 + 9$
$y = 19 \quad\quad\quad y = 5$
The solutions are (5, 19) and (−2, 5).

5. (1) $y^2 = -x + 3$
(2) $x - y = 1$
Solve equation (2) for x.
$x - y = 1$
$x = y + 1$
Use the substitution method.
$y^2 = -x + 3$
$y^2 = -(y+1) + 3$
$y^2 = -y - 1 + 3$
$y^2 = -y + 2$
$y^2 + y - 2 = 0$
$(y+2)(y-1) = 0$
$y + 2 = 0 \quad y - 1 = 0$
$y = -2 \quad y = 1$
Substitute into equation (2).
$x - y = 1 \quad x - y = 1$
$x - (-2) = 1 \quad x - 1 = 1$
$x + 2 = 1 \quad x = 2$
$x = -1$
The solutions are $(-1, -2)$ and $(2, 1)$.

7. (1) $y^2 = 2x$
(2) $x + 2y = -2$
Solve equation (2) for x.
$x + 2y = -2$
$x = -2y - 2$
Use the substitution method.
$y^2 = 2x$
$y^2 = 2(-2y - 2)$
$y^2 = -4y - 4$
$y^2 + 4y + 4 = 0$
$(y+2)(y+2) = 0$
$y + 2 = 0$
$y = -2$
The solution is a double root.
Substitute into equation (2).
$x + 2y = -2$
$x + 2(-2) = -2$
$x - 4 = -2$
$x = 2$
The solution is $(2, -2)$.

9. (1) $x^2 + 2y^2 = 12$
(2) $2x - y = 2$
Solve equation (2) for y.
$2x - y = 2$
$-y = -2x + 2$
$y = 2x - 2$
Use the substitution method.
$x^2 + 2y^2 = 12$
$x^2 + 2(2x - 2)^2 = 12$
$x^2 + 2(4x^2 - 8x + 4) = 12$
$x^2 + 8x^2 - 16x + 8 = 12$
$9x^2 - 16x - 4 = 0$
$(x - 2)(9x + 2) = 0$
$x - 2 = 0 \quad 9x + 2 = 0$
$x = 2 \quad 9x = -2$
$x = -\dfrac{2}{9}$
Substitute into equation (2).
$2x - y = 2 \quad\quad 2x - y = 2$
$2(2) - y = 2 \quad 2\left(-\dfrac{2}{9}\right) - y = 2$
$4 - y = 2 \quad\quad -\dfrac{4}{9} - y = 2$
$-y = -2$
$y = 2 \quad\quad -y = \dfrac{22}{9}$
$y = -\dfrac{22}{9}$
The solutions are $(2, 2)$ and $\left(-\dfrac{2}{9}, -\dfrac{22}{9}\right)$.

11. (1) $x^2 + y^2 = 13$
(2) $x + y = 5$
Solve equation (2) for y.
$x + y = 5$
$y = -x + 5$
Use the substitution method.
$x^2 + y^2 = 13$
$x^2 + (-x + 5)^2 = 13$
$x^2 + x^2 - 10x + 25 = 13$
$2x^2 - 10x + 12 = 0$
$2(x^2 - 5x + 6) = 0$
$2(x - 3)(x - 2) = 0$
$x - 3 = 0 \quad x - 2 = 0$
$x = 3 \quad x = 2$
Substitute into equation (2).
$x + y = 5 \quad x + y = 5$
$3 + y = 5 \quad 2 + y = 5$
$y = 2 \quad y = 3$
The solutions are $(3, 2)$ and $(2, 3)$.

13. (1) $4x^2 + y^2 = 12$
(2) $y = 4x^2$

Use the substitution method.
$$4x^2 + y^2 = 12$$
$$4x^2 + (4x^2)^2 = 12$$
$$4x^2 + 16x^4 = 12$$
$$16x^4 + 4x^2 - 12 = 0$$
$$4(4x^4 + x^2 - 3) = 0$$
$$4(4x^2 - 3)(x^2 + 1) = 0$$

$4x^2 - 3 = 0 \qquad x^2 + 1 = 0$
$4x^2 = 3 \qquad\qquad x^2 = -1$
$x^2 = \dfrac{3}{4} \qquad\qquad x = \pm\sqrt{-1}$
$x = \pm\dfrac{\sqrt{3}}{2}$

Substitute the real number solutions into equation (2).

$y = 4x^2 \qquad\qquad y = 4x^2$
$y = 4\left(\dfrac{\sqrt{3}}{2}\right)^2 \quad y = 4\left(-\dfrac{\sqrt{3}}{2}\right)^2$
$y = 4\left(\dfrac{3}{4}\right) \qquad y = 4\left(\dfrac{3}{4}\right)$
$y = 3 \qquad\qquad y = 3$

The solutions are $\left(\dfrac{\sqrt{3}}{2}, 3\right)$ and $\left(-\dfrac{\sqrt{3}}{2}, 3\right)$.

15. (1) $y = x^2 - 2x - 3$
(2) $y = x - 6$

Use the substitution method.
$$y = x^2 - 2x - 3$$
$$x - 6 = x^2 - 2x - 3$$
$$0 = x^2 - 3x + 3$$
$$x = \dfrac{-b \pm \sqrt{b^2 - 4ac}}{2a}$$
$$= \dfrac{-(-3) \pm \sqrt{(-3)^2 - 4(1)(3)}}{2(1)}$$
$$= \dfrac{3 \pm \sqrt{9 - 12}}{2}$$
$$= \dfrac{3 \pm \sqrt{-3}}{2}$$

Since the discriminant is less than zero, the equation has two complex number solutions. Therefore, the system of equations has no real number solution.

17. (1) $3x^2 - y^2 = -1$
(2) $x^2 + 4y^2 = 17$

Use the addition method.
Multiply equation (1) by 4.
$$12x^2 - 4y^2 = -4$$
$$x^2 + 4y^2 = 17$$
$$13x^2 = 13$$
$$x^2 = 1$$
$$x = \pm\sqrt{1} = \pm 1$$

Substitute into equation (2).

$x^2 + 4y^2 = 17 \qquad x^2 + 4y^2 = 17$
$1^2 + 4y^2 = 17 \qquad (-1)^2 + 4y^2 = 17$
$1 + 4y^2 = 17 \qquad 1 + 4y^2 = 17$
$4y^2 = 16 \qquad\qquad 4y^2 = 16$
$y^2 = 4 \qquad\qquad y^2 = 4$
$y = \pm\sqrt{4} \qquad\qquad y = \pm\sqrt{4}$
$y = \pm 2 \qquad\qquad y = \pm 2$

The solutions are (1, 2), (1, –2), (–1, 2), and (–1, –2).

19. (1) $2x^2 + 3y^2 = 30$
(2) $x^2 + y^2 = 13$

Use the addition method.
Multiply equation (2) by –2.
$$2x^2 + 3y^2 = 30$$
$$-2x^2 - 2y^2 = -26$$
$$y^2 = 4$$
$$y = \pm\sqrt{4} = \pm 2$$

Substitute into equation (2).

$x^2 + y^2 = 13 \qquad x^2 + y^2 = 13$
$x^2 + 2^2 = 13 \qquad x^2 + (-2)^2 = 13$
$x^2 + 4 = 13 \qquad x^2 + 4 = 13$
$x^2 = 9 \qquad\qquad x^2 = 9$
$x = \pm\sqrt{9} \qquad\qquad x = \pm\sqrt{9}$
$x = \pm 3 \qquad\qquad x = \pm 3$

The solutions are (3, 2), (3, –2), (–3, 2), and (–3, –2).

372 Chapter 12: Conic Sections

21. (1) $y = 2x^2 - x + 1$
(2) $y = x^2 - x + 5$
Use the substitution method.
$$y = 2x^2 - x + 1$$
$$x^2 - x + 5 = 2x^2 - x + 1$$
$$0 = x^2 - 4$$
$$0 = (x + 2)(x - 2)$$
$$x + 2 = 0 \quad x - 2 = 0$$
$$x = -2 \quad x = 2$$
Substitute into equation (2).
$$y = x^2 - x + 5 \quad y = x^2 - x + 5$$
$$y = (-2)^2 - (-2) + 5 \quad y = 2^2 - 2 + 5$$
$$y = 4 + 2 + 5 \quad y = 4 - 2 + 5$$
$$y = 11 \quad y = 7$$
The solutions are $(2, 7)$ and $(-2, 11)$.

23. (1) $2x^2 + 3y^2 = 24$
(2) $x^2 - y^2 = 7$
Use the addition method.
Multiply equation (2) by 3.
$$2x^2 + 3y^2 = 24$$
$$3x^2 - 3y^2 = 21$$
$$5x^2 = 45$$
$$x^2 = 9 = \pm\sqrt{9} = \pm 3$$
Substitute into equation (2).
$$x^2 - y^2 = 7 \quad x^2 - y^2 = 7$$
$$3^2 - y^2 = 7 \quad (-3)^2 - y^2 = 7$$
$$9 - y^2 = 7 \quad 9 - y^2 = 7$$
$$-y^2 = -2 \quad -y^2 = -2$$
$$y^2 = 2 \quad y^2 = 2$$
$$y = \pm\sqrt{2} \quad y = \pm\sqrt{2}$$
The solutions are $(3, \sqrt{2})$, $(3, -\sqrt{2})$, $(-3, \sqrt{2})$, and $(-3, -\sqrt{2})$.

25. (1) $x^2 + y^2 = 36$
(2) $4x^2 + 9y^2 = 36$
Use the addition method.
Multiply equation (1) by -4.
$$-4x^2 - 4y^2 = -144$$
$$4x^2 + 9y^2 = 36$$
$$5y^2 = -108$$
$$y^2 = -\frac{108}{5}$$
$$y = \pm\sqrt{-\frac{108}{5}}$$
The system of equations has no real number solution.

27. (1) $11x^2 - 2y^2 = 4$
(2) $3x^2 + y^2 = 15$
Use the addition method.
Multiply equation (2) by 2.
$$11x^2 - 2y^2 = 4$$
$$6x^2 + 2y^2 = 30$$
$$17x^2 = 34$$
$$x^2 = 2$$
$$x = \pm\sqrt{2}$$
Substitute into equation (2).
$$3x^2 + y^2 = 15 \quad 3x^2 + y^2 = 15$$
$$3(\sqrt{2}) + y^2 = 15 \quad 3(-\sqrt{2}) + y^2 = 15$$
$$3(2) + y^2 = 15 \quad 3(2) + y^2 = 15$$
$$6 + y^2 = 15 \quad 6 + y^2 = 15$$
$$y^2 = 9 \quad y^2 = 9$$
$$y = \pm\sqrt{9} \quad y = \pm\sqrt{9}$$
$$y = \pm 3 \quad y = \pm 3$$
The solutions are $(\sqrt{2}, 3)$, $(\sqrt{2}, -3)$, $(-\sqrt{2}, 3)$, and $(-\sqrt{2}, -3)$.

29. (1) $2x^2 - y^2 = 7$
(2) $2x - y = 5$
Solve equation (2) for y.
$$2x - y = 5$$
$$-y = -2x + 5$$
$$y = 2x - 5$$
Use the substitution method.
$$2x^2 - y^2 = 7$$
$$2x^2 - (2x - 5)^2 = 7$$
$$2x^2 - (4x^2 - 20x + 25) = 7$$
$$2x^2 - 4x^2 + 20x - 25 = 7$$
$$-2x^2 + 20x - 32 = 0$$
$$-2(x^2 - 10x + 16) = 0$$
$$-2(x - 2)(x - 8) = 0$$
$$x - 2 = 0 \quad x - 8 = 0$$
$$x = 2 \quad x = 8$$
Substitute into equation (2).
$$2x - y = 5 \quad 2x - y = 5$$
$$2(2) - y = 5 \quad 2(8) - y = 5$$
$$4 - y = 5 \quad 16 - y = 5$$
$$-y = 1 \quad -y = -11$$
$$y = -1 \quad y = 11$$
The solutions are $(2, -1)$ and $(8, 11)$.

31. (1) $y = 3x^2 + x - 4$
(2) $y = 3x^2 - 8x + 5$
Use the substitution method.
$y = 3x^2 + x - 4$
$3x^2 - 8x + 5 = 3x^2 + x - 4$
$-9x = -9$
$x = 1$
Substitute into equation (1).
$y = 3x^2 + x - 4$
$y = 3(1)^2 + (1) - 4$
$y = 3(1) + 1 - 4$
$y = 3 + 1 - 4$
$y = 0$
The solution is (1, 0).

33. An ellipse and a circle that intersect in two points.

35. A parabola and a straight line that intersect in one point.

Applying Concepts 12.4

37. $y = 2^x$
$x + y = 3$

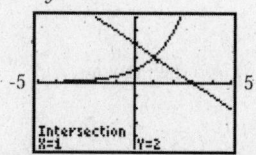

The solution is (1.000, 2.000).

39. $y = \log_2 x$
$\dfrac{x^2}{9} + \dfrac{y^2}{1} = 1$

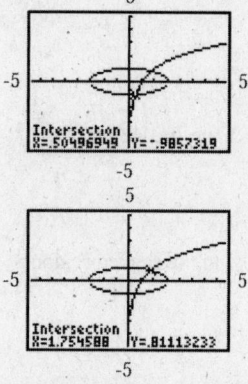

The approximate solutions are (1.755, 0.811) and (0.505, −0.986).

41. $y = -\log_3 x$
$x + y = 4$

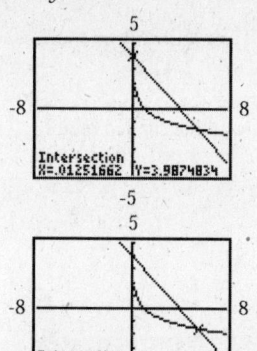

The approximate solutions are (0.013, 3.987) and (5.562, −1.562).

43. No. Because the center of each circle is the origin, it is not possible for two circles to have exactly two points of intersection. They can intersect at no points or at infinitely many points.

Section 12.5

Concept Review 12.5

1. Always true

3. Never true
$0 + 0 > 4$ is not a true statement. The point (0, 0) is not a solution of the inequality.

5. Sometimes true
If the inequality includes the symbol ≤ or ≥, the solution set will include the boundary.

Objective 12.5.1 Exercises

1.
a. parabola
b. dashed
c. is

3. $y \leq x^2 - 4x + 3$
Substitute the point (0, 0) into the inequality.
$0 \leq 0^2 - 4(0) + 3$
$0 \leq 3$ true
The point (0, 0) should be in the shaded region.

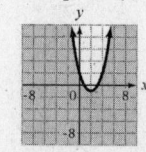

374 Chapter 12: Conic Sections

5. $(x-1)^2 + (y+2)^2 \leq 9$
Substitute the point (0, 0) into the inequality.
$(0-1)^2 + (0+2)^2 \leq 9$
$-1^2 + 2^2 \leq 9$
$3 \leq 9$ true
The point (0, 0) should be in the shaded region.

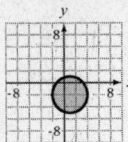

7. $(x+3)^2 + (y-2)^2 \geq 9$
Substitute the point (0, 0) into the inequality.
$(0+3)^2 + (0-2)^2 \geq 9$
$3^2 + (-2)^2 \geq 9$
$13 \geq 9$ true
The point (0, 0) should be in the shaded region.

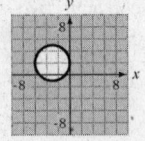

9. $\dfrac{x^2}{16} + \dfrac{y^2}{25} < 1$
Substitute the point (0, 0) into the inequality.
$\dfrac{0^2}{16} + \dfrac{0^2}{25} < 1$
$0 < 1$ true
The point (0, 0) should be in the shaded region.

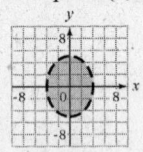

11. $\dfrac{x^2}{25} - \dfrac{y^2}{9} \leq 1$
Substitute the point (0, 0) into the inequality.
$\dfrac{0^2}{25} - \dfrac{0^2}{9} \leq 1$
$0 \leq 1$ true
The point (0, 0) should be in the shaded region.

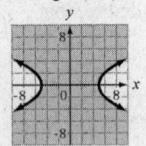

13. $\dfrac{x^2}{4} + \dfrac{y^2}{16} \geq 1$
Substitute the point (0, 0) into the inequality.
$\dfrac{0^2}{4} + \dfrac{0^2}{16} \geq 1$
$0 \geq 1$ false
The point (0, 0) should not be in the shaded region.

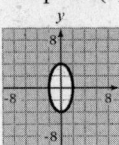

15. $y \leq x^2 - 2x + 3$
Substitute the point (0, 0) into the inequality.
$0 \leq 0^2 - 2(0) + 3$
$0 \leq 3$ true
The point (0, 0) should be in the shaded region.

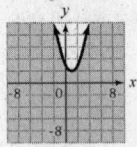

17. $\dfrac{y^2}{9} - \dfrac{x^2}{16} \leq 1$
Substitute the point (0, 0) into the inequality.
$\dfrac{0^2}{9} - \dfrac{0^2}{16} \leq 1$
$0 \leq 1$ true
The point (0, 0) should be in the shaded region.

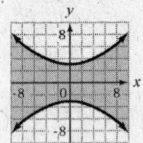

19. $\dfrac{x^2}{9} + \dfrac{y^2}{1} \leq 1$
Substitute the point (0, 0) into the inequality.
$\dfrac{0^2}{9} + \dfrac{0^2}{1} \leq 1$
$0 \leq 1$ true
The point (0, 0) should be in the shaded region.

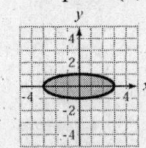

21. $(x-1)^2 + (y+3)^2 \leq 25$

Substitute the point (0, 0) into the inequality.

$(0-1)^2 + (0+3)^2 \leq 25$

$(-1)^2 + 3^2 \leq 25$

$10 \leq 25$ true

The point (0, 0) should be in the shaded region.

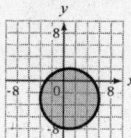

23. $\dfrac{y^2}{25} - \dfrac{x^2}{4} \leq 1$

Substitute the point (0, 0) into the inequality.

$\dfrac{0^2}{25} - \dfrac{0^2}{4} \leq 1$

$0 \leq 1$ true

The point (0, 0) should be in the shaded region.

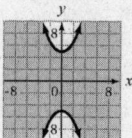

25. $\dfrac{x^2}{25} + \dfrac{y^2}{9} \leq 1$

Substitute the point (0, 0) into the inequality.

$\dfrac{0^2}{25} + \dfrac{0^2}{9} \leq 1$

$0 \leq 1$ true

The point (0, 0) should be in the shaded region.

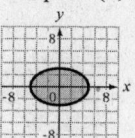

27. True

Objective 12.5.2 Exercises

29. intersection

31. $y \leq x^2 - 4x + 4$

$y + x > 4$

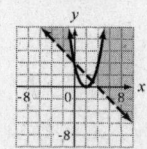

33. $x^2 + y^2 < 16$

$y > x + 1$

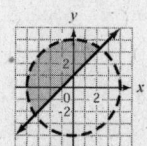

35. $\dfrac{x^2}{4} + \dfrac{y^2}{16} \leq 1$

$y \leq -\dfrac{1}{2}x + 2$

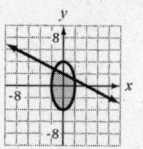

37. $x \geq y^2 - 3y + 2$

$y \geq 2x - 2$

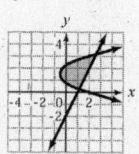

39. $x^2 + y^2 < 25$

$\dfrac{x^2}{9} + \dfrac{y^2}{36} < 1$

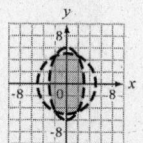

41. $x^2 + y^2 > 4$

$x^2 + y^2 < 25$

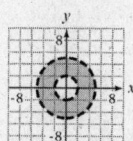

43. Region inside the circle $x^2 + y^2 = a$

45. Region between the circles $x^2 + y^2 = a$ and $x^2 + y^2 = b$

Applying Concepts 12.5

47. $y > x^2 - 3$

$y < x + 3$

$x \leq 0$

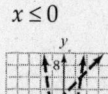

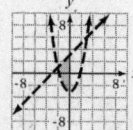

376 Chapter 12: Conic Sections

49. $x^2 + y^2 < 3$
$x > y^2 - 1$
$y \geq 0$

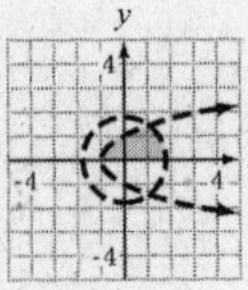

51. $\dfrac{x^2}{4} + \dfrac{y^2}{1} \leq 4$
$x^2 + y^2 \leq 4$
$x \geq 0$
$y \leq 0$

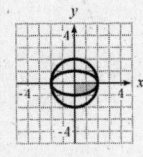

53. $y > 2^x$
$x + y < 4$

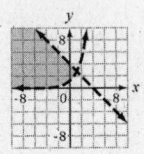

55. $y \geq \log_2 x$
$x^2 + y^2 < 9$

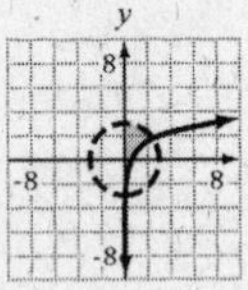

57. $y < 3^{-x}$
$\dfrac{x^2}{4} - \dfrac{y^2}{1} \geq 1$

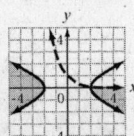

CHAPTER 12 REVIEW EXERCISES

1. $y = x^2 - 4x + 8$
$-\dfrac{b}{2a} = -\dfrac{-4}{2(1)} = \dfrac{4}{2} = 2$
$y = 2^2 - 4(2) + 8$
$= 4$
Vertex: $(2, 4)$
Axis of symmetry: $x = 2$

2. $y = -x^2 + 7x - 8$
$-\dfrac{b}{2a} = -\dfrac{7}{2(-1)} = \dfrac{7}{2}$
$y = -\left(\dfrac{7}{2}\right)^2 + 7\left(\dfrac{7}{2}\right) - 8$
$= -\dfrac{49}{4} + \dfrac{49}{2} - 8$
$= \dfrac{17}{4}$
Vertex: $\left(\dfrac{7}{2}, \dfrac{17}{4}\right)$
Axis of symmetry: $x = \dfrac{7}{2}$

3. $y = -2x^2 + x - 2$

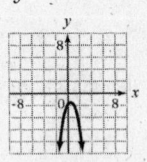

4. $x = 2y^2 - 6y + 5$

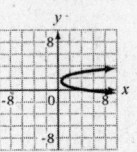

5. $(x_1, y_1) = (2, -1) \quad (x_2, y_2) = (-1, 2)$
$r = \sqrt{(x_2 - x_1)^2 + (y_2 - y_1)^2}$
$= \sqrt{(-1 - 2)^2 + (2 - (-1))^2}$
$= \sqrt{(-3)^2 + 3^2}$
$= \sqrt{9 + 9} = \sqrt{18}$
$(x - h)^2 + (y - k)^2 = r^2$
$[x - (-1)]^2 + (y - 2)^2 = (\sqrt{18})^2$
$(x + 1)^2 + (y - 2)^2 = 18$

6. $(x - h)^2 + (y - k)^2 = r^2$
$(x - (-1))^2 + (y - 5)^2 = 6^2$
$(x + 1)^2 + (y - 5)^2 = 36$

7. $(x + 3)^2 + (y + 1)^2 = 1$

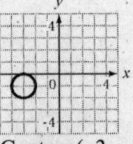

Center: $(-3, -1)$
Radius: 1

8. $x^2+(y-2)^2=9$

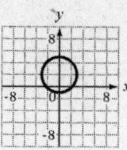

Center: (0, 2)
Radius: 3

9. $(x_1, y_1) = (4,6) \quad (x_2, y_2) = (0,-3)$
$r = \sqrt{(x_2-x_1)^2+(y_2-y_1)^2}$
$= \sqrt{(0-4)^2+(-3-6)^2}$
$= \sqrt{(-4)^2+(-9)^2}$
$= \sqrt{16+81} = \sqrt{97}$
$(x-h)^2+(y-k)^2 = r^2$
$(x-0)^2+(y-(-3))^2 = (\sqrt{97})^2$
$x^2+(y+3)^2 = 97$

10. $x^2+y^2+4x-2y=4$
$(x^2+4x)+(y^2-2y)=4$
$(x^2+4x+4)+(y^2-2y+1)=4+4+1$
$(x+2)^2+(y-1)^2=9$

11. $\dfrac{x^2}{1}+\dfrac{y^2}{9}=1$
x-intercepts: (1, 0) and (−1, 0)
y-intercepts: (0, 3) and (0, −3)

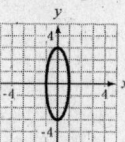

12. $\dfrac{x^2}{25}+\dfrac{y^2}{9}=1$
x-intercepts: (5, 0) and (−5, 0)
y-intercepts: (0, 3) and (0, −3)

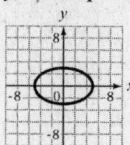

13. $\dfrac{x^2}{25}-\dfrac{y^2}{1}=1$
Axis of symmetry: x-axis
Vertices: (5, 0) and (−5, 0)
Asymptotes: $y=\dfrac{1}{5}x$ and $y=-\dfrac{1}{5}x$

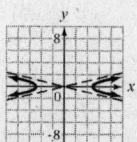

14. $\dfrac{y^2}{16}-\dfrac{x^2}{9}=1$
Axis of symmetry: y-axis
Vertices: (0, 4) and (0, −4)
Asymptotes: $y=\dfrac{4}{3}x$ and $y=-\dfrac{4}{3}x$

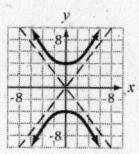

15. (1) $y=x^2+5x-6$
(2) $y=x-10$
Use the substitution method.
$x-10 = x^2+5x-6$
$0 = x^2+4x+4$
$0 = (x+2)^2$
$x+2=0 \quad x+2=0$
$x=-2 \quad x=-2$
Substitute into equation (2).
$y=-2-10 \quad y=-2-10$
$y=-12 \quad y=-12$
The solution is (−2, −12)

16. (1) $2x^2+y^2=19$
(2) $3x^2-y^2=6$
Use the addition method.
$2x^2+y^2=19$
$3x^2-y^2=6$
$5x^2=25$
$x^2=5$
$x=\pm\sqrt{5}$
Substitute into equation (1).
$2x^2+y^2=19 \quad 2x^2+y^2=19$
$2(\sqrt{5})^2+y^2=19 \quad 2(-\sqrt{5})^2+y^2=19$
$10+y^2=19 \quad 10+y^2=19$
$y^2=9 \quad y^2=9$
$y=\pm\sqrt{9} \quad y=\pm\sqrt{9}$
$y=\pm 3 \quad y=\pm 3$
The solutions are $(\sqrt{5}, 3), (\sqrt{5}, -3), (-\sqrt{5}, 3),$ and $(-\sqrt{5}, -3)$.

378 Chapter 12: Conic Sections

17. (1) $x = 2y^2 - 3y + 1$
    (2) $3x - 2y = 0$
    Use the substitution method.
    $3x - 2y = 0$
    $3x = 2y$
    $x = \dfrac{2}{3}y$
    $\dfrac{2}{3}y = 2y^2 - 3y + 1$
    $2y = 6y^2 - 9y + 3$
    $0 = 6y^2 - 11y + 3$
    $0 = (2y - 3)(3y - 1)$
    $0 = 2y - 3 \quad 0 = 3y - 1$
    $2y = 3 \quad\quad 3y = 1$
    $y = \dfrac{3}{2} \quad\quad y = \dfrac{1}{3}$
    Substitute into equation (2).
    $3x - 2\left(\dfrac{3}{2}\right) = 0 \quad\quad 3x - 2\left(\dfrac{1}{3}\right) = 0$
    $3x - 3 = 0 \quad\quad\quad\; 3x - \dfrac{2}{3} = 0$
    $3x = 3 \quad\quad\quad\quad\quad 3x = \dfrac{2}{3}$
    $x = 1 \quad\quad\quad\quad\quad\; x = \dfrac{2}{9}$
    The solutions are $\left(1, \dfrac{3}{2}\right)$ and $\left(\dfrac{2}{9}, \dfrac{1}{3}\right)$.

18. (1) $y^2 = 2x^2 - 3x + 6$
    (2) $y^2 = 2x^2 + 5x - 2$
    Use the addition method.
    Multiply equation (2) by –1.
    $y^2 = 2x^2 - 3x + 6$
    $-y^2 = -2x^2 - 5x + 2$
    $0 = -8x + 8$
    $8x = 8$
    $x = 1$
    Substitute into equation (1).
    $y^2 = 2x^2 - 3x + 6$
    $y^2 = 2(1)^2 - 3(1) + 6$
    $y^2 = 2 - 3 + 6$
    $y^2 = 5$
    $y = \pm\sqrt{5}$
    The solutions are $(1, \sqrt{5})$ and $(1, -\sqrt{5})$.

19. $(x-2)^2 + (y+1)^2 \leq 16$

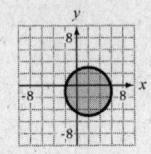

20. $\dfrac{x^2}{9} - \dfrac{y^2}{16} < 1$

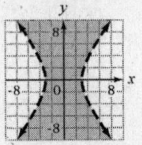

21. $y \geq -x^2 - 2x + 3$

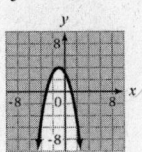

22. $\dfrac{x^2}{16} + \dfrac{y^2}{4} > 1$

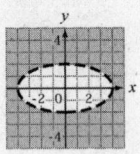

23. (1) $y \geq x^2 - 4x + 2$
    (2) $y \leq \dfrac{1}{3}x - 1$
    Write equation (1) in standard form.
    $y \geq (x^2 - 4x + 4) - 4 + 2$
    $y \geq (x-2)^2 - 2$

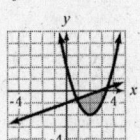

24. $\dfrac{x^2}{25} + \dfrac{y^2}{16} \leq 1$
    $\dfrac{y^2}{4} - \dfrac{x^2}{4} \geq 1$

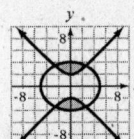

25. $\dfrac{x^2}{9} + \dfrac{y^2}{1} \geq 1$
    $\dfrac{x^2}{4} - \dfrac{y^2}{1} \leq 1$

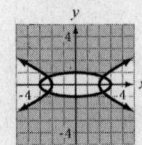

26. $\dfrac{x^2}{16}+\dfrac{y^2}{4}<1$
$x^2+y^2>9$

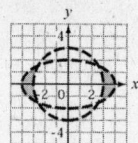

CHAPTER 12 TEST

1. $y=-x^2+6x-5$
$-\dfrac{b}{2a}=\dfrac{-6}{2(-1)}=\dfrac{-6}{-2}=3$
Axis of symmetry: $x=3$

2. $y=-x^2+3x-2$
$-\dfrac{b}{2a}=\dfrac{-3}{2(-1)}=\dfrac{3}{2}$
$y=-\left(\dfrac{3}{2}\right)^2+3\left(\dfrac{3}{2}\right)-2=-\dfrac{9}{4}+\dfrac{9}{2}-2=\dfrac{1}{4}$
Vertex: $\left(\dfrac{3}{2},\dfrac{1}{4}\right)$

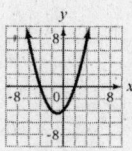

3. $y=\dfrac{1}{2}x^2+x-4$

4. $x=y^2-y-2$

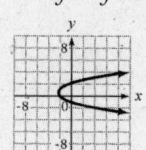

5. $(x-h)^2+(y-k)^2=r^2$
$[x-(-3)]^2+[y-(-3)]^2=4^2$
$(x+3)^2+(y+3)^2=16$

6. (1) $x^2+2y^2=4$
(2) $x+y=2$
Use the substitution method.
$x+y=2$
$y=2-x$
$x^2+2(2-x)^2=4$
$x^2+2(4-4x+x^2)=4$
$x^2+8-8x+2x^2=4$
$3x^2-8x+4=0$
$(3x-2)(x-2)=0$
$3x-2=0 \qquad x-2=0$
$x=\dfrac{2}{3} \qquad x=2$
Substitute into equation (2).
$x+y=2 \qquad x+y=2$
$\dfrac{2}{3}+y=2 \qquad 2+y=2$
$y=\dfrac{4}{3} \qquad y=0$

The solutions are $\left(\dfrac{2}{3},\dfrac{4}{3}\right)$ and $(2,0)$.

7. (1) $x=3y^2+2y-4$
(2) $x=y^2-5y$
Use the addition method.
Multiply equation (2) by -1.
$x=3y^2+2y-4$
$-x=-y^2+5y$
$0=2y^2+7y-4$
$0=(2y-1)(y+4)$
$2y-1=0 \qquad y+4=0$
$y=\dfrac{1}{2} \qquad y=-4$
Substitute into equation (2).
$x=y^2-5y \qquad x=y^2-5y$
$x=\left(\dfrac{1}{2}\right)^2-5\left(\dfrac{1}{2}\right) \qquad x=(-4)^2-5(-4)$
$x=\dfrac{1}{4}-\dfrac{5}{2} \qquad x=16+20$
$x=-\dfrac{9}{4} \qquad x=36$

The solutions are $\left(-\dfrac{9}{4},\dfrac{1}{2}\right)$ and $(36,-4)$.

8. (1) $x^2 - y^2 = 24$
(2) $2x^2 + 5y^2 = 55$
Use the addition method.
Multiply equation (1) by –2.
$-2(x^2 - y^2) = -2 \cdot 24$
$2x^2 + 5y^2 = 55$
$7y^2 = 7$
$y^2 = 1$
$y = \pm\sqrt{1}$
$y = \pm 1$
Substitute into equation (1).
$x^2 - y^2 = 24$ $x^2 - y^2 = 24$
$x^2 - (1)^2 = 24$ $x^2 - (-1)^2 = 24$
$x^2 - 1 = 24$ $x^2 - 1 = 24$
$x^2 = 25$ $x^2 = 25$
$x = \pm\sqrt{25}$ $x = \pm\sqrt{25}$
$x = \pm 5$ $x = \pm 5$
The solutions are (5, 1), (–5, 1), (5, –1), and (–5, –1).

9. $(x_1, y_1) = (2, 4)$ $(x_2, y_2) = (-1, -3)$
$r = \sqrt{(x_2 - x_1)^2 + (y_2 - y_1)^2}$
$= \sqrt{(-1-2)^2 + (-3-4)^2}$
$= \sqrt{(-3)^2 + (-7)^2}$
$= \sqrt{9 + 49}$
$= \sqrt{58}$
$(x - h)^2 + (y - k)^2 = r^2$
$[x - (-1)]^2 + [y - (-3)]^2 = (\sqrt{58})^2$
$(x + 1)^2 + (y + 3)^2 = 58$

10. $(x - 2)^2 + (y + 1)^2 = 9$

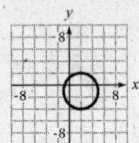

11. $(x - h)^2 + (y - k)^2 = r^2$
$(x - (-2))^2 + (y - 4)^2 = 3^2$
$(x + 2)^2 + (y - 4)^2 = 9$

12. $(x_1, y_1) = (2, 5)$ $(x_2, y_2) = (-2, 1)$
$d = \sqrt{(x_2 - x_1)^2 + (y_2 - y_1)^2}$
$= \sqrt{(-2-2)^2 + (1-5)^2}$
$= \sqrt{(-4)^2 + (-4)^2}$
$= \sqrt{32}$
$(x - h)^2 + (y - k)^2 = r^2$
$(x - (-2))^2 + (y - 1)^2 = \sqrt{32}$
$(x + 2)^2 + (y - 1)^2 = 32$

13. $x^2 + y^2 - 4x + 2y + 1 = 0$
$(x^2 - 4x) + (y^2 + 2y) = -1$
$(x^2 - 4x + 4) + (y^2 + 2y + 1) = -1 + 4 + 1$
$(x - 2)^2 + (y + 1)^2 = 4$
Center: (2, –1)
Radius: 2

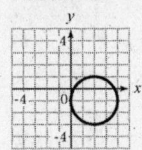

14. $\dfrac{y^2}{25} - \dfrac{x^2}{16} = 1$
Axis of symmetry: y-axis
Vertices: (0, 5) and (0, –5)
Asymptotes: $y = \dfrac{5}{4}x$ and $y = -\dfrac{5}{4}x$

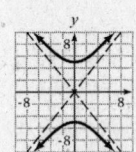

15. $\dfrac{x^2}{9} - \dfrac{y^2}{4} = 1$
Axis of symmetry: x-axis
Vertices: (3, 0) and (–3, 0)
Asymptotes: $y = \dfrac{2}{3}x$ and $y = -\dfrac{2}{3}x$

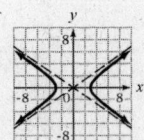

16. $\dfrac{x^2}{16} + \dfrac{y^2}{4} = 1$
x-intercepts: (4, 0) and (–4, 0)
y-intercepts: (0, 2) and (0, –2)

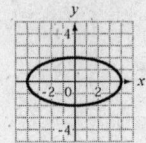

17. $\dfrac{x^2}{16} - \dfrac{y^2}{25} < 1$

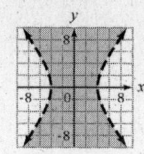

18. $x^2 + y^2 < 36$
    $x + y > 4$

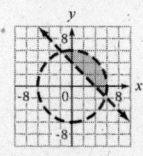

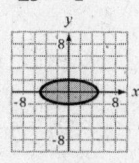

19. $\dfrac{x^2}{25} + \dfrac{y^2}{4} \le 1$

20. $\dfrac{x^2}{25} - \dfrac{y^2}{16} \ge 1$
    $x^2 + y^2 \le 9$

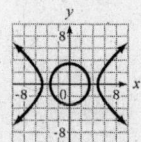

    The solution sets of these inequalities do not intersect, so the system has no real number solution.

FINAL EXAM

1. $12 - 8[3 - (-2)]^2 \div 5 - 3 = 12 - 8[5]^2 \div 5 - 3$
    $= 12 - 8(25) \div 5 - 3$
    $= 12 - 200 \div 5 - 3$
    $= 12 - 40 - 3$
    $= -31$

2. $\dfrac{a^2 - b^2}{a - b} = \dfrac{3^2 - (-4)^2}{3 - (-4)}$
    $= \dfrac{9 - 16}{3 + 4} = \dfrac{-7}{7}$
    $= -1$

3. $5 - 2[3x - 7(2 - x) - 5x] = 5 - 2[3x - 14 + 7x - 5x]$
    $= 5 - 2[5x - 14]$
    $= 5 - 10x + 28$
    $= 33 - 10x$

4. $\dfrac{3}{4}x - 2 = 4$
    $\dfrac{3}{4}x = 6$
    $\dfrac{4}{3} \cdot \dfrac{3}{4}x = \dfrac{4}{3} \cdot 6$
    $x = 8$
    The solution is 8.

5. $\dfrac{2 - 4x}{3} - \dfrac{x - 6}{12} = \dfrac{5x - 2}{6}$
    $12\left(\dfrac{2 - 4x}{3} - \dfrac{x - 6}{12}\right) = 12\left(\dfrac{5x - 2}{6}\right)$
    $4(2 - 4x) - (x - 6) = 2(5x - 2)$
    $8 - 16x - x + 6 = 10x - 4$
    $14 - 17x = 10x - 4$
    $-27x = -18$
    $x = \dfrac{2}{3}$

    The solution is $\dfrac{2}{3}$.

6. $8 - |5 - 3x| = 1$
    $-|5 - 3x| = -7$
    $|5 - 3x| = 7$
    $5 - 3x = 7 \qquad 5 - 3x = -7$
    $-3x = 2 \qquad\quad -3x = -12$
    $x = -\dfrac{2}{3} \qquad\quad x = 4$

    The solutions are $-\dfrac{2}{3}$ and 4.

7. $|2x + 5| < 3$
    $-3 < 2x + 5 < 3$
    $-3 - 5 < 2x < 3 - 5$
    $-8 < 2x < -2$
    $\dfrac{1}{2}(-8) < \dfrac{1}{2}(2x) < \dfrac{1}{2}(-2)$
    $-4 < x < -1$
    $\{x | -4 < x < -1\}$

8. $2 - 3x < 6 \quad$ and $\quad 2x + 1 > 4$
    $-3x < 4 \qquad\qquad\quad 2x > 3$
    $x > -\dfrac{4}{3} \qquad\qquad\quad x > \dfrac{3}{2}$
    $\{x | x > -\dfrac{4}{3}\} \cap \{x | x > \dfrac{3}{2}\} = \{x | x > \dfrac{3}{2}\}$

9.
$$3x - 2y = 6$$
$$-2y = -3x + 6$$
$$y = \frac{3}{2}x - 3$$
$$m_1 = \frac{3}{2}$$
$$m_1 \cdot m_2 = -1$$
$$\frac{3}{2}m_2 = -1$$
$$m_2 = -\frac{2}{3} \quad (x_1, y_1) = (-2, 1)$$
$$y - y_1 = m(x - x_1)$$
$$y - 1 = -\frac{2}{3}(x - (-2))$$
$$y - 1 = -\frac{2}{3}(x + 2)$$
$$y - 1 = -\frac{2}{3}x - \frac{4}{3}$$
$$y = -\frac{2}{3}x - \frac{1}{3}$$

The equation of the line is $y = -\frac{2}{3}x - \frac{1}{3}$.

10.
$$2a[5 - a(2 - 3a) - 2a] + 3a^2$$
$$= 2a[5 - 2a + 3a^2 - 2a] + 3a^2$$
$$= 2a[5 - 4a + 3a^2] + 3a^2$$
$$= 10a - 8a^2 + 6a^3 + 3a^2$$
$$= 6a^3 - 5a^2 + 10a$$

11.
$$\frac{3}{2+i} = \frac{3}{2+i} \cdot \frac{2-i}{2-i}$$
$$= \frac{6 - 3i}{4 - i^2}$$
$$= \frac{6 - 3i}{4 + 1}$$
$$= \frac{6 - 3i}{5}$$
$$= \frac{6}{5} - \frac{3}{5}i$$

12.
$$(x - r_1)(x - r_2) = 0$$
$$\left(x - \left(-\frac{1}{2}\right)\right)(x - 2) = 0$$
$$\left(x + \frac{1}{2}\right)(x - 2) = 0$$
$$x^2 - \frac{3}{2}x - 1 = 0$$
$$2\left(x^2 - \frac{3}{2}x - 1\right) = 0$$
$$2x^2 - 3x - 2 = 0$$

13.
$$8 - x^3y^3 = 2^3 - (xy)^3$$
$$= (2 - xy)(4 + 2xy + x^2y^2)$$

14.
$$x - y - x^3 + x^2y = x - y - x^2(x - y)$$
$$= 1(x - y) - x^2(x - y)$$
$$= (x - y)(1 - x^2)$$
$$= (x - y)(1 - x)(1 + x)$$

15.
$$\begin{array}{r} x^2 - 2x - 3 \\ 2x - 3 \overline{) 2x^3 - 7x^2 + 0x + 4} \\ \underline{2x^3 - 3x^2} \\ -4x^2 + 0x \\ \underline{-4x^2 + 6x} \\ -6x + 4 \\ \underline{-6x + 9} \\ -5 \end{array}$$
$$x^2 - 2x - 3 - \frac{5}{2x - 3}$$

16.
$$\frac{x^2 - 3x}{2x^2 - 3x - 5} \div \frac{4x - 12}{4x^2 - 4} = \frac{x^2 - 3x}{2x^2 - 3x - 5} \times \frac{4x^2 - 4}{4x - 12}$$
$$= \frac{x(x-3)}{(2x-5)(x+1)} \times \frac{4(x+1)(x-1)}{4(x-3)}$$
$$= \frac{x(x-3)4(x+1)(x-1)}{(2x-5)(x+1)4(x-3)} = \frac{x(x-1)}{2x-5}$$

17.
$$\frac{x-2}{x+2} - \frac{x+3}{x-3} = \frac{x-2}{x+2} \cdot \frac{x-3}{x-3} - \frac{x+3}{x-3} \cdot \frac{x+2}{x+2}$$
$$= \frac{x^2 - 5x + 6 - (x^2 + 5x + 6)}{(x+2)(x-3)}$$
$$= \frac{x^2 - 5x + 6 - x^2 - 5x - 6}{(x+2)(x-3)}$$
$$= -\frac{10x}{(x+2)(x-3)}$$

18.
$$\frac{\frac{3}{x} + \frac{1}{x+4}}{\frac{1}{x} + \frac{3}{x+4}} = \frac{\frac{3}{x} + \frac{1}{x+4}}{\frac{1}{x} + \frac{3}{x+4}} \times \frac{x(x+4)}{x(x+4)}$$
$$= \frac{3(x+4) + x}{x+4+3x} = \frac{3x + 12 + x}{4x + 4}$$
$$= \frac{4x + 12}{4x + 4} = \frac{4(x+3)}{4(x+1)} = \frac{x+3}{x+1}$$

19. $\dfrac{5}{x-2} - \dfrac{5}{x^2-4} = \dfrac{1}{x+2}$

$(x+2)(x-2)\left(\dfrac{5}{x-2} - \dfrac{5}{(x+2)(x-2)}\right) = (x+2)(x-2)\dfrac{1}{x+2}$

$5(x+2) - 5 = x - 2$
$5x + 10 - 5 = x - 2$
$5x + 5 = x - 2$
$4x = -7$
$x = -\dfrac{7}{4}$

The solution is $-\dfrac{7}{4}$.

20. $a_n = a_1 + (n-1)d$
$a_n - a_1 = (n-1)d$
$\dfrac{a_n - a_1}{n-1} = d$

21. $\left(\dfrac{4x^2 y^{-1}}{3x^{-1}y}\right)^{-2} \left(\dfrac{2x^{-1}y^2}{9x^{-2}y^2}\right)^3 = \dfrac{4^{-2} x^{-4} y^2}{3^{-2} x^2 y^{-2}} \cdot \dfrac{2^3 x^{-3} y^6}{9^3 x^{-6} y^6}$

$= 4^{-2} \cdot 3^{-(-2)} \cdot x^{-4-2} y^{2-(-2)} \cdot 2^3 9^{-3} x^{-3-(-6)} y^{6-6}$

$= 4^{-2} \cdot 3^2 x^{-6} y^4 \cdot 2^3 \cdot 9^{-3} x^3 y^0$

$= \dfrac{9 x^{-3} y^4 \cdot 8}{16 \cdot 729} = \dfrac{y^4}{162 x^3}$

22. $\left(\dfrac{3x^{2/3} y^{1/2}}{6x^2 y^{4/3}}\right)^6 = \dfrac{3^6 x^4 y^3}{6^6 x^{12} y^8}$

$= \dfrac{729 x^{4-12} y^{3-8}}{46656}$

$= \dfrac{1 x^{-8} y^{-5}}{64} = \dfrac{1}{64 x^8 y^5}$

23. $x\sqrt{18x^2 y^3} - y\sqrt{50x^4 y}$
$= x\sqrt{3^2 x^2 y^2 (2y)} - y\sqrt{5^2 x^4 (2y)}$
$= 3x^2 y\sqrt{2y} - 5x^2 y\sqrt{2y} = -2x^2 y\sqrt{2y}$

24. $\dfrac{\sqrt{16x^5 y^4}}{\sqrt{32xy^7}} = \sqrt{\dfrac{16x^5 y^4}{32xy^7}}$

$= \sqrt{\dfrac{x^4}{2y^3}}$

$= \sqrt{\dfrac{x^4}{y^2 (2y)}}$

$= \dfrac{x^2}{y} \sqrt{\dfrac{1}{2y}} \cdot \sqrt{\dfrac{2y}{2y}}$

$= \dfrac{x^2}{y} \sqrt{\dfrac{1 \cdot 2y}{(2y)^y}}$

$= \dfrac{x^2 \sqrt{2y}}{2y^2}$

25. $2x^2 - 3x - 1 = 0$
$a = 2, b = -3, c = -1$

$x = \dfrac{-b \pm \sqrt{b^2 - 4ac}}{2a}$

$= \dfrac{-(-3) \pm \sqrt{(-3)^2 - 4(2)(-1)}}{2(2)}$

$= \dfrac{3 \pm \sqrt{9 + 8}}{4} = \dfrac{3 \pm \sqrt{17}}{4}$

The solutions are $\dfrac{3 + \sqrt{17}}{4}$ and $\dfrac{3 - \sqrt{17}}{4}$.

26. $x^{2/3} - x^{1/3} - 6 = 0$
$(x^{1/3})^2 - x^{1/3} - 6 = 0$
Let $u = x^{1/3}$.
$u^2 - u - 6 = 0$
$(u - 3)(u + 2) = 0$

$u - 3 = 0 \qquad u + 2 = 0$
$u = 3 \qquad u = -2$
$x^{1/3} = 3 \qquad x^{1/3} = -2$
$(x^{1/3})^3 = 3^3 \qquad (x^{1/3})^3 = (-2)^3$
$x = 27 \qquad x = -8$

The solutions are 27 and -8.

27. $(x_1, y_1) = (3, -2), (x_2, y_2) = (1, 4)$

$m = \dfrac{y_2 - y_1}{x_2 - x_1} = \dfrac{4 - (-2)}{1 - 3} = \dfrac{6}{-2} = -3$

$y - y_1 = m(x - x_1)$
$y - (-2) = -3(x - 3)$
$y + 2 = -3x + 9$
$y = -3x + 7$

The equation of the line is $y = -3x + 7$.

28. $\dfrac{2}{x} - \dfrac{2}{2x + 3} = 1$

$x(2x + 3)\left(\dfrac{2}{x} - \dfrac{2}{2x + 3}\right) = x(2x + 3)(1)$

$2(2x + 3) - 2x = 2x^2 + 3x$
$4x + 6 - 2x = 2x^2 + 3x$
$2x + 6 = 2x^2 + 3x$
$0 = 2x^2 + x - 6$
$0 = (2x - 3)(x + 2)$

$2x - 3 = 0 \qquad x + 2 = 0$
$x = \dfrac{3}{2} \qquad x = -2$

The solutions are $\dfrac{3}{2}$ and -2.

29. (1) $3x - 2y = 1$
(2) $5x - 3y = 3$

Eliminate y.
Multiply equation (1) by -3 and equation (2) by 2.
Add the two new equations.

$-3(3x - 2y) = -3(1) \qquad -9x + 6y = -3$
$2(5x - 3y) = 2(3) \qquad 10x - 6y = 6$
$\qquad\qquad\qquad\qquad\qquad\qquad x = 3$

Substitute 3 for x in equation (1).
$3x - 2y = 1$
$3(3) - 2y = 1$
$9 - 2y = 1$
$-2y = -8$
$y = 4$

The solution is (3, 4).

30. $\begin{vmatrix} 3 & 4 \\ -1 & 2 \end{vmatrix} = 3(2) - (-1)(4)$

$= 6 + 4$
$= 10$

31. $\log_3 x - \log_3 (x - 3) = \log_3 2$

$\log_3 \left(\dfrac{x}{x - 3}\right) = \log_3 2$

Use the fact that if $\log_b u = \log_b v$, then $u = v$.

$\dfrac{x}{x - 3} = 2$

$(x - 3) \cdot \dfrac{x}{x - 3} = 2 \cdot (x - 3)$

$x = 2x - 6$
$-x = -6$
$x = 6$

The solution is 6.

32. $\displaystyle\sum_{i=1}^{5} 2y^i = 2y^1 + 2y^2 + 2y^3 + 2y^4 + 2y^5$

$= 2y + 2y^2 + 2y^3 + 2y^4 + 2y^5$

33. $0.5\overline{1} = 0.5 + 0.01 + 0.001 + 0.0001 + \cdots$

$= \dfrac{5}{10} + \dfrac{1}{100} + \dfrac{1}{1000} + \dfrac{1}{10,000} + \cdots$

$r = \dfrac{a_2}{a_1} = \dfrac{\frac{1}{1000}}{\frac{1}{100}} = \dfrac{1}{10}$

$S = \dfrac{a_1}{1 - r} = \dfrac{\frac{1}{100}}{1 - \frac{1}{10}} = \dfrac{\frac{1}{100}}{\frac{9}{10}} = \dfrac{1}{100} \cdot \dfrac{10}{9} = \dfrac{1}{90}$

$0.5\overline{1} = \dfrac{5}{10} + \dfrac{1}{90} = \dfrac{46}{90} = \dfrac{23}{45}$

34. $n = 9, a = x, b = -2y, r = 3$

$$\binom{9}{3-1}x^{9-3+1}(-2y)^{3-1} = \binom{9}{2}x^7(-2y)^2$$
$$= 36x^7 \cdot 4y^2$$
$$= 144x^7y^2$$

35. (1) $x^2 - y^2 = 4$
(2) $x + y = 1$

Solve equation (2) for y and substitute into equation (1).
$$x + y = 1$$
$$y = -x + 1$$
$$x^2 - y^2 = 4$$
$$x^2 - (-x+1)^2 = 4$$
$$x^2 - (x^2 - 2x + 1) = 4$$
$$x^2 - x^2 + 2x - 1 = 4$$
$$2x = 5$$
$$x = \frac{5}{2}$$

Substitute $\frac{5}{2}$ for x into equation (2).

$$\frac{5}{2} + y = 1$$
$$y = -\frac{3}{2}$$

The solution is $\left(\frac{5}{2}, -\frac{3}{2}\right)$.

36. $f(x) = \frac{2}{3}x - 4$
$$y = \frac{2}{3}x - 4$$
$$x = \frac{2}{3}y - 4$$
$$x + 4 = \frac{2}{3}y$$
$$\frac{3}{2}(x+4) = \frac{3}{2} \cdot \frac{2}{3}y$$
$$\frac{3}{2}x + 6 = y$$
$$f^{-1}(x) = \frac{3}{2}x + 6$$

37. $2(\log_2 a - \log_2 b) = 2\log_2 \frac{a}{b}$
$$= \log_2 \left(\frac{a}{b}\right)^2 = \log_2 \frac{a^2}{b^2}$$

38.
x-intercept	y-intercept
$2x - 3y = 9$	$2x - 3y = 9$
$2x - 3(0) = 9$	$2(0) - 3y = 9$
$2x = 9$	$-3y = 9$
$x = \frac{9}{2}$	$y = -3$
$\left(\frac{9}{2}, 0\right)$	$(0, -3)$

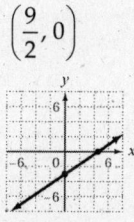

39. $3x + 2y > 6$
$$2y > -3x + 6$$
$$y > -\frac{3}{2}x + 3$$

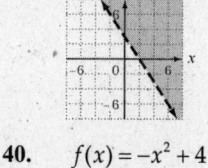

40. $f(x) = -x^2 + 4$
$$-\frac{b}{2a} = -\frac{0}{2(-1)} = 0$$
$$f(x) = -0^2 + 4 = 4$$
Vertex: $(0, 4)$
Axis of symmetry: $x = 0$

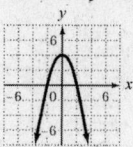

41. $\frac{x^2}{16} + \frac{y^2}{4} = 1$
x-intercepts: $(4, 0)$ and $(-4, 0)$
y-intercepts: $(0, 2)$ and $(0, -2)$

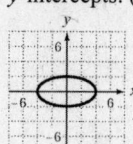

42. $f(x) = \log_2(x+1)$
$$y = \log_2(x+1)$$
$$2^y = x + 1$$
$$2^y - 1 = x$$

43. $f(x) = x + 2^{-x}$

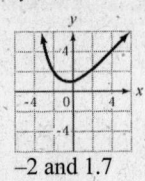

   −2 and 1.7

44. $f(x) = \ln x$
    $g(x) = \ln(x+3)$

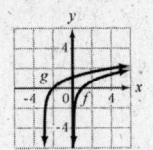

45. **Strategy**
    To find the range of scores on the 5th test, write and solve a compound inequality using x to represent the 5th test.

    Solution
    $70 \leq$ average of the 5 test scores ≤ 79

    $79 \leq \dfrac{64 + 58 + 82 + 77 + x}{5} \leq 79$

    $70 \leq \dfrac{281 + x}{5} \leq 79$

    $350 \leq 281 + x \leq 395$

    $69 \leq x \leq 114$

    The range of scores is 69 or better.

46. **Strategy**
    Average speed of jogger: x
    Average speed of cyclist: $2.5x$

	Rate	Time	Distance
Jogger	x	2	$2x$
Cyclist	$2.5x$	2	$2(2.5x)$

    The distance traveled by the cyclist is 24 more miles than the distance traveled by the jogger.
    $2x + 24 = 2(2.5x)$

    Solution
    $2x + 24 = 2(2.5x)$
    $2x + 24 = 5x$
    $24 = 3x$
    $8 = x$
    $2(2.5x) = 5x = 5(8) = 40$
    The cyclist traveled 40 mi.

47. **Strategy**
    Amount invested at 8.5%: x
    Amount invested at 6.4%: $12000 - x$

	Amount	Rate	Interest
8.5%	x	0.085	$0.085x$
6.4%	$12,000 - x$	0.064	$0.064(12,000 - x)$

    The sum of the interest earned by the two investments is $936.
    $0.085x + 0.064(12,000 - x) = 936$

    The distance traveled by the cyclist is 24 more miles than the distance traveled by the jogger.
    $2x + 24 = 2(2.5x)$

    Solution
    $0.085x + 0.064(12,000 - x) = 936$
    $0.085x + 768 - 0.064x = 936$
    $0.021x + 768 = 936$
    $0.021x = 168$
    $x = 8000$
    $12,000 - x = 4000$
    The amount invested at 8.5% is $8000.
    The amount invested at 6.4% is $4000.

48. **Strategy**
    The width of the rectangle: x
    The length of the rectangle: $3x - 1$
    Use the formula for the area of a rectangle ($A = LW$) if the area is 140 ft².

    Solution
    $A = L \cdot W$
    $140 = (3x - 1)x$
    $140 = 3x^2 - x$
    $0 = 3x^2 - x - 140$
    $0 = (3x + 20)(x - 7)$
    $3x + 20 = 0 \qquad x - 7 = 0$
    $x = -\dfrac{20}{3} \qquad x = 7$

    The solution $-\dfrac{20}{3}$ does not check because the width cannot be negative.
    $3x - 1 = 3(7) - 1 = 20$
    The length of the rectangle is 20 ft and the width is 7 ft.

49. **Strategy**
To find the number of additional shares, write and solve a proportion using x to represent the additional number of shares.

Solution
$$\frac{300}{486} = \frac{300+x}{810}$$
$$(810 \cdot 486) \cdot \frac{300}{486} = \frac{300+x}{810} \cdot (810 \cdot 486)$$
$$300(810) = (300+x)486$$
$$243{,}000 = 145{,}800 + 486x$$
$$97{,}200 = 486x$$
$$200 = x$$
The number of additional shares to be purchased is 200.

50. **Strategy**
Rate of car: x
Rate of plane: $7x$

	Distance	Rate	Time
Car	45	x	$\frac{45}{x}$
Plane	1050	$7x$	$\frac{1050}{7x}$

The total time traveled is $3\frac{1}{4}$ h.
$$\frac{45}{x} + \frac{1050}{7x} = 3\frac{1}{4}$$

Solution
$$\frac{45}{x} + \frac{1050}{7x} = 3\frac{1}{4}$$
$$\frac{45}{x} + \frac{150}{x} = \frac{13}{4}$$
$$\frac{195}{x} = \frac{13}{4}$$
$$4x\left(\frac{195}{x}\right) = 4x\left(\frac{13}{4}\right)$$
$$780 = 13x$$
$$60 = x$$
$$7x = 6(60) = 420$$
The rate of the plane is 420 mph.

51. **Strategy**
To find the distance of the object has fallen, substitute 75 ft/s for v in the formula and solve for d.

Solution
$$v = \sqrt{64d}$$
$$75 = \sqrt{64d}$$
$$75^2 = (\sqrt{64d})^2$$
$$5625 = 64d$$
$$87.89 \approx d$$
The distance traveled is 88 ft.

52. **Strategy**
Rate traveled during the first 360 mi: x
Rate traveled during the next 300 mi: $x + 30$.

	Distance	Rate	Time
First part of trip	360	x	$\frac{360}{x}$
Second part of trip	300	$x+30$	$\frac{300}{x+30}$

The total time traveled during the trip was 5 h.
$$\frac{360}{x} + \frac{300}{x+30} = 5$$

Solution
$$\frac{360}{x} + \frac{300}{x+30} = 5$$
$$x(x+30)\left(\frac{360}{x} + \frac{300}{x+30}\right) = 5x(x+30)$$
$$360(x+30) + 300x = 5(x^2+30x)$$
$$360x + 10800 + 300x = 5x^2 + 150x$$
$$660x + 10800 = 5x^2 + 150x$$
$$0 = 5x^2 - 510x - 10800$$
$$0 = 5(x^2 - 102x - 2160)$$
$$0 = (x+18)(x-120)$$
$$x+18 = 0 \qquad x-120 = 0$$
$$x = -18 \qquad x = 120$$
The solution -18 does not check because the rate cannot be negative.
The rate of the plane for the first 360 mi was 120 mph.

53. **Strategy**
To find the intensity:
Write the basic inverse variation equation, replace the variable by the given values, and solve for k. Write the inverse variation equation, replacing k by its value. Substitute 4 for d and solve for L.

Solution
$$L = \frac{k}{d^2}$$
$$8 = \frac{k}{(20)^2}$$
$$8 \cdot 400 = k$$
$$3200 = k$$
$$L = \frac{3200}{d^2}$$
$$L = \frac{3200}{4^2}$$
$$L = \frac{3200}{16} = 200$$
The intensity is 200 lumens.

54. Strategy
Rate of the boat in calm water: x
Rate of the current: y

	Rate	Time	Distance
With current	$x+y$	2	$2(x+y)$
Against current	$x-y$	3	$3(x-y)$

The distance traveled with the current is 30 mi. The distance traveled against the current is 30 mi.
$2(x+y) = 30$
$3(x-y) = 30$

Solution
$2(x+y) = 30 \quad \frac{1}{2} \cdot 2(x+y) = \frac{1}{2} \cdot 30$
$3(x-y) = 30 \quad \frac{1}{3} \cdot 3(x-y) = \frac{1}{3} \cdot 30$

$x+y = 15$
$x-y = 10$

$2x = 25$
$x = 12.5$
$x+y = 15$
$12.5 + y = 15$
$y = 2.5$

The rate of the boat in calm water is 12.5 mph.
The rate of the current is 2.5 mph.

55. Strategy
The find the value of the investment after two years, solve the compound interest formula for P. Use $A = 4000$, $n = 24$,
$i = \frac{9\%}{12} = \frac{0.09}{12} = 0.0075$.

Solution
$P = A(1+i)^n$
$P = 4000(1+0.0075)^{24}$
$P = 4000(1.0075)^{24}$
$P \approx 4785.65$
The value of the investment is $4785.65.

56. Strategy
To find the value of the house in 20 years, use the formula for the nth term of a geometric sequence.

Solution
$n = 20$, $a_1 = 1.06(180,000) = 190,800$, $r = 1.06$
$a_n = a_1(r)^{n-1}$
$a_{20} = 190,800(1.06)^{20-1}$
$\quad = 190,800(1.06)^{19}$
$\quad \approx 577,284$
The value of the house will be $577,284.